CHEMISTRY

A Project of the American Chemical Society

Volume II, Chapters 6–11

Beta Version

W. H. Freeman and Company
New York

ISBN 0-7167-9671-6

Here's what field testers are saying about...

CHEMISTRY

A Project of the
American Chemical Society

"For the first time in more than 20 years, I felt I was teaching my students something real."
    —*Jonathan Mitschele, Saint Joseph's College*

"By the end of the semester, I saw a dramatic improvement in my students' critical thinking skills."
    —*Kent Chambers,*
    Hardin-Simmons University

"What drew me to this book was the treatment of entropy in Chapter 8. It is absolutely without parallel and the chapter is supported by an equally marvelous interactive Web tutorial as well."
    —*Glenn Keldsen,*
    Purdue University North Central

"I have enjoyed working with a textbook that incorporates student activities directly into the text. This makes it much easier to enable students to be actively involved in the course . . . this is a unique feature. . . . I look forward to seeing the final product."
    —*Laura Eisen,*
    George Washington University

"The text emphasizes the concepts of chemistry . . . the how and why of chemical phenomena, not the usual teaching of algorithms and applied mathematics."
    —*Michelle Dose, Hardin-Simmons University*

"It is hard for an old professor to learn new teaching tricks, but it was worth it to engage in the trial of the ACS GenChem project. The promise of helping students gain a deeper and more intuitive understanding of the concepts of chemistry is very rewarding."
    —*Neil Rudolph, Adams State College*

"The text facilitates (even encourages) the use of lecture groups and in-class activities, both of which the students enjoyed. Working effectively in groups is one of the goals/objectives of Adams State's general education program, so this course helps to fulfill that specific goal."
    —*Marty Jones, Adams State College*

"I think this text has the potential to change both how we teach and how the students learn chemistry. The students are much more engaged with the material in class and appreciate the balance between activities and lecture. I'm growing in enjoyment of this also."
    —*Priscilla Bell, Whittier College*

". . . the experience provided me a challenge that made teaching more fun and exciting."
    —*Amina El-Ashmawy,*
    Collin County Community College

An Innovative Support Package for the Instructor . . .

The *Faculty Resource Online Guide (FROG)* for **Chemistry** is an Instructor's Manual that provides instructors with a set of options for integrating active learning into their curriculum. The teaching strategies presented in the *Faculty Resource Online Guide* cover the spectrum of general chemistry courses, from small studio sections to large, multilaboratory sections. The *Faculty Resource Online Guide* includes information for each type of activity students will encounter. Access the *Faculty Resource Online Guide* via our Web site:
www.whfreeman.com/acsgenchem

For *Investigate This* activities, the *Faculty Resource Online Guide* includes:

- Goals of activity
- Anticipated results
- Optional modifications and alternative activities
- Set-up and clean-up times

- Detailed lists of reagents, materials, procedures, and classroom options
- Digitized photographs of the activity set-ups and results
- Follow-up discussion and activities

For *Consider This* activities, the *Faculty Resource Online Guide* includes:

- Goals of activity
- Detailed solutions
- Follow-up discussion and activities

For *Check This* activities, the *Faculty Resource Online Guide* includes:

- Detailed solutions

For the end-of-chapter problems the *Faculty Resource Online Guide* includes:

- Detailed solutions

For the Student . . .

The *Personal Tutor*

The *Personal Tutor* is a print supplement for **Chemistry**. Its primary purpose is to give you more guidance and practice with problems and computations. Using the *Personal Tutor* is much like visiting the professor during office hours or going to a human tutor. The *Personal Tutor* is referenced where appropriate in the text.

First, go to the Web site **www.whfreeman.com/acsgenchemhome** click on "Personal Tutor," and take the diagnostic exam. Once you've completed the exam you will receive feedback on what sections of the *Personal Tutor* you should review.

You may need to take advantage of the *Personal Tutor* frequently or you may need its assistance only a few times during the course. The questions can be a good review when preparing for a test even if you are confident that you know how to solve the problems or do the computations.

The Student Web site

To complement your studies, visit our Web site at **www.whfreeman.com/acsgenchemhome** where you will find

"Web Companion" featuring:

- Molecular level animations
- Drag and drop activities
- Integrated review questions

Molecular Database: More than one hundred 3-D rotatable molecules

Images from the text: Every image in full color

The Molecular Structure Model Set
ISBN 0–7167–4822–3

Models of molecules help you to understand their physical and chemical properties by providing a way to visualize the three-dimensional arrangement of the atoms. A superb study aid, the Molecular Structure Model Set uses polyhedra to represent atoms and plastic connectors (scaled to correct bond lengths) to represent bonds. Plastic plates representing orbital lobes are included for indicating lone pairs of electrons, radicals, and multiple bonds—a feature unique to this set.

Student Survey

The feedback that you provide to the writers and editors of **Chemistry** will be analyzed and integrated into the final version of this text. We appreciate your thoughtful comments and insights. ***Please cut out this survey and return it to your instructor.***

Investigate This activities

    a. were useful and/or informative.

    b. were somewhat useful and/or informative.

    c. I tried using, but did not find useful.

    d. I did not try using.

Consider This activities

    a. were useful and/or informative.

    b. were somewhat useful and/or informative.

    c. I tried using, but did not find useful.

    d. I did not try using.

Check This activities

    a. were useful and/or informative.

    b. were somewhat useful and/or informative.

    c. I tried using, but did not find useful.

    d. I did not try using.

Web site tutorials at www.whfreeman.com/acsgenchem

    a. were useful and/or informative.

    b. were somewhat useful and/or informative.

    c. I tried using, but did not find useful.

    d. I did not try using.

What was your overall impression of this text?

How does this text compare with those you've used in other courses?

May we have your permission to use your comments in our marketing materials? If so, please print your name, and the name of your school below.

Name_____

School_____

Career aspiration_____

Year: Freshman ____
        Sophomore ____
        Junior ____
        Senior ____
        Other ____

Brief Contents

Table of Contents

Kinetic energy of an electron wave
Potential energy of an electron wave
Total energy of an atom

CHAPTER 5. Structure of Molecules

Introduction

Everything you hear, see, smell, taste, and touch involves chemistry and chemicals (matter). And hearing, seeing, smelling, tasting, and touching all involve intricate series of chemical reactions and interactions in your body. With such an enormous range of topics, chemistry offers you fascinating opportunities to explore and study. At the same time, all these possibilities make chemistry seem a daunting subject to study. Aware of both the fascination and the challenge of studying chemistry, the American Chemical Society chose a team of chemists to consider what concepts would help you open the doors to opportunities that require a knowledge of chemistry without being overwhelming. The team also took up the challenge to develop effective approaches to learning and teaching chemistry. The result of the team's efforts are this textbook, *Chemistry*, and its complementary materials, including project-based laboratory experiments, your molecular model kit, the *Web Companion*, and the *Personal Tutor*.

Learning chemistry, even with a limited range of concepts and content, requires a good deal of effort from both you and your instructors. To facilitate your efforts, we have written *Chemistry* in a conversational tone designed to be accessible and engaging. But you cannot learn chemistry only by reading about it, just as you cannot learn how to write a short story or how to find fossils simply by reading about how others do it. Learning how others do something you want to do is important, but you must also practice doing it yourself. Chemists and other scientists learn about the world through experimenting. They then try, often in collaborative efforts, to develop models of the world at the molecular level that explain their results and allow them to predict the outcomes of other possible experiments. We have tried to incorporate this same approach in this textbook.

Throughout *Chemistry*, we present activities and thought-provoking questions that are intended to promote active small-group and whole-class participation. To encourage your participation and collaborative learning efforts, four features appear often in each chapter:

Intro.1.   **Investigate This**

An *Investigate This* usually involves short experiments that introduce the chemical concepts explored in the following paragraphs. The investigations are designed to be carried out in small groups or in the whole class setting.

Intro.2.   **Consider This**

A *Consider This* follows each *Investigate This* and usually asks you to discuss and develop hypotheses or explanations for what you have observed. At other places a *Consider This* will ask you to think about and discuss the consequences of what has just been presented or to anticipate what is to come. The intent in all cases is to involve the class in a discussion.

Intro.3.   **Worked Example**

Each *Worked Example* guides you through the reasoning involved in solving a problem. Thinking about *how* to solve a problem is often more important and more challenging than actually carrying out the solution procedure, so we place an emphasis on this thinking. Almost all Worked Examples include the following components, after the statement of the problem:

Necessary information: What do you need to know, including the information from the problem statement, in order to solve the problem?

Strategy: How do you put the information together in order to solve the problem? What concepts are involved and how are they to be used? With an appropriate strategy (there is often more than one) in hand, the problem is essentially solved.

Implementation: Carry out the strategy using the needed information to obtain the solution to the problem. Calculations, if necessary, are done at this stage.

Does the answer make sense? Once you get an answer to a problem, you should always check to be sure it makes sense. You should also check to be sure you have carried out any numerical calculations correctly, but making *sense* of the answer is a distinct task (and can sometimes flag possible numerical problems). Is the answer about the size you would expect (based on other experiences, for example)? Does it have the expected direction (sign or change from some baseline)? And so on…

Intro.4.   **Check This**

At least one *Check This* follows each *Worked Example* and presents a similar problem or problems so that you can practice the strategy presented in the *Worked Example*. *Check This* problems also appear in other places, where you are asked to practice some technique or answer questions based on what has just been presented in the text. **Chemistry** is designed to be used with paper, pencil, calculator, and model kit at hand, so you can try each *Check This* as you come to it.

In addition to these features within each chapter, there is an *Outcomes Review* section near the end of each chapter and *End-of-Chapter Problems* you can use to test your problem-solving skills. Use the *Outcomes Review* to remind yourself of the important ideas from the chapter and the *End-of-Chapter Problems* to check your understanding of these ideas. Some of these problems will give you more practice with the kinds of problems you meet in the *Worked Example* and *Check This* activities throughout the chapter. Other *End-of-Chapter Problems* are included to stretch your thinking and engage you in problem-solving strategies that are combinations of strategies introduced in the chapter or that extend a bit beyond them. Most scientists work cooperatively, and we encourage you to try working on these problems collaboratively as well. Often a group can come up with more and better solutions than an individual working in isolation.

Throughout **Chemistry**, you will find an emphasis on understanding and reasoning, and on models of all kinds: physical, computer, and mathematical models, and analogies. We use models, because it is difficult to observe individual atoms or molecules as they undergo the changes and interactions that lead to the events we can easily observe in nature or in the laboratory. As we try to understand the physical and chemical properties of atoms and molecules and how they cause observable effects we will use three levels of description, which are exemplified by this page from the **Web Companion**:

$$Na^+(aq) + Cl^-(aq) + Ag^+(aq) + NO_3^-(aq) \longrightarrow AgCl(s) + Na^+(aq) + NO_3^-(aq)$$

- *Lab Level:* These are the observations you make on macroscopic systems, such as that shown in these frames from a movie of a reaction between two solutions in a test tube or in your Investigate This activities.

- *Molecular Level:* These are our models of what is going on among the particles (atoms, ions, and molecules) that gives rise to the effects observed in the laboratory. These models are animated in the *Web Companion* and are usually shown as less complicated still figures in the text. You will also often use models you build yourself with your molecular model kit.

- *Symbolic Level:* Intermediate between and connecting the Lab and Molecular Levels is the Symbolic Level of description. This is the descriptive level which you probably associate with chemistry. This level is essential, because it combines a great deal of information in a succinct format. It is also the most abstract level of representation, since all the symbols need to be interpreted to make sense of the description. The usual approach in *Chemistry* will be first to try to understand systems at the Lab and Molecular Levels and only then proceed to the Symbolic Level.

The *Web Companion*, from which the above figure is taken, is designed to provide you opportunities to use interactive animations, movies, and other resources that provide a visual (usually moving) means to examine many of the concepts included in the written text. When a

> **WEB Chap X, Sect X.5.3-4**
> A brief description of what you will find on these pages is given here.

Web Companion page (or pages) is available for some concept, this marginal box appears in the text. To access the *Web Companion*, visit www.whfreeman.com/acsgenchemhome and select "Web Companion." Find the appropriate location in the *Companion* by selecting the chapter and subsection referenced. The Companion may also be available on your institutional computers and, if so, your instructors will tell you how to access it. There are also *Consider This* and *Check This* problems, as well as *End-of-Chapter Problems,* based on the *Web Companion*. These are denoted by this symbol, WEB , and a reference to the chapter and section you will need to access.

A very large percentage of students who take the general chemistry course in a college or university have already had at least one year of a high school chemistry course. We assume that you are in this category and have probably been exposed to a good deal of the nomenclature and methods that are part of the study of chemistry. We take advantage of this background to move quickly into an examination of the properties of water that depends on some familiarity with the properties of atoms and molecules. You probably have also done some of the algebraic and arithmetic calculations that are a part of essentially all beginning chemistry courses. We take advantage of this experience as well, by providing a review of only necessary concepts and then using them to try to answer questions based on our initial studies of water. The brief reviews we provide in the text may, however, not be enough to make you comfortable with the problems we pose, so we have provided a *Personal Tutor*.

The primary purpose of the *Personal Tutor* is to give you more guidance and practice with problems and computations in the areas that seem to give students trouble. Using the Personal Tutor is much like visiting your instructor during office hours or going to a human tutor.

Marginal boxes like this one will alert you to a section in the *Personal Tutor* that might be helpful for the topic under discussion. Before using the *Tutor*, for the first time, visit

| **Personal Tutor** |
| A brief description directs you to the section you might find helpful for this part of the text. |

www.whfreeman.com/acsgenchemhome, select "Personal Tutor," and take the diagnostic exam. When you have completed the exam, you will get feedback on what sections of the *Personal Tutor* would be helpful for you to study. You may need to take advantage of it frequently or you may need its assistance few times or not at all during the course. The questions in the Tutor can be a good review when preparing for tests, even if you are confident that you know how to solve the problems or do the computations. We urge you to take advantage of this resource in whatever ways it can be helpful.

We have outlined above how we designed this textbook and its complementary materials to provide you, your classmates, and your instructors the resources to learn and teach chemistry actively and interactively. Now let us return to the first task the American Chemical Society team considered, what concepts to include in *Chemistry*. Several concepts or "big ideas" recur in one form or another through the book. Brief statements of these concepts are:

- Attractions between positive and negative centers hold matter together and are responsible for chemical reactions.

- The lower its energy, the more stable the system.

- During change, energy is conserved: $\Delta E_{net} = 0$.

- The properties of elements repeat periodically as the atomic number of their atoms increases.

- Electrons in atoms and molecules act like matter waves with quantized energies; the more spread out a matter wave, the lower (more favorable) its energy.

- Change occurs in the direction that increases the number of distinguishable arrangements of particles and/or energy quanta. Entropy, S, is a measure of this number, and in all spontaneous processes, net entropy increases, $\Delta S_{net} > 0$.

- Reactions are at equilibrium when $\Delta S_{net} = 0$ for the change from reactants to products.

- When a reaction at equilibrium is disturbed, the system reacts to minimize the disturbance. This is LeChatelier's principle. Reactions at equilibrium are quantitatively described by a temperature dependent equilibrium constant ratio.

- Electric current can produce reduction-oxidation chemical reactions. Reduction-oxidation chemical reactions can produce an electric current.

- The rate of a chemical reaction depends on the concentrations of species and the temperature of the system. These are a result of the reaction pathway.

Some of the concepts in this list may look familiar and others probably do not. Our goal in structuring **Chemistry** to emphasize active and collaborative learning has been to provide the means for you to understand these concepts. The understanding you gain will allow you to apply the concepts not only to the problems we and your instructors provide, but also to the problems and systems you meet in other courses, and most importantly to interesting and intriguing systems you meet in the world outside the classroom. It has been an enjoyable challenge to write **Chemistry** and develop the complementary materials. We hope it is an enjoyable challenge to use them and learn chemistry.

Chapter 6. Chemical Reactions

For millennia, humans have used natural polymers such as wool and silk to make cloth for clothing, shelter, and other purposes. Within the past century, our knowledge of atomic and molecular structure has led to the discovery of reactions to make synthetic polymers, such as nylon and polyester, with the same bonding as natural polymers and often with properties tailored to a particular purpose, such as clothing or containers. Spinnerets used to form nylon fibers are human versions of the spinnerets a spider uses to produce silk for her web.

Chapter 6. Chemical Reactions

Omnia mutantur, et nos mutamur in illis.
(All things change, and we change with them.)

attributed to the Emperor Lothar I (795-855)

More than seven thousand years have passed since humans learned to purposefully create new substances that better served their needs. Early artisans discovered that the green mineral malachite was almost magically transformed in the heat of crude ovens to a reddish, lustrous metal, which we now call copper. A material that scarcely existed in the natural world had been born in the flame. It could be worked into useful shapes for tools, weapons, and ornaments. A copper–tin alloy called bronze, and then iron, became the central metals upon which civilization depended. Civilizations are built on more than metal, so, while our ancestors were busily converting various stones into metals, they were just as busily converting grain into beer and bread. Taking advantage of their ability to shape their material world, hunter-gatherer societies became agrarian and mercantile societies. As the quotation above suggests, change seems to be a constant of nature and of ourselves.

Today, we continue to convert available resources into an almost unimaginable number of items that we use every day. The illustration on the facing page shows a few examples that emphasize materials and products containing natural or synthetic (manufactured) polymers as well as the production of two polymers, silk from spider spinnerets and nylon from industrial spinnerets. Our ancestors used natural polymers like cotton, wool, and silk to make cloth and decorated the cloth with various natural dyes. In all these processes, smelting, brewing, baking, dyeing, and many others, they learned a great deal of practical chemistry by tinkering with stuff. However, they had no underlying model of matter, so they had to rely on trial and error, guided by past experience, in their search for improvements or new processes.

Only within the past few hundred years, has our chemical knowledge become well enough organized to carry out studies directed at understanding the fundamental basis of the reactions upon which our civilization depends and developing entirely new materials and processes. Our predictive power has increased enormously as we have learned how molecules and ions are attracted to each other, and of how atoms and electrons redistribute themselves to form the products of reactions, including the polymers shown on the facing page.

The common characteristic of all these reactions is that they begin with the attraction of positive and negative centers for one another. From the beginning of the book, we have emphasized that the fundamental basis for atom formation, molecular bonding, and molecular interactions is the attraction of positive and negative charges (or partial charges). In this chapter, we will extend this fundamental idea to interactions that lead to chemical reactions. The objective is for you to develop the ability to use what you have learned about molecular structure to make predictions about what is likely to happen when chemicals are mixed with one another.

Section 6.1. Classifying Chemical Reactions

6.1.   **Consider This**   *What chemical reactions do you know?*

(a) Work in small groups for a few minutes listing all the changes you can think of that are chemical reactions. Share your list with the rest of the class and see how many different reactions the whole class has come up with. How many reactions were listed by more than one group?

(b) How did you decide whether a change is the result of a chemical reaction? List the criteria used to make this decision and to accept the reactions suggested by each group.

6.2.   **Investigate This**   *What changes do you observe?*

Do this as a class investigation and work in small groups to discuss and analyze the results. Prepare four 250-mL beakers or clear, colorless plastic tumblers with the contents shown in this table (A "tsp" is one rounded teaspoonful; "dried yeast" is a packet of dried baker's yeast; and, "$KHC_4H_4O_6$" is potassium hydrogen tartrate, also called "cream of tartar.").

Beaker	1	2	3	4
Contents	dried yeast	dried yeast + tsp glucose	tsp $CaCl_2$ + tsp $NaHCO_3$	tsp $NaHCO_3$ + tsp $KHC_4H_4O_6$

To each beaker, add about 20 mL of warm (50 °C) water and swirl to mix the ingredients. Observe the beakers for about two minutes and record any evidence of change(s) occurring. If there is no apparent change in one or more of the beakers, wait for a few minutes and check them again. Continue this checking until you are convinced no change is going to occur.

Your lists of changes in Consider This 6.1 show that you are already quite familiar with chemical reactions. Evidence that different substances are present before and after the change was probably one of the criteria you used for deciding whether a change is a chemical reaction. For example, the evidence might be that the reaction mixture foams or bubbles, like some of the mixtures in Investigate This 6.2, indicating that a new substance, a gas, has been produced. We can define a **chemical reaction** as a change that produces molecular structures or ionic combinations that are different from those in the reactants.

Identifying reaction products

Since chemical reactions produce new compounds, product identification is an important part of the process we use to find out whether a chemical reaction has occurred, and, if so, what kind.

6.3.   **Investigate This**   *How can you analyze the gas from a reaction?*

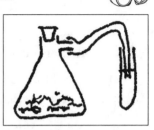

Do this as a class investigation and work in small groups to discuss and analyze the results. Repeat the reaction in the fourth beaker in Investigate This 6.2, but carry out the reaction in a stoppered, 250-mL filter flask with a length of rubber tubing attached to the filter arm, as shown in the diagram. Clamp the open end of the tubing below the surface of the liquid in a test tube about half full of limewater (a saturated solution of calcium hydroxide, $Ca(OH)_2$). Add the solid reactants to the flask and then stopper it as soon as the water has been added. Swirl the flask to mix the reactants and observe and record what happens for a few minutes.

6.4.   **Consider This**   *What is the gas from the $NaHCO_3 + KHC_4H_4O_6$ reaction?*

(a) What did you observe as the gas from the reaction in Investigate This 6.3 bubbled into the limewater? Recall the reactions of limewater in Chapter 2, Section 2.15. What is the gas? Give the reasoning for your identification.

(b) Do you think that this gas is a reasonable product to expect from these reactants? Explain the reasoning for your answer.

The three reactions in Investigate This 6.2 that produce a gaseous product all produce the same gaseous product. The reaction in the yeast-sugar mixture, the yeast fermentation of sugar, is complex, but the fermentation of glucose (also called dextrose) can be written as:

$$C_6H_{12}O_6(aq) \rightarrow 2CH_3CH_2OH(l) + 2CO_2(g) \tag{6.1}$$

Fermentation is central to the production of many alcoholic beverages and important in baking where the gaseous carbon dioxide product makes bread dough rise.

The net reaction of calcium chloride with sodium hydrogen carbonate in aqueous solution is:

$$Ca^{2+}(aq) + 2HCO_3^-(aq) \rightarrow CaCO_3(s) + CO_2(g) + H_2O(l) \tag{6.2}$$

We discussed the components of this reaction in Chapter 2, Section 2.15, in relation to the carbon cycle and the formation of limestone caves.

The net reaction of sodium hydrogen carbonate with potassium hydrogen tartrate is:

$$HC_4H_4O_6^-(aq) + HCO_3^-(aq) \rightarrow C_4H_4O_6^{2-}(aq) + CO_2(g) + H_2O(l) \tag{6.3}$$

Baking powders contain sodium hydrogen carbonate and a solid acid, such as potassium hydrogen tartrate. These react to give the carbon dioxide that makes cakes and cookies rise when they are baked.

6.5. Consider This   *How can you classify chemical reactions?*

(a) Think about your list of reactions from Consider This 6.1 as well as the reactions from Investigate This 6.2. Can you classify them into groups of reactions that have similar properties and/or behaviors? What are the properties and/or behaviors you use?

(b) If you were given a pair of reactants, could you use your classification scheme to predict the likely outcome of a reaction between them? Why or why not? If not, what other information would you need to make your prediction?

Classifying chemical reactions

The reactants in the above reactions are so different that you probably would not have predicted that each pair would produce the same product. In Consider This 6.5, you probably found that reaction classifications based on observable changes do not provide adequate information to make predictions about the likely reaction between a new set of reactants. In order to make predictions about the likely reaction, another essential piece of information you need is some idea what interactions can occur between the reactant molecules and/or ions. We have repeatedly emphasized the importance of electrical charge or partial charge (polarity) in explanations for chemical phenomena. ***Chemical reactions start when a center of positive charge in one reactant molecule or ion is attracted to a center of negative charge in another.*** Thus, understanding chemical reactions requires understanding the structure of molecules discussed in Chapter 5.

Once reactant molecules have come together, a chemical reaction between them involves rearrangements of electrons and/or atoms to form products that are different from the reactants. In Chapter 2, we introduced two kinds of chemical reactions: precipitation (and its reverse, dissolution) of solid ionic salts from solution and acid-base reactions. In this chapter we will introduce other kinds of chemical reactions. Using the background on atomic and molecular structure from Chapters 4 and 5, we organize chemical reactions into three broad classes: ionic precipitation, Lewis acid-base, and reduction-oxidation. Ionic precipitations occur as a result of the attraction between separated positive ions and negative ions to form ordered crystalline solids. Lewis acid-base reactions are characterized by the presence of a nonbonding pair of electrons on one of the reactants which ends up as an electron pair bond in one of the products. (Brønsted-Lowry acid-base reactions are a special case of Lewis acid-base reactions.) Reduction-oxidation reactions are characterized by the transfer of one or more electrons from one reactant to another. You will find that reactions (6.1), (6.2), and (6.3) fit into this classification scheme.

This organization of chemical reactions is based on the electron distributions and redistributions in and between molecules and ions when reactions occur. Ours is not the only possible organization or classification of chemical reactions. Each classification has its merits and utility. In most general classification schemes, there are exceptions and specific examples that fall into more than one class. You have to accept special cases as a necessary consequence of trying to describe an enormous number of different reactions with only a few variables. Keep in mind that there are exceptions, but focus on the overall organization and you will be able to understand the direction of many reactions and be able to predict their products.

Section 6.2.   Ionic Precipitation Reactions

6.6.   Investigate This   *How can you find the stoichiometry of a precipitate?*

Do this as a class investigation and work in small groups to discuss and analyze the results. Label five, clean, small centrifuge tubes from 1 to 5. Mix in these tubes the volumes of 0.10 M aqueous calcium nitrate, $Ca(NO_3)_2$, and sodium oxalate, $Na_2C_2O_4$, solutions specified in this table. Centrifuge the tubes for about a minute to settle the precipitates. Record all your observations.

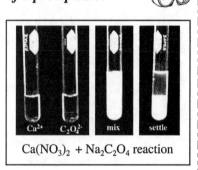

$Ca(NO_3)_2 + Na_2C_2O_4$ reaction

tube number	1	2	3	4	5
calcium nitrate, mL	0.5	1.5	2.5	3.5	4.5
sodium oxalate, mL	4.5	3.5	2.5	1.5	0.5

6.7.   Consider This   *What is the stoichiometry of the calcium oxalate precipitate?*

(a) In Investigate This 6.6, what is varying in the contents of the solutions in each tube as you go from the first to the sixth?

> **(b)** Is the amount of precipitate the same in each tube? If not, which has the most precipitate? which the least? Make a rough bar graph showing the amount of precipitate in each tube as a function of tube number.
>
> **(c)** What trends, if any do you see in the graph? How would you interpret them?

Ionic precipitation reactions are a result of attractions between cations that have lost one or more electrons and anions that have gained one or more electrons. The cations of interest to us are almost always metal ions. The anions may be monatomic (such as Cl^- and S^{2-}) or polyatomic (such as PO_4^{3-} and HCO_3^-). You learned from your analysis of Investigate This 2.30, Chapter 2, that singly-charged cations and anions form relatively soluble salts. The attractions of polar water molecules to the ions, hydration energy, is strong enough to overcome the attractions of the ions to one another in the solid crystal, the lattice energy. Multiply-charged anions and cations most often form relatively insoluble salts with each other. The larger charges attract one another strongly in the crystal and the hydration energy is not enough to overcome the lattice energy. The multiple charges on aqueous calcium and oxalate ions, $Ca^{2+}(aq)$ and $C_2O_4^{2-}(aq)$, help to explain the appearance of the precipitates in Investigate This 6.6.

$$Ca^{2+}(aq) + C_2O_4^{2-}(aq) \rightleftharpoons CaC_2O_4(s) \tag{6.4}$$

Table 6.1, gathers the solubility rules we have generated.

Table 6.1. General Solubility Rules for Ionic Salts

Cation	Anion	Solubility
singly-charged	singly-charged	soluble; Ag^+ is an exception
singly-charged	multiply-charged	alkali metal salts soluble
multiply-charged	singly-charged	halide and nitrate salts soluble
multiply-charged	multiply-charged	not soluble

6.8. **Consider This** *Does ionic precipitation fit the definition of a chemical reaction?*

We have said that a chemical reaction involves rearrangements of electrons and/or atoms to form products that are different from the reactants. Does reaction (6.4) fit this description? What rearrangements of electrons and/or atoms occurs? *Hint*: Consider what the designation *(aq)* on the ions means.

Calcium–oxalate reaction stoichiometry

There are different amounts of $Ca^{2+}(aq)$ and $C_2O_4^{2-}(aq)$ in each tube in Investigate This 6.6, so we would expect differences in the amount of precipitate that can be formed in each. The observed differences are a function of the stoichiometry of the precipitation reaction (6.4). The stoichiometry is based on charge balance. We have written the double arrows to represent the reversibility of the reaction; solid calcium oxalate does dissolve to a slight extent in water. Let's look quantitatively at the stoichiometry for one of the samples in Investigate This 6.6.

6.9.   **Worked Example**   *Stoichiometric calculation for calcium oxalate formation*

How much precipitate can be formed in sample 5 in Investigate This 6.6?

Necessary information: We are asked "how much" without any specification of units. We'll choose to calculate the number of moles of precipitate, since moles are easiest to get using the stoichiometric equation (6.4) and the volume and concentration data for the sample.

Strategy: The approach for solving problems of this kind was discussed in Chapter 2, Section 2.10, and is illustrated in the partial stoichiometry route map shown here. We calculate the number of moles of each reactant present in the solution and then, using the 1:1

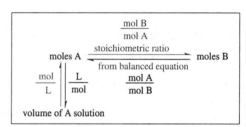

stoichiometry of the reaction, determine which is the limiting reactant. Finally, we convert the number of moles of limiting reactant to the number of moles of precipitate it can form.

Implementation:

$$\text{moles of } Ca^{2+}(aq) = (0.10 \text{ mol·L}^{-1})(4.5 \text{ mL})\left(\frac{1 \text{ L}}{1000 \text{ mL}}\right) = 4.5 \times 10^{-4} \text{ mol } Ca^{2+}(aq)$$

$$\text{moles of } C_2O_4^{2-}(aq) = (0.10 \text{ mol·L}^{-1})(0.5 \text{ mL})\left(\frac{1 \text{ L}}{1000 \text{ mL}}\right) = 0.5 \times 10^{-4} \text{ mol } C_2O_4^{2-}(aq)$$

The number of moles of $C_2O_4^{2-}(aq)$ required to react with 4.5×10^{-4} mol $Ca^{2+}(aq)$ is:

$$\text{mol } C_2O_4^{2-}(aq) \text{ required} = (4.5 \times 10^{-4} \text{ mol } Ca^{2+}(aq))\left(\frac{1 \text{ mol } C_2O_4^{2-}}{1 \text{ mol } Ca^{2+}}\right)$$

$$= 4.5 \times 10^{-4} \text{ mol } C_2O_4^{2-}(aq)$$

This amount of $C_2O_4^{2-}(aq)$ is more than is available, so $C_2O_4^{2-}(aq)$ is the limiting reactant. The number of moles of precipitate, $CaC_2O_4(s)$, formed by 0.5×10^{-4} mol $C_2O_4^{2-}(aq)$ is:

$$\text{mol } CaC_2O_4(s) = (0.5 \times 10^{-4} \text{ mol } C_2O_4^{2-}(aq))\left(\frac{1 \text{ mol } CaC_2O_4}{1 \text{ mol } C_2O_4^{2-}}\right)$$

$$= 0.5 \times 10^{-4} \text{ mol } CaC_2O_4(s)$$

Does the answer make sense? The concentrations of the reactants are the same and they react in a 1:1 ratio. Since we used less of the $C_2O_4^{2-}(aq)$ solution, it makes sense that it is the limiting reactant and that the amount of precipitate is the same as the amount of $C_2O_4^{2-}(aq)$. From its molar mass, you can calculate that about 6 mg of $CaC_2O_4(s)$ is formed.

6.10. Check This   *Stoichiometric calculation for calcium oxalate formation*

(a) Calculate the number of moles of precipitate that can be formed in the other four samples in Investigate This 6.6.

(b) How do your results compare with the bar graph you sketched in Consider This 6.7? Which sample should give the largest amount of precipitate? Is this what you observe?

(c) Is there anything different about the sample with the largest amount of precipitate? What is different?

Continuous variations

Notice that the *sum* of the numbers of moles of $Ca^{2+}(aq)$ and $C_2O_4^{2-}(aq)$ is the same for all five samples in Investigate This 6.6. What varies among the samples is the *ratio* of moles of $Ca^{2+}(aq)$ to moles of $C_2O_4^{2-}(aq)$. As you have found, the maximum amount of precipitate is formed in sample number 3, the sample with the 1:1 stoichiometric reaction ratio. In this sample, essentially all of the $Ca^{2+}(aq)$ and $C_2O_4^{2-}(aq)$ ions precipitate from the solution. In all the other samples, there is less of one or the other of these ions and less precipitate can be formed. This technique, in which the ratio of numbers of moles of two reactants is varied while their sum is held constant, is called the **continuous variation method**. Figure 6.1 is a representation of a continuous variation experiment, similar to the one you did in Investigate This 6.6.

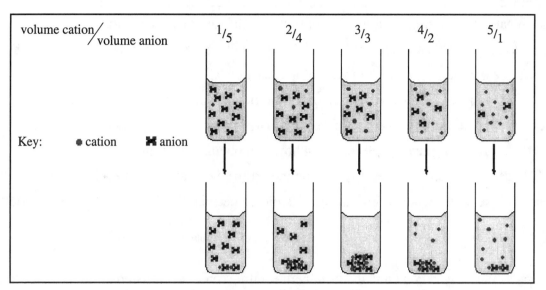

Figure 6.1. A continuous variations experiment.

Solutions of the same concentration of a cation and anion that react 1:1 are mixed in the volume ratios shown. The top row represents the mixture in the instant before reaction; the bottom row after reaction to form a precipitate. The nonprecipitating counter ions are not shown.

Continuous variation methods are often used to determine the stoichiometry of reactions. The maximum amount of reaction occurs when the ratio is stoichiometric. The method is applicable to any kind of reaction which produces a measurable effect that is proportional to the "number of moles of reaction." Measurable effects include formation of precipitates, production of heat, changes in color, and so on. Later in the chapter, we will introduce other techniques for determining reaction stoichiometries.

6.11. **Check This** *Nickel-dimethylglyoxime reaction stoichiometry*

Solutions containing nickel ion, $Ni^{2+}(aq)$, are a beautiful green. When a $Ni^{2+}(aq)$ solution is mixed with a slightly basic, clear, colorless alcoholic solution of **dimethylglyoxime** (dmg), $C_4H_8N_2O_2(alc)$, a deep red solid is formed:

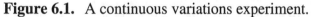

$$xNi^{2+}(aq) + yC_4H_8N_2O_2(alc) \rightleftharpoons \text{red solid} \qquad (6.5)$$

A continuous variations study was carried out on this reaction with the results shown here.

sample	1	2	3	4	5	6
$Ni^{2+}(aq)$, mol/10^{-4}	2.55	4.89	5.98	7.50	9.89	12.6
$C_4H_8N_2O_2(alc)$, mol/10^{-4}	18.2	15.8	14.8	13.1	10.9	8.19
mass solid g	0.074	0.141	0.173	0.189	0.158	0.118

(a) What is the *average* total number of moles of the two reactants in these samples?

(b) Assuming that x and y in equation (6.5) are small whole numbers (1, 2, or 3), use these data to figure out the likely stoichiometry of the reaction. Clearly explain your reasoning. *Note*: None of the samples contains the exact stoichiometric ratio of reactants.

(c) Based on the stoichiometry you found in part (b), how many moles of the red solid product are formed in each sample? Explain how you obtain your answers.

(d) Based on the stoichiometry you found in part (b), what amount of each reactant would you have to use in this study to give the exact stoichiometric ratio? What mass of solid would you expect from this sample? Explain.

Because they are so common and usually unexciting, precipitation and dissolution reactions seem straightforward and perhaps uninteresting. Consider, however, that you see all about you solids that do not dissolve well or rapidly in water. These include the wood, masonry, metals, and fibers that make up your shelter, your means of transportation, and your clothing. Many of these are complicated ionic solids, like concrete and brick. Most materials are, in fact, rather insoluble in water. That's why the great majority of ions in biological fluids are mostly singly-charged cations and anions, such as Na^+, K^+, and Cl^-, whose salts are water soluble. Your body also contains a good deal of Ca^{2+}, a small amount of which is an important component of fluids, but most of which is present as the solid salts that make up your bones and teeth. It's important to recognize possible precipitation or dissolution reactions as a way of understanding materials and the way nature and humans use them.

Reflection and projection

A chemical reaction is a change that involves rearrangements of electrons and/or atoms to form products that are different from the reactants. You listed several kinds of evidence you can use to tell whether a change produces new compounds, we tested one reaction, and saw how the method of continuous variations can be used to get information about reaction stoichiometry. Chemical reactions begin when a positive center in one molecule is attracted to a negative center in another. Subsequent rearrangements of electrons and/or atoms then lead to products. We are classifying reactions into three categories: ionic precipitation reactions, Lewis acid-base reactions, and reduction-oxidation reactions. Ionic precipitation reactions and the accompanying stoichiometric calculations have been a reminder of concepts you met previously in Chapter 2. Acid-base reactions, proton transfers, were also introduced in Chapter 2. In this chapter, based on our molecular model from Chapter 5, we will first look more deeply at proton transfers and then broaden the acid-base concept to include other reactions.

Section 6.3. Lewis Acids and Bases: Definition

The Brønsted-Lowry model of acids and bases introduced in Chapter 2, Section 2.12, is a special case of a more general model of acid-base reactions. G. N. Lewis, the same chemist who developed the Lewis bonding model that we have been using, proposed an acid-base model now called the **Lewis acid-base model**. The Lewis acid-base model, like his bonding model, is based on electron pairs. A **Lewis base** is an *electron pair donor*. A **Lewis acid** is an *electron pair acceptor*. One product of a Lewis acid-base reaction is a new molecule in which the electron pair from the Lewis base forms a covalent, electron-pair bond with the Lewis acid.

Any molecule or ion with one or more pairs of nonbonding electrons can react as a Lewis base. A molecule with a nonbonding pair of electrons is almost always polarized with the nonbonding pair as a partial negative center. Lewis bases have centers of negative charge. Therefore, cations or molecules with partial positive centers are most likely to react as Lewis acids. These generalizations help to classify a large number of reactions as Lewis acid-base reactions. But we need specific examples to make the model useful for predicting whether reactions between two molecules will occur and, if so, what the products will be.

6.12. Consider This   *How do ionic precipitation and Lewis acid-base reactions differ?*

Ionic precipitation reactions such as reaction (6.4) are usually not classified as Lewis acid-base reactions. The reactants in this reaction are an anion, $C_2O_4^{2-}$, that has several pairs of nonbonding electrons, a Lewis base. The cation, Ca^{2+}, seems that it could accept a pair of electrons and act as a Lewis acid. Why don't we call the precipitation of the ionic salt, CaC_2O_4, a Lewis acid-base reaction? Explain your response clearly.

Types of Lewis acid-base reactions

Many kinds of reactions can be classed as Lewis acid-base reactions; we are going to include three types. In the example reaction for each type, the Lewis base is the first reactant. The only electrons shown explicitly are the electron pair that is donated and the electron pair bond formed in the product. The reaction types are:

(1) Brønsted-Lowry acid-base reactions; proton transfer to a Lewis base:

$$:CN^-(aq) + H_2O(l) \rightleftharpoons H-CN(aq) + OH^-(aq) \tag{6.6}$$

(2) formation of metal ion complexes:

$$6 :CN^-(aq) + Fe^{3+}(aq) \rightleftharpoons \begin{bmatrix} NC-Fe \begin{smallmatrix} CN \\ | \\ \end{smallmatrix} \overset{CN}{\underset{CN}{\diagup}} \\ NC \end{bmatrix}^{3-} (aq)$$ (6.7)

(3) nucleophile (Lewis base)-electrophile (Lewis acid) reactions:

$$CH_3\ddot{O}H(l) + CH_3CH_2\overset{\delta+ \ \delta-}{CHO}(l) \rightleftharpoons CH_3CH_2\underset{OCH_3}{CHOH}(l)$$ (6.8)

In the next sections, we will go more deeply into each of these Lewis acid-base reaction types with further examples, investigations, and questions to help you identify and predict reaction products.

Section 6.4. Lewis Acids and Bases: Brønsted-Lowry Acid-Base Reactions

---

6.13. Investigate This *Do the acidities of acids differ?*

Do this as a class investigation and work in small groups to discuss and analyze the results. Use a pH meter and pH electrode (or short range pH indicator papers) to determine the pH (to the nearest 0.1 pH unit) of distilled (or deionized) water and the following 0.1 M aqueous solutions: hydrochloric acid, HCl; sodium chloride, NaCl; ethanoic (acetic) acid, $CH_3C(O)OH$; and, sodium ethanoate (acetate), $CH_3C(O)ONa$.

6.14. Consider This *How do the acidities of acids differ?*

List water and the solutions in Investigate This 6.13 in order of increasing pH. In which, if any, of the solutions is the concentration of hydronium ion, $H_3O^+(aq)$, about the same as in distilled water? Compared to the water, in which solutions does hydronium ion predominate? In which does hydroxide ion predominate? How might you explain these results?

Strong and weak Brønsted–Lowry acids and bases

In the Brønsted–Lowry definition of acids as proton donors and bases as proton acceptors, a **strong Brønsted–Lowry acid** is one that donates all of its acidic protons to water molecules in

WEB Chap 6, Sect 6.4
Interactive animations illustrate
proton transfer reactions and
strong and weak acids.

aqueous solution. With water as the base (proton acceptor), the amount of hydronium ion, $H_3O^+(aq)$, formed is equivalent to the amount of the acid added. In Investigate This 6.13, you found that a 0.1 M HCl solution has a pH ≈ 1. We can rewrite the definition, $pH \equiv -\log_{10}[H_3O^+(aq)]$, to get the molar concentration of hydronium ion:

$$[H_3O^+(aq)] = 10^{-pH} \text{ M} \tag{6.9}$$

For our solution with pH ≈ 1.0, equation (6.9) gives $[H_3O^+(aq)] = 10^{-1}$ M = 0.1 M. The hydronium ion concentration, $[H_3O^+(aq)]$, is the same as the concentration of HCl added to the solution. HCl is a strong Brønsted–Lowry acid, which we can represent by showing "complete" donation of its proton to water:

> Compare equation (6.9) to the pH scale in Figure 2.24, Chapter 2, Section 2.12.

$$HCl(aq) + H_2O(l) \rightarrow H_2OH^+(aq) + Cl^-(aq) \tag{6.10}$$

6.15. Check This   *[$H_3O^+(aq)$] in 0.1 M ethanoic acid solution*

  (a) Use your results from Investigate This 6.12 to calculate the $[H_3O^+(aq)]$ in a 0.1 M aqueous solution of ethanoic (acetic) acid. How does the $[H_3O^+(aq)]$ in this solution compare to the concentration of ethanoic acid present? What conclusion can you draw about the strength of ethanoic acid? Explain your reasoning.

  (b) WEB Chap 6, Sect 6.4.2. Explain clearly how the two molecular-level diagrams on this page are related to the hydrochloric acid and ethanoic acid solutions discussed in the preceding text and part (a).

    A **weak Brønsted–Lowry acid** is one that does not donate all of its acidic protons to water molecules in aqueous solution. As you have found in Check This 6.15, ethanoic acid in aqueous solution does not donate all its acidic protons to water to form hydronium ion. Ethanoic acid is a weak Brønsted–Lowry acid, which we can represent by using double arrows to indicate that its reaction with water reaches equilibrium (with the forward and reverse reactions balancing one another) before all its protons are donated:

$$CH_3C(O)OH(aq) + H_2O(l) \rightleftharpoons H_2OH^+(aq) + CH_3C(O)O^-(aq) \tag{6.11}$$

(The reaction of $HCl(aq)$ to donate its protons also reaches equilibrium, but only when there are essentially no $HCl(aq)$ molecules left; we use the one-way arrow to represent this condition.)

6.16. Check This   *Fraction of proton transfer between ethanoic acid and water*

    Use the reaction stoichiometry from expression (6.11) and your value for $[H_3O^+(aq)]$ in 0.1 M ethanoic (acetic) acid solution (from Check This 6.15) to determine the fraction of the ethanoic acid, $CH_3C(O)OH(aq)$, that transfers a proton to water. Explain your reasoning.

6.17.   **Consider This**   *What is the molecular level interpretation of expression (6.11)?*

  WEB   Chap 6, Sect 6.4.1. Explain how the animation represents the double arrows in reaction expression (6.11). Reaction expressions like (6.11) may suggest that the same two product molecules formed in the forward reaction (left to right) react in the reverse reaction (right to left). Is this what the animation shows? What is the interpretation of reaction expression (6.11) shown by the animation? Explain clearly.

In aqueous solution, a **strong Brønsted–Lowry base** accepts protons from water molecules to form an amount of hydroxide ion, $OH^-(aq)$, equivalent to the amount of base added. The amide ion, $NH_2^-(aq)$ (in solutions of sodium amide, $NaNH_2(s)$, for example), is a strong base:

$$HOH(l) + NH_2^-(aq) \rightarrow HNH_2(aq) + OH^-(aq) \tag{6.12}$$

A **weak Brønsted–Lowry base** does not accept an amount of protons equivalent to the amount of base added, so the $[OH^-(aq)]$ in a weak base solution is not equivalent to the concentration of base added. Recall from Chapter 2 that ammonia, $NH_3(g)$, dissolves readily in water but the solutions contain only modest amounts of hydroxide ion; ammonia is a weak base:

$$HOH(l) + NH_3(aq) \rightleftharpoons HNH_3^+(aq) + OH^-(aq) \tag{6.13}$$

6.18.   **Check This**   *$[OH^-(aq)]$ in an aqueous sodium amide solution*

  When 0.1 mole of sodium amide is dissolved in enough water to make one liter of solution, the pH of the solution is 13. What is the concentration of hydroxide ion, $[OH^-(aq)]$, in the solution? Is your result consistent with reaction expression (6.12)? Explain why or why not. *Note:* In aqueous solutions at 25 °C, the product of the hydronium and hydroxide ion concentrations is 10^{-14} M^2:

$$[H_3O^+(aq)][OH^-(aq)] = 10^{-14} \; M^2 \tag{6.14}$$

Check this on the pH scale, Figure 2.24, Chapter 2, Section 2.12. If you know the concentration of hydronium ion, you can calculate the hydroxide ion concentration, and *vice versa.*

Protons and electron pairs

    Reactions that can be characterized as either Brønsted-Lowry acid-base or Lewis acid-base, differ only in where we focus our attention. In Brønsted-Lowry acids and bases we focus on donating or accepting a proton. In reaction expressions (6.10) through (6.13), we have highlighted the proton that is transferred. A Brønsted-Lowry acid-base reaction always includes two conjugate acid-base pairs that differ only in the loss of a proton by the acid to form its

---

conjugate base. In expression (6.10), for example, the conjugate acid-base pairs are: {$HCl(aq)$ / $Cl^-(aq)$} and {$H_2OH^+(aq)$ / $H_2O(l)$}.

The Lewis model focuses on a nonbonding electron pair and the covalent bond it forms. We can rewrite the previous equations to highlight

> To emphasize the change of a nonbonding pair of electrons to a covalent bond, we show only one pair of valence electrons. In most cases, the Lewis structures have two or more nonbonding electrons.

these electron pairs and covalent bonds and show the Lewis acid-base conjugate pairs. For hydrochloric acid, reaction (6.10), we write:

$$H–Cl(aq) + H_2O:(l) \rightarrow H_2O–H^+(aq) + :Cl^-(aq) \hspace{2cm} (6.15)$$

The base, $H_2O:(l)$, has donated a pair of electrons to form a covalent bond with the proton from the H–Cl(aq). Expressions (6.10) and (6.15) represent the same reaction looked at from different viewpoints. The Lewis conjugate acid-base pairs are: {H–Cl(aq) / :Cl^-(aq)} and {$H_2O–H^+(aq)$ / $H_2O:(l)$}, which are identical to the Brønsted-Lowry conjugate pairs. But, in the Lewis case, our focus is on the electron pair, either nonbonding (in the base) or bonding (in the acid).

6.19. Check This   *Writing Lewis acid-base reactions*

Use reaction expression (6.15) as a model to rewrite (6.11), (6.12), and (6.13), highlighting the electron pairs and covalent bonds that characterize the reactions as Lewis acid-base reactions.

Relative strengths of Lewis bases

A **strong Lewis base** is one that has nonbonding electrons capable of forming a strong covalent bond with a Lewis acid. For the reactions in this section, this is a bond with a proton to form the conjugate Lewis acid. The more strongly the Lewis base holds the proton in its conjugate Lewis acid form, the less likely it is to give up this proton to another Lewis base. Thus, *the stronger the Lewis base, the weaker its conjugate acid.* We can use the results from Investigate This 6.13 to begin a ranking of the **basicity**, base strength, of several Lewis bases.

6.20. Worked Example   *Relative basicities of chloride ion and water*

In Investigate This 6.13, you found that 0.1 M hydrochloric acid has a pH ≈ 1. What is the basicity of the chloride ion, :Cl^-(aq), relative to water, $H_2O:(l)$?

Necessary information: We need our results from above which show that the hydrochloric acid solution contains essentially all :Cl^-(aq) and no H–Cl(aq).

Strategy: If two Lewis bases are present in a solution and in competition for protons, the stronger base will win the competition. ***The weaker base will be less protonated and will be present mainly in its base form; we will find more of the weaker Lewis base in the solution.***

We can also reason that we will find very little of its conjugate Lewis acid.

Implementation: Expression (6.15) represents the reaction in a solution containing $:Cl^-(aq)$ and $H_2O:(l)$ competing for protons. Since the solution contains $:Cl^-(aq)$ and $H_2O-H^+(aq)$, but no $H-Cl(aq)$, we know that $H_2O:(l)$ must have won the competition: $H_2O:(l)$ is a stronger base than $:Cl^-(aq)$. Since, we know that the stronger the base, the weaker its conjugate acid, we also know that $H_2O-H^+(aq)$ is a weaker acid than $H-Cl(aq)$.

Does the answer make sense? We almost never think of the chloride ion as a base, because solutions of simple chloride salts have little affect on the acid-base properties of their solutions. Therefore, to find that it is a weaker Lewis base than water makes sense.

6.21.   **Check This**   *Relative basicities of chloride and hydroxide ions*

    (a) In Investigate This 6.13, how does the pH of the 0.1 M sodium chloride solution compare to the pH of water? Which ion is present in higher concentration in the sodium chloride solution, $:Cl^-(aq)$ or $:OH^-(aq)$? Explain how you get your answer.

    (b) The reaction of $:Cl^-(aq)$ as a Lewis base with water as a Lewis acid, is:

$$H-OH(l) + :Cl^-(aq) \rightleftharpoons H-Cl(aq) + :OH^-(aq) \qquad (6.16)$$

Reaction (6.16) represents a competition for protons between two Lewis bases, $:Cl^-(aq)$ and $:OH^-(aq)$. Which one wins the competition? Which is the stronger Lewis base? Clearly explain the reasoning for your answers.

    (c) Which is the stronger acid, $H-Cl(aq)$ or $H-OH(l)$? Explain.

6.22.   **Worked Example**   *Relative basicities of ethanoate ion and water*

    In Investigate This 6.13, you found that 0.1 M ethanoic acid has a pH between 2 and 3. What is the basicity of the ethanoate (acetate) ion, $CH_3C(O)O:^-(aq)$, relative to water, $H_2O:(l)$?

Necessary information: We need your results from Check This 6.16 which show that, in this ethanoic acid solution, less than 10% of the acid, $CH_3C(O)O-H(aq)$, transfers a proton to water.

Strategy: The strategy for this problem is the same as Worked Example 6.20.

Implementation: In Check This 6.19, you wrote the Lewis acid-base equation for ethanoic acid:

$$CH_3C(O)O-H(aq) + H_2O:(l) \rightleftharpoons H_2O-H^+(aq) + CH_3C(O)O:^-(aq) \qquad (6.17)$$

Expression (6.17) represents the reaction in a solution containing $CH_3C(O)O:^-(aq)$ and $H_2O:(l)$ competing for protons. We know from your calculation (based on the pH measurement) that most of the $CH_3C(O)O:^-(aq)$ is protonated as ethanoic acid $CH_3C(O)O-H(aq)$. The ethanoate ion, $CH_3C(O)O:^-(aq)$, has won the competition: $CH_3C(O)O:^-(aq$ is a stronger Lewis base than $H_2O:(l)$. Since, we know that the stronger the Lewis base, the weaker its conjugate acid, we also know that $CH_3C(O)O-H(aq)$ is a weaker acid than $H_2O-H^+(aq)$.

Does this answer make sense? See Check This 6.23.

6.23.   **Check This**   *Relative basicities of ethanoate and hydroxide ions*

    (a) The reaction of $CH_3C(O)O:^-(aq)$ as a Lewis base with water as the Lewis acid is:

$$H-OH(l) + CH_3C(O)O:^-(aq) \rightleftharpoons CH_3C(O)O-H(aq) + :OH^-(aq) \qquad (6.18)$$

Do you have any evidence from Investigate This 6.13 that this reaction occurs? Explain your response.

    (b) The 0.1 M solution of sodium ethanoate in Investigate This 6.13 is more basic, higher pH, than water. In this solution, ethanoate and water are in competition for protons from water. Recall that the reaction of water with itself produces tiny amounts of hydronium and hydroxide ions. The reaction of ethanoate with water, expression (6.18) produces much more hydroxide. How does this observation show that the conclusion in Worked Example 6.22 makes sense?

    (c) Use the pH you measured for the 0.1 M sodium ethanoate solution to determine the concentration of hydroxide ion, $[:OH^-(aq)]$, in the solution. *Hint*: Use equation (6.14).

    (d) From your result in part (c) and the stoichiometry of reaction (6.18), calculate the concentration of unreacted ethanoate ion, $[CH_3C(O)O:^-(aq)]$, in the solution. Which is present in higher concentration, $CH_3C(O)O:^-(aq)$ ion or $:OH^-(aq)$ ion?

    (e) Which is the stronger Lewis base, $CH_3C(O)O:^-(aq)$ ion or $:OH^-(aq)$ ion? Clearly explain your reasoning.

The results from Worked Example 6.20 and Check This 6.21 show that chloride ion, $:Cl^-(aq)$, is a weaker Lewis base than either water, $H_2O:(l)$, or hydroxide ion, $:OH^-(aq)$. To get our relative Lewis basicities in order, we also need to know how the basicities of water and hydroxide ion compare. In Check This 6.23(b) we recalled the reaction of water with itself:

$$H-OH(l) + H_2O:(l) \rightleftharpoons H_2O-H^+(aq) + :OH^-(aq) \qquad (6.19)$$

In pure water, reaction (6.19) produces 10^{-7} M concentrations of hydronium and hydroxide ions. Almost all the water molecules remain unreacted. In the competition for protons in this solution, hydroxide has won. The hydroxide ion, $:OH^-(aq)$, is a stronger Lewis base than water, $H_2O:(l)$. Therefore, in order of *decreasing* Lewis basicity, we have:

$$\text{(strongest)} \quad :OH^-(aq) > H_2O:(l) > :Cl^-(aq) \quad \text{(weakest)} \tag{6.20}$$

Their conjugate Lewis acids, in order of *increasing* acidity, are:

$$\text{(weakest)} \quad HO-H(l) < H_2O-H^+(aq) < H-Cl(aq) \quad \text{(strongest)} \tag{6.21}$$

Other experiments like those in Investigate This 6.13 and more complex ones in non-aqueous solvents lead to the order of basicities and acidities for Lewis/Brønsted-Lowry bases and acids shown in Table 6.2.

6.24.   **Worked Example**   *pH of a sodium cyanide solution*

Use the information in Table 6.2 to predict whether the pH of a solution of sodium cyanide, NaCN, will be quite high, quite low, or moderately high or low.

Necessary information: We need to know that NaCN dissolves to give $Na^+(aq)$ and $CN^-(aq)$ ions and that $Na^+(aq)$ is a spectator ion in this solution.

Strategy: We write the acid and/or base reactions of $CN^-(aq)$ ions and use Table 6.2 to determine what species will predominate in the solution.

Implementation: The $CN^-(aq)$ ion is a Lewis base:

$$H-OH(l) + :CN^-(aq) \rightleftharpoons H-CN(aq) + :OH^-(aq) \tag{6.22}$$

Since $:OH^-(aq)$ is produced in this reaction, we predict that the solution will contain more hydroxide ion than hydronium ion, so the pH will be above 7. Table 6.2 shows that $:CN^-(aq)$ is a weaker base than $:OH^-(aq)$. Thus, in the solution, most of the $:CN^-(aq)$ will remain unreacted and the solution will be only moderately basic.

Does the answer make sense? In Table 6.2, we see that cyanide ion is near ammonia. We know that ammonia solutions are only moderately basic, so it makes sense that cyanide solutions should also be moderately basic.

6.25.   **Check This**   *pH of an aminium (ammonia-like) chloride solution*

The active ingredients in some eye drops are rather insoluble amines, RNH_2, that are made soluble by converting them to aminium chlorides, RNH_3Cl (called amine hydrochlorides), which dissolve in water to give, $RNH_3^+(aq)$ and $Cl^-(aq)$ ions. Use the information in Table 6.2 to predict approximately how high or low the pH of a solution of RNH_3Cl will be.

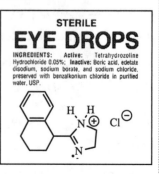

STERILE
EYE DROPS
INGREDIENTS: Active: Tetrahydrozoline Hydrochloride 0.05%; Inactive: Boric acid, edetate disodium, sodium borate, and sodium chloride, preserved with benzalkonium chloride in purified water, USP.

Table 6.2. Relative strengths of Lewis/Brønsted-Lowry bases and acids.

The strongest bases are at the top and the weakest at the bottom, as the blue arrow shows. The strengths of the acids, shown by the red arrow go in the reverse order. The bases and acids that can be compared in aqueous solutions are shaded. The other bases and acids are either too strong or too weak to be differentiated in aqueous solutions.

Base			Conjugate acid		
structure	name		structure	name	
$:CH_3^-$	methide	strong ↑	$H–CH_3$	methane	weak ↓
$:NH_2^-$	amide		$H–NH_2$	ammonia	
$:OCH_2CH_3^-$	ethoxide		$H–OCH_2CH_3$	ethanol	
$:OH^-$	hydroxide		$H–OH$	water	
$:NH_2R$	amine		$H–NH_2R^+ (RNH_3^+)$	"ammonium-like" ion	
$:OCO_2^{2-} (CO_3^{2-})$	carbonate		$H–OCO_2^- (HCO_3^{2-})$	hydrogen carbonate	
(phenolate structure) —O:$^-$	phenolate		(phenol structure) —O–H	phenol	
$:NH_3$	ammonia		$H–NH_3^+ (NH_4^+)$	ammonium	increasing acid strength
$:CN^-$	cyanide	increasing base strength	$H–CN$	hydrogen cyanide	
$:SH^-$	hydrogen sulfide		$H–SH (H_2S)$	hydrosulfuric acid	
$:OC(O)OH^- (HCO_3^-)$	hydrogen carbonate		$H–OC(O)OH (H_2CO_3)$	carbonic acid	
$CH_3C(O)O:^-$	ethanoate (acetate)		$CH_3C(O)O–H$	ethanoic (acetic) acid	
(2,4-dinitrophenolate structure) O_2N— —O:$^-$ with NO_2	2, 4-dinitrophenolate		(2,4-dinitrophenol structure) O_2N— —O–H with NO_2	2,4-dinitrophenol	
$:F^-$	fluoride		$H–F$	hydrogen fluoride	
$:OPO(OH)_2^- (H_2PO_4^-)$	dihydrogen phosphate		$H–OPO(OH)_2 (H_3PO_4)$	phosphoric acid	
$H_2O:$	water		$H_2O–H^+$	hydronium	
$:OSO_2OH^- (HSO_4^-)$	hydrogen sulfate		$H–OSO_2OH (H_2SO_4)$	sulfuric acid	
$:Cl^-$	chloride		$H–Cl$	hydrogen chloride	
$:Br^-$	bromide	weak ↓	$H–Br$	hydrogen bromide	strong ↓
$:I^-$	iodide		$H–I$	hydrogen iodide	
$:OClO_3^- (ClO_4^-)$	perchlorate		$H–OClO_3 (HClO_4)$	perchloric acid	

Note: The conventional representation of some of the molecules and ions is shown in parentheses; compare with Table 2.7, Chapter 2, Section 2.13.

Reflection and projection

What goes on in a chemical reaction does not depend on what we name the reaction. We can name reactions involving proton transfers as either Brønsted-Lowry or Lewis acid-base reactions and the only difference is where we place our emphasis in describing them. The reason we have introduced the Lewis acid-base model is that it is more general and can be extended to cover more reactions, as we will see in later sections. Our emphasis in the Lewis model is on a nonbonding pair of electrons on the Lewis base. We can rank Lewis bases (and, hence, also, their conjugate acids) in order of relative basicity by observing which Lewis base predominates in a solution in which two bases compete for protons: the weaker Lewis base predominates. We can use tables of relative Lewis base (and conjugate Lewis acid) strengths, like Table 6.2, to predict which species will predominate in a solution.

The list doesn't explain *why* one Lewis base is stronger than another. If you have an explanation, you can apply it when you meet Lewis bases and acids that aren't on the list. Our task in the next section is to provide at least some of that explanation.

Section 6.5.   Predicting Strengths of Lewis/Brønsted-Lowry Bases and Acids

Electronegativity and relative Lewis base strength

Electronegativity is one factor that plays a role in determining the strength of Lewis bases. The electronegativities of the elements increase going from left to right across a period of the periodic table:

$$\text{(least electronegative)} \quad C < N < O < F \quad \text{(most electronegative)} \tag{6.23}$$

Compare the order of electronegativities for the second-period elements with these basicities from Table 6.2:

$$\text{(most basic)} \quad :CH_3^- > :NH_2^- > :OH^- > :F^- \quad \text{(least basic)} \tag{6.24}$$

The order of basicities is the reverse of the order of electronegativities of the atom with the nonbonding electron pair. Nonbonding electrons on atoms with high electronegativity are held more tightly and are less available for donation to form a bond with a Lewis acid, including the

> The methide ion, $:CH_3^-$, may be unfamiliar, but methyl lithium, $LiCH_3$, is a reagent that is commonly used to attach a methyl group to a carbonyl carbon. You can think of methyl lithium as $(Li^+)(:CH_3^-)$.

proton. The differences between the basicities of these anions are quite large, more than 10 orders of magnitude. We'll return in Chapter 9 to look at acid-base systems more quantitatively.

6.26. **Check This**   *Relative Lewis base strengths*

(a) WEB Chap 6, Sect 6.5.1. Explain clearly how the charge density models shown on this page help explain the order of basicities in expression (6.24).

(b) If these ions, Cl^-, SH^-, and PH_2^-, could be tested for their relative base strengths, what do you predict the order would be? What would be the relative acid strengths of the conjugate acids? Clearly explain the reasoning for your answers. Do you have any evidence that your order is correct?

6.27. Consider This   *Is electronegativity a reliable predictor of Lewis acid-base strength?*

If Lewis base strength increases inversely with electronegativity, then the conjugate acids should increase in acid strength directly with electronegativity. Consider the hydrogen halides, whose charge density models are shown here (to the same scale):

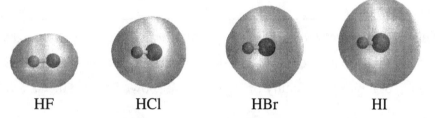

HF   HCl   HBr   HI

(a) How do the electronegativities of the halogens vary from F to I? Assuming there is a direct dependence of Lewis acid strength on the electronegativity of the halogen, what would you predict for the relative order of acid strength of the four hydrogen halides? Explain.

(b) Based on your relative order of acidity in part (a), what would you predict about the relative base strengths of the halide ions, $:F^-$, $:Cl^-$, $:Br^-$, and $:I^-$? Explain.

(c) Are your answers in parts (a) and (b) consistent with data in Table 6.2? Why or why not?

Atomic size and relative Lewis base strength

Electronegativities decrease going down a column (family) of the periodic table. Therefore, based on electronegativity, your prediction for the acid strengths of the hydrogen halides in Consider This 6.27 would be decreasing acidity from HF to HI. And for the halide ions, the conjugate Lewis bases, you would predict increasing Lewis base strength from $:F^-$ to $:I^-$. However, Table 6.2 shows that the halide ions follow exactly the reverse order of Lewis basicity:

$$\text{(most basic)}\quad :F^- > :Cl^- > :Br^- > :I^- \quad \text{(least basic)} \tag{6.26}$$

Electronegativity must not be the whole story. A second factor that affects the strength of a Lewis base is its size. The second-period elements we considered in the previous paragraph are all about the same size (within about 20%), so differences in electronegativity dominate their basicity.

The halide ions increase a great deal in size down the halogen family column of the periodic table; iodide is almost twice the size of fluoride. (You see this great size difference in the models for the hydrogen halides in Consider This 6.27.) The consequence of large size is that the valence electron charge is spread out in a much larger volume. As you have seen in Chapters 4 and 5, ions and molecules are more stable when their valence electrons occupy larger volumes. Iodide has the same number of valence electrons as fluoride, but is more stable and less reactive toward protons, because it is so much larger. The halide ions are such weak Lewis bases, except for fluoride, that their conjugate acids, the hydrogen halides, are essentially completely ionized to hydronium ions and halide ions in aqueous solutions. The order of the halide basicities (or hydrogen halide acidities) in Table 6.2 has to be determined from experiments in solvents other than pure water.

6.28.   Consider This   *Does electronegativity or size dominate Lewis acid-base strength?*

Charge density models for the hydrides (binary compounds of the elements with hydrogen) of the first four members of the oxygen family of elements are shown here (to the same scale):

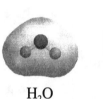  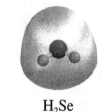

H_2O H_2S H_2Se H_2Te

(a) Based on the electronegativities of the oxygen family elements, what would you predict for the relative order of acid strength of these oxygen-family hydrides? Explain.

(b) Based on the sizes of the oxygen family elements, what would you predict for the relative order of acid strength of these oxygen-family hydrides? Explain.

(c) Which of these acids do you find in Table 6.2? What is the order of their acidities? Assuming this trend is maintained for all these acids, what is the relative order of their acid strengths? What is the relative order of base strengths for :OH⁻, :SH⁻, :SeH⁻, and :TeH⁻?

(d) What is the answer to the title question? Explain your reasoning.

6.29.   Check This   *Relative Lewis base strengths*

If phosphine, $:PH_3$, reacts as a Lewis base, would you expect it to be a stronger or weaker base than ammonia? Explain the reasoning for your prediction.

Oxyacids and oxyanions: carboxylic acids

The acids we have considered so far in this section are **binary acids**, compounds of hydrogen and one other element. These are interesting and important compounds, especially as a beginning

---

for understanding the factors that affect the strength of Lewis bases and acids, but there are only a small number of binary acids. The most common proton-donating acids are the oxyacids, which we introduced in Chapter 2, Section 2.13. Recall that the acidic protons of oxyacids are always bonded to an oxygen atom which is, in turn, bonded to another atom. The oxyanions of oxyacids are Lewis bases with one or more oxygen atoms that have nonbonded pairs of electrons that can bond to a proton. Almost all the acids in biological systems are oxyacids and most of these are carboxylic acids or derivatives of phosphoric acid.

6.30. Investigate This   *Do the acidities of alcohols and carboxylic acids differ?*

Use pH paper or a pH meter to measure the pH of water and 0.1 M solutions of ethanol and ethanoic (acetic) acid in water. Is the acidity of the alcohol greater than, less than, or the same as the acidity of the carboxylic acid? Show how you use your pH data as evidence for your answer.

The water, ethanol, and ethanoic acid in Investigate This 6.30 are oxyacids:

$$\text{H–OH}(l) + \text{H}_2\text{O:}(l) \rightleftharpoons \text{H}_2\text{O–H}^+(aq) + \text{:OH}^-(aq) \tag{6.19}$$

$$\text{CH}_3\text{CH}_2\text{O–H}(aq) + \text{H}_2\text{O:}(l) \rightleftharpoons \text{H}_2\text{O–H}^+(aq) + \text{CH}_3\text{CH}_2\text{O:}^-(aq) \tag{6.27}$$

$$\text{CH}_3\text{C(O)O–H}(aq) + \text{H}_2\text{O:}(l) \rightleftharpoons \text{H}_2\text{O–H}^+(aq) + \text{CH}_3\text{C(O)O:}^-(aq) \tag{6.17}$$

Reaction expressions (6.19) and (6.17) remind you of the acid-base properties of water and ethanoic acid discussed in the previous section. Reaction (6.27) is the corresponding reaction of ethanol as an oxyacid. (Only electrons involved in the Lewis acid-base reaction are shown.)

6.31. Consider This   *How do the acidities of alcohols and carboxylic acids differ?*

(a) Do your results in Investigate This 6.30 provide any evidence for or against reaction (6.27)? What is the evidence?

(b) Based on your results, how would you rank the acid strengths of water, ethanol, and ethanoic acid? How would you rank the base strengths of the hydroxide, :OH$^-$(aq), ethoxide, CH$_3$CH$_2$O:$^-$(aq), and ethanoate (acetate), CH$_3$C(O)O:$^-$(aq), anions? Explain your reasoning.

Water and aqueous solutions of ethanol have essentially the same pH, so you can conclude that reaction (6.27) does not proceed to a measurable extent to produce hydronium ions. You know that reaction (6.17) does proceed to a small extent to give the ethanoic acid solution a moderately acidic pH. Therefore, you can conclude that ethanoic acid is a stronger acid than ethanol (or, to look at these molecules from the point of view of their conjugate bases, ethanoate anion is a weaker base than ethoxide anion). Since reaction (6.27) does not proceed as written, we can conclude that ethanol and water must have approximately the same acid strength (and ethoxide anion and hydroxide anion about the same base strength).

6.32. Check This   *Relative basicities of ethanoate, ethoxide, and hydroxide anions*

(a) Are the rankings in Table 6.2 consistent with the conclusions about relative basicities in the previous paragraph? Explain why or why not.

(b) Write Lewis structures for these three ions. Which Lewis structure(s) can be written in more than one equivalent way? Can you think of any reason why ethanoate should be a weaker base than ethoxide or hydroxide? Explain.

Delocalized π bond in carboxylate

You have found that there is only one way to write the Lewis structures of ethoxide and hydroxide, but there are two equivalent ways to write the Lewis structure of the carboxylate group of the ethanoate anion: and . Recall, from Chapter 5, Section 5.7, that when two or more equivalent Lewis structures can be written for a molecule (or ion), none of the

> **WEB Chap 6, Sect 6.4.3-4.**
> View animations of ethanoic acid and ethanoate ion charge densities and proton transfers.

structures is correct. The pi electrons in the molecule are spread over several atoms in a delocalized orbital or orbitals. Figure 6.2 shows the overall charge density for the ethanoate (acetate) ion. Observe the equal charge density on the oxygens.

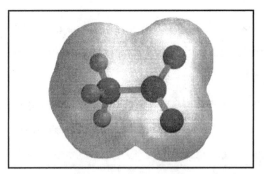

Figure 6.2. Charge density model for the ethanoate ion.

Energetics of proton transfer

Delocalization of the pi electrons in the ethanoate anion lowers the energy of the ion, that is, makes it more stable relative to similar molecules that lack such electron delocalization. The stability of the ethanoate anion is the major reason why it is such a weak base compared to the ethoxide anion. Figure 6.3 shows energy-level diagrams comparing the energies for proton transfer from ethanol to water and ethanoic acid to water, reactions (6.27) and (6.17), respectively. The lower total energy of a carboxylate ion makes a large difference in the energy required for proton transfer from the carboxylic acid and the alcohol. For carboxylic acids, transfer of a proton to water requires almost no energy; the reactants and products have just about the same energy.

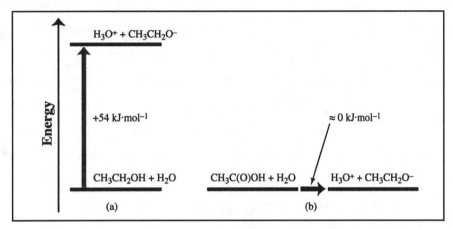

Figure 6.3. Energy-level diagrams for proton transfer by (a) ethanol and (b) ethanoic acid.

You might wonder why, if no energy is required, only a few percent of the ethanoic acid molecules in a 0.1 M aqueous solution transfer their proton to water. Recall that you have been confronted with this same kind of puzzle before. For example, in Chapter 2, you found that both exothermic and endothermic solubility reactions occur. Energy is often a guide for *comparing similar reactions*. In Figure 6.3, the comparison is between proton transfers from electrically neutral oxyacids to water to form hydronium ion and oxyanions with a −1 charge. Energy is not, however, a reliable guide for predicting the extent of *individual* acid-base (or solubility-precipitation) reactions. The missing piece of the puzzle is the entropy of these reactions, which we will consider in Chapters 8 and 9.

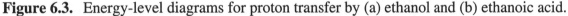

6.33. **Check This**   *Relative basicities of oxyanions and acidities of oxyacids*

(a) WEB Chap 6, Sect 6.5.2. How is the cyclohexanoxide anion on this page similar to the ethoxide anion? How is the phenoxide anion similar to the ethanoate anion? Is your selection of the weaker base consistent with these similarities? Explain your reasoning. Is your selection of the weaker base consistent with the information in Table 6.2? Explain.

(b) Write the reaction for the transfer of a proton from carbonic acid, $(HO)_2CO(aq)$, to water. Write the Lewis structure(s) for the hydrogen carbonate anion formed in this reaction.

(c) What would you predict about the relative acidities of carbonic acid and ethanoic acid? Give the reasoning for your prediction. Is your prediction consistent with the information in Table 6.2? Why or why not?

Oxyacids and oxyanions of third-period elements

Our reasoning about relative acid or base strengths based on delocalization of electrons can be extended to oxyacids and oxyanions whose molecules contain a third period element, phosphorus, sulfur, or chlorine, to which one or more –OH groups is bonded. These compounds

> Phosphates, sulfates, and their acids are vital components of living systems and are important in many industrial processes. Perchlorates are often used in studies where the very low basicity of the anion gives desirable solution properties.

also often have other oxygen atoms bonded to the third-period element. Figure 6.4 shows the four oxyacids of chlorine in order from least acidic, hypochlorous acid, HOCl, to most acidic,

perchloric acid, $HOClO_3$. Chloric and perchloric acids are stronger acids than the hydronium ion and transfer their protons completely to water in aqueous solution. Experiments in other solvents are required to determine the order of their acid strengths shown in the figure.

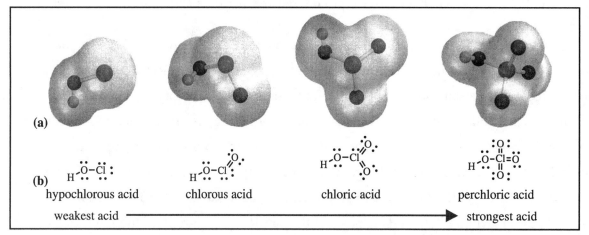

Figure 6.4. The oxyacids of chlorine.
Both (a) charge density models and (b) Lewis structures are shown for comparison.

6.34. Consider This   *How can you explain the relative acidities of the chlorine oxyacids?*

(a) Write the Lewis structures for the oxyanions formed when the chlorine oxyacids transfer a proton to a base. For which one(s) can you write two or more equivalent structures? How many more equivalent structures?

(b) Is there any correlation between your answers in part (a) and the relative strengths of the oxyacids? If so, suggest an explanation for the correlation.

You have found that there are, respectively, one, two, three, and four equivalent Lewis structures for the hypochlorite, ClO^-, chlorite, ClO_2^-, chlorate, ClO_3^-, and perchlorate, ClO_4^-, anions. The larger the number of equivalent Lewis structures, the greater the delocalization of the electrons and the more stable the oxyanion. The base strength of the oxyanions decreases as they become more stable. The perchlorate anion is the weakest base and, therefore, as we see in Figure 6.4, perchloric acid is the strongest acid. In Chapter 9, we will treat acid-base reactions more quantitatively and be able to assign numerical values to acid and base strength. The difference in acid strength going from one acid to the next in Figure 6.4 is almost a factor of a million, 10^6; perchloric acid is about 10^{17} times stronger than hypochlorous acid.

6.35. Check This   *Acid-base properties of bleach solutions*

Most liquid bleaches contain sodium hypochlorite, $NaOCl$, which ionizes in solution to give $Na^+(aq)$ and $ClO^-(aq)$ ions. Would sodium hypochlorite make bleach solutions acidic or basic? Give the reasoning for your answer.

6.36. Check This   *Another way to predict relative strengths of oxyacids and oxyanions*

(a) How many oxygen atoms are doubly bonded to chlorine in each of the oxyacids in Figure 6.4? Is there a correlation between these numbers and the relative strengths of the oxyacids? If so, formulate a rule that predicts the relative acid strengths of oxyacids (and relative base strengths of their oxyanions).

(b) What does your rule from part (a) predict about the relative basicity of the hydrogen sulfate, $HOSO_3^-$, and hydrogen sulfite, $HOSO_2^-$, oxyanions? Is this also what the delocalization of electrons in these ions predicts? Explain how you make your predictions.

Your results in Check This 6.36 show an alternative way to predict the order of acid or base strength for oxyacids and oxyanions that does not require you to write the Lewis structures for the oxyanions. Keep in mind, however, that this rule is a correlation, not an explanation. The explanation for the basicity and acidity is the delocalization of electrons and consequent greater stability of the oxyanions.

6.37. **Consider This**   *How do oxyacids/oxyanions of different central atoms compare?*

(a) Write the Lewis structure for phosphoric acid, $(HO)_3PO$. Write the reaction for transfer of one proton from phosphoric acid to water. What is the Lewis structure for the dihydrogen phosphate oxyanion formed in this reaction?

(b) Experimentally, we find that phosphoric and chlorous acids are about equal in acid strength. Is this what your rule from Check This 6.36(a) predicts? Is this what the delocalization of electrons in the oxyanions predicts? From this example, what conclusion might you draw about the influence of the central atom (phosphorus or chlorine here) on the strength of the oxyacids? State your reasoning clearly.

(c) Draw Lewis structures for the oxyanions of nitrous acid, HONO, and chlorous acid, HOClO. Experimentally, we find that the nitrite oxyanion is a somewhat stronger base than the chlorite oxyanion. Is this what your rule from Check This 6.36(a) predicts? Is this what the delocalization of electrons in the oxyanions predicts? From this example, what conclusion might you draw about the influence of the central atom (nitrogen or chlorine here) on the strength of the oxyacids? State your reasoning clearly.

Comparisons within and between periods

In Consider This 6.37(b), you find that two third-period oxyacids whose oxyanions have the same amount of electron delocalization have about the same acid strength. As a general rule, the relative acid (or base) strengths for oxyacids (or oxyanions) of elements in the same period of the periodic table depend on the amount of electron delocalization in the oxyanions and are little affected by the identity of the central atom. Comparisons between oxyacids/oxyanions of elements from different periods of the periodic table, as in Consider This 6.37(c), show that the period does make a difference. Second row oxyanions are stronger bases than third row oxyanions with the same electron delocalization (as indicated by the number of equivalent Lewis structures). This observation suggests that the third period oxyanions are more stable than their second period counterparts. Recall that third period atoms are larger than second period atoms, so third period oxyanions are somewhat larger than second period oxyanions. Thus delocalized electrons occupy a larger volume in the third-period oxyanions and are a bit lower in energy than the second-period oxyanions.

6.38. Check This   *Relative acid and base strengths*

(a) List these oxyanions in order from the weakest base to the strongest base. Explain the reasoning for your ranking.

$$CH_3CO_2^-, HOCO_2^-, NO_3^-, NO_2^-, OH^-$$

(b) The following oxyanions have at least one proton available to transfer to a base. List these ions in order from the strongest to the weakest *acid*. Explain your reasoning.

$$HOSO_3^-, HOSO_2^-, (HO)_2PO_2^-, HOPHO_2^- \text{ (one non-acidic H bonded to P)}$$

(The names of these oxyanions are hydrogen sulfate, hydrogen sulfite, dihydrogen phosphate, and hydrogen phosphite.)

(c) List the following acids (carbonic, phosphoric, and sulfuric) in order from the strongest acid to the weakest acid. Explain your reasoning.

$$(HO)_2CO, (HO)_3PO, (HO)_2SO_2$$

Predominant acid and base in a reaction

The point of our discussion of relative strengths of acids and bases is to enable you to tell, without doing any mathematics or making any measurements, what chemical species will predominate in aqueous solutions of acids and bases. You can use Table 6.2 or the reasoning from above, for example, to determine which of two bases in a solution is the stronger. The stronger will win the competition for available protons. The concentration of the weaker base will be larger than the concentration of the stronger base. This is the reasoning we used to find relative basicities in Section 6.4. As you have seen, we can reverse the process to predict which base (and acid) will predominate in a reaction between acids and bases. There is one stumbling block to watch out for in aqueous solutions. The predominant acid-base species in aqueous solutions is water; its concentration is close to 55 M. When you are asked about the predominant bases and acids in an aqueous solution, you are almost always being asked about the acids and bases *other than water* itself.

6.39. Worked Example   *Predicting the direction of reaction for acid-base reactions*

If equal volumes of 0.1 M solutions of hydrochloric acid and sodium cyanide, NaCN, are mixed, what species will predominate in the mixture? Will the mixture be acidic or basic?

Necessary information: We have to know that NaCN(s) dissolves in water to give $Na^+(aq)$ and $CN^-(aq)$ and that $Na^+(aq)$ is a spectator ion in the mixture. From Table 6.2 we find that $CN^-(aq)$ is a stronger base than either water or $Cl^-(aq)$. We also have to recall that a 0.1 M solution of hydrochloric acid is 0.1 M in hydronium ion, $H_3O^+(aq)$.

Strategy: Write the reaction between the reactants that are mixed and use its stoichiometry to determine the species that would be present, if no further reactions occurred. Then write the acid-base reaction(s) for these species and use the relative basicities (or acidities) of the reactants and products to determine which species will predominate.

Implementation: Since cyanide is a stronger base than chloride, protons from the hydronium ions in the hydrochloric acid solution will be transferred to cyanide when the solutions are mixed:

$$H_3O^+(aq) + Cl^-(aq) + Na^+(aq) + CN^-(aq) \rightleftharpoons HCN(aq) + H_2O(l) + Na^+(aq) + Cl^-(aq) \qquad (6.28)$$

Note that, because it is such a weak base, chloride ion is also a spectator ion in this mixture.

Reaction (6.28) shows that $H_3O^+(aq)$ and $CN^-(aq)$ react in a 1:1 ratio. Equal volumes of 0.1 M solutions of hydrochloric acid and sodium cyanide contain the same number of moles $H_3O^+(aq)$ and $CN^-(aq)$, so the reaction uses up both reactants to produce this number of moles of HCN(aq). In a solution of HCN(aq), the acid-base reaction is:

$$HCN(aq) + H_2O(l) \rightleftharpoons H_3O^+(aq) + CN^-(aq) \qquad (6.29)$$

Since $CN^-(aq)$ is a stronger base than water, we know that it will win the competition for protons. The reactants, HCN(aq) (and $H_2O(l)$), will predominate in the mixture, which also contains all the unreacted $Cl^-(aq) + Na^+(aq)$. Reaction (6.29) proceeds to some extent to produce small amounts of $H_3O^+(aq)$ (and $CN^-(aq)$), so the solution will be somewhat acidic.

Does the answer make sense? The stoichiometry of the reaction produces a solution that is identical to what we would get by dissolving HCN in salt water. The salt does not affect the acidity of the solution. HCN(aq) is a weak acid, so it will transfer only a few of its protons to water to produce a weakly acidic solution and a low concentration of $CN^-(aq)$. This is what we concluded above.

6.40. Check This   *Predicting the direction of reaction for acid-base reactions*

Phenol, Table 6.2, is a solid that is moderately soluble in water at room temperature. The phenolate anion is soluble in water to a much greater extent. If you want to dissolve a lot of phenol in an aqueous solution, should you use a solution of hydrochloric acid or a solution of sodium hydroxide to try to dissolve it? Clearly explain the reasoning for your choice.

Some cases can be a little complicated to sort out. For example, both the hydrogen tartrate anion, $HOOC(CHOH)_2COO^-(aq)$, and the hydrogen carbonate anion, $HOCO_2^-(aq)$, can accept and donate protons. When these anions are mixed, there are two possible acid-base reactions:

$$HOOC(CHOH)_2COO^-(aq) + HOCO_2^-(aq) \rightleftharpoons HOOC(CHOH)_2COOH(aq) + CO_3^{2-}(aq) \quad (6.30)$$

                                             tartaric acid   carbonate

$$HOOC(CHOH)_2COO^-(aq) + HOCO_2^-(aq) \rightleftharpoons {}^-OOC(CHOH)_2COO^-(aq) + (HO)_2CO(aq) \quad (6.31)$$

                                             tartrate   carbonic acid

The data in Table 6.2 show that the carbonate dianion is a stronger base than carboxylate anions like the hydrogen tartrate anion. Thus, reaction (6.30) is not likely to proceed too far toward products and can be eliminated as the predominant reaction in this mixture.

To analyze reaction (6.31), we have to compare the base strength of hydrogen carbonate anion with the base strength of a carboxylate anion, one of the ends of the tartrate dianion. In Check This 6.33, you considered this comparison from the point of view of the conjugate acids (carbonic and carboxylic acids) and found that they were of about equal strength. Thus, the bases are also about equal in strength. Without having more quantitative information about the relative base (or acid) strengths, we can reasonably conclude that reaction (6.31) is likely to proceed to yield the products, but, at equilibrium, there will be a good deal of the reactants left as well.

Recall that this is the reaction mixture from Investigate This 6.3 which produced gaseous carbon dioxide as a product. On that basis, we wrote this net equation for the reaction (with hydrogen tartrate and tartrate represented by molecular formulas):

$$HC_4H_4O_6^-(aq) + HCO_3^-(aq) \rightarrow C_4H_4O_6^{2-}(aq) + CO_2(g) + H_2O(l) \qquad (6.3)$$

In Chapter 2, Section 2.17, we saw that when carbonic acid is formed in a solution, it can react to release carbon dioxide gas:

$$(HO)_2CO(aq) \rightleftharpoons CO_2(g) + H_2O(l) \qquad (6.32)$$

Reaction (6.32) "uses up" the carbonic acid formed by reaction (6.31) and the system adjusts by producing more, which drives the reaction toward products and explains the single direction arrow in reaction expression (6.3).

This analysis is an example of Le Chatelier's principle, which we introduced in Chapter 2, Section 2.15. **Le Chatelier's principle** says that when systems at equilibrium are disturbed they respond in a way that minimizes the effect of the disturbance. The equilibrium represented by reaction expression (6.31) is disturbed when some of the carbonic acid reacts further to produce carbon dioxide, reaction (6.32). The system responds by making more carbonic acid to compensate for the loss. This process continues until the reactants are used up. The mixture of hydrogen tartrate and hydrogen carbonate anions represents the most complicated system we will consider. We usually will focus on acid-base interactions that are easier to analyze.

Reflection and projection

For binary Lewis/Brønsted-Lowry acids, two properties of the nonhydrogen atom in the acid explain the relative acidities and basicities shown in Table 6.2. Within a period of the periodic table, basicity decreases with increasing electronegativity and, conversely, the acidity of the conjugate acids increases. Within a family (column) of the periodic table, the basicity decreases with increasing size of the atom. In comparisons between two binary acids, the size effect, due to greater volume for the electron waves, usually dominates.

Most acids and bases are oxyacids and oxyanions in which the Lewis-base electron pairs are always on an oxygen atom. The relative basicity of the oxyanions is determined by electron delocalization which lowers the energy of the anion and makes it a weaker base. You get a qualitative measure of the relative amount of delocalization by counting the number of equivalent Lewis structures for the oxyanion. The identity of the central atom, for atoms in the same period, has little affect on the basicity of oxyanions. Oxyanions with central atoms from the third period are weaker bases than oxyanions of second period atoms, because of the increased size of the third period atoms.

Now you have the background to predict the direction of many Lewis/Brønsted-Lowry acid-base reactions. In the next sections, we will expand our view of Lewis acid-base reactions to include Lewis acids that are not proton donors.

Section 6.6.   Lewis Acids and Bases: Metal Ion Complexes

6.41. **Investigate This**   *Do Lewis bases react with metal ions?*

For this investigation, use a 0.1 M aqueous solution of nickel chloride, $NiCl_2$, and a 0.5 M aqueous solution of ammonia, NH_3. Record the colors of the solutions. Mix equal volumes of the two solutions. Record the color of the mixture. Does a reaction occur between the ingredients of the mixture? What is the evidence for your answer?

In Chapter 2, Section 2.5, we used ion–dipole attraction to explain how ionic solids dissolve. Water molecules were pictured as surrounding the cation with the negative end of the water dipoles oriented toward the cation. What that picture fails to convey is that, for many cations, a fixed number of water molecules arrange themselves *symmetrically* around the cation. The structure shown in Figure 6.5 is an example of a **metal ion complex**. Metal ion complexes form between cations and many different kinds of neutral molecules and ions. The molecules and ions that are attached to the central metal ion are called **ligands**, because they bind to the cation.

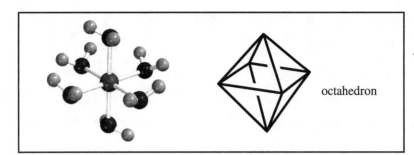

Figure 6.5. An octahedral complex ion, $Cr(H_2O)_6^{3+}$, and a regular octahedron.

All ligands have one or more atoms with nonbonding electron pairs that can be shared with

> Ligand (and the ligaments that connect your bones) are from the Latin *ligare* = to bind.

the central metal cation. ***All ligands are Lewis bases; the central metal ion is reacting as a Lewis acid.*** The structure shown in Figure 6.5 is the chromium, Cr^{3+}, ion in aqueous solution. It is often written as $Cr(H_2O)_6^{3+}$, rather than as $Cr^{3+}(aq)$, to show the stoichiometry of the complex ion. The complex is called an **octahedral complex ion** because the geometry of the atoms

>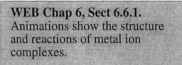
> **WEB Chap 6, Sect 6.6.1.**
> Animations show the structure and reactions of metal ion complexes.

bonded to the central metal ion is octahedral. The oxygens are at the vertices of an imaginary octahedron around the metal ion.

6.42. Consider This   *How do you know a Lewis acid-base complex ion has reacted?*

(a) Assume that $Ni^{2+}(aq)$ is an aqueous octahedral complex like the one shown for $Cr^{3+}(aq)$ in Figure 6.5. Make a molecular model of the nickel complex. What is one of the properties of this complex that you can infer from Investigate This 6.41? Explain.

(b) Which is the stronger Lewis base, water or ammonia? In a competition for a Lewis acid, which will win? How is your answer relevant to your observations in Investigate This 6.41? Explain.

(c) Make a molecular model of the product of the reaction in Investigate This 6.41. Explain why you think this is the product. What is one of the properties of this product?

Colors of metal ion complexes

Color is one of the most striking properties of complex ions. Most of the complexes formed by **transition metal** ions are colored. Transition metals are the metals in the middle of the periodic table shown inside the front cover. The "transition" is a result of electrons occupying a subshell (the *d* orbitals, Chapter 4, Section 4.11) with energies that are approximately the same as those of the elements at either end of the transition. The wavelengths of light these electrons absorb as they change energy within the subshell are in the visible region of the spectrum, so the ions are colored. The energies of the subshell electrons and the wavelengths absorbed are influenced by interactions with ligand electrons. You have seen that nickel(II) is green when complexed with water, $Ni(H_2O)_6^{2+}(aq)$, blue with ammonia, $Ni(NH_3)_6^{2+}(aq)$, and red with dimethylglyoxime, Check This 6.11. Much of the color you see around you in nature and in human decoration is due to metal ion complexes.

Note that we have introduced a new notation, nickel(II), for the nickel cation with a +2 charge, that is, a nickel atom that has lost two electrons. The Roman numeral in parentheses next to the name of the element denotes the charge on the metal ion. This notation is often used for metals, especially the transition metals, that form compounds in which the cation has two or more different charges. Iron, for example, commonly forms compounds in which the iron has a

> The older nomenclature, which you will still find in use, assigns different names to the cations with different charges. For example, iron(II) is called ferrous and iron(III) is called ferric, so FeO is ferrous oxide and Fe_2O_3 is ferric oxide.

2+ or 3+ charge. Two of the oxides of iron are iron(II) oxide, FeO, and iron(III) oxide, Fe_2O_3. This notation makes it easier to name compounds unambiguously, without having to remember the names formerly used for metal cations with different charges. We will use this notation when it is convenient to do so.

6.43. Investigate This   *Do calcium ions react with complexing ligands?*

Do this as a class investigation and work in small groups to discuss and analyze the results. You will use aqueous solutions of 0.1 M calcium nitrate, $Ca(NO_3)_2$, 0.1 M sodium oxalate, NaC_2O_4, 0.1 M sodium oleate, $NaO(O)C(CH_2)_7CH=CH(CH_2)_7CH_3$ (an ingredient in some soaps), and 0.1 M tetrasodium **ethylenediaminetetraacetic acid** (**EDTA**), $Na_4C_{10}H_{12}N_2O_8$ (used as a preservative in many foods and found in many other consumer products).

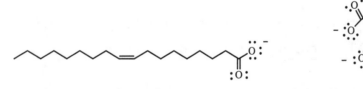

oleate anion   ethylenediaminetetraacetic acid (EDTA)$^{4-}$ anion

(a) Add about 2 mL of water and 4 drops of calcium nitrate solution to each of two test tubes. To one of the test tubes, add 5 drops of sodium oxalate solution, swirl gently to mix the solution, and record your observations. Add 5 drops of EDTA solution, gently swirl to mix the solution, and record your observations.

(b) Repeat with the other test tube, except use the oleate solution instead of the oxalate.

6.44.   Consider This   *How do calcium ions react with complexing ligands?*

   (a) In Investigate This 6.43, what happened when oxalate or oleate was added to calcium ion solutions? Did you expect these results? Explain why or why not.

   (b) What happened when EDTA solution was added to the mixtures? How can you account for these results? Does the same explanation apply to both mixtures? Explain why or why not.

Since you have seen it before, it isn't surprising to see a precipitate form when calcium and oxalate ions are mixed. Calcium oleate, $Ca[O(O)C(CH_2)_7CH=CH(CH_2)_7CH_3]_2$, is also insoluble in water, because the long nonpolar, hydrocarbon tail of oleate is not soluble in water. Anions with extended carbon frameworks often form insoluble salts. One of the ions responsible for "hard" water is $Ca^{2+}(aq)$. Soap scum and bathtub rings are precipitates of these metal ions with the ingredients of soap like sodium oleate.

Calcium-EDTA complex ion formation

   Perhaps you were surprised to observe that addition of $EDTA^{4-}(aq)$ to solutions containing precipitates of calcium oxalate and calcium oleate resulted in the disappearance of the precipitates. When the calcium oxalate precipitate goes back into solution, the calcium and oxalate ions must again be present in the solution from which they precipitated. This isn't possible (without forming a precipitate), so some reaction must have occurred to "use up" either the calcium ion or the oxalate ion or both. The extra ingredient in the solution, $EDTA^{4-}(aq)$, must be one of the reactants in this new reaction. It's not likely that the negatively charged $C_2O_4^{2-}(aq)$ and $EDTA^{4-}(aq)$ will react with one another; they strongly repel one another. The positive $Ca^{2+}(aq)$ cation is a likely candidate to react with the $EDTA^{4-}(aq)$.

6.45.   Investigate This   *What is the three-dimensional structure of EDTA?*

   Use your model kit to make a molecular model of the ethylenediaminetetraacetate tetraanion, $EDTA^{4-}$. To make it easier to handle and visualize the structure, leave off all the hydrogens, as they have been in the skeletal structure shown in Investigate This 6.43. Use the long yellow connectors in your kit to represent the nonbonding electron pairs on nitrogen and one of the nonbonding electron pairs on each charged oxygen. Twist your model about to show that all six of the yellow connectors (nonbonding electron pairs) can be directed toward a central point.

   What kind of complex ion can $EDTA^{4-}(aq)$ form with $Ca^{2+}(aq)$? In Investigate This 6.45, you manipulated your $EDTA^{4-}$ model to bring nonbonding electron pairs from six of the atoms close together around a central point. If there is a positive cation at the center of this array, the electron

---

pairs (Lewis bases) and the cation (Lewis acid) attract each other (bonds are made) and a stable

> **WEB Chap 6, Sect 6.6.2.**
> Observe animations of $EDTA^{4-}$ and the formation and structure of its metal ion complex.

structure is formed. Figure 6.6(a) shows this structure for a model of $EDTA^{4-}$ wrapped around and bonded to a central metal cation.

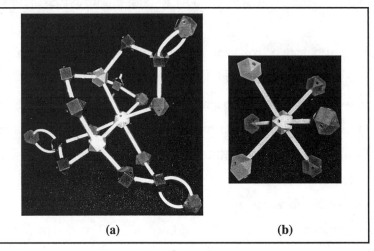

(a)   (b)

Figure 6.6. Structure of the $EDTA^{4-}$ complex with a central metal ion.

(a) The entire complex with none of the $EDTA^{4-}$ hydrogens shown. The metal ion is gray. (b) To make the geometry clearer, only the octahedral array of atoms directly bonded to the metal ion is shown.

$EDTA^{4-}$ is only one of many ligands that contain two or more Lewis base groups which can complex with metal ions. Usually, these ligand molecules can curl about, as shown in Figure 6.6(a), so more than one of the Lewis base groups can occupy bonding positions about a metal

> Chelate is from Latin *chele* = claw, like the pincer of a lobster.

ion. Such metal ion complexes are called **chelates** and the ligands are called chelating ligands. Chelating ligands like EDTA are particularly effective for dissolving metal ions, such as Pb^{2+}, that have few soluble salts. EDTA is used, for example, to treat people who have lead poisoning.

Complex ion formation and solubility

In a solution containing $Ca^{2+}(aq)$, $C_2O_4^{2-}(aq)$, and $EDTA^{4-}(aq)$, Investigate This 6.43(a), the anions are in competition for the calcium ion. Oxalate reacts to precipitate the calcium ion as calcium oxalate:

$$Ca^{2+}(aq) + C_2O_4^{2-}(aq) \rightleftharpoons CaC_2O_4(s) \tag{6.4}$$

$EDTA^{4-}(aq)$, reacts to complex (chelate) the calcium ion:

$$Ca^{2+}(aq) + EDTA^{4-}(aq) \rightleftharpoons Ca(EDTA)^{2-}(aq) \tag{6.33}$$

Evidently, under the conditions of Investigate This 6.43(a), $EDTA^{4-}(aq)$ wins the competition.

When Ca^{2+} cations react to form $Ca(EDTA)^{2-}(aq)$ complex ions by reaction (6.33), they are no longer present as $Ca^{2+}(aq)$ ions and are not available for reaction (6.4). As $EDTA^{4-}(aq)$ was added to the mixture containing $Ca^{2+}(aq)$, $C_2O_4^{2-}(aq)$, and $CaC_2O_4(s)$, the $Ca^{2+}(aq)$ in solution reacted to form $Ca(EDTA)^{2-}(aq)$. Because reaction (6.4) is reversible, some of the $CaC_2O_4(s)$ dissolved to replace the missing $Ca^{2+}(aq)$. These reactions continued, reaction (6.33) going in the forward direction and reaction (6.4) going in reverse, until all the precipitate dissolved. Figure 6.7 is a schematic representation of precipitate formation and then dissolution of the precipitate upon addition of a complexing reagent.

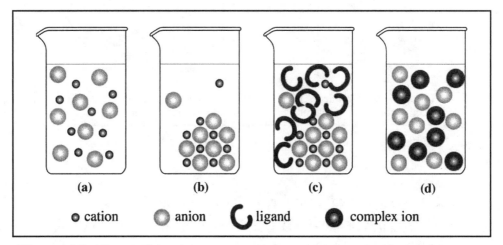

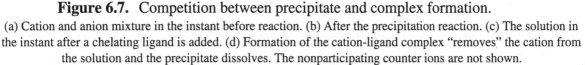

Figure 6.7. Competition between precipitate and complex formation.

(a) Cation and anion mixture in the instant before reaction. (b) After the precipitation reaction. (c) The solution in the instant after a chelating ligand is added. (d) Formation of the cation-ligand complex "removes" the cation from the solution and the precipitate dissolves. The nonparticipating counter ions are not shown.

The analysis of this competition is another application of Le Chatelier's principle. The reaction represented by equation (6.4) and Figure 6.7(b) is an ionic solid in equilibrium with its ions in solution. When the chelating ligand is added, it reacts with the cation, reaction (6.33), and reduces its concentration. This disturbs the solubility equilibrium and the solubility system responds to minimize the disturbance by adding more cation to the solution, that is, by dissolving. Le Chatelier's principle gives the direction but not the magnitude of these effects. In Chapter 9, we will examine competitive reactions like these more quantitatively.

6.46. Check This   *Keeping the shine in your hair*

Read the ingredients label on your shampoo (and see Check This 6.25). Many contain EDTA in some form (tetrasodium EDTA, trisodium HEDTA, Edetate, and so on). What is EDTA doing there? Is your answer related to Investigate This 6.43(b)? If so, clearly explain the connection.

Metal ion complexes with four ligands

The octahedral arrangement of ligands in metal ion complexes is the most common geometry you will find. Other geometries are possible and range from two to nine ligands. Complexes with four and six ligands are by far the most common. The geometry of complexes with four ligands is either square planar or tetrahedral. Examples of square planar complexes are the platinum, Pt^{2+}, complexes used in cancer treatment. One of these is $Pt(NH_3)_2Cl_2$, dichlorodiammine platinum(II); the structures of its *cis*- and *trans*-isomers are shown in Figure 6.8. Only the cis isomer, called cisplatin, is effective in cancer chemotherapy.

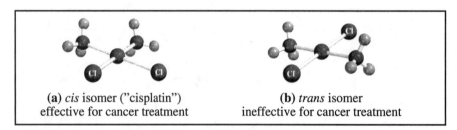

(a) *cis* isomer ("cisplatin")
effective for cancer treatment

(b) *trans* isomer
ineffective for cancer treatment

Figure 6.8. (a) *cis*- and (b) *trans*- isomers of $Pt(NH_3)_2Cl_2$.

6.47. Check This   *Names and structures of Pt(NH₃)₂Cl₂*

Explain why the structures of the $Pt(NH_3)_2Cl_2$ isomers are named as they are.

Tetrahedral metal ion complexes are less common, but are found in several classes of protein molecules. For example, many proteins that bind to nucleic acids are complexed to zinc ions through two sulfur atoms and two nitrogen atoms (Lewis bases with pairs of nonbonding electrons) from their amino acid side chains, as shown in Figure 6.9(a). This complex, called a *zinc finger*, holds the protein chain in a finger-like shape that can fit into the groove of a DNA helix, Figure 6.9(b).

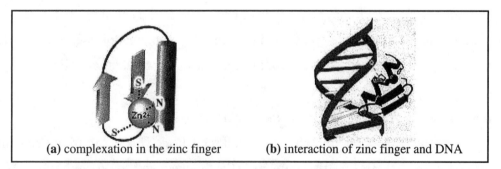

(a) complexation in the zinc finger   **(b)** interaction of zinc finger and DNA

Figure 6.9. Zinc finger complex in DNA binding proteins.
(a) Two sheet-like parts of the protein (the broad arrows) and a helical region (the blue cylinder) are held together by complexing with zinc(II). (b) The finger fits in the major groove of double helical DNA.

6.48. **Consider This**   *What is the structure of the nickel-dimethylglyoxime complex?*

In Check This 6.11, you had enough data to determine the ratio of moles of $Ni^{2+}(aq)$ that react with dimethylglyoxime, $C_4H_8N_2O_2$, to give a red solid. The red solid is an electrically neutral complex, which is part of the reason it comes out of solution. The metal ion and all of the second row element atoms in the complex lie in a plane. In addition to bonding to the central metal ion, hydrogen bonding between chelate ligands is an important factor in stabilizing the complex. Construct a model of the red solid formed in the nickel(II)-dimethylglyoxime reaction.

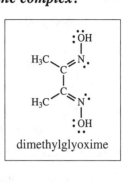

dimethylglyoxime

Porphine-metal ion complexes in biological systems

Zinc-fingers are but one example of a vast number of metal ion complexes in biological systems. Two of the most familiar are **chlorophyll**, Figure 6.10(a), the complex that gives plants their green color, and **heme**, Figure 6.10(c) the complex that gives blood its red color. The complexing molecules in both cases are derived from the same basic **porphine** structure, Figure 6.10(b). The square array of nitrogens held in place by the four joined rings is perfectly set up to complex a metal ion in its center, after two protons are lost to form the porphine dianion. The large number of π electrons in the double bonds of the planar porphine structure are in delocalized π orbitals that are spread over all the carbons and nitrogens in the molecule. The stability of this ring system may be one of the reasons that derivatives of porphine are found as metal ion complexing ligands in many biological roles.

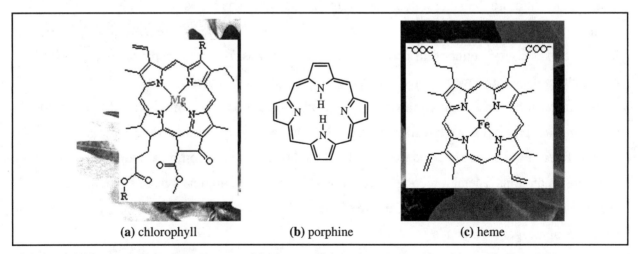

(a) chlorophyll   **(b)** porphine   **(c)** heme

Figure 6.10. Metal ion complexes with the porphine ring system in chlorophyll and heme.
There are no charged groups on chlorophyll. Differences in the R group at the top of the chlorophyll cause slightly different light absorptions. The R group at the bottom is a 20-carbon chain that makes chlorophyll insoluble in water but soluble in cell membranes, which also contain long hydrocarbon chains.

6.49. Check This   *The porphine ring system in chlorophyll and heme*

Trace the porphine ring system, Figure 6.10(b), in the structure of chlorophyll, Figure 6.10(a), and heme, Figure 6.10(c). How many π electrons are there in the porphine ring? How many π electrons are there in the porphine ring system in chlorophyll? in heme?

Chlorophyll is a Mg^{2+} *chlorin* complex and heme is a Fe^{2+} *porphyrin* complex. The ring structures are different for the two molecules, but you have found that the core of each is the porphine ring system. Chlorophyll molecules are stacked together in the membranes of chloroplasts, the organelles in green plants that absorb sunlight and use the captured energy to combine carbon dioxide and water into sugars. Heme is incorporated into a protein named globin to produce **hemoglobin**, the oxygen carrying molecule in your blood. (See the Chapter 9 opening illustration and Figure 9.7.) Iron(II) is usually found octahedrally complexed with six ligands. In hemoglobin, one of the side groups of the protein complexes to the fifth position around the iron(II), which bonds the heme to the protein. The sixth position is where an oxygen molecule is bound when the hemoglobin is carrying oxygen from the lungs. In this oxygenated form, the complex is red. After the oxygen is left with a cell, the sixth position is occupied by a water molecule for the return to the lungs. In this deoxygenated form, the complex is bluish-purple.

Reflection and projection

Metal ions in solution act as Lewis acids and can accept pairs of electrons donated by Lewis bases, ligands, to form metal ion complexes. Complex ion formation can compete with precipitation for metal ions and can dissolve precipitates or prevent their formation. Many ligands contain more than one functional group that act as Lewis bases; these chelating ligands form more than one bond to a metal ion to give chelate complexes. The great majority of metal ion complexes contain either four or six ligands. Four ligands form square planar or tetrahedral arrays about the central metal ion. Six ligands form an octahedral array.

All metal ions can form complexes with appropriate ligands. The complexes of the transition metal ions are particularly interesting, because they are so highly colored; their variation in color with different ligands can be used to learn about the bonding in the complex and the energies of electrons in the ions. Metal ion complexes are widespread in biological systems where they perform many functions, including capture of the sun's energy for photosynthesis by plants and transport of oxygen in animals.

You're used to thinking about protons or hydronium ions as acids and it isn't too much of a leap to include metal ions as Lewis acids, electron pair acceptors. It's a somewhat larger jump to think about centers of partial positive charge in molecules as electron pair acceptors; that's the jump we'll make in the next section.

Section 6.7. Lewis Acids and Bases: Electrophiles and Nucleophiles

6.50. Investigate This   *Does an acid react with an alcohol?*

Do this as a class investigation and work in small groups to discuss and analyze the results. Mix about 2 g of solid salicylic acid, 5 mL of methanol, and 2 small drops of concentrated sulfuric acid in each of two 20×200-mm test tubes. Place one of the test tubes in a water bath at about 70 °C. After 15 minutes, remove the test tube from the water bath, allow it to cool to room temperature, and then add 2 mL of a saturated aqueous solution of sodium hydrogen carbonate to each test tube. Carefully smell the contents of both test tubes. Record your observations.

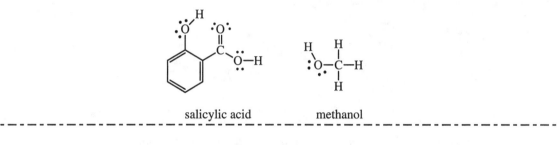

                     salicylic acid   methanol

6.51. Consider This   *What is the product of reaction of an acid with an alcohol?*

In Investigate This 6.50, what evidence, if any, do you have that a reaction has occurred between salicylic acid and methanol? Can you identify the product of the reaction? What conditions are required for the reaction?

Nucleophiles and electrophiles

Many of the reactions of the functional groups described in Chapter 5, Section 5.10, are Lewis acid-base reactions. Electronegative atoms in these groups always have nonbonding electrons and usually gather negative charge to themselves at the expense of less electronegative atoms. These centers of high electron density are Lewis bases. Some molecules have a significant positive center on carbon atoms that are bonded to oxygen or nitrogen and these act as Lewis acids. Another terminology for these reactions is in such common use that you will need to know it as well. The centers of negative charge density, Lewis bases, are called **nucleophiles** (*philos* = loving, hence "nucleus loving"). The positive centers, Lewis acids, are called **electrophiles** ("electron loving").

6.52. **Check This** *Centers of positive and negative charge in molecules*

(a) Redraw the Lewis structures for salicylic acid and methanol, Investigate This 6.50, and label the centers of positive charge, δ+, and negative charge, δ-. Explain how you decided where to place your labels.

(b) Draw Lewis structures for each of these molecules and label the positive and negative charge centers. Explain your labeling.

CO_2, H_2O, $(HO)_3PO$, $CH_3C(O)OH$, $HOCH_2CH_2OH$, $H_2N(CH_2)_6NH_2$

Alcohol–carboxylic acid reaction

To illustrate the nature of reactions between nucleophiles, Lewis bases, and electrophiles, Lewis acids, consider the reaction of a carboxylic acid, such as ethanoic acid, with an alcohol, such as ethanol, to form an ester and water:

(6.34)

The carbon that is attached to two oxygen atoms in the carboxylic acid is electrophilic (positive), while the oxygen in the alcohol is nucleophilic (negative). When the conditions are favorable, a nonbonding electron pair from the alcohol oxygen bonds to the carboxylic acid carbon. Following this initial Lewis acid-base reaction, the -OH group from the acid leaves (combined with a proton to form water). This second step in the overall reaction is the *reverse* of a Lewis acid-base reaction; the electrons in an electron pair bond become a nonbonding pair in the product, an ester, ethyl ethanoate. Chemists often use arrows to show where electrons are *imagined* to go in the course of such reactions (and we have used blue to show the atom centers and electrons that become part of the product water):

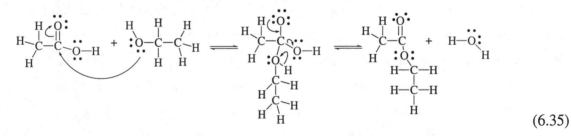

(6.35)

Our point here is not to concentrate on the specifics of this particular reaction, but to recognize that electrophilic and nucleophilic centers in molecules are attracted to each other. If conditions are favorable, they can react to produce new products. As you continue through the text, you will encounter more of these Lewis acid-base reactions in which mutually attracting

---

nucleophilic and electrophilic centers come together and undergo chemical reactions. In order to understand the interactions and the pathways of these reactions, you must recall the architecture of molecules introduced in Chapter 5.

6.53. Check This   *More nucleophile-electrophile reactions*

(a) In Investigate This 6.50, a carboxylic acid reacted with an alcohol. What functional group is formed as a product of this reaction? Write the Lewis structure for the reaction product, methyl salicylate. Write reaction expressions that parallel those of equation (6.35) that show how the reaction of salicylic acid (electrophile) and methanol(nucleophile) produces methyl salicylate.

(b) The 3rd and 4th reactions in Investigate This 6.2 produce carbonic acid, $(HO)_2CO(aq)$, which, as we have seen, reacts to form $CO_2(aq)$ and water:

$$(HO)_2CO(aq) \rightleftharpoons CO_2(aq) + H_2O(l) \tag{6.32}$$

CO_2 is only moderately soluble in water so it bubbles out of solution and accounts for the $CO_2(g)$ product shown in reaction expressions (6.2) and (6.3) for the reactions from Investigate This 6.2. Reaction (6.32) can be viewed as a nucleophilic-electrophilic reaction going in reverse. The *formation* of carbonic acid is the reaction of a nucleophile, water, with an electrophile, carbon dioxide. The first step in the reaction is analogous to the first step in reaction (6.35). Write this first step in the carbonic acid formation reaction and then show the rearrangement of electrons and protons that produces carbonic acid.

Condensation polymers

The ester formation reaction (6.35) is also called a **condensation reaction**, because the two reactants condense, come together into a single structure, in the first step of the reaction. In the overall reaction, a molecule of water is formed for every ester that is formed. Traditionally, the overall reaction is still called a condensation, because the focus is on the carbon-containing molecules. These reactions are particularly important when the reactant molecules are

polyfunctional, that is, contain more than one functional group that can take part in condensation reactions. An example is the reaction between terephthalic acid and ethylene glycol:

> Terephthalic acid is 1,4-benzenedicarboxylic acid and ethylene glycol (a common automobile antifreeze) is 1,2-ethanediol.

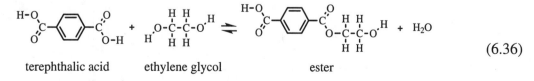

terephthalic acid   ethylene glycol   ester

$$\tag{6.36}$$

6.54. Check This   A condensation reaction

Write the condensation step, the first step in overall reaction (6.36). Show the positive and negative (electrophilic and nucleophilic) centers that attract one another to cause the reaction.

As far as it goes, reaction (6.36) is the same as reaction (6.35) with a different acid and alcohol. But reaction (6.36) isn't as far as these reactants can go. The ester product of the reaction still has an acid and an alcohol functional group which can react, respectively, with another molecule of ethylene glycol and terephthalic acid to form a polymer molecule. **Polymers** (from Greek *polys* = many + *meros* = part) are molecules that contain many repeating units of the same kind. The individual molecules that make up the repeating units are called **monomers** (*mono* = one, single). The reaction of terephthalic acid and ethylene glycol (the monomers) is a, **condensation polymerization**, that produces long chains of esters, **polyesters** (*poly*meric *ester*):

many of each of these   polyester (Dacron, Mylar, PET)

(6.37)

The structure in brackets in reaction expression (6.37) is the **polymer repeat unit** for this

Polyester drawn into fibers is named Dacron. Formed into thin sheets, it is Mylar (the familiar shiny material in party balloons). Molded into containers such as soft drink bottles, it is PET (polyethylene terephthalate).

condensation polymer. The repeat unit tells you what reactant or reactants condense to make the polymer and shows what the linkage is between the reactants.

6.55. Check This   Polyamide condensation polymers

(a) WEB Chap 6, Sect 6.7.1. Condensation polymerization is not limited to polyester formation between acid and alcohol groups. The condensation reaction of an amine with a carboxylic acid is analogous to the reaction we have shown for alcohols with carboxylic acids. Use the charge density models on this page to explain which molecule is the electrophile and which the nucleophile in this condensation reaction.

(b) The amine-carboxylic acid condensation product is an amide (Chapter 5, Table 5.4) and polymers with amide bonds are **polyamides**. Familiar synthetic polyamides are nylons, such as:

nylon-66

What are the reactants that condense to make this polymer? Write a reaction, analogous to (6.37), for the formation of this polymer from these reactants. Explain clearly how this structure is related to the product of the amide condensation reaction shown in the Web Companion. Why is this polymer called nylon-66?

(c) Is(are) the reaction(s) represented in Figure 1.30, Chapter 1, Section 1.9, related to the condensation reactions discussed in this section? Explain why or why not.

Biological condensation polymers

Proteins, starch, cellulose, and nucleic acids are all condensation polymers. Many of the linkages that build biomolecules from smaller precursors, for example, fats from glycerol and long chain carboxylic acids, are also made by condensation reactions like those shown above. Examples of products made from natural polymers are shown in the chapter opening illustration. Wool and silk are proteins (polyamides); cotton, paper, and cardboard are cellulose, a condensation polymer of glucose illustrated in Figure 6.11.

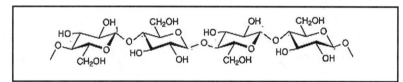

Figure 6.11. Structure of cellulose, a polymer of glucose.

Biological condensation reactions usually occur between reactants that are *derived* from acids, alcohols, and amines. However, you can think about the reactions as though they take place between the parent acid, alcohol, or amine. In these reactions, water is formed, in addition to the condensed product. We have shown all these nucleophile-electrophile condensation reactions as reversible. Indeed, all of the condensation products, including biological molecules, can be **hydrolyzed**, or broken down by water (*hydro* = water + *lyein* = loosen, break), under appropriate conditions to reform the reactant molecules.

6.56. **Check This**   *Water as a nucleophile*

Hydrolysis is a nucleophile-electrophile reaction with water acting as the nucleophile. Write out the steps for this hydrolysis of an amide. *Hint*: Think about the reverse of reaction (6.35).

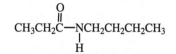

Condensation reactions can also take place with non-carbon-containing Lewis bases (nucleophiles) and acids (electrophiles). An example is reaction of a substituted deoxyribose (a polyalcohol) and phosphoric acid to form a **phosphate ester**:

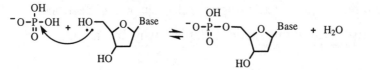

(6.38)

The electronegative oxygen atoms in phosphate draw electron density away from the central

> The "Base" shown on the deoxyribose represents one of the four nitrogen bases in DNA. The bond between the Base and the deoxyribose is also formed by a condensation reaction.

phosphorus atom and make it an electrophilic center, a Lewis acid. The nucleophilic electron pair on the alcohol is attracted to this Lewis acid, as shown by the curved arrow. Nucleic acids formed in replication (or transcription) of DNA are condensation polymers of phosphate esters:

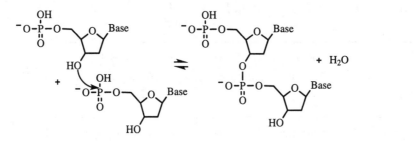

(6.39)

***

6.57. Check This *Polymeric structure of DNA*

Compare the product of reaction (6.39) with the structure of DNA shown in Figure 2.27, Chapter 2, Section 2.13. What is the polymer repeat unit for one strand of DNA? Why are the links between these repeat units often called "phosphodiester bonds?" Explain your answers.

***

Section 6.8. Formal Charge

Rules for formal charge

Formal charge is a useful concept to help understand the rearrangements that often occur following the condensation step in a nucleophile-electrophile reaction. **Formal charge** *is the charge an atom in a molecule or ion would have if all bonding electrons were shared equally between the bonded atoms*. Formal charges are calculated values, not experimental quantities that can be measured. The procedure for finding the formal charge on an atom in a molecule or ion is:

- draw the Lewis structure for the molecule.
- assign one electron from each electron pair bond to each of the bonded atoms.

- assign all nonbonding electrons on an atom to that atom.
- count all the valence electrons assigned to an atom.
- subtract this number of electrons from the positive charge on the atomic core.
- the difference, including its sign, is the formal charge on the atom.

The sum of all the formal charges in a molecule or ion has to be the overall charge on the molecule or ion. Use this sum to check your procedure.

6.58. Worked Example   *Assigning formal charge*

What are the formal charges on the atoms in the hydronium ion, H_3O^+, and in methanal, H_2CO?

Necessary information: We need to know the number of valence electrons and atomic core charges, 1, 4, and 6 for H, C, and O atoms, respectively.

Strategy: We apply the preceding rules to the two species.

Implementation: Lewis structures :

$$\left[H\text{-}\overset{\cdot\cdot}{\underset{\underset{H}{|}}{O}}\text{-}H \right]^+ \qquad\qquad\qquad \overset{H}{\underset{H}{}}C=\overset{\cdot\cdot}{O}\!:$$

Assign the bonding electrons evenly and the nonbonding electrons to their atom:

$$3\ H\cdot \quad \cdot\overset{\cdot\cdot}{\underset{\cdot}{O}}\cdot \qquad\qquad\qquad 2\ H\cdot \quad :\overset{\cdot}{\underset{\cdot}{C}}\cdot \quad :\overset{\cdot\cdot}{\underset{\cdot\cdot}{O}}:$$

Calculate the formal charges:

formal charge on H = 1 – 1 = 0   formal charge on H = 1 – 1 = 0

formal charge on O = 6 – 5 = +1   formal charge on C = 4 – 4 = 0

                                                formal charge on O = 6 – 6 = 0

To remind ourselves of the results, we can show the formal charges in the Lewis structures (omitting any zero formal charges):

$$\left[H\text{-}\overset{\cdot\cdot\ +1}{\underset{\underset{H}{|}}{O}}\text{-}H \right]^+ \qquad\qquad\qquad \overset{H}{\underset{H}{}}C=\overset{\cdot\cdot}{O}\!:$$

Does the answer make sense? We followed the rules to get the formal charges. The sum of all the formal charges is:

For H_3O^+: 0 + 0 + 0 + (+1) = +1   For H_2CO: 0 + 0 + 0 + 0 = 0

The sums give the correct overall charge on the ion and the electrically neutral molecule, so we can have some confidence that we have correctly assigned the formal charges.

6.59. **Consider This**   *How do you assign formal charge?*

(a) The isomeric ions, cyanate, NCO⁻, and isocyanate, CNO⁻, both exist. Write a Lewis structure for each ion. What are the formal charges on the atoms in each ion? Show that the calculated formal charges sum to the correct overall charge on the ions.

(b) The Lewis structure for the intermediate structure shown in equation (6.35) is redrawn here. Calculate the formal charges on all three oxygens and the carbon to which they are all bonded. All the rest of the formal charges in the structure are zero. Show that your calculated formal charges sum to the correct overall charge.

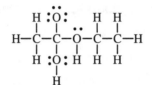

Interpretation of formal charges

The +1 formal charge on oxygen in the hydronium ion, Worked Example 6.58, does not mean that the oxygen has a positive charge in the ion. It means that the partial charge on the oxygen is less negative — more positive — than it is in the neutral molecule, where the formal charge on all atoms is zero. The electronic structures we showed for molecules in Chapter 5, Section 5.4, Figure 5.10, show identical bonding for all three hydrogen atoms in H_3O^+, with each proton in a σ bonding orbital. Only one nonbonding electron pair is left wholly for the oxygen atom, so it makes sense to consider the oxygen as being less negatively polarized than when it has two nonbonding pairs of electrons all to itself. The formal charge can be interpreted as an indicator of the *change* in polarization of an atom compared to its polarization in a different bonding situation.

Reactions reduce formal charges

Although there is no overall charge on the intermediate structure in equation (6.35), you found in Consider This 6.59(b) that there are non-zero formal charges on some of the oxygens in the structure. But stable molecular structures usually have as many atoms as possible with zero formal charge. In ions, the charge obviously has to be somewhere, so all atoms can't have zero formal charge. Electrically neutral molecules, however, rarely have formal charges. Reaction intermediates, which are postulated but usually not experimentally observable, often have formal charges on several atoms. The intermediates undergo electron and atomic core redistributions to reduce the number of formal charges; these redistributions lead to the observed products.

6.60. Check This   *Using formal charge*

(a) Use your answers for Consider This 6.59(a) to predict which isomeric ion, cyanate or isocyanate, is the more stable. Give the reasoning for your answer.

(b) In Consider This 6.59(b), are there any correlations you might make between formal charge and number of bonds to the oxygen atoms? Explain how you arrive at your response.

(c) In Section 6.5, we wrote oxyacid and oxyanion Lewis structures with as many double bonds to oxygen as possible [see Figure 6.4(b) and Consider This 6.34(a)]. There are alternative ways to write Lewis structures for species such as the chlorite ion:

$$\left[:\ddot{\text{O}}\text{-}\ddot{\text{C}}\text{l}\text{-}\ddot{\text{O}}: \right]^{-}$$

What are the formal charges on each atom in this structure? What are the formal charges on the atoms in the chlorite ion structure(s) you wrote for Consider This 6.34(a)? Is there a reason to favor one Lewis structure over the other? Explain why or why not.

(d) Using the model for chlorite ion in part (c), write Lewis structures for the chlorate, ClO_3^-, and perchlorate, ClO_4^-, ions. What are the formal charges on the atoms in these structures? Explain why we chose to correlate base strength with the structures you wrote for these ions in Consider This 6.34(a) rather than those you wrote here.

In reaction intermediates, oxygen is one of most common atoms with a formal charge. The three possible formal charges on oxygen, -1, 0, and $+1$, are all present in the intermediate structure you analyzed in Consider This 6.59(b). You can quickly recognize these three cases by just looking at the number of bonds to the oxygen. Table 6.3 shows the three possibilities and, to help you remember them, their comparison to the ions water forms. Typical kinds of electron and atomic core rearrangements that change the formal charge on oxygen to zero are also shown.

6.61. Consider This   *How are the correlations in Table 6.3 applied?*

(a) Which of the rearrangements in Table 6.3 occur when the intermediate in reaction (6.35) goes to products? Show and explain each one clearly.

(b) Which of the rearrangements in Table 6.3 occur when water reacts with carbon dioxide to produce carbonic acid? You wrote this reaction as a nucleophile-electrophile reaction in Check This 6.53(b). Show and explain each one clearly.

(c) Write reaction (6.8) as a nucleophile-electrophile reaction and show which rearrangements in Table 6.3 occur to produce the product. Show and explain each one clearly.

Table 6.3. Correlation between formal charge and reactions at oxygen atom centers.

The relationship between the number of bonds to oxygen and its formal charge in molecules and reaction intermediates is also demonstrated. As a comparison, water and its ions are shown.

formal charge on oxygen	bonds to oxygen	nonbonding pairs	water species	electron and atomic core rearrangements that bring the oxygen formal charge to zero
−1	1	3	HO^-	(a) change a nonbonding pair to bonding pair (makes a double bond) (b) gain a proton from an acid in solution to form an –OH group
0	2	2	H_2O	
+1	3	1	H_3O^+	change a bonding pair to nonbonding pair (most often with loss of H_2O or H^+ to a base*)

*   The base to which the proton is transferred is often not shown because it might be any of several Lewis bases, including water, in the solution mixture.

Reflection and projection

Lewis acid-base reactions of electrophilic and nucleophilic centers in molecules are the same as other Lewis acid-base reactions: a pair of nonbonding electrons becomes an electron pair bond in the initial product. Complications arise because the initial products of these reactions undergo rearrangements that form more stable products. The rearrangements can often be understood in terms of electron and atomic center movements that minimize formal charges in the final products. Because we are usually interested in the overall reaction, the simple, underlying Lewis acid-base reaction is sometimes obscured.

For this introduction, most of our reactions were, nucleophiles (alcohols and amines) reacting with the electrophilic carbon of the carboxyl group to produce condensation products (esters and amides). In these cases, the final products are a carbon-containing molecule that combines (condenses) all the carbons in both reactants into a single molecule and splits out a water molecule as a second product. When the reactants are polyfunctional, the condensations can continue adding to the ends of the condensation products to form condensation polymers. Polyesters and polyamides are used to make a vast number of products you use every day. Biological phosphate polyesters and polyamides make up your ribonucleic acids, DNA and RNA, and proteins. All these condensation polymers can be hydrolyzed, broken down by reaction with water, under appropriate conditions.

This completes our introduction to Lewis acid-base reactions that transform a nonbonding electron pair to a bonding pair. Now we will turn our attention to the third large class of reactions, those that usually involve complete transfer of one or more electrons from one reactant to another.

Section 6.9. Reduction-Oxidation Reactions: Electron Transfer

6.62. **Investigate This**   *Do ion-metal reactions between copper and silver occur?*

Do this as a class investigation and work in small groups to discuss and analyze the results. Use an overhead projector to view what happens when copper wire is immersed in an aqueous silver nitrate, $AgNO_3$, solution and silver wire is immersed in an aqueous copper sulfate, $CuSO_4$, solution in adjacent wells of a 6-well microtiter plate. Watch for several minutes and record your observations.

close-up
reaction view

6.63. **Consider This**   *What copper and silver ion-metal reactions occur?*

(a) In Investigate This 6.62, do you see any evidence for reaction in either or both of the wells? What is the evidence?

(b) WEB Chap 6, Sect 6.9.1. Is the movie related to what you observe in either of the wells in Investigate This 6.62? If so, explain how they are related.

(c) Try to write a reaction expression to describe what is happening in each well.

In precipitation reactions, the ions retain their electrons and their identity in the solid product. In Lewis acid-base reactions, a nonbonding electron pair in the Lewis base becomes a bonding pair shared by two atoms in the product. In **reduction-oxidation reactions (redox reactions –** pronounced "ree-dox") one or more electrons is completely transferred from one reactant to the other reactant. The reactant that gains electrons is **reduced**. The reactant that loses electrons is **oxidized**. The easiest reduction-oxidation reactions to recognize are probably those in which a metal is one of the reactants, as in Investigate This 6.62. The disappearance of the metal (or the formation of another) is a sure sign that

> The reactant that loses electrons is often called a **reducing agent,** since it provides electrons to the reactant that is reduced. The reactant that gains electrons is called the **oxidizing agent,** since it accepts electrons from the reactant that is oxidized. Thus, the reducing agent is oxidized and the oxidizing agent is reduced during the reaction. This can be a confusing nomenclature. We will not use it in this introduction, but will return to it in Chapter 10 when reduction-oxidation reactions are examined in more detail.

electrons have been transferred in the reaction. Metals have to lose electrons to become cations and "disappear" into solution. Metal cations have to gain electrons to get to their elemental metal form.

Direction of a reduction-oxidation reaction

In Investigate This 6.62, you observed needles of silvery metal growing on the copper wire when it was placed in a solution containing silver cation, $Ag^+(aq)$. You also observed the initially clear and colorless solution become a pale blue color. In the other sample, there was no evidence of formation of any metal or other solid and the initially clear, blue solution containing $Cu^{2+}(aq)$ did not change. It is difficult to tell whether any of the copper metal disappears in the first sample, but the change of the solution color shows that there is some new component in the solution. The second sample shows that $Cu^{2+}(aq)$ in aqueous solution is blue.

Taken together, all this evidence suggests that aqueous silver ion and copper metal react to produce silver metal and aqueous copper ion:

$$Ag^+(aq) + Cu(s) \rightarrow Ag(s) + Cu^{2+}(aq) \tag{6.40}$$

We have written reaction (6.40) as though it proceeds only in the forward direction. This is because, in Investigate This 6.62, there was no evidence of any reaction between $Ag(s)$ and $Cu^{2+}(aq)$, the reverse of reaction (6.40). As we have said before, most reactions are, to some extent reversible, but often greatly favor one direction over the other and we express such directionality with a single arrow.

There is no simple way to predict the direction of a reduction-oxidation reaction. The factors that affect the direction of the reaction (or if a redox reaction occurs at all) include pH, the presence of complexing ligands, and the exact nature of the solvent in which the reaction occurs. We will examine these more quantitative aspects of reduction-oxidation reactions in Chapter 10. In this chapter, we will focus on recognizing reduction-oxidation reactions and writing balanced equations to describe the reactions.

Charge balance

In the reaction between $Cu(s)$ and $Ag^+(aq)$, an electron has to be gained by $Ag^+(aq)$, in order to become $Ag(s)$. The $Ag^+(aq)$ ions are reduced: *electrons are gained.* Two electrons have to be lost

| WEB Chap 6, Sect 6.9.2-3. |
| Study animations of these |
| processes that occur in the |
| copper-silver ion reaction. |

by $Cu(s)$ to become $Cu^{2+}(aq)$. The $Cu(s)$ atoms are oxidized: *electrons are lost.* Reaction expression (6.40) is not balanced in terms of the electron transfer that must occur from copper to silver ion. The imbalance in electron transfer results in an imbalance in charge on each side of the expression. As written, there is one positive charge on the left and two on the right. Charges must be balanced.

6.64.   Worked Example   *A balanced chemical equation for the Cu/Ag⁺ reaction*

Write a balanced chemical equation for the reaction of copper metal, $Cu(s)$, with silver ion in solution, $Ag^+(aq)$.

Necessary information: Reaction expression (6.40) is unbalanced but tells us the reactants and products of the reaction.

Strategy: Balanced reaction expressions have to conserve both atoms and charge. In simple reduction-oxidation reactions like this one, we first balance the charge and then the atoms by inspection.

Implementation: To balance charge, we add a $Ag^+(aq)$ to the reactants:

$$2Ag^+(aq) + Cu(s) \rightarrow Ag(s) + Cu^{2+}(aq)$$

In this intermediate expression, the charges and copper atoms and ions balance, but silver atoms and ions do not. To balance the silvers, we add an atom of silver metal to the product side:

$$2Ag^+(aq) + Cu(s) \rightarrow 2Ag(s) + Cu^{2+}(aq) \tag{6.41}$$

Does the answer make sense? Equation (6.41) still describes the reaction we observed and it is balanced in atoms and charge. The two electrons lost by each $Cu(s)$ atom are transferred to two $Ag^+(aq)$ ions. Electron transfer is also balanced.

6.65.   Check This   *Balancing simple reduction-oxidation reaction expressions*

Balance these reduction-oxidation reactions. They are shown as reversible, since we have no evidence for the favored direction.

(a) $Mg^{2+}(aq) + Cu(s) \rightleftharpoons Mg(s) + Cu^{2+}(aq)$ $\qquad\qquad$ (6.42)

(b) $Fe^{3+}(aq) + Sn^{2+}(aq) \rightleftharpoons Fe^{2+}(aq) + Sn^{4+}(aq)$ $\qquad\qquad$ (6.43)

(c) $Zn(s) + H^+(aq) \rightleftharpoons Zn^{2+}(aq) + H_2(g)$ [$H^+(aq)$ is a way to represent $H_3O^+(aq)$.] $\quad$ (6.44)

(d) $Cu^{2+}(aq) + I^-(aq) \rightleftharpoons CuI(s) + I_2(s)$ [CuI contains the copper(I) ion.] $\qquad$ (6.45)

6.66.   Investigate This   *Does Cu(s) react with nitric and/or hydrochloric acid?*

Do this as a class investigation and work in small groups to discuss and analyze the results. **CAUTION:** concentrated nitric and hydrochloric acids are powerfully corrosive; handle them with respect and with rubber or plastic gloves. Some products of these reactions are toxic and must be kept in the plastic bag and disposed of properly in a fume hood.

(a) Place a 400-mL beaker of water, a 100-mL beaker containing 5 mL of concentrated nitric acid, HNO_3, and 0.25 gram of copper wire into a one-gallon zip-seal plastic bag. Seal the bag and then drop the copper wire into the beaker of nitric acid. Observe and record any changes that occur. When it appears that no further change is occurring, pour the entire contents of the 100-mL beaker into the 400-mL beaker of water. Observe and record any changes that occur.

(b) Repeat this procedure with concentrated hydrochloric acid, HCl, instead of nitric acid.

$Cu + HNO_3$

6.67. Consider This *What are products of Cu(s) reactions with nitric and hydrochloric acid?*

(a) In Investigate This 6.66. do you see any evidence for reaction of copper with either nitric or hydrochloric acid? What is the evidence? If there is a reaction, what reaction products can you identify in the reactions?

(b) What conclusions can you draw about the reaction between copper and nitric acid? between copper and hydrochloric acid?

The predominant ions in nitric acid are hydronium, $H_3O^+(aq)$, and nitrate, $NO_3^-(aq)$, formed by the acid-base reaction of HNO_3 with water. Nitrogen and oxygen form several gaseous compounds, one of which is reddish-brown nitrogen dioxide, $NO_2(g)$. In Investigate This 6.62, you observed that aqueous solutions containing $Cu^{2+}(aq)$ cation are blue. Your observations in Investigate This 6.66, together with the information in this paragraph suggest that we can write the reaction between copper and nitric acid as:

$$Cu(s) + H^+(aq) + NO_3^-(aq) \rightarrow Cu^{2+}(aq) + NO_2(g) \tag{6.46}$$

In expression (6.46) [and expression (6.44) in Check This 6.65], we have written the hydronium ion as $H^+(aq)$. You will usually find it easier to balance reduction-oxidation reactions, if you have fewer atoms to worry about. If you wish, you can add the "missing" H_2O [from $H_3O^+(aq)$] to the expressions after they are balanced.

6.68. Consider This *What is the reaction of Cu(s) with $H_3O^+(aq)$?*

Reaction (6.44) in Check This 6.65(c) proceeds as written: zinc metal reacts with hydronium ion to give hydrogen gas and zinc cation , $Zn^{2+}(aq)$, in solution. We can write an analogous (unbalanced) reaction for copper:

$$Cu(s) + H^+(aq) \rightarrow Cu^{2+}(aq) + H_2(g) \tag{6.47}$$

Should we include this possibility as part of reaction expression (6.46)? What evidence do you have that copper does or does not react with hydronium ion? Explain your reasoning.

Reaction expression (6.46) is not balanced in terms of either atoms or charge. The only change in charge that is evident from the expression is that copper metal has been oxidized (lost two electrons) to become $Cu^{2+}(aq)$ cation. But there is no obvious atom that has been reduced (gained electrons) in the reaction. Many reduction-oxidation reactions involve nonmetals and look something like expression (6.46). We need a way to keep track of electrons, so we can balance electron transfers in these reactions. And we need a way to tell whether a reaction *is* a reduction-oxidation. Both problems can be solved by defining and using oxidation numbers.

Oxidation numbers

The **oxidation number** for an atom is *defined* as the difference between the number of electrons in the neutral atom and the number of electrons in the atom as it exists in a compound or ion. If the atom has lost electrons, its oxidation number is positive. If the atom has gained electrons, its oxidation number is negative. The definition of oxidation number is easy to apply to elemental atoms and monatomic ions: the oxidation number is the charge on the atom or ion. In reaction (6.46), for example, copper metal, Cu, loses two electrons (is oxidized) in becoming Cu^{2+}. The number of electrons in Cu is two more than in Cu^{2+}, so Cu^{2+} is assigned an oxidation number of +2, which is equal to the charge on the ion. Sometimes, elemental atoms are written with a superscript zero, Cu^0, for example, to denote that their oxidation number is, by definition, zero.

6.69. Check This   *Oxidation numbers for elements and monatomic ions*

   (a) The oxidation numbers for Fe(s), $Fe^{3+}(aq)$, $Cl_2(g)$, and $Cl^-(aq)$ are 0, +3, 0, and –1, respectively. Show how the definition of oxidation number gives the preceding values.

   (b) What are the oxidation numbers for: $V^{4+}(aq)$, $H^-(g)$, $P_4(s)$, $N_2(g)$, and $Na^+(aq)$. Explain.

Assigning oxidation numbers to atoms in molecules and polyatomic ions, requires a procedure for dividing the valence electrons among the atoms so we can apply the oxidation number definition. The procedure we will use is:

- write a Lewis structure for the molecule or ion.
- for each electron pair bond between identical atoms, assign one electron to each atom. For the purposes of these assignments, carbon and hydrogen are assumed to be identical.

- for each electron pair bond between unlike atoms, assign all the electrons to the more electronegative atom.
- assign all nonbonding valence electrons on an atom to that atom.
- count all the valence electrons assigned to an atom, n_e.
- subtract the number of valence electrons, n_e, from the positive charge on the atomic core, Z_{core}. The difference, including its sign, is the oxidation number, ON, for the atom.

$$ON = Z_{core} - n_e \qquad\qquad (6.48)$$

As a check on the application of this procedure, the sum of the oxidation numbers on all the atoms has to be the overall charge on the molecule or ion. Note that, like formal charge, these oxidation numbers are calculated values, not experimentally measurable quantities. There is no universally accepted procedure for assigning oxidation numbers in molecules and polyatomic ions. As long as you use one procedure consistently, your results will be internally consistent and you will be able to keep track of electron transfers.

6.70. Check This   ***The procedure and the definition of oxidation number***

Note that our procedure for assigning oxidation numbers in multiatomic species ignores inner shell electrons, which never take part in electron transfer reactions and always are associated with their atomic core. Explain how our procedure, which accounts only for valence electrons and the atomic core charges, is consistent with the definition of oxidation number.

6.71. Worked Example   ***Assigning oxidation numbers***

What are the oxidation numbers for the atoms in NO_3^- and NO_2?

Necessary information: We need to know that the core charges (= number of valence electrons) for N and O atoms are 5 and 6, respectively. It is useful to know that two molecules of NO_2 can react to form O_2NNO_2, which is bonded through the nitrogens. For molecules with more than one equivalent Lewis structure, any one of them can be used to assign oxidation numbers.

Strategy: Apply the procedure for assigning oxidation numbers.

Implementation: Lewis structures for NO_3^- and NO_2:

NO_2 has an odd number of valence electrons, 17, so there is no way to write a structure that satisfies the octet rule on all the atoms. We have shown the odd electron on the N atom, because

---

this explains how two NO_2 molecules can combine to form O_2NNO_2 using the odd electron from each for a bond between the N atoms.

Assign the bonding electrons to the more electronegative atom and nonbonding electrons to their atom:

Calculate the oxidation numbers:

ON for N = 5 – 0 = +5   *ON* for N = 5 – 1 = +4

ON for O = 6 – 8 = –2   *ON* for O = 6 – 8 = –2

The structures with oxidation numbers shown on each atom are:

Does the answer make sense? Check the procedure by summing oxidation numbers:

for NO_3^-: (–2) + (–2) + (–2) + 5 = –1   for NO_2: (–2) + (–2) + 4 = 0

The sums give the correct overall charges so we have some assurance that we have assigned the oxidation numbers correctly.

6.72.   Check This   *Assigning oxidation numbers*

Assign oxidation numbers to each of the atoms in these molecules and ions:

H_2O, NH_3, $(HO)_3PO$, PO_4^{3-}, $HOCl$, Cl_2, C_2H_6, CH_3OH, $HC(O)OH$

Interpreting oxidation numbers

The oxidation number for an atom in a molecule or ion is a measure of its share of the valence electrons compared to what it has as a neutral atom. If the oxidation number of an atom increases, it has been **oxidized** (has lost electrons). If the oxidation number of an atom decreases, it has been **reduced** (has gained electrons). Changes in oxidation numbers signal electron transfers in a reaction. *In all reduction-oxidation reactions, one atom is oxidized (oxidation number increases going from reactants to products) and another atom is reduced (oxidation number decreases).* Table 6.4 compares the descriptions of oxidation and reduction in terms of electron loss or gain and oxidation number increase or decrease.

Table 6.4. Comparison of electron and oxidation number change in oxidation and reduction.

Process	Electrons	Oxidation Number
oxidation of an atom	*loss*	*increase*
reduction of an atom	*gain*	*decrease*

6.73. Check This   *Reduction-oxidation reactions*

(a) Show that reaction (6.46) is a reduction-oxidation reaction by showing that one of the atoms in the reactants is oxidized and another is reduced. *Hint*: Use the results from Worked Example 6.71.

(b) Assign oxidation numbers to all the atoms in this reaction expression.

$$HCl(aq) + H_2O(l) \rightleftharpoons H_3O^+(aq) + Cl^-(aq) \tag{6.49}$$

Is reaction (6.49) a reduction-oxidation reaction? Explain the reasoning for your answer.

(c) Assign oxidation numbers to the iodine atoms(ions) in this reaction expression:

$$H^+(aq) + H_2O_2(aq) + I^-(aq) = H_2O(aq) + I_2(aq) \tag{6.50}$$

Is reaction (6.50) an reduction-oxidation reaction? Which atom(ion) is reduced? which oxidized? Explain the reasoning for your answer.

6.74. Consider This   *Are there patterns in oxidation numbers?*

(a) List the oxidation number for the oxygen atoms in Worked Example 6.71 and Check This (6.72) and (6.73). What conclusion can you draw about the oxidation number of oxygen atoms in compounds? Explain how you reach your conclusion. Can you think of a reason why this pattern is observed?

(b) List the oxidation number for the hydrogen atoms in Check This (6.72) and (6.73). What conclusion can you draw about the oxidation number of hydrogen atoms in compounds? Explain how you reach your conclusion. Can you think of a reason why this pattern is observed?

An alternative way to assign oxidation numbers

The procedure we have been using to assign oxidation numbers is based on electronegativity differences (except for hydrogen bonded to carbon) and always works to give consistent assignments. Your analyses in Consider This 6.74 show that oxygen in its compounds has an oxidation number of –2, except in peroxides, when its oxidation number is –1. Hydrogen, in the compounds we have considered so far, always has an oxidation number of +1, except when it is bonded to carbon, when its oxidation number is zero. These patterns lead to the set of rules, shown in Table 6.5, for assigning oxidation numbers without writing a Lewis structure and dividing valence electrons between the atoms. The rules do not work in every case you meet, but they are often useful for a quick appraisal of whether a reaction is a reduction-oxidation.

Table 6.5. An alternative set of rules for assigning oxidation numbers, *ON*.

Rule
1.   The *ON* for atoms in elements is 0.
2.   The *ON* for monatomic ions is the ionic charge.
3.   The *ON* for oxygen in most compounds is –2.
The *ON* for oxygen in peroxides is –1.
4.   The *ON* for hydrogen in many compounds is +1.
The *ON* for hydrogen bound to carbon is zero.
5.   For neutral compounds, *ON*'s must sum to zero.
6.   For polyatomic ions, *ON*'s must sum to charge.

6.75.   Check This   *Oxidation number for oxygen in peroxides*

Use electronegativies and the definition of oxidation number to explain how the bonding in peroxides, such as hydrogen peroxide, H_2O_2, gives these oxygen atoms a –1 oxidation number.

6.76.   Worked Example   *Assigning oxidation numbers*

Use the rules in Table 6.5 to assign oxidation numbers for the atoms in NO_3^- and NO_2.

Necessary information: All the information we need is in Table 6.5.

Strategy: Apply the applicable rules in Table 6.5. For these cases, this means rule 3 and either rule 5 or 6.

Implementation: The sum of the oxidation numbers in NO_3^- has to be –1:

$(ON$ for N$) + 3(ON$ for O$) = -1$

Substitute –2 for (ON for O) and solve for (ON for N):

$(ON$ for N$) = -1 -3(-2) = +5$

The sum of the oxidation numbers in NO_2 has to be 0:

$(ON$ for N$) + 2(ON$ for O$) = 0$

$(ON$ for N$) = 0 - 2(-2 =) = +4$

Does the answer make sense? The oxidation numbers based on the rules are the same as those assigned from the definition in Worked Example 6.71, so they make sense.

6.77. Check This   *Assigning oxidation numbers*

Use the rules in Table 6.5 to assign oxidation numbers for each of the atoms in these molecules and ions: H_2O, NH_3, $(HO)_3PO$, PO_4^{3-}, $HOCl$, Cl_2, C_2H_6, CH_3OH, $HC(O)OH$. How do your answers compare to what you got in Check This 6.72? Can you explain any differences?

Reflection and projection

Our third large class of reactions, reduction-oxidation reactions, involves complete transfer of one or more electrons from one reactant to another. The reactant that loses electrons is oxidized and the reactant that gains electrons is reduced. In simple cases, usually with monatomic ions and elements, you can easily tell that reactants have gained and lost electrons and thus identify the reaction as a reduction-oxidation. In more complex cases, usually those with molecules and/or polyatomic ions, you usually will need to assign oxidation numbers to all the atoms in the reactants and products, in order to identify a reaction as a reduction-oxidation. The atom that decreases in oxidation number in the reaction is reduced and the atom that increases in oxidation number is oxidized. You can often use the rules in Table 6.5 to make the decision whether a reaction is a reduction-oxidation. If the rules lead to problems or ambiguities, go back to the definition of oxidation numbers to assign them appropriately.

You can usually balance simple reduction-oxidation reaction equations by inspection, as we demonstrated early in this section. You have probably noticed, however, that we have not yet balanced the more complex reaction of copper with nitric acid. Balancing reduction-oxidation reaction equations is the topic of the next section and we will finish the chapter with a brief look at reduction-oxidation reactions of carbon-containing molecules, including sugar fermentation.

Section 6.10. Balancing Reduction-Oxidation Reaction Equations

In a balanced reduction-oxidation reaction equation, the total number of electrons lost by one reactant must equal the total number of electrons gained by another. Many such reactions are not easy to balance by inspection, but require a systematic procedure. Two procedures are in wide use: the oxidation-number method and the half-reactions method. We will introduce both and you can choose to use whichever is easier for you. Initially, we will restrict our consideration to reactions carried out in acidic solutions. Reaction equations in acidic solutions can always incorporate water and/or hydronium ion, two species that are present in high concentration in these solutions. Balancing equations for reactions in basic solutions requires a slightly modified procedure, which we will introduce at the end of the section.

The oxidation-number method

Table 6.6 presents the systematic procedure for balancing reduction-oxidation reaction equations by the **oxidation-number method**. The method uses the change in oxidation numbers for the atom that is oxidized and the atom that is reduced to determine the number of electrons lost and gained, respectively.

Table 6.6. Oxidation-number method for balancing reduction-oxidation equations.
This procedure applies to acidic and neutral solutions.

Step	Procedure
1.	Write all reactants and products, except H^+ and H_2O, in the form of a chemical equation.
2.	Assign oxidation numbers to those atoms that undergo changes in oxidation number.
3.	Adjust coefficients for the reactant and product containing the atom being oxidized to balance that atom. Adjust coefficients for the reactant and product containing the atom being reduced to balance that atom.
4.	Compare the number of electrons being released by the atom undergoing oxidation to the number of electrons being gained by the atom undergoing reduction. While maintaining the coefficient ratios from Step 3, adjust coefficients to equalize the number of electrons associated with oxidation and reduction.
5.	If the number of oxygen atoms differs between the left and right sides of the equation, add water molecules to the side needing more oxygen atoms.
6.	If the number of hydrogen atoms differs between the left and right sides of the equation, add hydrogen ions (H^+) to the side needing more hydrogen atoms.
7.	Check the equation to be certain that all atoms are balanced, and that the net charge is the same on both sides of the equation.

6.78. Worked Example   *Balancing redox equations by the oxidation-number method*

Use the oxidation-number method to balance the equation for the reaction of copper with concentrated nitric acid.

$$Cu(s) + H^+(aq) + NO_3^-(aq) \rightarrow Cu^{2+}(aq) + NO_2(g) \tag{6.46}$$

Necessary information: We need to know, from Section 6.9, that the oxidation numbers for Cu, Cu^{2+}, N in NO_3^-, and N in NO_2 are 0, 2, +5, and +4, respectively.

Strategy: Follow the steps outlined in Table 6.6.

Implementation:

Step 1: The reactants containing atoms that are oxidized and reduced are Cu and NO_3^-; these atoms appear in the products as Cu^{2+} and NO_2.

$$Cu(s) + NO_3^-(aq) \rightarrow Cu^{2+}(aq) + NO_2(g) \qquad (6.51)$$

Step 2: Oxidation numbers are shown in red above the atoms.

$$\overset{0}{Cu}(s) + \overset{+5}{NO_3^-}(aq) \rightarrow \overset{+2}{Cu^{2+}}(aq) + \overset{+4}{NO_2}(g) \qquad (6.52)$$

Step 3: The number of copper atoms in the reactants and products is the same. The number of nitrogen atoms in the reactants and products is the same. No adjustment of the coefficients in equation (6.52) is necessary at this step.

Step 4: We can show the gain and loss of electrons as:

gain 1 electron

$$\overset{0}{Cu}(s) + \overset{+5}{NO_3^-}(aq) \longrightarrow \overset{+2}{Cu^{2+}}(aq) + \overset{+4}{NO_2}(g)$$

lose 2 electrons

$$(6.53)$$

To balance the gain and loss of electrons, two nitrate ions need to be reduced to two nitrogen dioxide molecules:

gain 2 electrons

$$\overset{0}{Cu}(s) + 2\overset{+5}{NO_3^-}(aq) \longrightarrow \overset{+2}{Cu^{2+}}(aq) + 2\overset{+4}{NO_2}(g)$$

lose 2 electrons

$$(6.54)$$

Step 5: In expression (6.54), there are six oxygen atoms in the reactants and four in the products. We add two water molecules to the products to balance the oxygens:

$$Cu(s) + 2NO_3^-(aq) \rightarrow Cu^{2+}(aq) + 2NO_2(g) + 2H_2O(l) \qquad (6.55)$$

Step 6: In expression (6.55), there are four hydrogen atoms in the products, but none in the reactants. We add four hydrogen ions (hydronium ions) to the reactants to balance hydrogens:

$$Cu(s) + 4H^+(aq) + 2NO_3^-(aq) \rightarrow Cu^{2+}(aq) + 2NO_2(g) + 2H_2O(l) \qquad (6.56)$$

Step 7: In reaction equation (6.56), atoms are balanced and charge is balanced. There is a net charge of +2 [= 4(+1) + (−2)] on the reactant side and +2 on the product side.

Does the answer make sense? Reaction equation (6.56) describes the observed changes in the reaction of copper with concentrated nitric acid and is balanced in atoms, charge, and electrons transferred. The answer makes sense.

6.79.   **Check This**   *Balancing redox equations by the oxidation-number method*

(a) In less concentrated nitric acid solutions, Cu^0 is oxidized to Cu^{2+}, but the gaseous reaction product is nitric oxide, NO, rather than NO_2. Assign the oxidation number for N in NO and use the oxidation-number method to write the balanced equation for the reaction.

(b) Use the oxidation-number method to balance the reaction of hydrogen peroxide with iodide:

$$H_2O_2(aq) + I^-(aq) \rightarrow H_2O(l) + I_2(aq) \tag{6.57}$$

(c) Use the oxidation-number method to balance the reaction of thiosulfate anion, $S_2O_3^{2-}$, with iodine [a reaction often used in analyses of reactants that produce iodine by oxidizing iodide, as in reaction (6.57)]:

$$\left[\begin{array}{c}:O:\\ \| \\ :O{-}S{-}S: \\ \| \\ :O:\end{array}\right]^{2-}_{(aq)} + I_{2\,(aq)} \longrightarrow \left[\begin{array}{c}:O: \quad\quad :O:\\ \| \quad\quad\quad \| \\ :O{-}S{-}S{-}S{-}S{-}O: \\ \| \quad\quad\quad \| \\ :O: \quad\quad :O:\end{array}\right]^{2-}_{(aq)} + I^-_{(aq)} \tag{6.58}$$

As you first use the oxidation-number method for reduction-oxidation equation balancing, you will probably find it helpful to show the loss and gain of electrons explicitly, as we did in expressions (6.53) and (6.54) in Worked Example 6.78. With more experience, you can probably do the gain and loss in your head and adjust the coefficients without showing the intermediate expressions.

6.80.   **Investigate This**   *What happens when bleach and iodide are mixed?*

Do this as a class investigation and work in small groups to discuss and analyze the results. Mix 1 mL of 6 M aqueous sulfuric acid, $(HO)_2SO_2$, solution with 20 mL of 20% aqueous potassium iodide, KI, solution in a small flask. Record your observations. Add a few drops of household bleach solution to the flask and swirl to mix. Record your observations.

6.81.   **Consider This**   *What is the reaction between bleach and iodide?*

(a) Did you observe any evidence for reaction between acid and potassium iodide solution in Investigate This 6.80? between bleach and acidic potassium iodide solution? What is the evidence in each case?

(b) Can you identify any product or products of the reaction? Explain your response. Can you tell whether the reaction is a reduction-oxidation? Why or why not?

(c) The active ingredient in most common household bleaches is hypochlorite ion, OCl^-. If the reaction in Investigate This (6.80) is a reduction-oxidation, is hypochlorite probably reduced or oxidized in the reaction? What is the reasoning for your answer? What might be the product(s) formed from the hypochlorite? Could you see any evidence for this(these) product(s) in the investigation? Why or why not?

The half-reactions method

The **half-reactions method** for balancing reduction-oxidation reaction equations is based on splitting the reaction in two, considering the reduction reaction and oxidation reaction separately, and then recombining them so that the number of electrons lost in the oxidation is the same as the number of electrons gained in the reduction. You do not have to consider oxidation numbers explicitly in the balancing procedure outlined in Table 6.7, but you may have to use them to determine which species is reduced and which oxidized, in order to write the appropriate half reactions.

Table 6.7. Half-reactions method for balancing reduction-oxidation reactions.

This procedure applies to acidic and neutral solutions.

Step	Procedure
1.	Write all reactants and products except H^+ and H_2O in the form of separate chemical equations for the oxidation and reduction half-reactions.
2.	For each half reaction, adjust coefficients for atoms of all elements other than hydrogen and oxygen.
3.	For each half-reaction, if the number of oxygen atoms differs between the left and right sides of the equation, add water molecules to the side needing more oxygen atoms.
4.	For each half-reaction, if the number of hydrogen atoms differs between the left and right sides of the equation, add hydrogen ions (H^+) to the side needing more hydrogen atoms.
5.	For each half-reaction, add the number of electrons, e^-, needed to balance the charge. One half-reaction (the reduction) will need electrons added to the left side, the other half-reaction (the oxidation) will need electrons added to the right side.
6.	Multiply each half-reaction by the minimum factor required to equalize the number of electrons in each half-reaction.
7.	Add the equations for the half-reactions together, canceling electrons and excess water molecules or hydrogen ions.

6.82.   Worked Example   *Balancing redox equations by the half-reactions method*

In Investigate This 6.80, the red-brown color formed in the reaction is due to elemental iodine, I_2, formed by oxidation of I^-. (The I_2 combines with excess I^- to form a complex ion, I_3^-, which is what you observe, but we will focus on the I_2.) Hypochlorite ion, OCl^-, is the species that is reduced and the product of the reduction is chloride ion, Cl^-. (There is no visible evidence for this product; other experiments are required to confirm that it is formed.) Thus, we can write the reactants and products as:

$$I^-(aq) + OCl^-(aq) \rightarrow I_2(aq) + Cl^-(aq) \tag{6.59}$$

Use the half-reactions method to balance this reaction.

Necessary information: We need to know that $I^-(aq)$ reacts to give $I_2(aq)$ and $OCl^-(aq)$ reacts to give $Cl^-(aq)$.

Strategy: Follow the steps outlined in Table 6.7.

Implementation:

Step 1: The half reactions are:

$$I^-(aq) \rightarrow I_2(aq) \tag{6.60}$$
$$OCl^-(aq) \rightarrow Cl^-(aq) \tag{6.61}$$

Step 2: There are two iodine atoms on the product side of expression (6.60) and only one on the reactant side, so the coefficient on $I^-(aq)$ must be 2. The number of chlorine atoms in the reactants and products is the same in expression (6.61).

$$2I^-(aq) \rightarrow I_2(aq) \tag{6.62}$$
$$OCl^-(aq) \rightarrow Cl^-(aq) \tag{6.61}$$

Step 3: In expression (6.61), there is one oxygen atom in the reactants and none in the products. Add a water molecule to the products to balance the oxygens.

$$2I^-(aq) \rightarrow I_2(aq) \tag{6.62}$$
$$OCl^-(aq) \rightarrow Cl^-(aq) + \mathbf{H_2O}(l) \tag{6.63}$$

Step 4: In expression (6.63), there are two hydrogen atoms in the products and none in the reactants. Add two hydrogen ions (hydronium ions) to the reactants to balance hydrogens.

$$2I^-(aq) \rightarrow I_2(aq) \tag{6.62}$$
$$\mathbf{2H^+}(aq) + OCl^-(aq) \rightarrow Cl^-(aq) + H_2O(l) \tag{6.64}$$

Step 5: In expression (6.62), there are two negative charges on the reactant side and no charges on the product side. Add two electrons, e^-, to the product side.

$$2I^-(aq) \rightarrow I_2(aq) + \mathbf{2e^-} \tag{6.65}$$

In expression (6.64), the net charge on the reactant side is +1 [= 2(+1) + (−1)] and on the product side is −1. Add two electrons to the reactant side to make the net charge on each side −1.

$$2H^+(aq) + OCl^-(aq) + 2e^- \rightarrow Cl^-(aq) + H_2O(l) \qquad (6.66)$$

Step 6: The number of electrons in each half reaction is the same, so no multiplicative factor is required to make them the same.

Step 7: Add the two half-reaction equations (6.65) and (6.66) to cancel the electrons. There are no excess water or hydrogen ions to be cancelled.

$$2I^-(aq) \rightarrow I_2(aq) + \cancel{2e^-} \qquad (6.65)$$
$$\underline{2H^+(aq) + OCl^-(aq) + \cancel{2e^-} \rightarrow Cl^-(aq) + H_2O(l)} \qquad (6.66)$$
$$2H^+(aq) + OCl^-(aq) + 2I^-(aq) \rightarrow Cl^-(aq) + H_2O(l) + I_2(aq) \qquad (6.67)$$

Does the answer make sense? Reaction equation (6.67) describes the observed changes in the reaction of bleach (OCl^-) with an acidic solution of iodide ion and is balanced in atoms, charge, and electrons transferred. The answer makes sense.

6.83.   Check This   *Balancing redox equations by the half-reactions method*

(a) One way to analyze iron(II) ion in solution is by reaction with dichromate ion, $Cr_2O_7^{2-}$:

$$Fe^{2+}(aq) + Cr_2O_7^{2-}(aq) \rightarrow Fe^{3+}(aq) + Cr^{3+}(aq) \qquad (6.68)$$

Explain how you know this is a reduction-oxidation reaction and balance it by the half-reactions method.

(b) Household bleach, hypochlorite solutions, always have a warning label that says something like: "Hazard. Do not mix with other household chemicals such as toilet bowl cleaners, rust removers, acids…" All of these are often quite acidic. When hypochlorite ion solution is mixed with hydronium ion solution the products are chlorine gas, Cl_2 and oxygen gas, O_2. Explain how you know this is a reduction-oxidation reaction, write the reactants and products, and balance the reaction equation by the half reactions method. *Hint*: Water has an oxygen atom with a -2 oxidation number and can decompose to O_2 with a zero oxidation number.

(c) In some reduction-oxidation reactions the same element, in different molecules, is both reduced and oxidized. One example that is commonly used in chemical analysis is the reaction of iodate ion, IO_3^-, with iodide ion, I^-, in acidic solution:

$$IO_3^-(aq) + I^-(aq) \rightarrow I_2(aq) \qquad (6.69)$$

Balance this reaction equation by the half reactions method.

Reduction-oxidation reactions in basic solutions

In basic solutions the concentration of hydronium ion is so small that we cannot consider it as a reactant. Hydroxide ions, $OH^-(aq)$ and water are the predominant species that we can use as sources of oxygen and hydrogen, respectively, to balance reduction-oxidation reaction equations. There are several ways to account for this change. By far the easiest is first to balance the reaction equation as if it occurs in acidic solution, using either of the two methods we have just introduced. Then add the same number of $OH^-(aq)$ to both sides of the equation so that the added $OH^-(aq)$ is equivalent to the $H^+(aq)$ and "reacts" with it to produce water. Finish up by eliminating any excess water molecules and then rechecking for atom and charge balance.

6.84.   **Worked Example**   *Balancing redox equations in basic solution*

To make household bleach chlorine, $Cl_2(g)$, is reacted with an aqueous solution of sodium hydroxide [$Na^+(aq)$ and $OH^-(aq)$]. The reaction products are $OCl^-(aq)$ and $Cl^-(aq)$:

$$Cl_2(g) \rightarrow OCl^-(aq) + Cl^-(aq) \tag{6.70}$$

Write the balanced equation for this reaction in basic solution.

Necessary information: The reactants and products are given in expression (6.70).

Strategy: Use the half-reaction method to balance the equation as though it occurs in acidic solution and then add enough $OH^-(aq)$ to both sides of the equation to convert any $H^+(aq)$ to water.

Implementation: We will go through the steps of the half-reaction procedure without comment. Be sure you can explain each step for yourself.

Step 1:	$Cl_2(g) \rightarrow OCl^-(aq)$	$Cl_2(g) \rightarrow Cl^-(aq)$
Step 2:	$Cl_2(g) \rightarrow 2OCl^-(aq)$	$Cl_2(g) \rightarrow 2Cl^-(aq)$
Step 3:	$Cl_2(g) + 2H_2O(l) \rightarrow 2OCl^-(aq)$	$Cl_2(g) \rightarrow 2Cl^-(aq)$
Step 4:	$Cl_2(g) + 2H_2O(l) \rightarrow 2OCl^-(aq) + 4H^+(aq)$	$Cl_2(g) \rightarrow 2Cl^-(aq)$
Step 5:	$Cl_2(g) + 2H_2O(l) \rightarrow 2OCl^-(aq) + 4H^+(aq) + 2e^-$	$Cl_2(g) + 2e^- \rightarrow 2Cl^-(aq)$
Step 6:	unnecessary	
Step 7:	$Cl_2(g) + 2e^- \rightarrow 2Cl^-(aq)$	

$$\underline{Cl_2(g) + 2H_2O(l) \rightarrow 2OCl^-(aq) + 4H^+(aq) + 2e^-}$$

$$2Cl_2(g) + 2H_2O(l) \rightarrow 2OCl^-(aq) + 2Cl^-(aq) + 4H^+(aq) \tag{6.71}$$

There are four $H^+(aq)$ among the products of reaction equation (6.73), so we will add four $OH^-(aq)$ to both sides of the equation to convert these $H^+(aq)$ in the product to an equivalent amount of water:

$$2Cl_2(g) + 2H_2O(l) + 4OH^-(aq) \rightarrow 2OCl^-(aq) + 2Cl^-(aq) + \underline{4H^+(aq) + 4OH^-(aq)}$$
$$= 4H_2O(l)$$

Eliminating excess water gives the final balanced reaction equation in basic solution:

$$2Cl_2(g) + 4OH^-(aq) \rightarrow 2OCl^-(aq) + 2Cl^-(aq) + 2H_2O(l) \tag{6.72}$$

Does the answer make sense? Reaction equation (6.72) describes the reaction discussed in the presentation of the problem and is balanced in atoms, charge, and electrons transferred. The answer makes sense.

6.85. Check This   ***Balancing redox equations in basic solution***

(a) Balance reaction expression (6.70) in basic solution by the oxidation-number method. *Hint*: Write Cl_2 twice on the reactant side and let one of them be reduced and the other oxidized. Add all the Cl_2 molecules together after the equation is balanced.

(b) Hydrogen peroxide, H_2O_2, can oxidize manganous ion, Mn^{2+}, to manganese dioxide, MnO_2 in basic solution:

$$H_2O_2(aq) + Mn^{2+}(aq) \rightarrow MnO_2(s) \tag{6.73}$$

Balance this equation for reaction in basic solution. *Hint*: Assume that the product of reduction of the peroxide is water.

Section 6.11. Reduction-Oxidation Reactions of Carbon-containing Molecules

6.86. Investigate This   ***Does methanal react with silver ion?***

Do this as a class investigation and work in small groups to discuss and analyze the results. Place 4 mL of a basic solution of silver diammine complex, $Ag(NH_3)_2^+(aq)$, in a clean 200-mm test tube. CAUTION: This basic solution is caustic and can harm clothes and flesh. Silver ion solutions can stain clothing and flesh. Handle with care and rubber or plastic gloves. Add to the test tube a few drops of an aqueous solution of methanal (formaldehyde), H_2CO. Swirl the mixture in the test tube for a minute or two. Record your observations.

6.87. Consider This   ***What is the reaction of methanal with silver ion?***

In Investigate This 6.86, what is the evidence, if any, that methanal reacts with silver ion? What products of the reaction can you identify? How would you classify the reaction? Explain.

Oxidation of methanal by silver ion

The most obvious observation in Investigate This 6.86 is the formation of silver metal. It was also a product of the reaction with copper in Investigate This 6.62. Once again, the silver metal signals the presence of a reactant that can provide the electrons required to produce the Ag^0 from the Ag^+. The silver is formed when methanal is added to the mixture, so methanal is probably the source of electrons and must itself be oxidized. What are the possible oxidation products of methanal?

6.88.   Consider This   *What oxidation numbers are available for carbon?*

(a) What is the oxidation number of each of the atoms in each of these compounds?

methane, CH_4 methanol, CH_3OH methanal, CH_2O

methanoic (formic) acid, $HC(O)OH$ carbon dioxide, CO_2

(b) Is there a correlation between the number of bonding pairs of electrons between carbon and oxygen and the oxidation number of carbon? If so, state a rule you can use to assign oxidation numbers to a carbon atom, based on the number of bonds to oxygen atoms.

(c) When methanal is oxidized in the reaction with silver ion in Investigate This 6.86, what are the possible oxidation products? Do you have any evidence for or against any of the possible products? Explain your reasoning.

We can use the results of your analysis from Consider This 6.88 to write the reactants and products for methanal oxidation by silver ion. Although, silver ion is present in the solution as its diammine complex, $Ag(NH_3)_2{}^+(aq)$, we will write it as $Ag^+(aq)$. The ammonia is not oxidized or reduced and it makes the equations simpler, if it is left out. You saw no gas produced by the reaction, so the most likely oxidized product from methanal is methanoic acid:

$$Ag^+(aq) + CH_2O(aq) \rightarrow Ag(s) + HC(O)OH(aq) \tag{6.74}$$

In basic solution, a possible balanced reaction is:

$$2Ag^+(aq) + CH_2O(aq) + 2OH^-(aq) \rightarrow 2Ag(s) + HC(O)OH(aq) + H_2O(l) \tag{6.75}$$

6.89.   Check This   *Balancing the methanal-silver ion reaction equation*

(a) Show how to balance expression (6.74) to get (6.75) in basic solution.

(b) What will happen to methanoic acid in basic solution? Write the reaction equation for this reaction. *Hint*: See Section 6.4.

(c) Combine your equation from part (b) with the balanced equation (6.75) to get the overall reaction of methanal with silver ion that accounts for the fate of methanoic acid in basic solution.

Yeast fermentation of glucose

In Investigate This 6.2 you observed the formation of foam as glucose was fermented by yeast to produce carbon dioxide and ethanol:

$$2CH_3CH_2OH(aq) + 2CO_2(g) \tag{6.76}$$

Equation (6.76) is equation (6.1) rewritten to show the connectivity of the atoms in glucose.

6.90. Check This *Oxidation numbers for carbons in glucose fermentation*

Assign oxidation numbers for all the carbons in the reactant and products in the glucose fermentation reaction. Use your rule from Consider This 6.88(b), if this makes the task easier.

The carbons in glucose all have oxidation numbers of +1 or +2, but two carbons in the fermentation products have an oxidation number of +4 and two others an oxidation number of zero. The fermentation of glucose represents an *internal reduction-oxidation reaction*. Electrons are transferred from one carbon to another to yield an oxidized product, carbon dioxide, and a reduced product, ethanol. Equation (6.76) is the net result of two series of biochemical reactions:

> In 1897, Hans and Eduard Buchner were making cell-free extracts of yeast for medicinal purposes and discovered that these extracts rapidly fermented sugar. This was the first demonstration that living cells are not necessary to carry out a metabolic reaction sequence.

glycolysis and fermentation. **Glycolysis** (*glyco* = sugar + *lyein* = loosen, break; sugar breaking) is a metabolic pathway that oxidizes glucose to two molecules of pyruvic acid. The balanced half reaction for glycolysis is:

$$\longrightarrow 2H-C-C-C + 4H^+(aq) + 4e^- \tag{6.77}$$

The molecule that accepts the electrons lost by glucose in equation (6.77) is **nicotinamide dinucleotide, NAD⁺**, an oxidant used by almost all organisms. The reduction of NAD⁺ to **NADH**, reduced nicotinamide dinucleotide, can be represented by this half reaction (where the reactive part of the molecule is enclosed in the rectangle and R is the rest of the molecule, which is not oxidized or reduced):

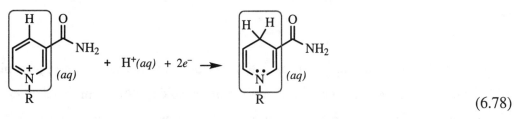

$$(6.78)$$

6.91. Check This   *Net glycolysis reaction*

(a) Check to see that equations (6.77) and (6.78) are balanced with respect to atoms and electrons.

(b) Write the balanced equation for oxidation of glucose to pyruvic acid by NAD^+. You may substitute the abbreviations NAD^+ and NADH for the structures in equation (6.78).

Other changes occur during glycolysis, including the production of adenosine triphosphate, ATP, the compound that organisms use to provide energy for most life processes. We will say more about energy and ATP in the next chapter. As glycolysis proceeds, organisms face a problem; NAD^+ gets used up, because there is only a tiny amount of it in a cell. If oxygen is available, other metabolic processes oxidize the NADH back to NAD^+ and glycolysis can keep going to oxidize more glucose and produce more ATP. Yeast can live in the absence of oxygen; in order to do so, they need a way to regenerate their NAD^+ from NADH. **Fermentation**, the reduction of pyruvic acid to produce ethanol and carbon dioxide is the pathway yeast use to regenerate their NAD^+:

$$(6.79)$$

6.92. Check This   **Reduction-oxidation of glucose**

(a) Write the balanced equation for the reduction of two molecules of pyruvic acid by NADH to produce ethanol and carbon dioxide.

(b) Show that the sum of the balanced equation from Check This 6.91(b) and the one you wrote in part (a) gives equation (6.76), the overall fermentation (reduction-oxidation) reaction for glucose.

(c) What happens to the NAD^+ and NADH in the summation you made in part (b)? Does this explain why the cell needs only a small amount of these molecules for its reduction-oxidation reactions.

All life depends upon reduction-oxidation reactions to provide the energy required to survive and to carry out processes such as silk production by spiders shown in the chapter opening illustration. The fermentation pathway is an inefficient use of glucose for energy. Total oxidation of glucose can produce about twenty times more energy than is available from fermentation. In order to understand these energy considerations, you need to know more about energy, the subject of the next chapter. Then, in Chapter 10, we will look again at the energy released by reduction-oxidation reactions and the way that energy is captured by organisms.

Section 6.12. Outcomes Review

We have divided chemical reactions into three large classes: precipitation reactions, Lewis acid-base reactions, and reduction-oxidation reactions. In precipitation reactions, ions lose their bonding to waters of hydration and form crystals held together by attractions between cations and anions. The electrons on the cations and anions remain with the same atomic core throughout the reaction.

In Lewis acid-base reactions, a nonbonding electron pair on a Lewis base is attracted to a positive center on a Lewis acid and the electrons form a new covalent bond between the two reactants. Brønsted-Lowry acid-base reactions are a special class of Lewis acid-base reactions in which the new covalent bond is to a proton. We showed how to account for and predict the relative strengths of these acids and bases. Lewis acid-base reactions also occur between ligands, Lewis bases, and metal ions, Lewis acids, to which the bases bond to form metal ion complexes, many of which are important in biological systems. Finally Lewis acid-base reactions occur between nucleophiles, Lewis bases, and electrophiles, Lewis acids, to form a variety of final products from the rearrangements that occur after the initial bond formation. The examples we chose were condensation reactions, many of which are used by both man and nature to create polymers.

Reduction-oxidation reactions involve complete transfer of one or more electrons from one reactant to another. The reactant that loses electrons is oxidized and the reactant that gains electrons is reduced. Oxidation numbers identify reduction-oxidation reactions and can be used to balance their reaction equations. Another method for balancing is to break the reaction into half reactions, a reduction and an oxidation, balance these separately, and then recombine them.

Check your understanding of the ideas in the chapter by reviewing these expected outcomes of your study. You should be able to:

• classify a chemical change as a precipitation, Lewis acid-base, or reduction-oxidation reaction based on known reactants and products, or from experimental observations on the change [Sections 6.2, 6.4, 6.6, 6.7, 6.9, and 6.11].

• predict probable precipitation reactions based on the cation and anion charges [Section 6.2].

• use the stoichiometry of continuous variations studies to determine the formula of a reaction product or the ratio in which reactants react [Section 6.2].

• define and give examples of three classes of Lewis acid-base reactions: Brønsted-Lowry proton transfers, metal ion complexation, and nucleophile-electrophile reactions [Sections 6.3, 6.4, 6.6, and 6.7].

• use the observed pH and stoichiometry of a solution to determine the relative basicity (acidity) of the species in the solution [Section 6.4].

• predict the relative basicities (acidities) of a series of Lewis/Brønsted-Lowry bases (acids) of known structure [Section 6.5].

• use relative basicities (acidities) to predict the predominant species in solutions of Lewis/Brønsted-Lowry bases (acids) [Sections 6.4 and 6.5].

• recognize the formation of a metal ion complex between a Lewis base and a metal ion in solution by observations on mixtures of the reactants (or in competitions between the Lewis base and another reactant for the metal ion) and suggest a structure for the complex [Section 6.6].

• identify the nucleophilic and electrophilic sites in a pair of reactants that react to form a condensation product, including condensation polymers [Section 6.7].

• predict the product of a nucleophile-electrophile reaction, including condensation polymerization, between reactants with the functional groups we have introduced [Sections 6.7 and 6.8].

• use formal charge to explain the rearrangements that some reaction intermediates undergo or to explain the relative stability of different isomeric Lewis structures [Section 6.8].

• identify the molecules or ions that are reduced and oxidized and their respective products in a reduction-oxidation reaction [Sections 6.9, 6.10, and 6.11].

• assign oxidation numbers to all the atoms in a given molecule or ion [Sections 6.9 and 6.11].

• balance a given reduction-oxidation reaction in acidic or basic solution by inspection, the oxidation-number method, or the half-reactions method [Sections 6.9, 6.10, and 6.11].

- use your knowledge of the oxidation numbers of atoms in various molecules or ions to predict possible reduced or oxidized products from a reaction [Sections 6.10 and 6.11].

- use oxidation numbers to show that a given reaction is an internal reduction-oxidation [Section 6.11].

Section 6.13. EXTENSION — Titration

6.93. Investigate This   *How can you analyze an acid-base reaction?*

Do this as a class investigation and work in small groups to discuss and analyze the results. Add two drops of bromocresol green acid-base indicator solution to about 20 mL of a 0.10 M aqueous solution of ethanoic (acetic) acid, $CH_3C(O)OH$, and swirl to mix well. Use a calibrated pipet to add 1.0 mL of this solution to each of eight wells in a 24-well microtiter plate. Leave the solution in the first well as it is. Add 0.20 mL of 0.10 M aqueous sodium hydroxide, NaOH, solution to the second well, 0.40 mL to the third, 0.60 mL to the fourth, and so on to 1.40 mL to the eighth well. Use an overhead projector to view the plate. Record the color in each of the wells.

6.94. Consider This   *How can you follow an acid-base reaction?*

What is the pattern of colors that you observe in the wells in Investigate This 6.93? How can you explain the pattern?

Ethanoic acid reaction with hydroxide

In Investigate This 6.93, hydroxide ion (from the NaOH solution) is added to ethanoic acid. The acid-base reaction that can occur is:

$$CH_3C(O)OH(aq) + OH^-(aq) \rightleftharpoons H_2O(l) + CH_3C(O)O^-(aq) \qquad (6.80)$$

Table 6.2 shows that hydroxide ion is a stronger base than ethanoate. If equal numbers of moles of ethanoic acid and hydroxide are mixed, the predominant base in the mixture, after reaction occurs, is the ethanoate ion, $CH_3C(O)O^-(aq)$. What is the predominant base, if *unequal* numbers of moles of ethanoic acid and hydroxide are mixed?

If fewer moles of hydroxide than ethanoic acid are mixed, reaction (6.80) proceeds until the hydroxide is used up, but then no more ethanoate can be formed. Ethanoate will be the predominant base in the mixture and there will also be ethanoic acid left over. If more moles of hydroxide than ethanoic acid are mixed, reaction (6.80) proceeds until the ethanoic acid is used up. This leaves a solution containing substantial concentrations of both hydroxide and ethanoate ions.

6.95.   **Check This**   *Acid-base equivalence in Investigate This 6.93*

In which well in Investigate This 6.93 is the number moles of hydroxide added equal to the number of moles of ethanoic acid at the start? Show how you get your answer. Where is this well in the pattern of colors you observed in the investigation?

Titration

When base is added to an acid (or *vice versa*), as in Investigate This 6.93, the **equivalence point** of the addition is when the number of moles of base added is equal to the number of moles of acid originally present. If you have some way to recognize the equivalence point, you can determine the number of moles of acid in an unknown sample by keeping track of the number of moles of base you need to add to get to the equivalence point. This kind of analysis is called a **titration**. Titrations are another way to determine reaction stoichiometry, if the concentrations of both reactants are known.

In Investigate This 6.93, you observed that the **acid-base indicator** (bromocresol green) is yellow in acetic acid solution, the first well. You also observed that the indicator is blue in solutions that contain more hydroxide than acid, the wells after the equivalence point. In wells to which some base had been added, but before the equivalence point, the indicator is green (a mixture of blue and yellow). These wells contain ethanoate ion formed by reaction (6.80) as well as unreacted ethanoic acid; the wells are at some intermediate acidity. At the equivalence point the indicator color in the well is blue. The equivalence point with this indicator is signaled by the appearance of the blue color (the disappearance of the green color). More accurate experiments are required to prove this assertion, but the proper choice of indicator makes it possible to determine the equivalence point in many acid-base titrations.

> Acid-base indicators are highly colored dyes whose color is sensitive to the pH of the solution. There are many different acid-base indicators each of which changes color at a different pH. The pH test paper you have used contains these dyes that give the color you use to get the pH.

A common use of titrations is to find the concentration of a reactant in a solution of unknown concentration. The traditional laboratory apparatus for carrying out titrations is illustrated in Figure 6.12. The long, graduated glass tube with a valve at one end to control the outflow of liquid is a **buret**. The buret usually contains a reactant solution of accurately known concentration, called the **titrant**. The idea is to add the titrant in small portions to a vessel containing a *measured volume* of the other reactant, as you modeled in Investigate This 6.93, whose concentration is not known. When the equivalence point is reached, the volume of titrant required, its concentration, and the volume of the unknown solution are used to calculate the unknown concentration. For acid–base titrations, the equivalence point is often signaled by an indicator that changes color.

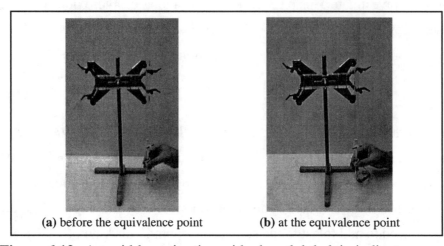

(a) before the equivalence point   (b) at the equivalence point

Figure 6.12. An acid-base titration with phenolphthalein indicator.

The flask contains ethanoic acid of unknown concentration and a few drops of phenolphthalein indicator which is colorless in acid and red in base. The titrant is sodium hydroxide solution of known concentration. The titrant is added slowly while the flask is swirled. The addition is stopped just as the solution turns light pink.

6.96. **Worked Example**   *An acid-base titration*

25.00 mL of an ethanoic acid solution of unknown concentration are titrated with 0.105 M sodium hydroxide with phenolphthalein as the indicator. The buret reading at the beginning of the titration is 3.45 mL and the reading at the end (just as the titrated solution turns pink) is 22.67 mL. What is the concentration of the ethanoic acid?

Necessary information: All the volume and molarity information we need are available from the problem statement. Burets are graduated beginning with zero at the top, so the volume delivered (in milliliters) is the final reading minus the initial reading. We know from equation (6.80) that hydroxide and ethanoic acid react in a 1:1 ratio.

Strategy: Determine the volume of hydroxide added and use its molarity to calculate the number of moles of hydroxide that react. Since the acid and base react in a 1:1 ratio, the number of moles hydroxide used is the number of moles of ethanoic acid in the original sample. Use the number of moles ethanoic acid, the volume of the sample, and the definition of molarity to get the concentration.

Implementation: The volume of base used in the titration is:

$$\text{vol base} = (22.67 \text{ mL}) - (3.45 \text{ mL}) = 19.22 \text{ mL}$$

$$\text{moles base} = (19.22 \text{ mL})\left(\frac{1 \text{ L}}{1000 \text{ mL}}\right)(0.105 \text{ M}) = 2.02 \times 10^{-3} \text{ mol}$$

$$\text{moles acid} = \text{moles base} = 2.02 \times 10^{-3} \text{ mol}$$

$$\text{concentration of acid} = \frac{\text{mol acid}}{\text{vol acid}} = \frac{2.02 \times 10^{-3} \text{ mol}}{0.02500 \text{ L}} = 0.0807 \text{ M}$$

Does the answer make sense? The volumes of acid and base that react are similar so the concentrations of the solutions should be similar, as we found. The volume of base used is smaller than the volume of acid, so the acid is less concentrated, as we found.

6.97. **Check This**   *An acid-base titration*

   (a) If two drops of phenolphthalein are added to 50.0 mL of an aqueous solution of ethylenediamine, $H_2NCH_2CH_2NH_2$, will the solution be colored or colorless? Explain the reasoning for your answer.

   (b) When the solution in part (a) is titrated with a 0.500 M solution of hydrochloric acid, a color change of the solution occurs when 25.0 mL of the acid have been added. What is the color change observed? What is the concentration of the ethylenediamine solution? The reaction of ethylenediamine with hydronium ion is:

$$2H_3O^+(aq) + H_2NCH_2CH_2NH_2(aq) \rightleftharpoons {}^+H_3NCH_2CH_2NH_3{}^+(aq) + 2H_2O(l) \qquad (6.81)$$

6.98. **Investigate This**   *What other titrations are possible?*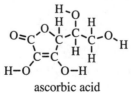

   Work in small groups to do this investigation and to discuss and analyze the results. Each group will use two small test tubes and three, labeled thin-stem plastic pipets containing, respectively, water, 0.0050 M aqueous **ascorbic acid** (vitamin C) solution, and 0.0050 M aqueous iodine, I_2, solution.

ascorbic acid

   (a) In one test tube, mix 20 drops of water and 3 drops of iodine solution. Record your observations. In the second test tube, mix 20 drops of ascorbic acid solution and 3 drops of iodine solution. Record your observations.

(b) Continue to mix iodine solution, 2-3 drops at a time, to the ascorbic acid solution while keeping a count of the *total* number of drops added. Stop adding at the first sign that the ascorbic acid has all reacted. Record the number of drops required.

6.99. Consider This   *What is the stoichiometry of the iodine-ascorbic acid reaction?*

(a) In Investigate This 6.97, what evidence do you have that a reaction occurs between ascorbic acid and iodine? Explain.

(b) How many drops of the iodine solution were required to react with all the ascorbic acid in 20 drops of the ascorbic acid solution? Assuming that the size of drops is the same for both solutions, what can you conclude about the stoichiometric reaction ratio between iodine and ascorbic acid? Clearly explain the reasoning for your answer.

Ascorbic acid-iodine reaction

Iodide ion in water, $I^-(aq)$, is colorless. If the iodine in the iodine solution is reduced to iodide, the color disappears:

$$I_2(aq) + 2e^- \rightarrow 2I^-(aq) \tag{6.82}$$

The reaction with ascorbic acid in Investigate This 6.98 must furnish the two electrons. Your results from Consider This 6.99 show that iodine reacts 1:1 with ascorbic acid. Each ascorbic acid molecule must transfer two electrons to reduce a molecule of I_2 to $2I^-$. The product of this oxidation of ascorbic acid is **dehydroascorbic acid** ("dehydro" means "without hydrogen"):

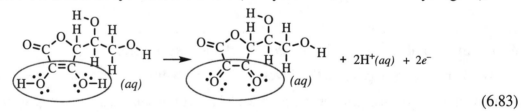

$$\tag{6.83}$$

Titration with iodine, as in Investigate This 6.98, is one of the methods for determining the amount of ascorbic acid (vitamin C) in foods.

6.100. Check This   *Ascorbic acid oxidation*

(a) Show how two electrons are lost in going from reactant to product(s) in reaction (6.83).

(b) Write the balanced chemical equation for the reaction of ascorbic acid with iodine.

In principle, any reaction that goes essentially to "completion" with well-defined stoichiometry can be used in a titration to analyze one of the reactants. For efficiency, we also want reactions that are rapid, so the titration can be done quickly. The usual limitation on possible titration reactions is finding a way to detect the equivalence point. Color changes are most convenient, but there are also instrumental methods that can often be used when appropriate indicators or other color changes are not available.

Index of Terms

Chapter 6 Problems

Section 6.1.   Classifying Chemical Reactions

6.1.   In each of the following cases, indicate whether a chemical reaction (or reactions) occurs and explain how you know.

(a) Water is boiled in a teakettle.

(b) Boiling water is poured into a bowl containing a package of instant oatmeal and stirred to make a hot cereal breakfast.

(c) A glass of a carbonated soft drink is left overnight and tastes "flat" the next morning.

(d) A glass of ice cubes and water is left overnight and there are no ice cubes in the water the next morning.

6.2.   In each of the following cases, indicate whether a chemical reaction (or reactions) occurs and explain how you know.

(a) A match is dropped on the floor.

(b) A match is struck and used to start a barbeque.

(c) A piece of paper is folded to make a paper airplane.

(d) A piece of paper is torn into many small pieces to make confetti.

6.3.   In each of the following cases, indicate whether a chemical reaction (or reactions) occurs and explain how you know.

(a) An acorn buried and forgotten by a squirrel grows into an oak tree.

(b) A bottle of milk left too long in the refrigerator turns sour.

(c) Equal volumes of solutions of blue food coloring and yellow food coloring are mixed and the resulting solution is green.

(d) A few drops of bromocresol green solution are added to 20 mL of a colorless 0.1 M solution of sodium acetate and the resulting solution is blue. When 10 mL of a colorless solution of 0.1 M hydrochloric acid are added, the mixture is green. When a further 10 mL of the hydrochloric acid are added, the mixture is yellow.

6.4.   A few small pieces of dry ice (solid carbon dioxide, –78 °C) are dropped into a tall glass cylinder containing a red solution of dilute ammonia to which has been added a few drops of phenol red acid-base indicator solution. Bubbles of gas are rapidly evolved, a white fog forms above the solution, and, after a short time, the solution in the cylinder turns yellow. What evidence, if any, do you have for physical and chemical reactions occurring in this system? Explain your reasoning carefully.

Section 6.2.   Ionic Precipitation Reactions

6.5.   Which of the following would you predict to be insoluble ionic solids when placed in water?

     (a) $NaCl$ **(d)** $BaCl_2$

     (b) $CaCO_3$ **(e)** $BaSO_4$

     (c) Na_2CO_3

6.6.   25.0 mL of 0.100 M sodium sulfate, Na_2SO_4, is added to 25.0 mL of 0.200 M barium chloride, $BaCl_2$. The mixing of these two solutions results in the formation of a white precipitate. How many grams of the precipitate can be formed? What is the limiting reactant? *Note*: You can assume that the two volumes are additive when the solutions are mixed, that is, the final volume of solution mixture is 50.0 mL.

6.7.   Precipitation reactions often play a role in quantitative analysis. Consider these procedural steps and measurements taken to determine the mass percent of iron in a dietary iron tablet.

Procedural Steps	Measurements Taken
1. The mass of 20 tablets is determined. Then the tablets are ground into a fine powder.	Total mass of 20 tablets = 22.131 g
2. The mass of a sample of the powder is measured. Then the powder is dissolved in nitric acid, converting all iron present to soluble Fe(III) ions.	Mass of sample of powder = 2.998 g
3. The iron is precipitated from the solution through the addition of aqueous ammonia. Then the solid is separated and strongly heated to drive off all water. The iron is now present as the dry solid $Fe_2O_3(s)$.	Mass of dry, solid Fe_2O_3 = 0.264 g

     (a)   Use these data to determine the average mass of iron in a dietary iron tablet.

     (b)   What is the mass percent of iron in a dietary iron tablet?

     (c)   What are some of the assumptions made in using this procedure to determine the average mass of iron in a dietary iron tablet?

6.8. A 0.649 g sample containing only potassium sulfate, K_2SO_4, and ammonium sulfate, $(NH_4)_2SO_4$, are dissolved in water and then treated with excess barium nitrate, $Ba(NO_3)_2$, solution to precipitate all of the sulfate ion as barium sulfate, $BaSO_4$.

(a) Write the net ionic equation for the precipitation reaction.

(b) If 0.977 g of precipitate are formed, how many moles of sulfate ion were in the original sample?

(c) What is the mass percent of K_2SO_4 in the original sample? *Hint:* Let w = mass of K_2SO_4 in the sample and write expressions for the number of moles of sulfate present as K_2SO_4 and $(NH_4)_2SO_4$.

6.9. Use Figure 6.1, an idealized representation of a continuous variations experiment, to see how continuous variations experiments are analyzed by graphical procedures.

(a) Each sample contains the same total number of moles of the cation and anion. Calculate the mole fraction of cation in each sample, that is, the decimal fraction of the total moles that is cation in each sample.

(b) Assume that the "precipitate" shown in each sample is proportional to the amount of precipitate you would weigh from that sample. Make a graph of the amount of precipitate formed in each sample as a function of the mole fraction of cation.

(c) What does your graph in part (b) look like? Does it have straight line segments? If you draw the straight lines, where do they intersect? How is the intersection related to the 1:1 stoichiometry of the reaction?

(d) If you did not know the stoichiometry, could you use the graphical analysis to find it? Explain the procedure you would use to do so.

6.10. Analyze the data in Check This 6.11 by the graphical method outlined in Problem 6.9. What is the stoichiometry of the reaction between nickel ion and dimethylglyoxime? Is this the same result you got in Check This 6.11? Why or why not?

Section 6.3.   Lewis Acids and Bases: Definition

6.11. Identify the ions or molecules in the following list that you predict to be Lewis bases. Explain your reasoning for each selection.

(a) NH_3

(b) H^+

(c) NH_4^+

(d) OH^-

(e) CN^-

(f) CH_3NH_2

(g) H_3O^+

(h) CH_3^-

6.12.   **(a)** Do all Lewis bases share a common characteristic? If so, explain what it is.

        (b) Do Lewis acids also share a common characteristic? If so, explain what it is.

Section 6.4.   Lewis Acids and Bases: Brønsted-Lowry Acid-Base Reactions

6.13.   Identify which reactant can be classified as a Brønsted-Lowry acid or base, and as a Lewis acid or base in this reaction. In each case, give the reason for your choice.

$$\left[\underset{\displaystyle H}{\overset{\displaystyle H}{H-\overset{..}{\underset{..}{O}}-H}} \right]^{+} + \;\; \underset{\displaystyle H}{\overset{\displaystyle H}{:\!N\!-\!H}} \;\; \longrightarrow \;\; \underset{\displaystyle H}{H-\overset{..}{\underset{..}{O}}:} \;\; + \left[\underset{\displaystyle H}{\overset{\displaystyle H}{H-\overset{\displaystyle H}{N}-H}} \right]^{+}$$

6.14.   Lactic acid, $CH_3CH(OH)C(O)OH$, accumulates in our muscle tissue when we exercise strenuously. This acid transfers only a fraction of its acidic protons to water in a Brønsted-Lowry acid-base reaction. Write a chemical equation for this reaction and identify the Brønsted-Lowry acids and bases.

6.15.   A 0.05 M solution of perchloric acid, $HOClO_3$, in water has a pH of 1.3. If sodium perchlorate, $NaClO_4$, is dissolved in water, will the solution pH be higher, lower, or the same as the water before the salt was added? Explain your response.

6.16.   A 0.1 M solution of sodium cyanide, NaCN, in water has a pH of 11. If hydrogen cyanide gas, HCN, is dissolved in water to give a 0.1 M solution, will the solution pH be higher, lower, or about the same as the water before the gas was dissolved? Approximately what pH would you expect the solution to have? Explain your responses.

6.17.   Hydronium ions are able to pass only very slowly from one side to the other of biological membranes. A difference in pH from one side to the other of membranes is produced by metabolism and is the basis for synthesis of the ATP cells need to keep functioning. Any substance that can carry protons from one side to the other of the membrane can interrupt the synthesis of ATP. The requirement for such a substance is that it be soluble in water in a form that can react with hydronium, but less soluble in water (more soluble in nonpolar solvents like the inside of a membrane) after it has reacted.

    (a) 2,4-dinitrophenol (see Table 6.2) is one of the substances that has been used for this purpose in many studies of metabolism. Why is 2,4-dinitrophenol a good candidate for this purpose?

    (b) In the 1930s, some people used 2,4-dinitrophenol as a weight-reducing drug. (This use was discontinued after several deaths occurred.) What is the basis for this use of the compound?

6.18. Consider the reaction between boric acid and water:

$$B(OH)_3(s) + 2H_2O(l) \rightleftharpoons [B(OH)_4]^-(aq) + H_3O^+(aq)$$

(a) Rewrite this equation using Lewis structures. Then analyze the reactants in terms of Lewis acid-base definitions.

(b) Explain why the formula for boric acid is more correctly written as $B(OH)_3$ instead of H_3BO_3.

(c) Some acids, such as sulfuric, $(HO)_2SO_2$, and phosphoric, $(HO)_3PO$, can donate more than one proton to a Lewis (or Brønsted-Lowry) base. These are called polyprotic acids. Explain why boric acid acts as a monoprotic (one proton) rather than polyprotic acid.

6.19. Examine the following information and rank the following bases from weakest to strongest: $CH_3C(O)O^-$ (ethanoate or acetate), NO_3^- (nitrate), OH^- (hydroxide). Clearly explain the reasoning, based on this experimental evidence, for your ranking.

> 0.1 M $CH_3C(O)OH$ has pH = 2.9
> 0.1 M HNO_3 has a pH = 1.0
> 0.1 M $NaNO_3$ has a pH = water in which it was dissolved
> 0.1 M $CH_3COO^-Na^+$ (sodium ethanoate) has a pH = 8.9
> 0.1 M $NaOH$ has a pH = 13.0

Section 6.5.   Predicting Strengths of Lewis/ Brønsted-Lowry Bases and Acids

6.20. Select the member of each pair that you predict to be the stronger base.

(a) NH_3 *or* PH_3

(b) HP^{2-} *or* S^{2-}

(c) HP^{2-} *or* O^{2-}

(d) CH_3O^- *or* CH_3S^-

6.21. If equal volumes of 0.1 M aqueous solutions of sodium hydroxide, NaOH, and ammonium chloride, NH_4Cl, are mixed, what Lewis acid and what Lewis base will predominate in the mixture? Explain your reasoning. *Note:* Ammonium chloride dissolves to give a solution of ammonium cations and chloride anions.

6.22. What ions predominate in the solution formed when equal volumes of 0.10 M NH_3 and 0.10 M HCl are mixed? Explain your reasoning.

6.23. How do you think the basicity of the dihydrogen arsenate anion, $(HO)_2AsO_2^-$, compares to the basicity of the dihydrogen phosphate anion, $(HO)_2PO_2^-$? Explain the basis for your response.

6.24. Compare the Lewis structures of $CH_3C(O)O^-$ and $CH_3C(O)S^-$.

(a) Are there delocalized electrons in these ions? If so, where? Explain.

(b) Which ion do you predict to be the stronger base? Explain the basis for your prediction.

6.25. (a) A 0.1 M solution of potassium amide, KNH_2, in water has a pH of 13. Write a chemical equation for the acid-base reaction that occurs in this solution. What are the predominant Brønsted-Lowry acid and base in the solution? How would you characterize the base strengths of the bases in this solution. Explain your responses.

(b) A 0.1 M solution of ethylamine, $CH_3CH_2NH_2$, in water has a pH of about 11. Write a chemical equation for the acid-base reaction that occurs in this solution. What are the predominant Brønsted-Lowry acid and base in the solution? How would you characterize the base strengths of the bases in this solution. Explain your responses.

(c) How are the reactions you write and your interpretations of the experimental evidence in parts (a) and (b) the same and different? Explain clearly.

6.26. Consider these structural reaction expressions showing the stepwise formation of carbonic acid, H_2CO_3, from water and carbon dioxide:

(a) For each step analyze the Lewis acid-base reaction that is represented. Identify which reactant (or part of the intermediate structure) is acting as the Lewis acid and which as the Lewis base in each step.

(b) We have previously written the overall chemical equation for this reaction as:

$$H_2O(l) + CO_2(g) \rightleftharpoons H_2CO_3(aq)$$

What advantages or disadvantages does this equation have relative to the structural reactions for understanding the Lewis acid-base reaction? Explain your reasoning.

6.27. Consider this reaction: $H_2O(l) + HCO_3^-(aq) \rightleftharpoons H_3O^+(aq) + CO_3^{2-}(aq)$

(a) Use the structural representations shown in Table 6.2 to rewrite the equation for this reaction. Identify all Lewis acids and bases in the reaction and the conjugate pairs of Lewis acids and bases.

(b) Does the position of the equilibrium for this reaction lie mostly towards the product side or mostly towards the reactant side? Explain your reasoning.

Section 6.6.   Lewis Acids and Bases: Metal Ion Complexes

6.28. At pH = 7, aluminum hydroxide, $Al(OH)_3$, is quite insoluble. As the pH increases, $Al(OH)_3$ slowly dissolves into solution. Write a chemical reaction to explain this observation. *Hint:* Note that aluminum is in the boron family (column) of the periodic table and see Problem 6.18.

6.29. When a solution of ethylenediamine, $H_2NCH_2CH_2NH_2$, en, is added slowly to a solution of copper(II) ion, the original light blue color becomes a much deeper blue and then changes to a beautiful magenta. A group of students prepared 12 samples for a continuous variations study of these color changes. They mixed the volumes of 0.020 M ethylenediamine and 0.020 M copper sulfate solutions given in the table below. They measured the amount of light absorbed (the absorbance, A) by each solution at two wavelengths of light, 640 nm (red) and 560 nm (green). Their experimental results are also given in the table.

en, mL	0	1	2	3	4	5	6	7	8	9	10	11
Cu, mL	12	11	10	9	8	7	6	5	4	3	2	1
A_{640}	0.044	0.095	0.150	0.200	0.256	0.301	*	0.261	0.187	**	**	**
A_{560}	0.001	0.026	0.050	0.075	0.107	0.146	*	0.307	0.411	0.312	0.208	0.105

*   The group spilled some of one sample and didn't record absorbance readings for that sample.
**   No readings taken.

(a) Use graphical analyses to determine all you can about the complexes formed by copper(II) ion with ethylenediamine. In particular, try to figure out how many ethylenediamine molecules chelate to each copper ion. Clearly explain the reasoning you use in your analyses and to reach your conclusions. *Hint*: See Problems 6.9 and 6.10.

(b) If the 6:6 solution had been mixed before some was spilled, would it have been appropriate for the students to have used what was left to get the data for this sample? Why or why not? If it had not been spilled, what would the absorbance readings have been for this sample? Explain how you get your answers.

(c) Use your model kit to make a model of the copper-ethylenediamine complex(es).

6.30. As a 4.0 M aqueous solution of ethylenediamine, en, is added slowly to a stirred 0.10 M aqueous solution of nickel(II) ion, the original green color of the nickel(II) ion changes through light blue to purple. The color stops changing after 15 mL of the en solution has been added. What conclusion(s) can you draw about the complex(es) formed between the nickel(II) and en? Clearly state the reasoning for your conclusions.

6.31. One method of treating lead or mercury poisoning is to administer chelating agents that effectively sequester (bind and "hide") metal ions, preventing their incorporation into bone or blood. One such agent is EDTA, discussed in Section 6.6. Another is the ligand called "BAL", which stands for "British AntiLewisite". Its chemical name is 2,3-dimercaptopropanol and this is its structure:

$$\begin{array}{ccc} H & H & H \\ | & | & | \\ H-C- & C- & C-H \\ | & | & | \\ O & S & S \\ | & | & | \\ H & H & H \end{array}$$

(a) Use your model kit to make a molecular model of this molecule and then sketch the model showing the bond angles revealed by the model. *Hint:* As you did in Investigate This 6.45, leave off the hydrogen atoms to make the model easier to handle, visualize, and sketch.

(b) Use the model to explain how this molecule can engage in a Lewis acid-base reaction to sequester metal ions such as Pb^{2+}.

Section 6.7. Lewis Acids and Bases: Electrophiles and Nucleophiles

6.32. **(a)** Explain how both Brønsted-Lowry acid-base reactions and Lewis acid-base reactions illustrate the importance of charge in understanding chemical reactions.

(b) Why are *both* definitions necessary and useful for understanding chemical reactions?

6.33. Consider the reaction: $CH_3C(O)OH + NH_3 \rightleftharpoons CH_3C(O)NH_2 + H_2O$

(a) Which reactant is the electrophile? Explain.

(b) Which reactant is the nucleophile? Explain.

6.34. Proteins are important biological condensation polymers that result from the linking of amino acids.

(a) Draw the structural formula for the tripeptide (a short polymer of amino acids) that is formed when three glycine molecules, $H_2NCH_2C(O)OH$, are linked together by amide bonds.

(b) Write the reaction expression for formation of the glycine tripeptide.

6.35. WEB Chap 6, Sect 6.7.2. Explain in detail how the movie animation describing the formation of an amide bond is related to the formation of the ester bond shown in reaction (6.35). Does the animation help to clarify the steps in the reaction sequence? Explain why or why not.

6.36. Identify the nucleophile, the electrophile, and any spectator ions in each of these balanced equations. If necessary, rewrite the reactants showing all the valence electrons.

(a) $NaOH + CH_3–I \rightarrow NaI + CH_3–OH$

(b) $(CH_3)_3N + CH_3–I \rightarrow (CH_3)_4N^+ + I^-$

(c) $(CH_3)_3C–Br + CH_3CH_2OH \rightarrow (CH_3)_3C–O–CH_2CH_3 + HBr$

(d) $NH_3 + CH_3C(O)OCH_2CH_3 \rightarrow CH_3C(O)NH_2 + CH_3CH_2OH$

(e) $CH_3NH_2 + CH_3C(O)Cl \rightarrow CH_3C(O)NHCH_3 + HCl$

6.37. Many esters have pleasant odors. Examine the formulas of these esters (and their odors) and write the formula of the alcohol and the acid from which each ester is made. *Note:* C_6H_5 represents the phenyl group, a benzene ring with one of the hydrogens replaced by another atom or group, as in phenol (*phen*yl alcoh*ol*), Table 6.2.

(a) $CH_3C(O)OCH_2CH_2CH(CH_3)_2$ (oil of banana)

(b) $CH_3CH_2CH_2C(O)OCH_2CH_3$ (pineapple)

(c) $CH_3C(O)O(CH_2)_7CH_3$ (orange)

(d) $C_6H_5C(O)OCH_3$ (ripe Kiwi)

(e) $CH_3C(O)OCH_2C_6H_5$ (jasmine)

6.38. Many esters have pleasant odors. Write the formula of the ester that is formed by reaction of each of these acid and alcohol combinations.

(a) $CH_3(CH_2)_2C(O)OH + CH_3(CH_2)_4OH$ (apricot)

(b) $CH_3C(O)OH + CH_3CH_2OH$ (some nail polish removers)

(c) $CH_3C(O)OH + (CH_2)_3CH(CH_2)_2OH$ (pear)

(d) $CH_3CH_2CH_2C(O)OH + CH_3OH$ (apple)

(e) $HC(O)OH + CH_3CH_2OH$ (rum)

6.39. Alcohols can react with carbonyl compounds, aldehydes and ketones, as well as with carboxylic acids. The products of these reactions are called hemiacetals or hemiketals. Consider the reaction of methanol with propanal:

a hemiacetal

Describe the similarities and differences between these reaction steps and those shown in reaction (6.35) and in Problem 6.26. What do you think it is about the structures of a carboxylic acid and an aldehyde that makes their reactions with an alcohol somewhat different?

6.40. If a molecule contains both an alcohol and an aldehyde functional group, then an *intra*molecular reaction can occur, that is, a reaction between electrophilic and nucleophilic sites on the same molecule.

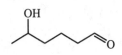

Intramolecular reactions like this are often much faster than intermolecular reactions. In part, this is because the nucleophile and electrophile are held close together; they are parts of the same molecule.

(a) Draw the product you would get from the intramolecular reaction of the alcohol and aldehyde in the molecule above. If this skeletal structure is confusing, draw out the structure, showing all the carbons and hydrogens. *Hint:* See Problem 6.39. Use your molecular models to model the reaction, if that helps you visualize what will happen.

(b) Compare the structure of the product you get in part (a) with the structure of glucose shown in reaction (6.76) and in Problem 6.41. What structure(s) in the two molecules is(are) the same? Explain.

6.41. Cellulose and starch are condensation polymers of glucose. The condensation reaction between two glucose molecules is represented in this equation:

OH   OH   OH   OH

HO—⟨O⟩—OH + HO—⟨O⟩—OH ⇌ HO—⟨O⟩—O—⟨O⟩—OH + H_2O

HO   OH   HO   OH   HO   OH HO   OH

glucose 1   glucose 2   glucose dimer

(a) Many of the condensation reactions we wrote in Sections 6.7 and 6.8 involved an alcohol or an amine, Lewis bases, reacting with a positive carbon, a Lewis acid, that was double bonded to an oxygen. Is there a carbon in the glucose molecule that might react as such a Lewis acid? Which one? Explain the reasoning for your choice.

(b) Based on your choice in part (a) show which pair of nonbonding electrons in the reactants is attracted to this positive carbon to start the reaction toward the product shown. On the basis of this interaction, show where the oxygen that is highlighted in the reactants will end up in the products. Explain how you make your choice.

6.42. Occasionally, there is more than one possible nucleophilic (or electrophilic) reaction site in a molecule or ion. This means that more than one possible product is possible. An example is the reaction of sodium nitrite ($NaNO_2$) with isoamyl bromide (3-methyl-1-bromobutane) to form isoamyl nitrite, the active ingredient in "smelling salts."

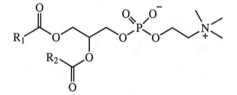

(a) Write the Lewis structures for the nitrite anion and the product, isoamyl nitrite.

(b) Identify the nucleophile, the electrophile, and any spectator ions in this reaction. *Another product is also formed.*

(c) Examine the Lewis structure you wrote for the nitrite anion in part (a). Can you identify more than one nucleophilic site?

(d) Draw the structure for the other product formed in the reaction of sodium nitrite with isoamyl bromide.

(e) Is there any reason to expect the oxygen atoms in the nitrite ion to be less nucleophilic than the oxygen atom in hydroxide, for example? Explain why or why not. *Hint:* Consider electron delocalization.

6.43. Phosphatidyl choline is a common component of your cell membranes.

$$\text{(structure of phosphatidyl choline)}$$

The units that make up this molecule are glycerol, $OHCH_2CH(OH)CH_2OH$, two carboxylic acids, $R_1C(O)OH$ and $R_2C(O)OH$, choline, $HOCH_2CH_2N(CH_3)_3^+$, and phosphoric acid. R_1 and R_2 are hydrocarbon chains, usually containing 15, 17, or 19 carbons.

(a) Write the overall reaction that forms phosphatidyl choline from its five units.

(b) Four bonds are required to hold the five units together. What kind of bond is each one of these four? Explain your answers.

Section 6.8.   Formal Charge

6.44. Calculate the formal charge for each atom in the following substances.

(a) NH_4^+ (c) NH_3

(b) NO_3^- (d) H_2O

6.45. Write Lewis structures with formal charges for each atom in these anions:

    (a) CO_3^{2-} **(d)** HPO_4^{2-}

    (b) NO_2^- **(e)** $H_2PO_4^-$

    (c) PO_4^{3-}

6.46. Is there more than one possible way to write Lewis structures for any of the ions in Problem 6.44? If so, calculate the formal charges for each atom in each structure. Which structure(s) are most favored? Explain why.

6.47. The formal charges are shown for ozone and nitric acid. Fill in the missing electron pairs and show that the formal charges are correct.

                     ozone   nitric acid

Section 6.9. Reduction-Oxidation Reactions: Electron Transfer

6.48. Consider the reaction: $3Sn^{2+}(aq) + 2Bi^{3+}(aq) \rightarrow 3Sn^{4+}(aq) + 2Bi(s)$

    (a) How could you tell that a reaction had occurred upon mixing solutions of the reactants? How would you know that it is a reduction-oxidation reaction? Explain.

    (b) Which element is oxidized?

    (c) Which element is reduced?

    (d) How many electrons are lost by each ion of the element that is oxidized?

    (e) How many electrons are gained by each ion of the element that is reduced?

6.49. A piece of sodium metal, Na(s), is placed in water. A vigorous reaction occurs in which hydrogen gas is evolved and the metal disappears. Upon testing the resulting solution with phenolphthalein, the solution turns pink. (See Figure 4.3(a).)

    (a) How could you tell that a reaction had occurred upon placing the sodium in the water? How would you know that it is a reduction-oxidation reaction? Explain.

    (b) Write the net ionic equation for the reaction described above.

    (c) Which reactant is oxidized?

    (d) Which reactant is reduced?

6.50. Identify the following changes in charge as oxidation or reduction.

    (a) $Zn \rightarrow Zn^{2+}$ **(c)** $Fe^{3+} \rightarrow Fe^{2+}$

    (b) $S^{2-} \rightarrow S$ **(d)** $Ag^+ \rightarrow Ag$

6.51. Given the following observations:

    (i) Mg metal reacts with $Zn^{2+}(aq)$ to yield $Mg^{2+}(aq)$ and Zn metal.

    (ii) Zn metal reacts with $Cu^{2+}(aq)$ to yield $Zn^{2+}(aq)$ and Cu metal.

    (iii) Mg metal reacts with $Cu^{2+}(aq)$ to yield $Mg^{2+}(aq)$ and Cu metal.

  (a) Write the balanced oxidation-reduction reaction for each observation.

  (b) Which metal is most likely to undergo an oxidation? Explain.

  (c) Which metal is most likely to undergo a reduction? Explain.

6.52. For the following reactions: (i) assign oxidation numbers to each atom; (ii) identify the element that is oxidized; (iii) identify the element that is reduced.

  (a) $V_2O_5 + 2H_2 \rightarrow V_2O_3 + 2H_2O$

  (b) $2K + Br_2 \rightarrow 2K^+ + 2Br^-$

  (c) $N_2 + 3H_2 \rightarrow 2NH_3$

6.53. **(a)** What is the oxidation number for the sulfur atom in sulfur trioxide, SO_3?

  (b) What is the oxidation number for the sulfur atom in the sulfite ion, SO_3^{2-}? in the sulfate ion, SO_4^{2-}?

  (c) Sulfur trioxide dissolves in water to give an acidic solution. This is not a redox reaction. What is the acid that is produced? Explain the reasoning for your answer.

6.54. Determine the oxidation numbers of the nitrogen atoms in the following:

  (a) NH_3 **(c)** NO_2^-

  (b) NO **(d)** NO_3^-

  (e) Does the location of nitrogen near the middle of the second period of the periodic table help to explain your results in parts (a) through (d)? Explain why or why not.

6.55. What is the oxidation number of the metal in the following metal-ion complexes?

  (a) $PtCl_4^{2-}$ **(d)** MoS_4^{2-}

  (b) $Cu(NH_3)_4^{2+}$ **(e)** $Zn(OH)_4^{2-}$

  (c) $Fe(CN)_6^{3-}$

Section 6.10. Balancing Reduction-Oxidation Reaction Equations

6.56. For each of the following oxidation-reduction reactions, balance the reaction and identify the reactant oxidized and the reactant reduced.

  (a) $MnO_4^-(aq) + SO_3^{2-}(aq) + H^+(aq) \rightarrow Mn^{2+}(aq) + SO_4^{2-}(aq) + H_2O$

  (b) $NO_3^-(aq) + Zn(s) + H^+(aq) \rightarrow NH_4^+(aq) + Zn^{2+}(aq) + H_2O$

  (c) $Cl_2 + OH^-(aq) \rightarrow ClO_3^-(aq) + Cl^-(aq) + H_2O$

6.57. Zinc-air cells take oxygen from the air to create electrochemical energy. Ambient oxygen gas, O_2, is converted in aqueous solution to hydroxide ion, OH^-, and zinc metal is converted to zinc oxide, ZnO, during discharge of a zinc-air cell.

(a) What is being oxidized in this reaction and what is being reduced as this cell discharges? What are the changes in oxidation number taking place?

(b) Write an overall chemical equation for the reaction that takes place in this cell.

(c) How many moles of electrons are being transferred per mole of zinc converted to zinc oxide?

6.58. These redox reactions occur in aqueous acidic solutions. Identify which element is oxidized and which reduced in each case. Explain how you make these identifications. Balance each reaction by both the oxidation-number and half-reactions methods and compare the results. Which method do you prefer? Explain the reason(s) for your preference.

(a) $S^{2-}(aq) + NO_3^-(aq) \rightarrow NO_2(g) + S_8(s)$ (S_8 rings are the common form of elemental sulfur.)

(b) $Hg^{2+}(aq) + NO_2^-(aq) \rightarrow Hg(l) + NO_3^-(aq)$

(c) $Fe^{2+}(aq) + NO_3^-(aq) \rightarrow Fe^{3+}(aq) + NO(g)$

6.59. These redox reactions occur in aqueous basic solutions. Identify which element is oxidized and which reduced in each case. Explain how you make these identifications. Balance each reaction by either the oxidation-number or half-reactions method, whichever you prefer.

(a) $S^{2-}(aq) + I_2(s) \rightarrow SO_4^{2-}(aq) + I^-(aq)$

(b) $MnO_4^-(aq) + C_2O_4^{2-}(aq) \rightarrow MnO_2(s) + CO_2(g)$

(c) $Bi(OH)_3(s) + Sn(OH)_3^-(aq) \rightarrow Bi(s) + Sn(OH)_6^{2-}(aq)$

6.60. The reaction that occurs when solid cobalt(II) sulfide dissolves in nitric acid can be represented as:

$$CoS(s) + NO_3^-(aq) \rightarrow Co^{2+}(aq) + NO(g) + S_8(s)$$ (S_8 rings are the common form of elemental sulfur.)

(a) Identify which element is oxidized and which reduced in this reaction. Explain how you make these identifications.

(b) Balance the reaction. *Hint:* You might find it easier to treat elemental sulfur as S to balance the equation and then multiply by eight to get S_8 as the product.

Section 6.11. Reduction-Oxidation Reactions of Carbon-containing Molecules

6.61. Lactic acid, $CH_3CH(OH)COOH$, produced in your muscles during strenuous exercise is converted by your cells into pyruvic acid, $CH_3C(O)COOH$, as the muscles rest. Is this change an oxidation or a reduction reaction? Explain your reasoning.

6.62. Cider (apple juice) can ferment to form some ethanol and carbon dioxide in solution. This "hard" (ethanol-containing) cider can react with oxygen from the air to form cider vinegar, a dilute solution of ethanoic (acetic) acid. Write a balanced equation for the vinegar-forming reaction. It might be helpful to consider that an oxygen molecule can gain four electrons and produce two oxide ions.

6.63. When wood is burned in a wood stove, water is always one product. However, carbon dioxide, carbon monoxide, and/or free carbon particles are among the carbon-containing products that form when wood cellulose burns. Wood cellulose is a polymer of glucose, $C_6H_{12}O_6$.

(a) Write three separate equations showing the different carbon-containing products (plus water) that can form if one mole of glucose in wood burns with oxygen gas from the air.

(b) Use these three equations to help explain why it is dangerous to use a wood stove in an area that is not properly ventilated.

6.64. *With respect to the carbon atom on the right in each structure,* identify each of the following reactions as an oxidation, a reduction, or neither.

(a) $H_2C{=}CH_2 \longrightarrow H_3C{-}C\overset{\displaystyle O}{\underset{\displaystyle OH}{\Big<}}$

(b) $\overset{\displaystyle HO}{H_2C}{-}\underset{\displaystyle OH}{CH_2} \longrightarrow H_2C{=}CH_2$

(c) $H_3C{-}\overset{\displaystyle O}{\underset{\displaystyle H}{C\Big<}} \longrightarrow H_2C{=}CH_2$

(d) $H_2C{=}CH_2 \longrightarrow H_3C{-}\overset{\displaystyle OH}{CH_2}$

Section 6.13. EXTENSION — Titration

6.65.   Potassium permanganate, $KMnO_4$, dissolves to form an intensely colored pink solution in water. $KMnO_{4(aq)}$ is often used to oxidize other substances. Depending on the pH of the solution, the permanganate ion, $MnO_4^-{}_{(aq)}$, may be converted to the colorless $Mn^{2+}{}_{(aq)}$ ion, to brown $MnO_{2(s)}$ in neutral or slightly alkaline solution, or to pale green $MnO_4^{2-}{}_{(aq)}$ in highly alkaline solutions.

(a) Identify the oxidation number of manganese in each of these species and explain why the permanganate ion gains electrons in each of these possible reactions.

(b) Write the reduction half-reaction for each of these changes.

(c) In each of these three cases, what visual clues might you have to determine if the reduction reaction was completed? These visual changes can make potassium permanganate useful in titration reactions to determine the concentration of species that reduce the permanganate ion. What limitations are there to these visual methods for finding the endpoint of a titration?

6.66.   The reaction of dichromate ion with iron(II) ion is a standard method for analyzing iron in samples such as ores:

$$Cr_2O_7^{2-}(aq) + Fe^{2+}(aq) \rightarrow Cr^{3+}(aq) + Fe^{3+}(aq)$$

A chromium atom has six valence electrons; dichromate is

(a) What is the oxidation number of the chromium atoms in the dichromate anion? Explain how you get your answer.

(b) Balance the reaction equation (in acidic solution).

(c) A 0.178 g sample of iron ore is dissolved in acid and treated to convert all the iron present to the iron(II) oxidation state. When titrated with a 0.0100 M solution of dichromate, 34.57 mL of the dichromate solution is required to reach the equivalence point. What is the mass percent of iron in the ore?

General Problems

6.67.   When potassium dichromate, $K_2Cr_2O_7$, an orange solid, is dissolved in an aqueous basic solution, a yellow solution containing chromate ion, $CrO_4^{2-}(aq)$, is formed. Write a balanced chemical equation that describes the reaction forming the chromate from dichromate. Is this a precipitation, Lewis acid-base, or oxidation-reduction reaction? Clearly state the reasoning for your response. *Hint:* See Problem 6.66 for the structure of the dichromate anion.

6.68. Consider this table of data.

Metal Ion	Ratio of Ionic Charge/Ionic Radius
Li^+	1.5
Ca^{2+}	2.1
Mg^{2+}	3.1
Al^{3+}	6.7

(a) Use these data to predict which ion is expected to act as the strongest Lewis acid toward water in aqueous solution. Explain your reasoning.

(b) How can you rationalize the relationship between the ratio of ionic charge to ionic radius for Ca^{2+} and Mg^{2+}?

(c) Given that sodium ion acts as a spectator ion in aqueous solution and has no significant acid-base properties, what do you predict will be the ratio of its ionic charge to its ionic radius?

6.69. **(a)** WEB Chap 6, Sect 6.7.2. To make nylon fibers, the solid polymer is melted and then extruded through very fine holes in spinnerets, as shown in the chapter opening illustration. The fine strands from the spinnerets are wound into threads that are used to make nylon fabric. The strength of nylon fibers is largely a result of the many hydrogen bonds between long individual polymer molecules aligned parallel to one another as illustrated in the Web Companion movie. How do you think the extrusion process works to produce these aligned, hydrogen-bonded polymers?

(b) Most silk is obtained from silkworms, which produce essentially one kind of protein polymer to make their cocoons. Spiders, on the other hand, need silk for a variety of tasks and produce several different kinds of silk, some of which are stronger than an equivalent steel wire. They do this using aqueous suspensions of different protein polymers and extruding different mixtures of these polymers, depending on the task for which it is used. The fibers are formed as the mixtures are extruded through their spinnerets, as shown in the chapter opening illustration. Why do you think spiders have evolved to use mixtures like this, rather than producing each type of silk as a separate unique polymer?

Chapter 7.   Chemical Energetics: Enthalpy

What do all these products have in common? They are fuels that provide energy for heating. lighting, and life. The overall reaction of food molecules with oxygen that is carried out in your body can be carried out in the laboratory by burning the food. In your body, a good deal of the energy released in this reaction is captured to produce adenosine triphosphate, ATP, molecules which you use to provide the energy for most of your life processes.

Chapter 7. Chemical Energetics: Enthalpy

"Watt and Stephenson whispered in the ear of mankind
their secret, that *a half-ounce of coal will draw two
tons a mile...*"

Ralph Waldo Emerson (1803–1882),
The Conduct of Life, "Wealth"

In the mid-19th century, Emerson marveled that "a half-ounce of coal [would] draw two tons a mile..." He was referring to the steam locomotive, Figure 7.1, that burned a small amount of coal to boil water to drive pistons to produce the power to move large loads. From the perspective of the early 21st century, it is difficult to imagine the excitement that accompanied the development of engines that could replace waterpower and draft animals to do humanity's heavy lifting and hauling. Engines that transformed chemical energy into mechanical energy fueled the Industrial Revolution, and transformed everyday life for most people on Earth.

Figure 7.1. George Stephenson's steam locomotive, *The Rocket.*
Invented in 1829, this locomotive introduced innovations shared by all steam locomotives since that time.

Now we take for granted vehicles that use the energy released when hydrocarbons burn and nuclear power plants that use the energy released from the fission of uranium–235 (Chapter 3, Section 3.5) to produce electricity. It is easy to lose sight of the fact that *everything we do requires a source of energy.* A variety of these energy sources, including some that supply the metabolic energy you need to stay alive, are shown in the illustration on the facing page. Every change that occurs involves the exchange of energy among substances. The same principles of energy exchange govern everything from the movement of locomotives to the movement of your eyes as you read this page.

These principles were discovered by scientists and engineers who were trying to understand what governed the amount of movement that heat (steam) engines could produce. The branch of science they developed to quantify the relationships between thermal and mechanical energy is

| The name, "thermodynamics" (*therme* = heat + *dynamis* = power), reflects its practical origin. |

called **thermodynamics**. Our emphasis in this chapter will be on the law of conservation of energy, also known as the first law of thermodynamics, which provides us with powerful tools for investigating heat and work and relating them to energy changes in chemical reactions.

Section 7.1.   Energy and Change

7.1.   Investigate This   *What happens when foods are put in a flame?*

Do this as a class investigation and work in small groups to discuss and analyze the results. Investigate several foods, like those shown in the opening illustration, that can provide you with energy. Use several pieces of uncooked spaghetti, one mini-marshmallow, a large potato chip, a pan partially filled with water, and a flame (matches, lighter, candle, or burner).

Take a full-length piece of spaghetti and hold it by one end over the pan of water. Bring the flame under the free end of the spaghetti and hold it there until the appearance of the spaghetti begins to change — then remove the flame. If necessary, extinguish the spaghetti by immersing it in water. Record your observations.

You might investigate other foods as well.

Repeat this procedure with a marshmallow held on the end of another full-length piece of spaghetti and then again with a large potato chip held in the flame.

7.2.   Consider This   *Is energy involved when foods are burned?*

(a) What changes did you observe in Investigate This 7.1? Was there any evidence that energy was involved in these changes? What evidence? What kinds of energy? Do your observations help you answer the question posed with the opening illustration?

(b) One function of food is to provide metabolic energy. Based on your observations for burning different foods, do you think the results are in any way indicative of their different food energy values? Compare your observations and interpretations with those of your classmates.

Your body does not burn food for energy in the way you burned foods in Investigate This 7.1, but the result is the same. When carbon–hydrogen–oxygen compounds, such as those in the carbohydrates and fats in the foods used in the investigation, are burned completely, the final products are carbon dioxide and water. When carbon–hydrogen–oxygen compounds are metabolized completely for energy, the final products are carbon dioxide and water. The *change is the same* in each case.

Measuring energy

Though the caloric content of a food, as displayed on its Nutrition Facts label, is a measure of the energy contained in the food, it is *not* determined by feeding the food to someone and measuring her or his increase in energy. The energy value is determined by burning the food under carefully controlled conditions. The thermal energy (heat) released (in Calories per gram) during complete combustion is the value that typically appears on the food label.

In chemistry, the international energy unit, the joule, J, is most often used to report energy measurements. We will express all energies in joules or kilojoules, kJ (1 kJ = 1000 J). An older metric unit, the calorie (lower case "c"), is also still widely used, especially in the life sciences. The **calorie** is now defined as 4.184 J. The nutritional Calorie (upper case "C") on food labels, which is what you count when dieting, is 1000 cal = 4184 J.

Energy conservation and chemical changes

Overall chemical reactions, especially those in living systems, typically proceed through several steps in the complete journey from reactants to products. Intermediate reactions are often difficult to study directly, because the transitional compounds are so rapidly converted to the next products. However, it is possible to measure accurately, the overall energy change for a reaction without knowing anything about the individual steps, and therein lies much of the power of thermodynamics.

During the Industrial Revolution, scientists, engineers, and inventors were seeking the most efficient means of converting thermal energy (heat) to mechanical energy. Scientist-inventors, including James Watt and George Stephenson (heroes of the Emerson quotation that opens the chapter) had begun to build locomotives and engines by improving the atmosphere-powered steam engine invented in 1705 by Thomas Newcomen. These practical thermodynamicists wanted to convert as much thermal energy to mechanical energy as possible.

Along the way, these scientists discovered a great deal about heat, work, and other forms of energy. The British physicist James Joule discovered the fundamental principle that *when any change occurs, energy is always conserved.* Energy does not appear from nowhere nor does any of it disappear. Only the *form* of the energy changes (as between mechanical energy and thermal energy). This principle, the **first law of thermodynamics**, applies to chemical reactions as well as other kinds of change.

The results of many measurements over many years have been gathered, tabulated, and published, making it possible for you to compare quantitatively the energy changes associated with different processes. Using only tabulated energy data, for example, you can determine the amount of energy associated with the bonds in molecules and you can predict energy changes for chemical reactions that have never been carried out in the laboratory. With the conceptual and mathematical tools presented in this chapter, you will learn how to do these analyses.

7.3.   Consider This   *How is release of thermal energy related to chemical bond energy?*

In Investigate This 7.1, you observed energy released in the form of heat and light from burning foods. What conclusion(s) can you draw about the strengths of the chemical bonds of the reactants and products in these reactions? Explain your reasoning. Remember that breaking bonds requires energy and forming bonds releases energy.

You have already begun your analyses by considering the energy released when foods are burned. The products are the same, water and carbon dioxide, when fats and carbohydrates are burned. However, fatty foods generally burn more steadily and release more energy than carbohydrates, starches and sugars. The greater the energy released, the greater the difference between the energy released in product bond formation and the energy required for reactant bond breaking. Breaking the bonds in fats must take less energy than breaking the bonds in carbohydrates, since more net energy is released when fats burn. You can conclude that the bonds in fats are weaker than the bonds in carbohydrates. We will extend this discussion of bond energies in Section 7.7. First, let us consider two forms of energy, heat and work, and their interconnections.

Section 7.2.   Thermal Energy (Heat) and Mechanical Energy (Work)

7.4.   Investigate This   *What happens when a pinwheel is held over a flame?*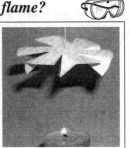

Do this as a class investigation and work in small groups to discuss and analyze the results. Use a candle, an open-ended metal cylinder clamped vertically, and a pinwheel.

(a) Hold your hand about 15 cm above the flame and note how warm the combustion gases coming from the flame feel. Hold the pinwheel about

15 cm above the flame and observe its motion, if any.

(b) Move the metal cylinder over the candle so that it serves as a chimney. Wait a few moments for the metal to become warm. Hold your hand above the top of the chimney and note how warm the combustion gases coming from the flame feel. Hold the pinwheel above the top of the chimney and observe its motion, if any.

7.5.   Consider This   *What causes a pinwheel over a flame to turn?*

(a) In Investigate This 7.4, did you feel a difference in the temperature of the combustion gases without and with the chimney in place over the flame? If so, which felt warmer? Try to explain why there was a temperature difference or why there was not.

(b) Did the pinwheel turn faster without or with the chimney in place over the flame? Try to explain why there was a difference or why there was not.

(c) Is there any correlation between your answers in parts (a) and (b)? If so, are your explanations consistent with the correlation? Explain how.

Two familiar forms of energy are **thermal energy** or **heat** (objects feel hot or cold) and **mechanical energy** or **work** (objects move from one place to another). The usual symbols we use for quantities of heat and work are q and w, respectively. In this chapter we will consider the thermal energy and mechanical energy associated with chemical reactions. Transfers of thermal and mechanical energy from one substance or place to another can be measured in the laboratory and related to the chemical reactions being studied.

Directed and undirected kinetic energy

Thermal energy and mechanical energy are two types of *kinetic* energy. Thermal energy results from the *undirected motions* of individual atoms, ions, and molecules. The molecules are moving chaotically in every direction. Mechanical energy results from *directed motion*, the net movement of the atoms, ions, and molecules in an object in the same direction. The chimney and pinwheel that you used in Investigate This 7.4 illustrate the difference between undirected kinetic energy and directed kinetic energy. We can use Figure 7.2 to help understand this example.

7.6.   Check This   *The difference between warm and cool gases*

If we had not labeled the cooler gas in Figure 7.2, would you still be able to tell that it is cooler? Explain why or why not.

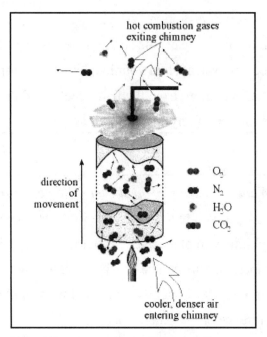

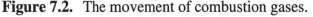

Figure 7.2. The movement of combustion gases.
The velocity (speed and direction) of each molecule is shown by an arrow whose length indicates the speed. The molecules rising in the chimney are a combination of combustion products and molecules from the air. The molecules are moving in all directions, but there is net upward motion as they are buoyed up by the denser gases that enter the chimney at the bottom.

The energy released by the flame is mostly thermal energy. (Energy is also released as light, but we are not considering light here because it represents only a small part of the total energy.) The thermal energy is contained in the undirected (chaotic) molecular motions of the hot combustion gases. Thermal energy is also transferred through collisions to the other gases in the air. The expanding hot gases are buoyed up the chimney by cooler, denser air that enters at the bottom of the chimney. The net upward movement of the gases depicted in Figure 7.2 is directed kinetic energy and can do work. The rising gases do work on the pinwheel, causing it to turn.

7.7. Consider This   ***How do thermal and mechanical energies interact?***

Clearly, the rising combustion gases from the flame produce a change in the pinwheel when it is placed above the chimney in Investigate This 7.4. If the gases produce a change in the pinwheel, would you expect the spinning pinwheel, in turn, to produce any change in the gases? If so, what kind of change? How might the properties of the emerging gases differ between when the pinwheel is there and when it is not? Discuss the possibilities in small groups and then share your ideas with the class.

The pinwheel turns because the moving gas molecules strike the vanes and transfer some of their kinetic energy to them. These gas molecules now have lower kinetic energy, so they are cooler than before they struck the pinwheel and did work on it. Such interplays of heat and work interest engineers who concern themselves with how to build machines that generate maximum work (w, directed kinetic energy) while producing minimum heat (q, undirected kinetic energy). Chemists, on the other hand, concern themselves with understanding why chemical reactions occur as they do. That understanding, gained through the study of heat and work in chemical systems, provides the keys to making better materials, understanding life processes, producing energy more efficiently, minimizing environmental damage, and much more.

Chemical reaction energies

In Chapter 5, you learned that the total energy of a molecule is a complex interplay of potential and kinetic energies that involves attractions and repulsions of electrical charges and the sizes of electron waves (orbitals). When molecules react to form different molecules, the total energy of the products is almost always different than the total energy of the reactants. Our problem is that the total energy of a molecular system cannot be measured directly. What we have to do is measure the energy *changes* that occur and try to infer from these changes something about the total energies of the molecules involved. We measure the energy changes in chemical reactions by measuring their effects on the properties of substances that are easy to observe, such as the temperature of a mass of water. We will concentrate on thermal energy and start by considering how it is exchanged between one substance and another.

Section 7.3. Thermal Energy (Heat) Transfer

7.8.   Investigate This   *Can radiation change the temperature of water?*

Do this as a class investigation and work in small groups to discuss and analyze the results. Use a bright incandescent light, two flat-sided containers filled with water, and two thermometers set up as shown in the photograph. Record the temperature of the water in the containers at the beginning and after 10 minutes in the light beam. Remove the container nearer the light source and read the temperature in the remaining container after another 10 minutes.

7.9.   Consider This   *How does radiation change the thermal energy of water?*

(a) How did the temperature of the water in the two containers in Investigate This 7.8 compare after 10 minutes in the light beam? Is there evidence that the thermal energy of the water in one or both of the containers changed? Explain.

(b) After 10 further minutes in the light beam (without the nearer container), what happened to the temperature of the water in the remaining container? Is there evidence for further change in the thermal energy of the water in this container? Explain why you think you obtained the results you did.

The early thermodynamicists conceived of heat as a fluid that flowed from one substance to another. The fluid was called "caloric" from the Latin word for heat. This imaginary fluid is the obvious source of the name calorie and of such terminology as "heat flow." Our present understanding is that thermal energy, the chaotic motion of atoms and molecules, can be transferred from one substance to another in two ways: by electromagnetic radiation emitted by the warmer body and absorbed by the cooler, or by contact of the two substances. To be consistent with our present model, we will use terminology such as "heat transfer" or "thermal energy transfer" and not use the term "flow" when we discuss heat effects and measurements.

Thermal energy transfer by radiation

You may have had the experience of sitting in front of bonfire or a fire in a fireplace and

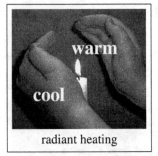

radiant heating

having your face feel hot while your back felt cool. What you were feeling on your face was the **radiant energy** from the glowing fire. Thermal energy was being transferred to you by the long wavelength, **infrared electromagnetic radiation**, from the flame. The radiant energy was absorbed by the molecules in your skin and caused them to move faster (increased their temperature). The thermal energy of your face was greater than it would have been if it was sensing just the temperature of the air.

Review Chapter 4, Sections 4.3 and 4.4, and Figures 4.13 and 4.14, for electromagnetic radiation wavelengths and energy emission by glowing bodies.

Radiation travels in straight lines and, since your body was in the way, couldn't reach your back, which simply sensed the temperature of the air.

7.10.   Consider This   *How does radiation transfer energy to water?*

(a) Explain your results from Investigate This 7.8 in terms of absorption of infrared radiation from the light beam being absorbed by the water. Can you see any difference in the light beam before and after it enters the water? Why or why not?

(b) If you were to place your hand in the light beam before it passed through the containers of water and then in the beam after it passed through the water, would you feel any difference? Explain your response. Try the experiment.

Thermal energy transfer by contact

The molecules in a warmer substance are moving more rapidly, on the average, than those in a cooler substance. In collisions between molecules of the two substances the more energetic, warmer molecules will cause the molecules of the cooler substance to start moving faster. For example, we said that the hot combustion product gases in Investigate This 7.4 transfer some of their thermal energy to the surrounding air by collisions with the air molecules. In this energy transfer process, the cooler substance will get warmer. And, since energy is conserved, the warmer substance will get cooler. Two substances at the same temperature exchange thermal energy, but there is no *net* energy transfer and thus no changes in temperature.

These energy transfers take place by two pathways: conduction and convection. You have experienced thermal energy transfer by conduction and convection, Figure 7.3, even if these are not familiar terms. When you touch a pan of water heated on a hot stove, thermal energy from the burner is *conducted* to your skin by the metal or glass walls of the pan. In **conduction**, Figure 7.3(a), thermal energy from a hot source causes particles in the material in contact with the

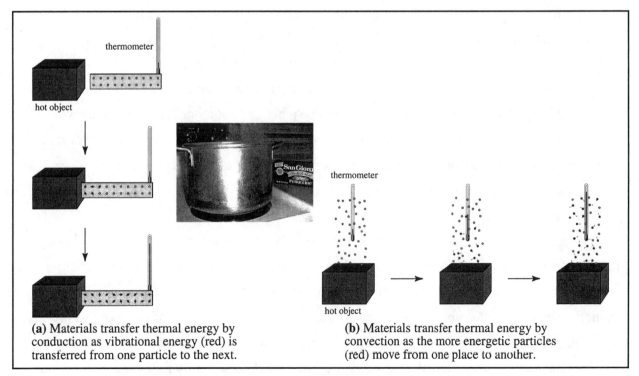

(a) Materials transfer thermal energy by conduction as vibrational energy (red) is transferred from one particle to the next.

(b) Materials transfer thermal energy by convection as the more energetic particles (red) move from one place to another.

Figure 7.3. Conduction and convection of thermal energy.

source to vibrate faster. These particles, in turn, collide with adjacent particles and start them

| Conduction is from Latin that means "to *lead* together." Convection is from Latin that means "to *carry* together." |

vibrating faster and so on through the conductor. The particles stay fixed in place as the energy is transferred.

In a fluid medium (gas or liquid), the particles can move about and **convection**, Figure 7.3(b), can transfer energy from one place to another. The particles initially in contact with the hot source move away from it and set up bulk movement in the fluid. As the pan of water is heated, the cooler denser water from the top sinks to the bottom displacing the warmed less dense water; *convection* currents are set up in the water which help transfer thermal energy from the burner throughout the liquid.

***7.11.* Consider This** *How are conduction and convection the same? different?*

(a) What is(are) the difference(s) between conduction and convection? Explain how the difference(s) are illustrated at the molecular level in Figure 7.3.

(b) Could the results in Investigate This 7.8 be caused by convection and/or conduction of thermal energy? What tests might you do to find out?

A similarity between conduction and convection is that atoms or molecules have to contact one another to transfer kinetic energy (motion) from one to another. In conduction, the motions are vibrations of the particles in the lattice and in convection the motions are mainly translations through space. In a metal pan on a stove, Figure 7.3, thermal energy is transferred from the burner to the entire pan without any atoms moving from one lattice position to another in the electric heating element or the pan. In the water in the pan and in the air and combustion gases in Figure 7.2, faster moving molecules move from one place to another and, if they collide with other molecules, can get them moving faster as well. Transfer of thermal energy does not require a mysterious substance (caloric); the transfer occurs by way of electromagnetic radiation or particles actually contacting one another.

Reflection and projection

The concepts of thermodynamics were discovered or formulated mainly by scientists and inventors seeking ways to convert as much thermal energy (heat) as possible to mechanical energy (work). Heat and work are forms of kinetic energy, undirected and directed molecular motion, respectively. An important thermodynamic concept, the first law of thermodynamics, is that energy is conserved in every process/change that occurs, including chemical reactions.

Thermal energy can be transferred from one substance to another by radiation, conduction, and/or convection. The temperature of the substance that loses the thermal energy decreases. The temperature of the substance that gains the thermal energy increases. Energy is conserved in these transfers. Often, however, we can bring about a change in the thermal energy of a substance without transferring thermal energy from another substance. One way to do this is by doing work on the substance of interest. Since there is more than one way to bring about changes in the thermal energy of a substance, we need to examine how overall changes depend upon the pathway for change. We also have to be clear about what parts of the environment we need to take into account when analyzing a change. These are the topics of the next sections.

Section 7.4. State Functions and Path Functions

7.12. **Investigate This**   *What happens when you rub your hands together?*

Put the palms of your hands together and rub them together briskly for 3–4 seconds. Besides the rubbing, what sensation do you feel on your palms?

7.13. **Consider This**   *What change occurs when you rub your hands together?*

(a) Rubbing your hands together, as you did in Investigate This 7.12 and as you have done when your hands are cold, warms your skin. What change must have occurred to the molecules in your skin to cause this sensation? Explain your reasoning.

(b) How do you think rubbing your hands together brings about the change in part (a)?

(c) Is there another way you could get the same sensation without rubbing your hands on something? What would bring about the change in the molecules in this second case? Explain.

One of the characteristics of thermodynamics is careful definition of the properties of thermodynamic variables such as temperature, thermal energy (heat), mechanical energy (work) and so forth. Usually we will take an approach that depends on what seems to make sense and can be described in terms of everyday experiences or easily observable effects. But there are some distinctions that will help you understand why we use different variables that seem to stand for the same quantity. One fundamental distinction is between variables that are state functions and variables that are path functions.

State functions

State functions are variables whose value depends only on the state (present condition) of the system under study and not at all on how it got into that state. Some important state functions are temperature, pressure, and volume; we will meet others as we go along. A physical analogy

> The term *state function* can be confusing, since we also commonly refer to different physical *states* (solid, liquid, or gas) of matter. The description of its present condition, the thermodynamic state of a system, has to take into account the physical states of its components, but, more fundamentally, it must account for the energy relationships among the components.

for state functions is altitude above sea level. If you are on top of Pike's Peak, Figure 7.4, your altitude above sea level, 3.8 km, is the same whether you drove to the top, walked to the top, or took the cog railway. State functions are like altitude. It makes no difference how the collection of molecules arrived at their particular state. What is important is that they are in that state.

Figure 7.4. Two ways to get to the top of Pike's Peak.
An auto road winds up the mountain in a series of switchbacks. The cog railway climbs more steeply and more directly.

If the state of a collection of molecules changes, the changes in its state functions depend only on the initial and final state of the molecules. Again using the altitude analogy, if you descend Pike's Peak from 3.8 km above sea level (initial state: height = h_i) and end your descent 2.1 km above sea level (final state: height = h_f), your change in state (altitude), Δh, is:

$$\Delta h = h_f - h_i = 2.1 \text{ km} - 3.8 \text{ km} = -1.7 \text{ km} \tag{7.1}$$

When the final state (h_f in this analogy) has a less positive value than the initial state (h_i), the numeric value for the change of state (Δh) has a negative sign, just as you have often seen for changes in energy in previous chapters.

7.14.   Consider This   How do different pathways for descent of Pike's Peak compare?

Imagine that you descend Pike's Peak 1.7 km in altitude by car and a friend takes the cog railway down. Will you both travel the same linear distance on the ground, as measured, for example, by the odometer on your car? Explain why or why not.

Path functions

Numeric values for some variables are dependent on the path taken to get from the initial to the final state; these are **path functions**. In Consider This 7.14, for example, to descend 1.7 km, you will travel a longer distance in a car down the winding road than your friend will coming down the straighter cog railway. The actual distance traveled is a path function, whereas the difference in altitude is a state function. In thermodynamic systems, thermal energy transfers (heat) and mechanical energy (work) are path functions. When you rubbed your hands together in Investigate This 7.12, they felt warmer. The molecules in your skin were made to move more rapidly by the mechanical work you did rubbing your hands together. You could bring about this same change by holding your hands near a source of thermal energy, such as a flame. These two possible pathways from the initial state of skin at about 37 °C to warmer skin at about 40 °C are shown in Figure 7.5.

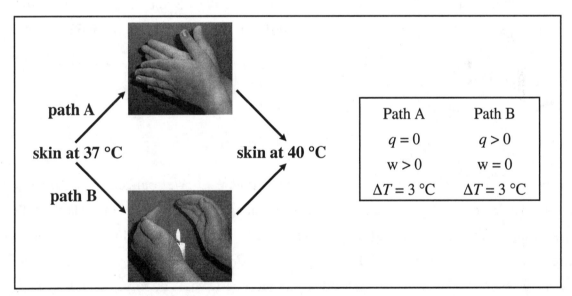

Path A	Path B
$q = 0$	$q > 0$
$w > 0$	$w = 0$
$\Delta T = 3\ ^\circ C$	$\Delta T = 3\ ^\circ C$

skin at 37 °C — path A / path B — skin at 40 °C

Figure 7.5. Two pathways that make the same change in the temperature of the skin.

The change in the temperature of the skin, $\Delta T = 3\ ^\circ C$, is the same for both pathways, but the work and heat are not. Along path A, work (moving your hands) was done on the skin to raise its temperature. There was no thermal energy transferred to the skin; q is zero for path A. Along path B, the temperature of the skin changes as thermal energy is transferred by radiation,

---

convection, and conduction from the flame to the skin. No mechanical energy is transferred to the skin; w is zero for path B. The chart in Figure 7.5 compares the two pathways.

7.15. Consider This   *How does the energy of the skin molecules change in Figure 7.5?*

We have said that temperature is a measure of the average energy of the molecules in a system. When the energy of the molecules increases, the temperature increases. How do the energy changes of the skin molecules compare for the two pathways in Figure 7.5? Explain how you reach your conclusion.

In Section 7.1, we said that burning foods produces the same change as complete metabolism of the foods. The initial and final states are the same in both cases. However, the pathways are quite different and the amounts of heat and work are different, as we will find in Section 7.10.

Section 7.5. System and Surroundings

You may have noticed that we often refer to "the system of interest," "the chemical reaction system," or some other reference to the "system." In Consider This 7.15, we referred to the "average energy of the molecules in a system." A thermodynamic analysis of a change always requires that we distinguish the collection of molecules that we are studying, the **system**, from its **surroundings**, everything else. Let's look at the different types of systems we will encounter.

7.16. Check This   *The system in Figure 7.5*

What is the system in Figure 7.5 that is referred to in Consider This 7.15?

Open, closed, and isolated systems

If we were studying the contents of the aquaria in Figure 7.6, we could consider the aqueous solution, fish, plants, and gravel as our system. Everything else (the glass of the aquarium, the surrounding air, the table upon which the aquarium rests, the room, the building, the planet, the entire universe) is the surroundings. In the aquarium in Figure 7.6(a), energy and matter can move between the system and the surroundings. Atmospheric gases dissolve in the water. Thermal energy is exchanged by conduction and convection with the table and the air. Light (radiant) energy enters the system. Work could be done on the water by stirring it. These must be accounted for as part of the surroundings. This aquarium is an **open system** because matter and energy can be exchanged with the surroundings. All living things are open systems. They must take in matter from their surroundings to survive.

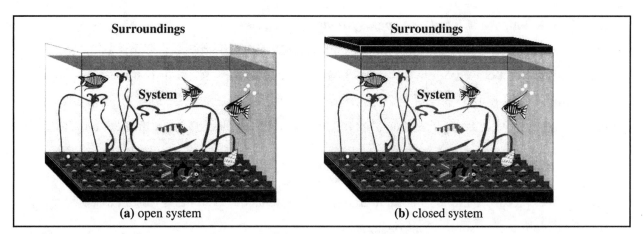

Figure 7.6. System (the contents of the aquaria) and surroundings (everything else).

If we were to seal the top of the aquarium, as shown in Figure 7.6(b), no exchange of matter could occur with the surroundings. (We would now include the small amount of air trapped above the water as part of the system.) Thermal energy could still enter or leave the system, and light energy could still be absorbed and, if we had sealed a stirrer inside, we could still do work on the system by stirring the water. The sealed aquarium is a **closed system**. Closed systems do not exchange matter with their surroundings. They can exchange work, if their walls are not rigid (a moving stirrer paddle counts as a moving "wall"). They can exchange thermal energy (heat), if their walls are not thermally insulated.

If we were to wrap the entire sealed aquarium in Figure 7.6(b) in an opaque thermally insulating blanket, no thermal energy could enter or leave the system. No exchange of matter could occur between the sealed system and the surroundings. If there is no stirrer in the thermally insulated and sealed aquarium, then no work could be done on the system. A system that can exchange neither work nor thermal energy with its surroundings is called an **isolated system**. Table 7.1 summarizes the matter, thermal energy, and mechanical energy transfers that can occur between the surroundings and the different kinds of systems.

Table 7.1. Transfers that can occur between system and surroundings.

type of system	matter	thermal energy (heat)	mechanical energy (work)
open	yes	yes	yes
closed	no	yes	yes
isolated	no	no	no

7.17. Check This *Closed and open systems*

(a) Think about the flame in Investigate This 7.4. Consider the fuel that burns as the system. Is the burning fuel a closed or open system? Explain the reasoning for your response.

(b) Ammonium nitrate, $NH_4NO_3(s)$, an important fertilizer, has also been responsible for some devastating explosions. The reactions that occur in the explosion are complex, but the overall reaction can be represented as:

$$2NH_4NO_3(s) \rightarrow 2N_2(g) + 4H_2O(g) + O_2(g)$$

Consider an exploding sample of solid ammonium nitrate as the system. Is this a closed or open system? Explain the reasoning for your response.

A fuel burning or a sugar undergoing aerobic metabolism in an organism are examples of processes occurring in open systems. At least one of the reactants, oxygen, has to be supplied from the surroundings in both cases. By contrast, explosions generally can be classified as occurring in closed systems. All the atoms necessary for the reaction are together in a single compound (ammonium nitrate or trinitrotoluene, TNT, for example) or a mixture of compounds (a hydrogen-oxygen mixture, for example). We will meet further examples of both kinds of systems in this and the succeeding chapters.

7.18. Consider This *How do systems and surroundings interact?*

(a) If you hold a beaker containing a sample of ammonium chloride, $NH_4Cl(s)$, dissolving in water, your hand feels cold. (Recall Investigate This 2.20 in Chapter 2.) If the ammonium chloride and water are the system, what are the relevant surroundings? How do the surroundings interact with the system? Is the system open, closed, or isolated? Explain your responses.

(b) If the same process as in part (a) is carried out in a well-insulated container, what are the relevant surroundings? How do the surroundings interact with the system? Is the system open, closed, or isolated? Explain your responses.

(c) Imagine that you drop a hot piece of metal into a well-insulated container of room temperature water. If the metal is the system, what are the surroundings? How do the surroundings interact with the system? What do you expect will happen to the system and the surroundings? Is the system open, closed, or isolated? Explain your responses.

(d) Imagine carrying out the same process as in part (c), but now take the metal plus the water as the system. How do the surroundings interact with the system? What changes do you expect will occur in the system and the surroundings? Is the system open, closed, or isolated? Explain your responses.

(e) WEB Chap 7, Sect 7.6.1. Compare with Figure 7.7 below. Click on the cups and classify the parts of the set up as system or surroundings.

Reflection and projection

In the last two sections we have made some important distinctions between different kinds of thermodynamic functions, between a system and its surroundings, and among the interactions of the system of interest with its surroundings. We found that the values for some functions, heat and work, in particular, can be different when a change occurs by different pathways. Functions that depend upon how a change is carried out are called *path functions*. The values of other functions depend only on the initial and final states of the system undergoing change and not on the pathway for the change. Functions that depend only on the initial and final state of the system are called *state functions*.

We will define what the systems are, as we use and analyze them in the remainder of this chapter and the rest of the text. We do this so that we can also be clear about what the surroundings are that we have to account for in our analysis of changes in the system. The matter, thermal energy, and mechanical energy transfers that can occur between the surroundings and open, closed, and isolated systems (Table 6.1) become clearer as they are applied to real chemical systems. We will begin by examining thermal energy transfers in more detail. The change in temperature of one substance can be used determine the amount of thermal energy it has gained (or lost) from another source, a chemical reaction, for example. This application of the first law is called calorimetry and is the topic of the next section.

Section 7.6.   Calorimetry and Introduction to Enthalpy

7.19.   **Investigate This**   *What happens when an acid and base are mixed?*

For this investigation, use a small test tube, and separate plastic pipets containing 1 M sodium hydroxide (NaOH), solution and 1 M hydrochloric acid (HCl) solution. Add about 1 mL of the NaOH solution to the test tube and then about 1 mL of the HCl solution. Gently touch the test tube near the bottom where the solution is. Record all your observations.

7.20.   **Consider This**   *What causes the changes when an acid and base are mixed?*

What was(were) your observation(s) when you touched the test tube after the acid and base were mixed in Investigate This 7.19? What process or reaction could be responsible for your observations? Explain how you reach your conclusion.

In Investigate This 7.19, we can designate the acid and base molecules (ions) that react with one another as the system. The surroundings are the water in which the acid and base are dissolved, the test tube, the air around it, and, when you touch the test tube, your finger. Everything else in the universe is also a part of the surroundings, but you use your common sense to make a judgment about what parts of it can have an effect on the system and with which parts the system exchanges thermal energy.

Since the amount of thermal energy transferred depends on the pathway for the process, we often use subscripts on q to designate the conditions under which the thermal energy transfer is measured. We are mostly interested in reactions that take place in open containers on the laboratory bench, as in Investigate This 7.19, or in biological systems. Since these reactions occur under conditions of constant atmospheric pressure, we designate the associated constant-pressure thermal energy transfer as q_P.

Enthalpy

The 19th-century thermodynamicists introduced a useful new energy function called **enthalpy, H**. A change in enthalpy is symbolized by ΔH for a process or reaction; *ΔH has the same numerical value as q_P for the process.* ΔH, which can be determined by measuring the thermal energy transfer at constant pressure, is the difference between the final value for the enthalpy, H_{final}, and the initial value of the enthalpy, $H_{initial}$:

$$\Delta H = H_{final} - H_{initial} = q_P \tag{7.2}$$

Note that the quantity q_P is the *difference* between enthalpies, not enthalpy itself. As you will see in Section 7.10, ΔH is not just another name for constant-pressure thermal energy transfer. The enthalpy of a chemical system corresponds to a reservoir of thermal energy. It might be helpful to think about enthalpy as "heat content," a term that was once used for this thermodynamic property.

When heat (thermal energy) is transferred *to* a system, q_P is positive — heat is added to the system. The enthalpy (heat content) of the system has increased, $H_{final} > H_{initial}$; therefore, from equation (7.2), ΔH is positive. Positive values for ΔH and q_P are characteristic of **endothermic** changes. On the other hand, if heat *leaves* the system we are studying, q_P is negative — heat is subtracted from the system. The enthalpy of the system decreases, $H_{final} < H_{initial}$, and ΔH is negative. Negative values for ΔH and q_P are characteristic of **exothermic** changes. Although total enthalpies cannot be measured directly, the amount of heat transferred can be experimentally determined, as you will see in the rest of this section. A measured value of q_P enables us to determine the enthalpy change, ΔH, for the system under investigation.

The thermal energy effect of a reaction can often be determined by touching the reaction vessel to determine whether the reaction is exothermic (it will be warm to the touch, because heat is leaving the reacting system and is being added to the surroundings, including your skin) or endothermic (it will be cool to the touch, because heat is being added to the reacting system and is being taken from the surroundings). If a reaction produces or consumes only a small amount of heat, your sense of touch may not be able to detect it. Several kinds of devices, called **calorimeters**, have been developed to measure thermal energy quantitatively.

7.21.   Check This   *Is the reaction of an acid with a base exothermic or endothermic?*

(a) Write the net ionic reaction that occurs in Investigate This 7.19.

(b) Is this reaction exothermic or endothermic? Give your reasoning.

Constant-pressure calorimetry

7.22.   Investigate This   *What happens when urea dissolves in water?*

Do this as a class investigation and work in small groups to discuss and analyze the results. Set up a simple constant-pressure calorimeter, like the one illustrated in Figure 7.7, containing 100. mL of room temperature water. While gently stirring the water, record its temperature every 15 seconds for about two minutes. Add 6.0 g of urea, $H_2NC(O)NH_2(s)$, to the water and continue stirring and recording the temperature of the solution for another three minutes.

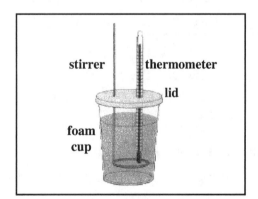

Figure 7.7.   A simple constant–pressure calorimeter.

7.23.   Consider This   *What is ΔT when urea dissolves in water?*

What were the initial and final temperatures of the water and solution in the calorimeter in Investigate This 7.22? What was $\Delta T = T_f - T_i$ for the urea dissolution reaction?

Figure 7.7 illustrates a simple **constant-pressure calorimeter**, a device that can be used to measure thermal energy changes in aqueous reactions. The setup requires only an insulated container, such as a Styrofoam® cup, an insulating lid, a thermometer, and perhaps a stirrer. Thermal energy produced (exothermic) or consumed (endothermic) by the reaction is transferred to or from the aqueous solution. The resulting temperature change of the solution is measured by the thermometer. If the calorimeter is perfect, no thermal energy will leave the calorimeter and none will enter from the outside. Under these conditions, the sum of all the thermal energy changes occurring in the calorimeter must be zero:

$$q_{P(reaction)} + q_{P(solution)} = 0 \qquad (7.3)$$

Energy is conserved; the transfer of thermal energy *from* one component of the calorimeter contents must result in the transfer of thermal energy *to* another component. If the reaction is exothermic, then $q_{P(reaction)}$ has a negative value. Consequently, in order to satisfy equation (7.3), $q_{P(solution)}$ must have a positive value. A positive value for $q_{P(solution)}$ means that thermal energy is added to the solution, that is, it gets warmer.

7.24. Consider This *What is observed for an endothermic reaction in a calorimeter?*

(a) If an endothermic reaction is carried out in a calorimeter, is $q_{P(reaction)}$ positive or negative? Explain.

(b) If the reaction is endothermic, what is the sign of $q_{P(solution)}$? Will you observe the water to become warmer or cooler? Explain your reasoning.

(c) Is the dissolution of urea in water exothermic or endothermic? Explain how you can use your observations from Investigate This 7.22 to answer this question.

In practice, calorimeters are never perfect. The solution absorbs most of the heat from an exothermic reaction and provides most of the heat for an endothermic reaction, but some of the heat remains unaccounted for. This heat goes to heating (or cooling) the calorimeter itself (the container, thermometer, and stirrer). For our purposes, however, we are going to neglect this deviation from ideality. We will assume that all the thermal energy released or absorbed by the reaction in our calorimeter is absorbed or furnished by the solution, as in equation (7.3). The error we introduce by assuming that the calorimeter is ideal is usually only a few percent.

7.25. Check This *Energy transfers in a calorimeter*

WEB Chap 7, Sect 7.6.2. Carry out the animated reaction and the analysis of changes. State in your own words what is going on in this calorimeter, the reasons for the observed changes, and the roles, if any, that thermal energy transfer by radiation, conduction, and convection play in these changes.

Temperature and thermal energy change

In a calorimetric experiment, you measure the temperature change of the liquid. You then have to relate the temperature change to the thermal energy changes in the calorimeter. The equation linking the thermal energy transferred to a substance to the temperature change of the substance was derived in Chapter 1, Worked Example 1.57, though we didn't use these symbols there:

$$q_{P(substance)} = (m)(c)(\Delta T) \tag{7.4}$$

Here, m is the mass of the substance (in grams), c is the specific heat of the substance (in joules per gram per degree Celsius temperature change, $J \cdot g^{-1} \cdot {}^{\circ}C^{-1}$), and ΔT is the temperature change in degrees Celsius, $T_{final} - T_{initial}$. Recall that specific heat is defined as the energy required to raise the temperature of 1 g of a substance by 1 °C. The specific heat of a substance is a characteristic of that substance, just as its boiling point and melting point are. The specific heat of water is 4.18 $J \cdot g^{-1} \cdot {}^{\circ}C^{-1}$. In most of our calculations, we will assume that the specific heat of dilute aqueous solutions is the same as that for water.

7.26.   Worked Example   *Determination of $q_{P(reaction)}$ for a reaction in a calorimeter*

100.0 mL of 0.105 M acetic acid solution (0.0105 mol) and 100.0 mL of 0.12 M ammonia solution (0.012 mol) were mixed in a calorimeter like the one in Figure 7.7. The reaction that occurs is:

$$CH_3C(O)OH(aq) + NH_3(aq) \rightarrow NH_4^+(aq) + CH_3C(O)O^-(aq) \tag{7.5}$$

The solutions were both at 22.50 °C before mixing. After mixing, the temperature rose to 23.15 °C. What is $q_{P(reaction)}$ for the reaction in the calorimeter? Assume that the specific heat of the solution is the same as the specific heat of water, 4.18 $J \cdot g^{-1} \cdot {}^{\circ}C^{-1}$, and that the density of the solutions is 1.00 $g \cdot mL^{-1}$.

Necessary Information: We need equations (7.3) and (7.4).

Strategy: Substitute the experimental quantities into equation (7.4) to find $q_{P(solution)}$ and then substitute $q_{P(solution)}$ in equation (7.3) to get $q_{P(reaction)}$.

Implementation: The change in temperature is:

$$\Delta T = (23.15 \text{ °C}) - (22.50 \text{ °C}) = 0.65 \text{ °C}$$

The mass of the mixed solution in which reaction occurs is the sum of the masses of the original solutions, each of which is 100. g [= (100.0 mL)(1.00 $g \cdot mL^{-1}$)]; the mass of the mixed

---

solution is 200. g. Thus, $q_{P(solution)}$ is:

$$q_{P(solution)} = (m)(c)(\Delta T) = (200.\ g)(4.18\ J \cdot g^{-1} \cdot °C^{-1})(0.65\ °C) = 5.4 \times 10^2\ J$$

Rearrange equation (7.3) to get $q_{P(reaction)}$:

$$q_{P(reaction)} = -q_{P(solution)} = -5.4 \times 10^2\ J$$

Does the Answer Make Sense? To increase the temperature of 200 g of water one degree, requires 836 J [= (200. g)(4.18 J·g⁻¹·°C⁻¹)(1 °C)]. Here we found that a smaller amount of thermal energy, about 550 J, warms 200 g of solution somewhat less than one degree, so the answer is in the right direction and of the right magnitude.

7.27. Check This *Determination of $q_{P(reaction)}$ for dissolution of urea in a calorimeter*

Use your data from Investigate This 7.22 to determine $q_{P(reaction)}$ for the dissolution of 6.0 g of urea in 100. mL of water. *Hint:* Assume that the density of water is 1.00 g·mL⁻¹ and that the specific heat of the solution is the same as the specific heat of water. Remember that the mass of the solution is the mass of the water plus the solute.

Calculating molar enthalpy change

In Worked Example 7.26 and Check This 7.27, values for $q_{P(reaction)}$ were determined from experimental calorimetric data. We have said that $\Delta H_{reaction} = q_{P(reaction)}$, so we also have $\Delta H_{reaction}$ values for these reactions under the conditions studied. Usually, however, we express enthalpy changes per mole of reaction, kJ·mol⁻¹, where the coefficients in the balanced reaction equation are taken to be molar quantities. We can convert our measured values for $q_{P(reaction)}$ (in kJ) to $\Delta H_{reaction}$ (in kJ·mol⁻¹) by accounting for the number of moles of reactants that reacted in our experimental systems.

7.28. Worked Example *Determination of $\Delta H_{reaction}$ for a reaction in a calorimeter*

Use the result from Worked Example 7.26 to calculate $\Delta H_{reaction}$ when one mole of acetic acid reacts with one mole of ammonia by reaction (7.5).

Necessary Information: We need the volume and concentration data from Worked Example 7.26 to determine the number of moles of acetic acid and ammonia that reacted to give $q_{P(reaction)} = -5.4 \times 10^2$ J.

Strategy: Use the data in Worked Example 7.26 to find the number of moles of reactants that react and then figure out what $q_{P(reaction)} = \Delta H_{reaction}$ would be if one mole reacted.

Implementation: Acetic acid and ammonia react in a one-to-one mole ratio and the data in

Worked Example 7.26 show that there was an excess of ammonia ($0.12 \ mol \cdot L^{-1} \times 0.100 \ L = 0.012 \ mol$) compared to acetic acid ($0.0105 \ mol$) in the mixture. The acetic acid is the limiting reactant, so $0.0105 \ mol$ of each reactant reacts to yield $q_{P(reaction)} = -5.4 \times 10^2 \ J$.

$$\Delta H_{reaction} = q_{P(reaction)} \ (per \ mole) = \left(\frac{-5.4 \times 10^2 \ J}{0.0105 \ mol} \right) = -51 \ kJ \cdot mol^{-1}$$

Does the Answer Make Sense? About one hundredth of a mole of reaction occurs in the calorimeter and transfers about 540 J to the solution and calorimeter. One mole of reaction (one hundred times the amount of reaction in the calorimeter) would produce one hundred times as much thermal energy, about 54 kJ. This is approximately the result we got.

7.29. Check This   *Determination of $\Delta H_{reaction}$ for dissolution of urea in a calorimeter*

Use your result from Check This 7.27 and the data in Investigate This 7.22 to calculate $\Delta H_{reaction}$ when one mole of urea is dissolved to yield a solution of the same molarity as the final solution in the investigation. *Hint:* Assume that the volume of the final solution in the investigation is 100. mL.

Handling significant figures properly in experimental measurements is essential, if the results are to be trusted. Worked Examples 7.26 and 7.28 are good exercises in one of the more troublesome aspects of determining the number of significant figures to include in the results of calculations. In both cases, the numerical values for the experimental measurements, were quite accurate, but the final result depended on a small difference between two large values (the measured temperatures). As you review the calculations to check your understanding of the concepts, also check your understanding of why the significant figures are reported as they are.

7.30. Consider This   *How can you get more accurate values for $\Delta H_{reaction}$?*

Suppose you decide to try to get a more accurate value of $\Delta H_{reaction}$ for the reaction between aqueous acetic acid and ammonia solutions by doubling the amount of reactants used in the experiment described in Worked Example 7.26. When you carry out the reaction with 200. mL of each reactant solution you find that the temperature change is 0.66 °C.

(a) Is this the result you would expect? Why or why not?

(b) Use these data to determine $\Delta H_{reaction}$ (Watch your significant figures.) and compare it to the value from Worked Example 7.28. Is this the result you would expect? Why or why not?

(c)   WEB   Chap 7, Sect 7.6.3. Is this *Web Companion* page related to parts (a) and (b)?

Improving calorimetric measurements

Trying to improve the precision of a calorimetric measurement by increasing the volumes of reactant solutions used does not work. As you discovered in Consider This 7.30, the amount of thermal energy produced increases, but, at the same time, the amount of solution increases by the same proportion. The increased thermal energy produces essentially the same temperature change in this greater mass of solution. You could work with more concentrated solutions, but they can also produce problems. There might be thermal effects from diluting the solutions when they are mixed or the specific heats of the solutions may be significantly different from water. If you do not account for such factors, your answer may be more precise (more significant figures) but less accurate (further from the true value).

The factor that can make the biggest improvement in calorimetric measurements is more precise temperature measurement, that is, use of a more sensitive temperature measuring device. For the simple calorimeters we have been discussing, more sensitive thermometers are not justified. If the temperature measurements are improved, losses of thermal energy to the surroundings are large enough to show up as important factors. Better temperature measurements are justified and required in more sophisticated calorimeters,

---

7.31.   **Check This**   *Effect of increasing the amount of another calorimetric reaction*

If the calorimetric experiment in Investigate This 7.22 is carried out by dissolving 12 g of urea in 100 mL of water, the final temperature of the solution is about 7 °C lower than the initial temperature of the water.

(a) Is this the result you would expect? Why or why not?

(b) How does this experiment differ from the one suggested in Consider This 7.30? Does the difference help explain the result? Why or why not?

Reflection and projection

Calorimetry is an important application of the first law of thermodynamics. One common use is the determination of the thermal energy released or gained by chemical reactions in aqueous solution. The thermal energy transfers are between the system (the reacting molecules) and their surroundings (the liquid solution). The thermal energy transfer to the surroundings can be calculated from the specific heat of the liquid and the measured values for the mass and temperature change of the liquid. If the reaction is carried out in an insulated container (calorimeter) the sum of the thermal energy transfers in the calorimeter is zero.

Calorimetric measurements on reactions in solution are almost always carried out in open containers at a constant pressure of one atmosphere. The thermal energy (heat) transferred in the reaction is usually called q_P to indicate that the process is carried out at constant pressure. A special name (and symbol), enthalpy (H), is used to designate the "heat content" of a system. A change in the system usually results in a change in the enthalpy, ΔH, which can be equated to the measured q_P for the change at constant pressure. Enthalpy changes in chemical reactions are a result of the breaking and making of bonds in the reactions. The enthalpies required to break bonds can be determined in various ways, including calorimetric experiments. The next sections show how these enthalpies can be used to calculate the enthalpy changes for reactions.

Section 7.7.   Bond Enthalpies

7.32.   Investigate This   *What happens when yeast is added to hydrogen peroxide?*

Do this as a class investigation and work in small groups to discuss and analyze the results. Put 2-3 g of dried baker's yeast into a 100-mL graduated cylinder. Add about 10 mL of 3% hydrogen peroxide, H_2O_2, solution to the cylinder. *CAUTION: 3% hydrogen peroxide can damage both skin and clothes. Wear protective gloves and avoid splashing the liquid.* Add about 1 mL of liquid dishwashing detergent and swirl gently to mix the contents of the cylinder. Gently touch the outside of the cylinder near the bottom, so you can feel any thermal effects. Observe any changes in either the yeast or the solution. Record your observations. Set the cylinder and its contents aside to use in Investigate This 7.35.

7.33.   Consider This   *What reaction occurs when yeast is added to hydrogen peroxide?*

What evidence do you have from Investigate This 7.32 that a reaction occurs when yeast is added to hydrogen peroxide? What do you think the product(s) of the reaction could be? How could you test for the presence of this(these) product(s)?

Chemical reactions: bond breaking and bond making

One of the goals in studying chemistry is to gain the ability to predict likely outcomes for chemical reactions and to calculate accompanying energy changes. In Chapter 5 we introduced covalent bonds, focusing mainly on the role of valence electrons in holding the atoms together. In chemical reactions of covalent compounds, at least some of the covalent bonds in the reactants break, the atoms rearrange themselves, and new covalent bonds form in the products. *The bond-breaking process requires an input of energy and the bond-forming process releases energy.*

- *If weaker bonds are replaced by stronger ones, the overall reaction is exothermic.*

- *If stronger bonds are replaced by weaker ones, the overall reaction is endothermic.*

Experimental determinations of enthalpy changes for thousands of reactions provide the basis for enormous predictive power. Using only enthalpy data tables (introduced below), you can figure out whether any reaction you might imagine would release energy or consume energy.

We will use the reaction of hydrogen peroxide, Investigate This 7.32, as an example for analysis. You probably observed the formation of gas as a product of the reaction when the yeast was added to the peroxide. Some of the gas was trapped in the foam that filled the rest of the cylinder. You could also feel that the solution warmed up as the reaction occurred. The reaction is exothermic. If we can find out what the gaseous product(s) is(are), we will know more about the bond-breakings and bond-formations that occur in the reaction.

Hydrogen and oxygen are the only elements in the aqueous hydrogen peroxide, $H_2O_2(aq)$, solution; the most likely gases that can be formed are molecular hydrogen, $H_2(g)$, and/or molecular oxygen, $O_2(g)$. Possible reactions that could form these gases are:

$$2H_2O_2(aq) \rightarrow 2H_2O(l) + O_2(g) \tag{7.6}$$

$$H_2O_2(aq) \rightarrow H_2(g) + O_2(g) \tag{7.7}$$

The bond rearrangements in reactions (7.6) and (7.7) are depicted in Figure 7.8.

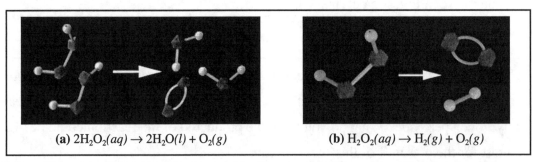

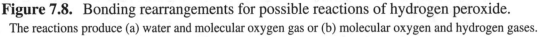
(a) $2H_2O_2(aq) \rightarrow 2H_2O(l) + O_2(g)$ **(b)** $H_2O_2(aq) \rightarrow H_2(g) + O_2(g)$

Figure 7.8. Bonding rearrangements for possible reactions of hydrogen peroxide.
The reactions produce (a) water and molecular oxygen gas or (b) molecular oxygen and hydrogen gases.

7.34. Check This   *Properties of the possible H_2O_2 reaction products*

   (a) What is the composition of the gas that is produced by reaction (7.6)? What are its properties?

   (b) What is the composition of the gas that is produced by reaction (7.7)? What are its properties?

(c) How could you test the product gas to distinguish between these possibilities?

7.35. Investigate This   *What is the gas formed in the hydrogen peroxide reaction?* 👓

Do this as a class investigation and work in small groups to discuss and analyze the results. Test the flammability of the gas trapped in the foam formed in Activity 7.32. *WARNING: Place the cylinder behind a safety shield that protects others as well as yourself when the gas is tested.* Light a long fireplace match, blow it out, and quickly insert the glowing match tip a few centimeters into the mouth of the cylinder while carefully observing the results.

7.36.   Consider This   *Which bonds break and form in the hydrogen peroxide reaction?*

(a) What bonds are broken and what bonds are formed as the reactant is converted to products in reaction (7.6)? Make models of the reactants and then take them apart and reassemble them to products to illustrate the reaction. How does the number of bonds broken compare with the number of bonds formed? How many two-electron bonds are present in the reactants and products.

(b) Do the same kind of model exercise and answer the same questions for reaction (7.7).

Bond enthalpy

To continue the analysis of the hydrogen peroxide reaction, you will need to know more about the energies in chemical bonds. The model for covalent bonding developed in Chapter 5 is based on the idea that most chemical bonds consist of shared pairs of electrons being closely held between pairs of atoms (atom cores). The carbon atom in methane, CH_4, for example, has four electron-pair bonds to the four hydrogen atoms. The question for us here is: How energetically independent are the four bonds? We characterize the strength of a chemical bond in terms of its **bond enthalpy**, *the enthalpy required to break the bond*. If one carbon–hydrogen bond is broken, how are the bond enthalpies of the remaining three carbon–hydrogen bonds affected?

This is an important question. If the bond enthalpy for each bond is independent of all the other bonds in the molecule, we should be able to find enthalpy changes for chemical reactions using nothing but bond enthalpies. We would add together the enthalpies required to break the bonds in the reactants and subtract the enthalpies released when the bonds in the products form. If bond enthalpies are independent of one another, this calculation should give the same enthalpy change as that obtained by measuring the enthalpy change for the reaction with a calorimeter.

Homolytic bond cleavage

The independence of bond enthalpies can be tested by measuring the energy (enthalpy) it takes to break covalent bonds one after another in a gaseous molecule. (We study molecules in the gas phase because they are relatively far apart and independent of one another.) These experiments are carried out in such a way that each atom keeps one electron from each pair of valence electrons in the bond that is broken. This symmetric bond breaking is called **homolytic bond cleavage**. Table 7.2 shows the homolytic bond cleavage reactions and enthalpies for successive removal of each hydrogen atom from methane.

> **Free radicals** are molecules with one or more one-electron, σ-nonbonding orbitals. Free radicals are quite reactive species. All of the reaction products in Table 7.2 are free radicals. One important free radical in biological systems is nitric oxide, NO, $\cdot \ddot{N} = \ddot{O} :$. NO plays a great many roles as a signaling molecule in our bodies. The reaction of oxygen with C=C and C≡C bonds often produces free radicals like $R_2COO\cdot$, which are responsible, for example, for spoiling food (rancid butter), for some of the tissue damage in arthritis, and for the aging of cells and organisms.

> Homolytic is from *homos* = same + *lyein* = loosen.

Table 7.2. Enthalpy required for successive homolytic bond cleavages in methane.

All reactants and products are gases.

Homolytic bond cleavage reactions	ΔH, $kJ \cdot mol^{-1}$
H−C(H)(H)−H ⟶ H−C(H)(H)• + H•	439
H−C(H)(H)• ⟶ H−C(H)• + H•	465
H−C(H)• ⟶ H−C• + H•	421
H−C• ⟶ •C• + H•	339
H−C(H)(H)−H ⟶ •C• + 4H•	1664

7.37. Check This   *Overall homolytic bond cleavage reaction of methane*

Show how to combine the four individual homolytic bond cleavages for methane, Table 7.2, to give the overall reaction and overall ΔH of reaction. Are all the bonding electrons in methane accounted for in the products? Explain your answer.

Average bond enthalpies

If bond enthalpies were independent of one another, all the values in Table 7.2 would be the same. As you see, the enthalpy required for breaking the C–H bonds in methane is not the same for each bond, which means that the bond enthalpies are not totally independent of one another. But the **average bond enthalpy**, 416 kJ·mol^{-1} (= $\dfrac{1664}{4}$), is within 20% of all the individual bond enthalpies. When similar experiments are done to remove hydrogen atoms from many other carbon-containing compounds, the experimental bond enthalpies average 414 kJ·mol^{-1}. Average bond enthalpies for several pairs of atoms in a variety of compounds are tabulated in Table 7.3.

Table 7.3. Average bond enthalpies for homolytic bond cleavage.
Values are in kJ·mol^{-1} of bonds for gaseous compounds at 25 °C.

	H	C	N	O	P	S
H–	*436.4*	414	393	460	326	368
C–	414	347	276	351	263	255
C=		620	615	745*		477
C≡		812	891	*1071*		
N–	393	276	193	176	209	
N=		615	418			
N≡		891	*941.4*			
O–	460	351	176	142	502	
O=		745*		*498.7*		469
P–	326	263	209	502	197	
S–	368	255				268
S=		477		469		352
Cl–	*431.9*	338				
Br–	*366.1*	276				
I–	*298.3*	238				

The bond enthalpy for a particular pair of atoms is in the cell where the row of the atom on the left (showing the correct number of bonds) intersects the column of the atom to which it is bound. The C=N bond enthalpy, for example, is read where the C= row intersects the N column, a value of 615 kJ·mol^{-1}. The same value is read beginning with the N= row and moving across to the C column.

Note: Values for diatomic molecules (in bold italics) have four significant figures, because they are not averaged over different compounds.

* The C=O bond enthalpy in CO_2 is 799 kJ·mol^{-1} of bonds. Recall from Chapter 5 that CO_2 has delocalized pi bonds that lead to stronger bonding; the bond enthalpy is larger than for a simple C=O bond.

7.38. Consider This *How do bond enthalpies for single and multiple bonds compare?*

(a) Examine the average bond enthalpies in Table 7.3. How do the bond enthalpies of multiple bonds (such as C=C, and C≡C) compare to the bond enthalpies of the corresponding single bonds (C–C)? Are all double bonds less than twice as strong as single bonds between the same two atoms or are they all more than twice as strong? Give examples to justify your answer.

(b) Do you see any consistent patterns in the strengths of multiple bonds relative to the strengths of single bonds? Give examples to justify your answer.

Patterns among average bond enthalpies

Before we go on to use average bond enthalpies to examine reactions, let's analyze the data in Table 7.3 to see what kinds of patterns there are among the bond enthalpies. These might help us gain further insight into the structure and bonding in molecules. As we do this analysis, we will focus on bond formation instead of bond breaking. For example, Table 7.3 shows that breaking an oxygen-oxygen single bond requires an input of 142 kJ·mol^{-1}. The reverse process, forming an oxygen-oxygen single bond releases 142 kJ·mol^{-1}; ΔH for the bond formation is –142 kJ·mol^{-1}.

The covalent compounds that usually interest us the most are those found in living systems and all of them involve bonds among carbon, nitrogen, and/or oxygen, so we will limit our analysis to these atoms. Table 7.4 shows the enthalpies of bond formation for carbon-carbon, nitrogen-nitrogen, and oxygen-oxygen bonds. These are values from Table 7.3 with negative signs because bond formations are exothermic.

Table 7.4. Enthalpies of formation for C–C, N–N, and O–O bonds.
All energies are in kJ·mol^{-1} of the specified bond.

bond order	carbon–carbon	nitrogen–nitrogen	oxygen–oxygen
single	–347	–193	–142
double	–620	–418	–498.7
triple	–812	–941.4	

As we expect from our bonding model, the enthalpy released in bond formation increases with bond order for all three atoms. The higher the bond order, the stronger the bond between the same atoms. There is a significant difference, however, between carbon and the other two atoms when we compare the *relative* enthalpy release for single bond formation with that for multiple bond formation. Notice, for example, that forming an oxygen-oxygen double bond releases almost twice as much enthalpy as the *sum* of *two* oxygen-oxygen single bonds. Replacing two oxygen-oxygen single bonds by a double bond is an exothermic reaction. For carbon, forming the double bond releases only about 90% as much enthalpy as the sum of two carbon-carbon single bonds. Replacing two single carbon-carbon bonds with a double bond is an endothermic reaction. These and other comparisons are shown graphically in Figure 7.9 for carbon, oxygen, and nitrogen.

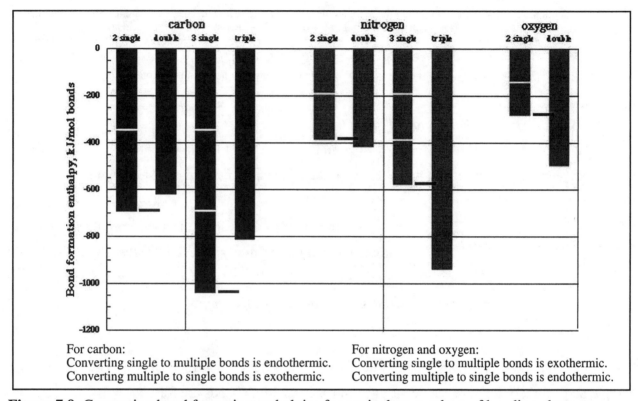

For carbon:
Converting single to multiple bonds is endothermic.
Converting multiple to single bonds is exothermic.

For nitrogen and oxygen:
Converting single to multiple bonds is exothermic.
Converting multiple to single bonds is endothermic.

Figure 7.9. Comparing bond formation enthalpies for equivalent numbers of bonding electrons.
Comparisons are for single versus multiple bonds for C, N, and O. Single bond enthalpy bars are stacked to make the comparisons clear and the horizontal black bars call attention to the sum of the single bond enthalpies.

In Figure 7.9, note that, for carbon, the enthalpy released in forming a multiple bond is always less than the sum of the enthalpies released in forming an equivalent number of single bonds. Just the reverse is true for nitrogen and oxygen. The comparisons in Table 7.4 and Figure 7.9 can help us understand the direction of chemical change without doing any calculations. For example, under the proper conditions, ethene (ethylene) polymerizes to give polyethylene chains:

$$2nH_2C=CH_2 \rightarrow -(-CH_2-CH_2-CH_2-CH_2-)_n- \tag{7.8}$$

The product chain continues in both directions. Count the number of bonding electrons between carbons pictured in Figure 7.10; the electrons from the double bond in ethene form two single bonds in polyethylene. This is an exothermic process; more stable single bonds are formed.

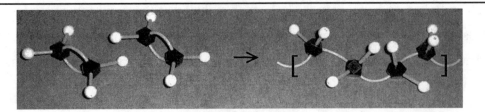

The reactants represent a large number of ethene molecules; only two are shown. The "bent" bonds are retained in the product polymer to show how each double bond in the ethenes has become two single bonds in the polymer.

Figure 7.10. Ethene polymerization.
The brackets enclose the same part of the chain as the parentheses in equation (7.8).

7.39. Check This   *Formation of polyethylene*

Make several models of ethene molecules and "react" them to form polyethylene. How does your polyethylene "product" compare to the product shown in reaction (7.8) and Figure 7.10?

Photo of an automobile airbag showing a label with ingredients.

Another example comes from the airbag used in automobiles. Solid sodium azide, NaN_3, is the reactant that produces the gas that fills the airbag. The reaction of the azide gives nitrogen gas as a product:

$$2Na^+ + 2\ :\overset{-}{\underset{}{N}}=\overset{+}{N}=\overset{-}{N}: \longrightarrow 2Na\cdot + 3\ :N\equiv N: \tag{7.9}$$

In this case, three triple bonds replace four double bonds between nitrogen atoms. The reaction is rapid and exothermic; more stable nitrogen-nitrogen multiple bonds are formed.

7.40. Check This   *Sulfur-sulfur bonds*

The most stable form of elemental oxygen is the diatomic molecule with a multiple bond between the oxygen atoms. The most stable form of sulfur, the third period element beneath oxygen in the periodic table, is the S_8 molecule, a ring of eight atoms bonded by single bonds.

(a) Write Lewis structures for S_2 and S_8.

(b) Make four molecular models of S_2 using oxygen or carbon atom centers to represent sulfur atom centers. Convert these four S_2 molecules to an S_8 molecule. How many two-electron bonds are there in the reactants? in the products?

(c) Sketch a bond formation enthalpy diagram (using the data in Table 7.3 and modeled on Figure 7.9) to show the relative enthalpies of bond formation for four moles of S_2 compared to one mole of S_8. Is the reaction you modeled in part (b) exothermic or endothermic?

(d) Does your result in part (c) help explain why oxygen and sulfur molecules are so different? Explain why or why not.

Bond enthalpy calculations

Now we will go on to use bond enthalpies to analyze reactions that involve bonds between unlike atoms and to estimate numeric values for enthalpies of reaction. Although average bond enthalpies cannot provide precise values for reaction enthalpies, they are a good starting point, if other reaction enthalpy data are not available. When the enthalpy change for a reaction that has never been run can be estimated to within 10-20%, for example, the information can help researchers decide whether to pursue an investigation or to abandon it.

Let's go back to estimate the energetics of the hydrogen peroxide reaction of Investigate This 7.32. In order to do this, we have to recognize a problem. All the average bond enthalpies in Table 7.3 refer to gas-phase reactions, but the reactions we want to analyze, reactions (7.6) and (7.7), involve reactants and products in solution as well as gases. For our estimates of the energetics of the hydrogen peroxide reaction we will analyze two analogous gas phase reactions:

$$2H_2O_2(g) \rightarrow 2H_2O(g) + O_2(g) \tag{7.10}$$

$$H_2O_2(g) \rightarrow H_2(g) + O_2(g) \tag{7.11}$$

We cannot expect the results of our calculations to give an accurate value for the actual reaction, but they will probably help us explain the direction of the reaction. In Worked Example 7.41, we will use average bond enthalpy data to determine the enthalpy change for reaction (7.10) and, in Check This 7.42, ask you do the same for reaction (7.11). Then you can compare the reaction enthalpies to see if they help you to understand why the reaction goes the way you observed.

7.41. Worked Example   *Reaction enthalpy from average bond enthalpies*

Use the average bond enthalpies from Table 7.3 to calculate the enthalpy change for the decomposition of hydrogen peroxide to give water and oxygen by reaction (7.10).

Necessary Information: We need to know which bonds are broken in the reactants and which bonds are formed in the products; this is what you found in Check This 7.36(a).

Strategy: Use average bond enthalpies to calculate the enthalpy required to break the bonds in the reactants and the enthalpy released when the bonds in the products are formed. To keep track of which bond enthalpy is under scrutiny, bond enthalpies will be symbolized as *BH*, with a subscript denoting the bond being analyzed. BH_{H-O}, for example, symbolizes the bond enthalpy for *breaking* a hydrogen–oxygen bond and $-BH_{H-O}$ symbolizes the bond enthalpy for *forming* a hydrogen–oxygen bond. One way to proceed is to imagine all the bonds in the products being broken homolytically to produce atoms, for which the enthalpy change is $\Sigma BH_{reactants}$:

$$2\,H\!-\!\ddot{\underset{..}{O}}\!-\!\ddot{\underset{..}{O}}\!-\!H \longrightarrow 4\,H\!\cdot\; +4\cdot\ddot{\underset{..}{O}}\cdot$$

$$2H_2O_2(g) \rightarrow 4H(g) + 4O(g) \qquad\qquad \Sigma BH_{\text{reactants}} \qquad\qquad (7.12)$$

The summation symbol, Σ, indicates that the bond enthalpies for breaking all the bonds in the reactants are added together. Then the atoms are recombined to form the products, for which the enthalpy change is $\Sigma(-BH_{\text{products}})$:

$$4\,H\!\cdot\; +4\cdot\ddot{\underset{..}{O}}\cdot \longrightarrow 2\,H\!-\!\ddot{\underset{..}{O}}\!-\!H\; +\; \ddot{\underset{..}{O}}\!=\!\ddot{\underset{..}{O}}$$

$$4H(g) + 4O(g) \rightarrow 2H_2O(g) + O_2(g) \qquad \Sigma(-BH_{\text{products}}) \qquad (7.13)$$

The sum of reactions (7.12) and (7.13) is the reaction of interest, reaction (7.10); $\Delta H_{\text{reaction}}$ is the sum of the enthalpy changes for reactions (7.12) and (7.13):

$$\Delta H_{\text{reaction}} = (\Sigma BH_{\text{reactants}}) + \Sigma(-BH_{\text{products}}) \qquad\qquad (7.14)$$

Figure 7.11 summarizes these equations and the strategy in an enthalpy-level diagram (which is exactly the same as an energy diagram, but with enthalpy as the energy function).

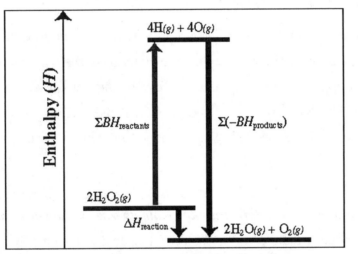

Figure 7.11. Enthalpy-level diagram for: $2H_2O_2(g) \rightarrow 2H_2O(g) + O_2(g)$
The reaction is assumed to be exothermic, as you found experimentally in Investigate This 7.32.

Implementation: Table 7.5 shows the combinations of average bond-breaking and bond-forming enthalpy changes for reactions (7.12) and (7.13).

Table 7.5. Bond-breaking and bond-forming enthalpy changes for: $2H_2O_2(g) \rightarrow 2H_2O(g) + O_2(g)$

Reactant bond breaking				Product bond formation			
bond	number of bonds	BH, kJ·mol^{-1}	ΣBH, kJ	bond	number of bonds	BH, kJ·mol^{-1}	$\Sigma(-BH)$, kJ
O–O	2	142	284	O=O	1	499	–499
H–O	4	460	1840	H–O	4	460	–1840
$\Sigma BH_{reactants} = 2BH_{O\text{-}O} + 4BH_{H\text{-}O} = 2124$				$\Sigma(-BH_{products}) = -BH_{O=O} - 4BH_{H\text{-}O} = -2339$			

Table 7.5 gives us the numerical values for the sums in equation (7.14) and substituting these values into the equation gives $\Delta H_{reaction}$:

$$\Delta H_{reaction} = (2124 \text{ kJ}) + (-2339 \text{ kJ}) = -215 \text{ kJ}$$

Does the Answer Make Sense? We set up an *imaginary* pathway for reaction (7.10) which involved enthalpy changes that could be calculated from average bond enthalpies. The sum of the enthalpy changes along this pathway from products to reactants is the same as the enthalpy change for the direct reaction. The calculated enthalpy change is negative; we predict that the reaction is exothermic. This makes sense, since you found experimentally that the reaction is exothermic. Note that this enthalpy change is for the reaction as written, with two moles of hydrogen peroxide reacting. The enthalpy change for one mole reacting is half this value:

$$H_2O_2(g) \rightarrow H_2O(g) + \tfrac{1}{2}O_2(g) \qquad \Delta H_{reaction} = -108 \text{ kJ (for one mole of } H_2O_2) \qquad (7.15)$$

7.42. Check This   *Reaction enthalpy from average bond enthalpies*

(a) Use average bond enthalpy data from Table 7.3 to determine the enthalpy change, $\Delta H_{reaction}$ for reaction (7.11):

$$H_2O_2(g) \rightarrow H_2(g) + O_2(g) \qquad\qquad\qquad\qquad (7.11)$$

(b) Draw an enthalpy-level diagram, modeled after Figure 7.11, for this reaction.

(c)   WEB   Chap 7, Sect 7.7.3. Describe how the interactive exercise and resulting enthalpy-level diagram on this web page are related to the calculations and enthalpy-level diagrams in parts (a) and (b) and in Worked Example 7.42 and Figure 7.11.

7.43. Consider This   *What is the hydrogen peroxide decomposition reaction?*

(a) How does the enthalpy change for reaction (7.11) in Check This 7.42, compare to the enthalpy change for reaction (7.10) in Worked Example 7.41?

(b) Is reaction (7.6) or reaction (7.7) responsible for your observations in Investigate This 7.32 and 7.35? Summarize all the experimental and calculated results that lead you to this conclusion.

Accuracy of bond enthalpy calculations

The results of the experiments and the calculated enthalpies of reaction in Worked Example 7.41 and Check This 7.42 are all consistent with reaction (7.6) being the exothermic decomposition reaction for hydrogen peroxide in the presence of yeast. The question remains whether the quantitative results of calculations based on average bond enthalpies for gas phase reactions are accurate enough to justify their use as more than a qualitative indicator of exothermicity or endothermicity, especially for reactions that involve solutions, liquids, and solids, as well as gases.

Hydrogen peroxide, hydrogen, oxygen, and water are compounds that have been extensively studied, and we can obtain accurate values for the enthalpy changes for reaction (7.6) (written here for one mole of hydrogen peroxide), and reaction (7.7):

$$H_2O_2(aq) \rightarrow H_2O(l) + \tfrac{1}{2}O_2(g) \qquad \Delta H_{reaction} = -95 \text{ kJ·(mol } H_2O_2)^{-1} \qquad (7.6)$$

$$H_2O_2(aq) \rightarrow H_2(g) + O_2(g) \qquad \Delta H_{reaction} = +191 \text{ kJ·(mol } H_2O_2)^{-1} \qquad (7.7)$$

The bond enthalpy calculations for these reactions (in the gas phase) produced reaction enthalpies that were in the right direction (exothermic and endothermic) and within about 50% of these experimental values.

Our calculated results could have been much better, if we had accounted for the enthalpy required to get hydrogen peroxide from solution into the gas phase and the enthalpy released when gaseous water condenses to a liquid. The calculations would not have been a great deal more difficult. However, we set out to find out how useful reaction enthalpy calculations based solely on average bond enthalpies can be. The answer appears to be that bond enthalpy calculations are quite useful in giving us the direction of a reaction enthalpy change, but that we cannot depend on them to give accurate numerical values, especially if the reactants and products are not all gases. The great advantage of bond enthalpies is that the modest set of values in Table 7.3 provides all the information you need to calculate reaction enthalpies for an enormous number of reactions, even ones that you only imagine — like reaction (7.7). In the next section, we will discuss another method for obtaining reaction enthalpies that gives accurate numerical values, but requires a much more extensive data table.

Reflection and projection

The two-electron, covalent bonds in molecules are not completely independent of one another. However, the energies (enthalpies) required to break similar bonds (H–C, for example) in the same or different gaseous molecules are not too different. We can, therefore, use the average of these enthalpies as a reasonable approximation of the bond enthalpy, tabulate the results for different kinds of bonds, and use these values to calculate enthalpy changes for reactions. Enthalpies of reaction are calculated by summing up the bond enthalpies for all the reactant bonds that are broken, summing up all the bond enthalpies for product bonds that are formed (negative values), and combining these positive and negative sums to get the enthalpy change for the reaction:

$$\Delta H_{reaction} = (\Sigma BH_{reactants}) + \Sigma(-BH_{products}) \tag{7.14}$$

You can't expect enthalpy changes calculated from average bond enthalpies to be as accurate as direct calorimetric measurements (when these are possible). However, these calculated values can almost always be used to determine whether a reaction will be exothermic or endothermic. Calculated values of reaction enthalpies are all you can get if the reactants and/or products are only imagined molecules. You also saw how bond enthalpies help to understand why oxygen and nitrogen tend to form simple diatomic molecules with double and triple bonds, respectively, whereas carbon preferentially forms extended chains of single bonds.

Useful as bond enthalpies are, calculations based on them are not accurate enough for all purposes, especially for analyzing complex reactions in solution. A different approach that is not dependent on assumptions about the bonding properties of molecules is required.

Section 7.8. Standard Enthalpies of Formation

Bond enthalpies are extraordinarily useful for predicting enthalpy changes for reactions that are experimentally inaccessible, but they are limited by the requirement that reasonable enthalpy estimates require reactants and products to be in the gas phase. The alternative approach is to use standard enthalpies of formation. The **standard enthalpy of formation** of a compound is the enthalpy change when one mole of the compound is formed at a pressure of 1 bar (= 10^5 kg·m^{-1}·s^{-2} = 100 kPa) from its elements in their standard states. The **standard state** of an element is its most

> The standard state pressure used to be defined as one atmosphere pressure, about 101 kPA. For most purposes, tiny differences between enthalpies of formation tabulated for one atm and one bar are negligible; values found in older tables and handbooks are still usable.

stable form at 1 bar and a specified temperature (usually 25 °C). We have to specify the element in its most stable form, because many elements exist in more than one structure at 1 bar and 25 °C. Different structures of an element are called **allotropes**. Carbon, for example, is an element that exists in several allotropic solid forms, Figure 7.11, that differ in enthalpy. **Graphite** is the most stable allotrope, and it is chosen as the standard state for carbon.

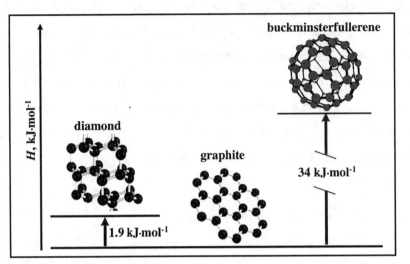

Figure 7.12. Enthalpy-level diagram for carbon allotropes.

ΔH values are for one mole of carbon atoms in each form. Graphite is made up of many layers; a tiny area of one layer is shown here. The diamond structure extends in three dimensions and can form large crystals. **Buckminsterfullerene**, C_{60}, is a discrete molecular allotrope of carbon. A family of these carbon-cage, fullerene molecules ("buckyballs"), with larger numbers of carbon atoms, has been made and analyzed.

7.44. Worked Example   *Formation of a compound from its standard-state elements*

Write the reaction equation for the formation of one mole of methanol, CH_3OH, in its standard state from its elements in their standard states.

Necessary Information: We need to know the standard states of all the reacting elements and the product. The standard states for the elements are: C*(graphite)*, $H_2(g)$, and $O_2(g)$. Methanol is a liquid at 25 °C and 1 bar, $CH_3OH(l)$.

Strategy: We write an expression for the formation reaction, using the elements and product in their standard states and then balance the equation for one mole of product.

Implementation:

unbalanced: C*(graphite)* + $O_2(g)$ + $H_2(g)$ → $CH_3OH(l)$

balanced: 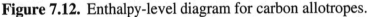 C*(graphite)* + $1/2 O_2(g)$ + $2H_2(g)$ → $CH_3OH(l)$

Does the Answer Make Sense? The reactants are elements in their standard states and the product is methanol in its standard state. The balanced reaction equation corresponds to the definition of the reaction whose enthalpy change is the enthalpy of formation of methanol. Note that the coefficients are chosen so that one mole of product is formed, even when fractional coefficients are necessary for reactant elements.

7.45.   **Check This**   *Formation of compounds from their standard-state elements*

Write the reaction equations for the formation of one mole of dimethylamine, $(CH_3)_2NH(g)$, and one mole of ethylamine, $CH_3CH_2NH_2(g)$, in their standard states (shown with their formulas) from their elements in their standard states. How do the reactants for the two cases compare? What can you conclude about the reaction equations for the formation of isomers?

In Worked Example 7.44, we wrote the reaction equation for the formation of one mole of methanol in its standard state from its elements in their standard states. The standard enthalpy of formation for methanol is the enthalpy change for the reaction we wrote:

$$C_{(graphite)} + 1/2O_2(g) + 2H_2(g) \rightarrow CH_3OH(l) \qquad \Delta H_f^\circ = -238.7 \text{ kJ·mol}^{-1} \qquad (7.16)$$

Thermodynamic tables use ΔH_f° to designate ΔH values for reactions under standard conditions. The superscript "°" in ΔH_f° designates the standard state pressure of one bar, and the subscript "f" reminds us that the value is for an enthalpy of formation.

Appendix XX lists standard enthalpies of formation for many compounds, including those used in problems in this text. Although temperature is not specified as part of the standard state, tables of enthalpies of formation are almost always compiled for compounds at 298 K (25 °C), as in Appendix XX. The bond enthalpies included in Table 7.3 are also all given for 25 °C. In order to set up enthalpy tables that everyone can use, scientists have agreed on the reference point for enthalpy measurements. The choice of reference point is the elements in their standard states which are all arbitrarily *assigned* an enthalpy of formation of zero. Since only *differences* in enthalpy are measurable, the choice of zero for the enthalpy of formation of the elements is made as a matter of convenience.

In addition to the accuracy of the data, enthalpies of formation greatly simplify calculations involving complex molecules because the particulars of chemical bonding are not factors in the calculations. The following Worked Examples illustrate the use of standard enthalpies of formation to obtain the standard enthalpy changes for chemical reactions and the Check This problems provide an opportunity for you to practice the procedures.

7.46.   **Worked Example**   *Standard enthalpy change for an isomerization reaction*

Use the data in Appendix XX to calculate the standard enthalpy of reaction, $\Delta H^\circ_{reaction}$, for the **isomerization**, change of one isomer to another, of one mole of cyclobutane, cyclo-$C_4H_8(g)$ (a four-carbon ring) to one mole of 1-butene, $CH_2=CHCH_2CH_3(g)$ (an alkene isomer of cyclobutane) at 25 °C:

$$\text{cyclo-C}_4\text{H}_8(g) \rightarrow \text{CH}_2=\text{CHCH}_2\text{CH}_3(g) \qquad \Delta H^\circ_{\text{reaction}} = ? \qquad (7.17)$$

Necessary Information: We need standard enthalpies of formation for the reactant and product. The values from Appendix XX are 26.65 and 1.17 kJ·mol^{-1}, respectively, for cyclobutane and 1-butene. Note that formation of these isomers from the elements is endothermic.

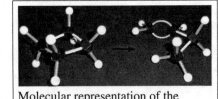

Molecular representation of the cyclobutane isomerization.

Strategy: Write the enthalpy of formation reaction equations for the reactant and product and combine them appropriately to obtain the desired reaction equation (7.17). The enthalpies of formation will be combined in the same way to get the standard reaction enthalpy.

Implementation:

$$4\text{C}(graphite) + 4\text{H}_2(g) \rightarrow \text{cyclo-C}_4\text{H}_8(g) \qquad \Delta H_f^\circ(\text{cyclo}) = 26.65 \text{ kJ·mol}^{-1} \qquad (7.18)$$

$$4\text{C}(graphite) + 4\text{H}_2(g) \rightarrow \text{CH}_2=\text{CHCH}_2\text{CH}_3(g) \qquad \Delta H_f^\circ(\text{butene}) = 1.17 \text{ kJ·mol}^{-1} \qquad (7.19)$$

Reverse reaction equation (7.18), including the sign of the enthalpy change, and add it to reaction equation (7.19) to get reaction equation (7.17):

$$\text{cyclo-C}_4\text{H}_8(g) \rightarrow 4\text{C}(graphite) + 4\text{H}_2(g) \qquad \Delta H^\circ = (1 \text{ mol})[-\Delta H_f^\circ(\text{cyclo})] \qquad -(7.18)$$

$$\underline{4\text{C}(graphite) + 4\text{H}_2(g) \rightarrow \text{CH}_2=\text{CHCH}_2\text{CH}_3(g) \qquad \Delta H^\circ = (1 \text{ mol})[\Delta H_f^\circ(\text{butene})]} \qquad (7.19)$$

$$\text{cyclo-C}_4\text{H}_8(g) \rightarrow \text{CH}_2=\text{CHCH}_2\text{CH}_3(g) \qquad \Delta H^\circ_{\text{reaction}} \qquad (7.17)$$

$$\Delta H^\circ_{\text{reaction}} = (1 \text{ mol})[-\Delta H_f^\circ(\text{cyclo})] + (1 \text{ mol})[\Delta H_f^\circ(\text{butene})]$$

Which can be rewritten as:

$$\Delta H^\circ_{\text{reaction}} = (1 \text{ mol})[\Delta H_f^\circ(\text{butene})] - (1 \text{ mol})[\Delta H_f^\circ(\text{cyclo})] \qquad (7.20)$$

$$\Delta H^\circ_{\text{reaction}} = (1 \text{ mol})[1.17 \text{ kJ·mol}^{-1}] - (1 \text{ mol})[26.65 \text{ kJ·mol}^{-1}] = -25.48 \text{ kJ}$$

Figure 7.13 is an enthalpy-level diagram showing another way to combine standard enthalpies of formation to get the standard enthalpy of the reaction, as in equation (7.20).

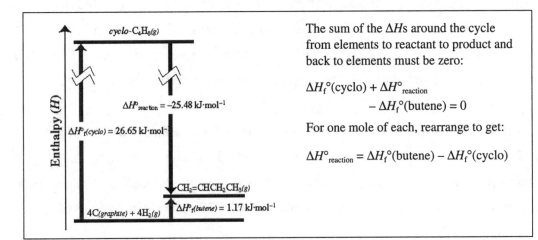

The sum of the ΔHs around the cycle from elements to reactant to product and back to elements must be zero:

$$\Delta H_f^\circ(\text{cyclo}) + \Delta H^\circ_{\text{reaction}} - \Delta H_f^\circ(\text{butene}) = 0$$

For one mole of each, rearrange to get:

$$\Delta H^\circ_{\text{reaction}} = \Delta H_f^\circ(\text{butene}) - \Delta H_f^\circ(\text{cyclo})$$

Figure 7.13. Enthalpy-level diagram for cyclobutane to 1-butene isomerization.

Does the Answer Make Sense? The enthalpy level diagram, Figure 7.13, clearly illustrates that 1-butene is a lower energy, more stable compound than cyclobutane. The reaction to form the 1-butene isomer should be exothermic, as we have calculated it to be.

7.47.   **Check This**   *Standard enthalpy change for an isomerization reaction*

(a) Make molecular models to illustrate the isomerization of cyclopentane, cyclo-$C_5H_{10(g)}$ (a five-carbon ring) to 1-pentene, $CH_2=CHCH_2CH_2CH_{3(g)}$ (an alkene isomer of cyclopentane):

$$cyclo\text{-}C_5H_{10(g)} \rightarrow CH_2=CHCH_2CH_2CH_{3(g)} \qquad \Delta H°_{reaction} = ? \qquad (7.21)$$

(b) Use the data in Appendix XX to calculate the standard enthalpy of reaction, $\Delta H°_{reaction}$, for reaction (7.21) at 25 °C.

(c) Sketch an enthalpy level diagram, modeled after Figure 7.13, for this reaction to show how your combination of standard enthalpies of formation gives the standard enthalpy of the reaction.

We have pointed out many times that bond formation is always exothermic: the atoms bonded together are lower in energy than the separated atoms. You might, therefore, have wondered about the positive standard enthalpy of formation of the isomers, cyclobutane and 1-butene, in Worked Example 7.46. How can the formation of these compounds from their elements be endothermic, if bond formation is always exothermic? The answer lies in the difference between formation of a molecule from its separated atoms (always exothermic) and from the atoms in their elemental standard states, as illustrated for cyclobutane in Figure 7.14.

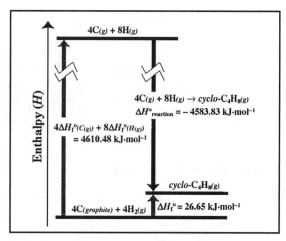

Figure 7.14. Comparison of formation of *cyclo*-C_4H_8 from its atoms and its elements.
Energy is always released when a molecule is formed from its separated gas-phase atoms.

Figure 7.14 shows that the formation of cyclobutane from its separated atoms is a highly exothermic process. However, the formation of the atoms from their elements requires a bit more energy (enthalpy) than we get back from compound formation, so the enthalpy of formation of cyclobutane from its *elements* is endothermic.

***7.48.* Check This** *Comparison of formation of 1-butene from its atoms and its elements*

(a) Use the data in Appendix XX to construct an enthalpy-level diagram, modeled after Figure 7.14, comparing the enthalpy changes in formation of 1-butene from its atoms and its elements.

(b) Explain why there are similarities and differences between your enthalpy-level diagram in part (a) and the one shown in Figure 7.14.

Equation (7.20), Figure 7.13, and your equivalent formulations in Check This 7.48 show that the standard enthalpy change for isomerization reactions is the difference between the standard enthalpy of formation of the product and the standard enthalpy of formation of the reactant (with the number of moles of reactant and product accounted for):

$$\Delta H°_{reaction} = (\text{mol product})[\Delta H_f°(\text{product})] - (\text{mol reactant})[\Delta H_f°(\text{reactant})] \qquad (7.22)$$

Let's see how this approach works for more complicated reactions.

***7.49.* Worked Example** *Standard enthalpy change for a reaction*

Use the data in Appendix XX to calculate the standard enthalpy of reaction, $\Delta H°_{reaction}$, for the complete oxidation of methanol, $CH_3OH(l)$ at 25 °C:

$$CH_3OH(l) + \tfrac{3}{2}O_2(g) \rightarrow CO_2(g) + 2H_2O(l) \qquad \Delta H°_{reaction} = ? \qquad (7.23)$$

Necessary Information: From Appendix XX, the standard enthalpies of formation for methanol, oxygen, water, and carbon dioxide are, respectively, -238.9, 0.0 (element in its standard state), -393.5, and -285.8 kJ·mol^{-1}.

Molecular representation of the methanol oxidation.

Strategy: Use the standard enthalpies of formation and the stoichiometry of the reaction to get the overall standard enthalpy changes for formation of the reactants, $\Delta H_f°(\text{reactants})$, and the products, $\Delta H_f°(\text{products})$. Combine these to get the standard enthalpy change for the reaction, $\Delta H°_{reaction}$.

Implementation:

For the reactants ($O_2(g)$ is included to be complete):

$C_{(graphite)} + 2H_2(g) + \frac{1}{2}O_2(g) \rightarrow CH_3OH(l)$ $\Delta H^\circ = (1 \text{ mol})[\Delta H_f^\circ(CH_3OH)]$

$\frac{3}{2}O_2(g) \rightarrow \frac{3}{2}O_2(g)$ $\Delta H^\circ = (\frac{3}{2} \text{ mol})[\Delta H_f^\circ(O_2)]$

$C_{(graphite)} + 2H_2(g) + 2O_2(g) \rightarrow CH_3OH(l) + \frac{3}{2}O_2(g)$ $\Delta H_f^\circ(\text{reactants})$ (7.24)

$\Delta H_f^\circ(\text{reactants}) = (1 \text{ mol})[\Delta H_f^\circ(CH_3OH)] + (\frac{3}{2} \text{ mol})[\Delta H_f^\circ(O_2)]$ (7.25)

$\Delta H_f^\circ(\text{reactants}) = (1 \text{ mol})[-238.9 \text{ kJ·mol}^{-1}] + (\frac{3}{2} \text{ mol})[0.0 \text{ kJ·mol}^{-1}] = -238.9 \text{ kJ}$

For the products:

$C_{(graphite)} + O_2(g) \rightarrow CO_2(g)$ $\Delta H^\circ = (1 \text{ mol})[\Delta H_f^\circ(CO_2)]$

$2H_2(g) + O_2(g) \rightarrow 2H_2O(l)$ $\Delta H^\circ = (2 \text{ mol})[\Delta H_f^\circ(H_2O)]$

$C_{(graphite)} + 2H_2(g) + 2O_2(g) \rightarrow CO_2(g) + 2H_2O(l)$ $\Delta H_f^\circ(\text{products})$ (7.26)

$\Delta H_f^\circ(\text{products}) = (1 \text{ mol})[\Delta H_f^\circ(CO_2)] + (2 \text{ mol})[\Delta H_f^\circ(H_2O)]$ (7.27)

$\Delta H_f^\circ(\text{products}) = (1 \text{ mol})[-393.5 \text{ kJ·mol}^{-1}] + (2 \text{ mol})[-285.8 \text{ kJ·mol}^{-1}] = -965.1 \text{ kJ}$

Figure 7.13 is an enthalpy-level diagram, which shows how the standard enthalpies of formation of the reactants and products combine to give the standard enthalpy of reaction:

$$\Delta H^\circ_{\text{reaction}} = \Delta H_f^\circ(\text{products}) - \Delta H_f^\circ(\text{reactants}) \qquad (7.28)$$

$$= -965.1 \text{ kJ} - (-238.9 \text{ kJ}) = -726.2 \text{ kJ}$$

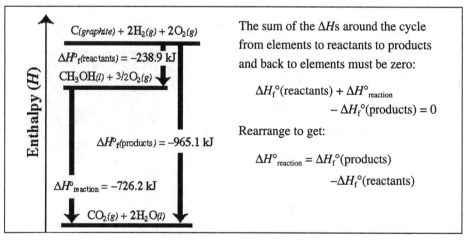

Figure 7.15. Enthalpy-level diagram for oxidation of methanol.

Does the Answer Make Sense? Without further calorimetric measurements, we do not know whether the calculated *numerical* value is correct. However, we know that methanol burns in air and gives off thermal and light energy, so the reaction must be exothermic, as our result shows.

7.50. Check This   *Standard enthalpy change for a reaction*

(a) Use the data in Appendix XX to calculate the standard enthalpy of reaction, $\Delta H°_{reaction}$, for the complete oxidation of glucose, $C_6H_{12}O_6(s)$ at 25 °C:

$$C_6H_{12}O_6(s) + 6O_2(g) \rightarrow 6CO_2(g) + 6H_2O(l) \qquad \Delta H°_{reaction} = ? \qquad (7.29)$$

(b) Draw an enthalpy-level diagram, modeled after Figure 7.15, for this reaction.

Look back at Worked Examples 7.46 and 7.49 at your work in Check This 7.47 and 7.50, and at equation 7.22 and examine the final form of the equations used to calculate the standard enthalpies of reaction. In all cases, the standard enthalpy of reaction was obtained by subtracting the standard enthalpy of formation of the reactants from the standard enthalpy of formation of the products. This result can be expressed as:

$$\Delta H°_{reaction} = \Sigma[n_j(\Delta H_f°)_j]_{products} - \Sigma[(n_j(\Delta H_f°)_j]_{reactants} \qquad (7.30)$$

In equation (7.30), the subscripts "j" refer to the individual reactant or product compounds; n_j is the coefficient in the chemical equation for the jth compound and $(\Delta H_f°)_j$ is the standard enthalpy of formation for that compound. The reactant and product sums include all the reactant and product compounds, respectively.

7.51. Check This   *Bond enthalpy and enthalpy of formation comparison*

Compare the enthalpy change for a reaction calculated from bond enthalpies and standard enthalpies of formation.

(a) The bond dissociation enthalpy of bromine is 193 kJ·mol⁻¹. Use this knowledge and bond enthalpies to estimate the standard enthalpy change, $\Delta H°_{reaction}$, for the gas-phase reaction of bromine with ethene to yield 1,2-dibromoethane:

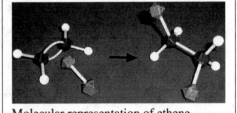

Molecular representation of ethene bromination.

$$\underset{H}{\overset{H}{\diagdown}} C = C \underset{H}{\overset{H}{\diagup}} {}_{(g)} + \ :\!\ddot{Br}\!-\!\ddot{Br}\!: {}_{(g)} \longrightarrow H-\underset{H}{\overset{:\ddot{Br}:}{\underset{|}{\overset{|}{C}}}}-\underset{:\ddot{Br}:}{\overset{H}{\underset{|}{\overset{|}{C}}}}-H \quad {}_{(g)} \qquad \Delta H_{reaction} = ? \qquad (7.31)$$

(b) The standard enthalpy of formation of 1,2-dibromoethane is –43.1 kJ·mol⁻¹. Use this value and others from Appendix XX to calculate $\Delta H°_{reaction}$ for reaction (7.31). *Hint:* Remember that bromine is not in its standard state in this reaction.

(c) How do your results for parts (a) and (b) compare? What might account for any difference?

Section 7.9.   Harnessing Energy in Living Systems

When you set a marshmallow on fire in Investigate This 7.1, you were rearranging the chemical bonds in the sugar(s) of the marshmallow and the oxygen from the air. The carbon in the sugar was oxidized to carbon dioxide and the hydrogen combined with more oxygen to form water. Equation (7.29) in Check This 7.50 showed this overall reaction for one sugar, glucose. In the combustion process the enthalpy difference between the reactants and products was released as thermal energy to the environment around the burning marshmallow. The pathway for conversion of

reactants to products, although complex, is relatively direct and involves no other reactants. This is not the case when that same glucose, perhaps eaten as a toasted marshmallow, is metabolized in your body. Glucose metabolism is a long, complicated, but highly organized, process that requires a large number of enzymes and several biochemical pathways. Energy from the oxidation of glucose is released piecemeal as the oxidation process proceeds.

Coupled reactions. During the oxidation process some of the energy released is "lost" as thermal energy that helps maintain your body temperature at 37 °C. Some of the energy is captured by coupling a reaction that *releases* energy to one that *requires* energy. The basis for these **coupled reactions** is that the reaction providing the energy does not proceed unless the reaction needing the energy input also occurs. In our bodies, the exothermic oxidation of glucose is coupled at several points along the biochemical oxidation pathway with the endothermic formation of adenosine triphosphate, ATP^{4-}, from adenosine diphosphate, ADP^{3-} (Figure 7.16):

$$ADP^{3-}(aq) + HOPO_3{}^{2-}(aq) + H^+(aq) \rightarrow ATP^{4-}(aq) + H_2O(l) \qquad \Delta H°_{ATP\ form} \approx +21\ kJ\cdot mol^{-1}$$

$$(7.32)$$

For simplicity, we are writing the hydronium ion as $H^+(aq)$, as we did in reduction-oxidation reactions in Chapter 6, Sections 6.9, 6.10, and 6.11.

The links between phosphate groups in ATP^{4-} and ADP^{3-} (labeled in red in Figure 7.16) are acid anhydride functional groups. An **acid anhydride** (from *anhydrous* = without water) is a functional group formed when two acid molecules condense to form a single unit with the loss of water, as shown in Figure 7.16. Formation of acid anhydrides from the acids is always an endothermic process. Furthermore, the grouping of four negative charges that repel one another in ATP^{4-} helps to explain why an input of energy is required to form it from ADP^{3-} and $HOPO_3{}^{2-}$.

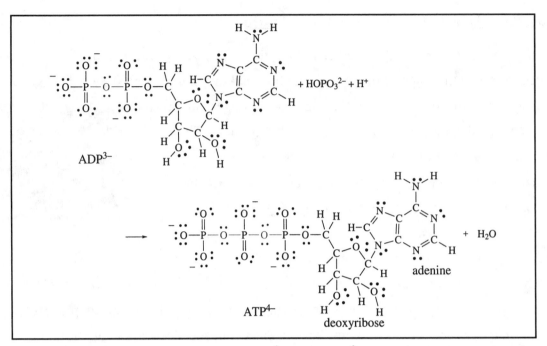

Figure 7.16. Reaction to form ATP^{4-} from ADP^{3-} and phosphate.

The most direct way that coupling occurs to form ATP^{4-} is by the transfer of a phosphate group from one of the oxidized products of glucose oxidation to an ADP^{3-}. Two consecutive reactions in the glucose oxidation pathway are:

$$^{-2}O_3POCH_2CHOHCHO + NAD^+ + HOPO_3{}^{2-} \rightarrow$$

$$^{-2}O_3POCH_2CHOHC(O)OPO_3{}^{2-} + NADH + H^+ \tag{7.33}$$

$$^{-2}O_3POCH_2CHOHC(O)OPO_3{}^{2-} + ADP^{3-} \rightarrow {}^{-2}O_3POCH_2CHOHC(O)O^- + ATP^{4-} \tag{7.34}$$

In reaction (7.33), an enzyme catalyzes the oxidation of an aldehyde group to a carboxylic acid and its conversion to a mixed acid anhydride in which two different acids (phosphoric and carboxylic) have condensed with the loss of water and formation of an anhydride bond between the acids. The compound that is reduced is NAD$^+$, nicotinamide dinucleotide (see Chapter 6, Section 6.11). The energy released by the reduction-oxidation reaction provides the energy required for endothermic formation of the anhydride.

In reaction (7.34) another enzyme catalyzes the transfer of a phosphate group from the mixed acid anhydride to form a phosphoric acid anhydride. Since one acid anhydride bond is broken and another is formed, the reaction overall involves very little enthalpy change. The *combination* of reactions (7.33) and (7.34) represents a coupling of the energy released in one of the glucose oxidation steps to the formation of ATP^{4-} from ADP^{3-} using the mixed anhydride formed during the oxidation to couple the reactions.

7.52.   **Check This**   *The phosphate bond to deoxyribose in ATP^{4-} and ADP^{3-}*

The phosphate groups in ATP^{4-} and ADP^{3-} are bonded by acid anhydride links. What kind of bond (functional group) links the first phosphate group to deoxyribose? *Hint:* See Chapter 6, Section 6.7.

Energy captured as ATP^{4-}

In our bodies, coupling of the glucose oxidation pathway to the formation of ATP^{4-} can produce about 36 mol of ATP^{4-} for each mole of glucose oxidized. We can estimate what this means in terms of enthalpy changes by combining reaction (7.29) and reaction (7.32) taken 36 times, and their standard enthalpy changes:

$$C_6H_{12}O_6(s) + 6O_2(g) \rightarrow 6CO_2(g) + 6H_2O(l) \qquad\qquad \Delta H°_{\text{glucose oxidation}} \quad (7.29)$$

$$\underline{36ADP^{3-}(aq) + 36HOPO_3^{2-}(aq) + 36H^+(aq) \rightarrow 36ATP^{4-}(aq) + 36H_2O(l) \qquad 36\Delta H°_{\text{ATP form}} \quad (7.32)}$$

$$C_6H_{12}O_6(s) + 6O_2(g) + 36ADP^{3-}(aq) + 36HOPO_3^{2-}(aq)\,) + 36H^+(aq) \rightarrow$$
$$6CO_2(g) + 42H_2O(l) + 36ATP^{4-}(aq) \qquad\qquad \Delta H°_{\text{coupled reaction}} \quad (7.35)$$

$$\Delta H°_{\text{coupled reaction}} = \Delta H°_{\text{glucose oxidation}} + 36\Delta H°_{\text{ATP form}} \qquad\qquad (7.36)$$

$$\Delta H°_{\text{coupled reaction}} = -2801 \text{ kJ} + (36 \text{ mol}) (21 \text{ kJ·mol}^{-1}) = -2045 \text{ kJ}$$

The $\Delta H°_{\text{glucose oxidation}}$ comes from your calculation in Check This 7.50. Overall, the enthalpy change for the coupled reaction is −2045 kJ for every mole of glucose oxidized to produce 36 moles of ATP^{4-}. This result means that, under standard conditions, 756 kJ out of the 2801 kJ from glucose oxidation would go into making ATP^{4-}. The rest of the energy is released as heat to the surroundings. Under these conditions, the percentage of enthalpy converted to the ATP^{4-} required for other biological processes is about 27% ($\left(\dfrac{756 \text{ kJ}}{2801 \text{ kJ}} \right) \times 100\% = 27\%$). It is interesting to compare this biological oxidation with combustion. Steam engines that evolved from *The Rocket*, Figure 7.1, were rarely more than about 10% efficient in converting the thermal energy of combustion to work and present-day standard internal combustion engines are about 30% efficient.

7.53.   **Check This**   *Other biological fuels*

What other biological fuel compounds did you investigate in Investigate This 7.1? What others are represented in the chapter opening illustration? Which seem to be the best fuels? Explain your answer.

Using the energy captured as ATP

The reverse of reaction (7.35), the **hydrolysis** (breaking down by water; see Chapter 6, Section 6.7) of ATP^{4-} to ADP^{3-}, is exothermic:

$$ATP^{4-}(aq) + H_2O(l) \rightarrow ADP^{3-}(aq) + HOPO_3^{2-}(aq) + H^+(aq) \qquad \Delta H^\circ_{\text{ATP hydrolysis}} \approx -21 \text{ kJ·mol}^{-1}$$

$$(7.37)$$

Figure 7.17 is a mechanical analogy showing how the enthalpy released by glucose oxidation is coupled through the synthesis and hydrolysis of ATP^{4-} to other energy-requiring biological reactions that provide such things as locomotion, information processing, and synthesis of new biological molecules.

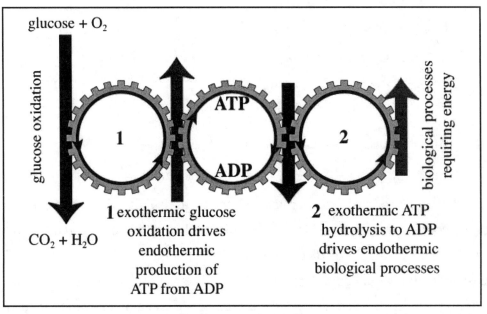

Figure 7.17. Glucose oxidation coupled via ATP-ADP to energy-requiring reactions.
Enthalpy change arrows are not to scale, but show the direction and relative magnitudes of the changes. Thermal energy is "lost" to the surroundings at each transformation, glucose-to-ATP and ATP-to-biological processes.

For example, joining amino acids to form a protein is an endothermic process. Consider the simple case of bonding two glycines, $H_2NCH_2C(O)OH$, to form diglycine, $H_2NCH_2C(O)NHCH_2C(O)OH$, and water:

$$2\text{Gly}(aq) \rightarrow \text{Gly–Gly}(aq) + H_2O(l) \qquad \Delta H^\circ_{\text{Gly-Gly}} = +8 \text{ kJ·mol}^{-1} \qquad (7.38)$$

If this reaction is coupled to the hydrolysis of ATP^{4-}, we can write the coupled reaction as the sum of reactions (7.37) and (7.38):

$$2\text{Gly}(aq) + ATP^{4-}(aq) + \rightarrow \text{Gly–Gly}(aq) + ADP^{3-}(aq) + HOPO_3^{2-}(aq) + H^+(aq)$$

$$\Delta H^\circ_{\text{reaction}} = -13 \text{ kJ·mol}^{-1} \qquad (7.39)$$

Figure 7.18 shows how this exothermic coupling is represented on an enthalpy-level diagram.

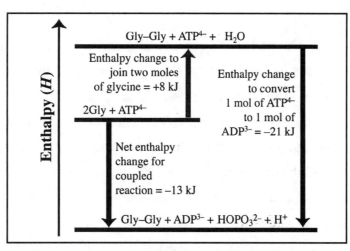

Figure 7.18. Enthalpy coupling of diglycine synthesis to ATP^{4-} hydrolysis.

Protein synthesis in organisms is much more complicated than reactions (7.38) and (7.39) suggest, but the overall energetics are reasonably represented here. The enzymes and ribonucleic acids that catalyze protein synthesis do not work unless ATP available to be hydrolyzed as part of the overall process. This is, as we said above, the essence of reaction coupling. The energy from glucose oxidation (and the oxidation of other fuel molecules, like fats and proteins) sustains all life processes, but the energy is furnished indirectly through ATP^{4-}. Knowledge of energy relationships in living systems is crucial to understanding those systems, and that knowledge is built, in part, on measuring and interpreting reaction enthalpies. We will return to this discussion in the next chapters as we add to our understanding of thermodynamics and coupled reactions.

7.54. **Check This**   *Other pathways for ATP hydrolysis*

(a) In some coupling reactions ATP^{4-} hydrolysis takes another pathway:

$$ATP^{4-}(aq) + H_2O(l) \rightarrow AMP^{2-}(aq) + (O_3POPO_3)^{4-}(aq) + 2H^+(aq)$$
$$(O_3POPO_3)^{4-}(aq) + H_2O(aq) \rightarrow 2HOPO_3^{2-}(aq)$$

The enthalpy change for each of these hydrolysis reactions is about the same as the ATP^{4-} hydrolysis to give ADP^{3-} and phosphate, reaction (7.37). The sum of these two reactions is:

$$ATP^{4-}(aq) + 2H_2O(l) \rightarrow AMP^{2-}(aq) + 2HOPO_3^{2-}(aq) + 2H^+(aq)$$

What is $\Delta H°$ for this reaction combination? What advantage (if any) is there for an organism to use this combination pathway for ATP^{4-} hydrolysis compared to reaction (7.37). Explain.

(b) How might you explain why the enthalpy change for reaction (7.37) and for each of the two individual hydrolysis reactions in part (a) are about the same?

Reflection and projection

The standard enthalpy of formation of a compound is defined as the enthalpy of reaction for formation of the compound in its standard state from the most stable form of its elements in their standard states. The standard state is one bar pressure and the tabulated values are usually given for 298 K (25 °C). Enthalpies of formation are experimental values independent of any bonding model and can be combined to give accurate values for enthalpies of reaction. An example from the complex metabolism of glucose coupled to the formation of ATP^{4-} shows how much information you can get from enthalpies of formation and reaction, even in the absence of detailed knowledge of the actual reactions.

Sometimes, however, the way a reaction is carried out is a determinant of the changes we observe. This is particularly the case for reacting systems involving both heat and work. Up to this point, we have quantitatively analyzed only thermal energy transfers; next we will briefly consider systems that also involve mechanical energy transfers.

Section 7.10. Pressure-Volume Work, Internal Energy, and Enthalpy

The *sum of the potential and kinetic energy* in a collection of molecules is a state function called its **internal energy**. We have been using the symbol E to represent internal energy. When a change takes place in the collection of molecules, the resulting collection has a new value for its internal energy. Using the symbols E_i to represent the initial (beginning) state and E_f to represent the final (ending) state, we write the change in internal energy, ΔE, as:

$$\Delta E = E_f - E_i \tag{7.40}$$

When the internal energy of a collection of molecules changes, in a chemical reaction, for example, the energy difference ($E_f - E_i$ or ΔE) can always be expressed as some combination of thermal energy (heat) and mechanical energy (work). To satisfy the law of conservation of energy, all the energy must be accounted for. This leads directly to the mathematical statement for the **first law of thermodynamics**:

$$\Delta E = q + w \tag{7.41}$$

Thermal and mechanical energy that enter a system are positive; they increase the internal energy. Thermal and mechanical energy that leave a system are negative; they decrease the internal energy.

A practical example of the way the first law is applied is the internal combustion engine in an automobile, which is a reminder of why thermodynamics was developed in the first place. When the mixture of gasoline and air explodes inside an engine cylinder, as in Figure 7.19, the rearrangement of atoms to form new molecules produces a large decrease in internal energy. The value for ΔE is negative when energy is released in the reaction, as is the case here. Expanding gases move the piston; work (w) leaves the system of reacting gases to move the automobile. The rest of the energy released by the exploding gasoline–air mixture is converted to heat (q), which does not contribute to the motion of the automobile. One task for automotive engineers is to design engines that convert the highest possible percentage of the energy released to work.

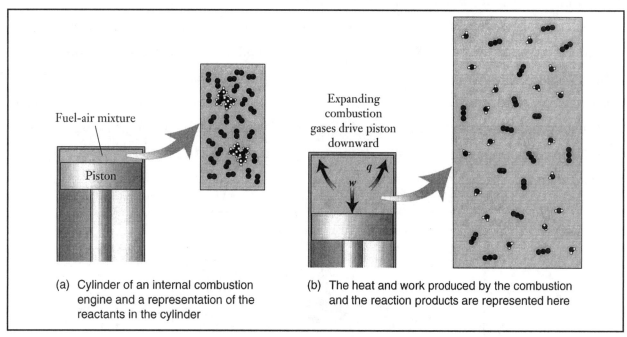

(a) Cylinder of an internal combustion engine and a representation of the reactants in the cylinder

(b) The heat and work produced by the combustion and the reaction products are represented here

Figure 7.19. Heat and work in the cylinder of an internal combustion engine.
Overall combustion reaction: $2C_8H_{18}(g) + 25O_2(g) \rightarrow 16CO_2(g) + 18H_2O(g)$

7.55.   Consider This   *How is the first law of thermodynamics used?*

Combustion in the cylinder of a small engine is found to produce 5 kJ of heat and 2 kJ of work during each power stroke. After re-designing the engine, engineers find that they can produce 3 kJ of work during each power stroke between the same initial and final states as in the original design. How much heat is produced during each power stroke in the re-designed engine? Explain the signs you assign to all the variables in solving this problem.

Definition of work

From physics, we have the definition of **work** (w) as the product of the force (F) acting on an object multiplied by the distance (d) the object is moved in the direction of the force:

$$w = F \times d \tag{7.42}$$

As we noted above, work, like thermal energy, has directionality. *Work entering a system of interest, is given a positive sign*, since it increases the energy of the system. The kind of work that most interested the 18th and 19th century thermodynamicists was the work of gases, steam in particular, pushing on pistons. In this section and in Section 7.13, we will discuss the work done by or on gases produced or consumed in chemical reactions and return to another important form of work, electrical work, in Chapter 10.

Constant volume and constant pressure reactions

When a reaction that produces gases is carried out in a closed container with rigid walls, the added gas increases the pressure inside the container, but nothing moves, because the gas is trapped in the container. This reaction is carried out at *constant volume*. In a constant volume system, no work is done by the gas, $w = 0$. By contrast, when the same reaction is carried out in a container with a movable piston as one wall, Figure 7.20, work is done pushing back against the constant external atmospheric pressure on the piston. The reaction is carried out at *constant pressure*; for the cases of most interest to us, this is the pressure of the atmosphere.

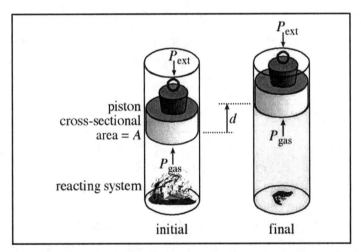

Figure 7.20. Constant pressure reaction apparatus.
A reaction that produces gas is carried out at constant pressure in a cylinder with a movable piston. A constant force on the piston, the external pressure, is represented by a weight sitting on the piston.

In order to determine the amount of work done in a constant pressure reaction, we'll analyze the changes that occur when the reaction occurs in the experimental set-up in Figure 7.20. The

> **WEB Chap 7, Sect 7.11.1**
> Work with an animation of a reaction carried out at constant P and at constant V.

pressure of the atmosphere, P_{ext}, acts like a weight resting on top of the piston. Before the reaction occurs, the absolute value of the pressure of gas inside the piston, $|P_{gas}|$, is equal to the

absolute value of the external pressure, $|P_{gas}| = |P_{ext}|$. When the reaction occurs, gas is produced in the cylinder, it pushes against the piston and raises the weight until, once again, $|P_{gas}| = |P_{ext}|$.

Pressure-volume work

To relate the pressure of the atmosphere, P_{ext}, to the force, F_{ext}, acting on the piston we need to know that pressure is defined as force per unit area:

$$P_{ext} = \frac{F_{ext}}{A}$$

Rearranging the definition gives the force:

$$F_{ext} = P_{ext} \times A \tag{7.43}$$

Before we can use this force to calculate the work, we need to define the directionality of the work we will calculate. We are interested in how much work is done *on* our reacting system, since work done on the system will increase its energy. The force acting on the system is the atmospheric pressure, P_{ext}. You can see in Figure 7.20 that P_{ext} is acting in a direction *opposite* to the movement of the piston when gas is produced. Therefore, the work done on the reacting system is negative:

$$w \text{ (on the reacting system)} = -F_{ext} \times d \tag{7.44}$$

Substituting the force from equation (7.43) into equation (7.44) gives:

$$w = -P_{ext} \times A \times d \tag{7.45}$$

The cross sectional area of the piston, A, times the distance the plunger moves, d, is the change in volume, ΔV, of the gas in the cylinder:

$$\Delta V = A \times d \tag{7.46}$$

Finally, we substitute ΔV from equation (7.46) into equation (7.45) to get the **pressure-volume work**:

$$w = -P_{ext} \times \Delta V \tag{7.47}$$

The sign of the pressure-volume work done on or by a system can be confusing. It takes work to raise(lift) a weight; work is being done *on* the weight to raise it. The only place for that work to come from is the reacting system. Therefore, *work has to leave the reacting system in order to raise the weight*. Indeed, equation (7.47) shows that when gas is produced, $\Delta V > 0$ (positive change in volume), work leaves the reacting system, $w < 0$. If a reaction uses up gas, $\Delta V < 0$ (negative change in volume), work enters the reacting system, $w > 0$. Next, we will look at how heat, pressure-volume work, and internal energy are related to enthalpy.

7.56. Worked Example   *Pressure-volume work at constant pressure*

A reaction was carried out at constant atmospheric pressure in a cylinder like the one in Figure 7.20. The volume of gas in the cylinder before reaction was 0.756 L and the volume of gas after the reaction was 0.983 L. The atmospheric pressure was one bar, 100. kPa. How much work was done on the reacting system during this change?

Necessary Information: To keep energy units straight, we need to know that 1 kPa·L = 1 J.

Strategy: Calculate the change in volume, ΔV, and use equation (7.47) to get the work.

Implementation:

$$\Delta V = (0.983\ \text{L}) - (0.756\ \text{L}) = 0.227\ \text{L}$$

$$w = -P_{\text{ext}} \times \Delta V = -(100.\ \text{kPa})(0.227\ \text{L}) = -22.7\ \text{J}$$

Does the Answer Make Sense? Since the volume is increased by the reacting system, work *leaves* the system, that is, the system does work in order to push back the piston. Work leaving the system is negative, so the sign of our result is correct and shows that work is done *by* not *on* the reacting system.

7.57. Check This   *Pressure-volume work at constant pressure*

For a reaction carried out at constant atmospheric pressure in a cylinder like the one in Figure 7.20, the final volume was found to be 0.521 L *less* than the initial volume. If the atmospheric pressure on the piston was 103 kPa, how much work was done on the reacting system?

Enthalpy revisited and defined

Whether a reaction is carried out at constant volume or at constant pressure, the change in internal energy, ΔE, is the same. Using the subscripts V and P to designate constant volume and constant pressure, respectively, we have:

$$\text{At constant volume:} \quad \Delta E = q_V + w_V = q_V + (-P\Delta V) = q_V - 0 = q_V \qquad (7.48)$$

$$\text{At constant pressure:} \quad \Delta E = q_P + w_P = q_P + (-P\Delta V) = q_P - P\Delta V \qquad (7.49)$$

For a constant volume system, $\Delta V = 0$, so $(-P\Delta V) = 0$. Equation (7.48) shows that the value of q_V is a direct measure of the change in internal energy. However, we have said that we are usually interested in reactions at constant pressure. When the thermal energy transfer, q_P, is measured for a reaction at constant pressure, a pressure-volume work term must be taken into account, if gases are involved as reactants and/or products. If gases are not involved, then ΔV will be nearly zero and the $-P\Delta V$ term in equation (7.49) will also be close to zero. In these cases $q_V \approx q_P$.

Recall that when we introduced enthalpy in Section 7.6, we said that we would explain why scientists devised this special function. The reason is that, even when gases are involved, chemists are almost never interested in pressure-volume work. The focus of chemistry is on how the energy stored in chemical bonds changes in the course of a chemical reaction. The work term

---

in the first law only makes things more complicated. The way to avoid dealing with the work term is to define a new thermodynamic quantity, and that is what was done with **enthalpy**:

$$H \equiv E + PV \tag{7.50}$$

The change in enthalpy, ΔH, for a reaction at constant pressure is:

$$\Delta H = \Delta E + P\Delta V \tag{7.51}$$

Equations (7.49) and (7.51) can be combined to give:

$$\Delta H = q_P \tag{7.52}$$

Equation (7.52) shows that q_P is a measure of the *change* in enthalpy. We get ΔH directly from an experimental measurement of the thermal energy transfer at constant pressure, as we have been doing throughout the chapter.

Also, we have been treating enthalpy as a state function (as when we sum the enthalpy changes around a cycle to zero in Figures 7.13 and 7.15). Now we see that enthalpy is defined in terms of variables that are state functions (internal energy, pressure, and volume), so enthalpy is a state function. We measure the enthalpy change in a system that undergoes a change by measuring q_P, the thermal energy transferred in a constant pressure pathway that brings about this change. The enthalpy change is the same no matter how the specified change in the system is carried out; ΔH is a function only of the initial and final states of the system. On the other hand, the thermal energy transferred in the change, q, is a path function and does depend upon the way the change is carried out. Only for a constant pressure pathway are the enthalpy change and thermal energy transfer the same.

7.58.   **Consider This**   *How can ΔE (or ΔH) be the same for a reaction at constant volume and constant pressure?*

Consider an exothermic reaction that produces a gaseous product from reactants in solution. When the reaction is run at constant volume, you find that the increase in temperature of the solution is greater than when the same amount of reaction occurs at constant pressure.

(a) Is q, the thermal energy transfer *to the reaction*, positive or negative? Explain the reasoning for your answer.

(b) Is q larger (more positive) for the reaction in the constant volume or constant pressure system? Explain the reasoning for your answer.

(c) What is the sign of the work, w, done *on* the system in the constant pressure reaction system? Explain the reasoning for your answer.

(d) Show how your answers in parts (b) and (c) can combine to make ΔE the same, whether the reaction is carried out at constant volume or at constant pressure. Show the same for ΔH.

Recall that the internal energy change is the change in the sum of potential and kinetic energies in going from reactants to products. The enthalpy change represents the thermal energy component of this change. To see this, consider that, if you subtract the pressure-volume work done on the system, $-P\Delta V$, from ΔE in equation (7.49), you get the enthalpy change, ΔH:

$$\Delta E - (-P\Delta V) = [q_P + (-P\Delta V)] - (-P\Delta V)$$

$$\Delta E + P\Delta V = q_P = \Delta H$$

Thus, you can understand why enthalpy is sometimes called heat content. To make these ideas more concrete, let's consider some chemical reaction systems.

7.59. Worked Example   *Comparing ΔE and ΔH*

Are ΔE and ΔH the same for the oxidation of methanol? If not, is work done *on* the reaction system or *by* the reaction system?

$$CH_3OH(l) + {}^3\!/_2O_2(g) \rightarrow CO_2(g) + 2H_2O(l) \tag{7.23}$$

Necessary Information: In addition to the balanced chemical equation (7.23) with the phases of the reactants and products given, we need the definition of work done on the system, $-P\Delta V$.

Strategy: Does a volume change occur in the reaction? If it does, then ΔE and ΔH are not the same and we use the sign of the volume change to determine whether work was done on or by the system.

Implementation: In reaction equation (7.23) there are more moles of gaseous reactants, ${}^3\!/_2$ mole, than moles of gaseous products, 1 mole. Thus, there is a change in volume: ΔE and ΔH are not the same.

Since the number of moles of gas decreases, ΔV is negative and $-P\Delta V$ is positive. Work is done on the system in this change.

Does the Answer Make Sense? In a reaction with a change in the number of moles of gas, the volume changes and ΔE and ΔH are not the same. A decrease in volume of the reacting system means work is done on it, as you saw earlier in this section.

7.60. Check This   *Comparing ΔE and ΔH*

From the following list, identify those chemical reactions where ΔE and ΔH are the same.

Where pressure–volume work is done, indicate whether work is done on the reaction system or by the reaction system:

   (a) $2H_2(g) + O_2(g) \rightarrow 2H_2O(l)$

   (b) $C_6H_{12}O_6(s) + 6O_2(g) \rightarrow 6CO_2(g) + 6H_2O(l)$

   (c) $CaCO_3(s) \rightarrow CaO(s) + CO_2(g)$

   (d) $H_3O^+(aq) + OH^-(aq) \rightarrow 2H_2O(l)$

7.61.   Worked Example   *Comparing ΔE and ΔH quantitatively*

   When one mole of methanol is oxidized, reaction equation (7.23), under standard conditions at a constant pressure of 1 bar, 1.2 kJ of work is done on the system. How do $\Delta E°_{reaction}$ and $\Delta H°_{reaction}$ compare for this reaction?

   Necessary information: We need $\Delta H°_{reaction} = -726.2$ kJ from Worked Example (7.49) and equation (7.49).

   Strategy: Substitute the heat, q_p $(= \Delta H°_{reaction})$ and the work, w_p, into equation (7.49) to get $\Delta E°_{reaction}$.

   Implementation:

   $$\Delta E°_{reaction} = q_p + w_p = \Delta H°_{reaction} + w_p = -726.2 \text{ kJ} + 1.2 \text{ kJ} = -725.0 \text{ kJ}$$

   We see that $\Delta E°_{reaction} \approx \Delta H°_{reaction}$. The values differ by less than 0.2%.

   Does the answer make sense? The thermal energy transfer from the reacting system is quite large compared to the work done on the system. The work term has little affect on the relative values of the energy and enthalpy changes. This is a general result for most highly exothermic or endothermic reactions, even when gases are involved: $\Delta E°_{reaction} \approx \Delta H°_{reaction}$.

7.62.   Check This   *Comparing ΔE and ΔH quantitatively*

   (a) When one mole of 1,2-dibromoethane decomposes to give ethene and gaseous bromine, under standard conditions at a constant pressure of 1 bar, 2.5 kJ of work is done by the system. How do $\Delta E°_{reaction}$ and $\Delta H°_{reaction}$ compare for this reaction?

   $$CH_2BrCH_2Br(g) \rightarrow CH_2CH_2(g) + Br_2(g)$$

   Hint: This is the reverse of reaction (7.31) in Check This 7.51.

   (b) How good is the approximation, $\Delta E°_{reaction} \approx \Delta H°_{reaction}$, for this reaction system? Under what conditions will the approximation not be valid? Explain your reasoning.

When there is no change in gas volume, $\Delta V = 0$, little pressure–volume work is done either on or by the system. In this case, $\Delta E \approx \Delta H$; the internal energy change and the enthalpy change are essentially the same. Also, when the thermal energy transfer is large, we can usually equate the energy and enthalpy changes. In Section 7.13, you can investigate the difference between a chemical change that is carried out under constant volume and constant pressure conditions.

Reflection and projection

Changes in the internal energy of a system are a result of heat and work transfers to the system (E increases) or from the system (E decreases). The sum of the heat and work transferred to or from a system is its change in internal energy, ΔE, which can be determined by measuring the transfer of thermal energy to or from the system in a constant volume change. For constant pressure systems, we must also account for the pressure-volume work to determine the changes in E. Pressure-volume work transfers are positive if the system is compressed or the number of moles of gas decreases (the volume decreases) and negative if the system expands or the number of moles of gas increases (the volume increases).

Enthalpy, H, was introduced in order to avoid dealing with pressure-volume work. ΔH can be determined by measuring the transfer of thermal energy to or from the system in a constant pressure change. Enthalpy, like energy, pressure, volume, and temperature, is a state function; the change in enthalpy for a change in a system depends only on the initial and final states of the system. For most reactions of practical importance, like combustions or reactions in solution, the difference between ΔH and ΔE is so small that it usually can be neglected; that is, $\Delta H \approx \Delta E$.

The energetics of reactions are critically important for understanding and using them, but the energy (or enthalpy) doesn't tell us everything, as we will remind you in the next section.

Section 7.11. What Enthalpy Doesn't Tell Us

Is it really possible to obtain enough energy from half an ounce of coal (14 g) to move two tons a mile? The citation from Emerson at the beginning of the chapter says that it is. We don't have enough information about *The Rocket*, Stephenson's locomotive and the loads it pulled to answer this question directly, but we can use what we know about modern engines to get some idea whether it was really possible. This is what we often have to do, substitute a problem (pathway) we know how to solve for another that we don't have enough information to solve.

There are small automobiles that can travel about 50 miles on the energy of combustion from one gallon (3.7 L) of gasoline or one mile on 74 mL. Assume that the car is a two-ton load (an overestimate) and its engine is our "locomotive." *The engine is our system*. Engines run in cycles and return to the same state at the end of each cycle. The internal energy of an engine is the same at the end of a mile of travel as it was at the beginning; $\Delta E = 0$ for the engine. Therefore, from the first law of thermodynamics, the maximum amount of work we can get *out* of an engine to move a load is equal to the amount of heat that we put into the engine from the fuel it burns, $q = w$. Table 7.6 provides a comparison between the heat available to *The Rocket* and the automobile.

Table 7.6. Comparison of thermal energy available to *The Rocket* and an automobile.

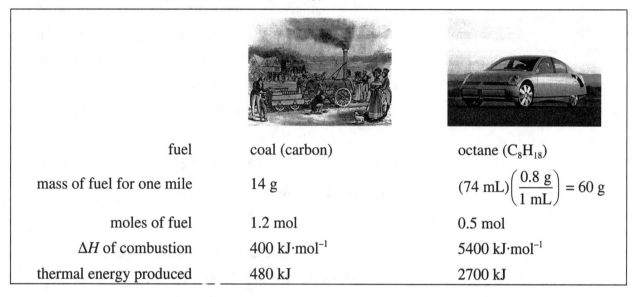

fuel	coal (carbon)	octane (C_8H_{18})
mass of fuel for one mile	14 g	$(74 \text{ mL})\left(\dfrac{0.8 \text{ g}}{1 \text{ mL}}\right) = 60 \text{ g}$
moles of fuel	1.2 mol	0.5 mol
ΔH of combustion	400 kJ·mol^{-1}	5400 kJ·mol^{-1}
thermal energy produced	480 kJ	2700 kJ

If *The Rocket* had been as fuel efficient as a modern car, it would have required $(2700/480) \times 0.5$ oz = 3 oz of coal to move the load a mile. Steam engines, especially the earliest ones were not nearly this efficient, but even if the difference was a factor of ten, only 30 oz, about two pounds, of coal would have been required to move two tons, 4000 pounds, a mile. The steam engine was a marvelous invention. Forgive Emerson for stretching the marvel a bit.

Our comparison of thermal energies in Table 7.6 is valid, but the statement that the amount of work available is equivalent to the thermal energy of fuel combustion is not. Not all the heat (undirected molecular motion) that goes into an engine can be converted to work (directed motion). The first law of thermodynamics, conservation of energy, does not provide any explanation for this limitation on conversion of one form of energy to another. What other puzzles are posed by the first law?

In Section 7.9, we emphasized coupling endothermic to exothermic reactions, in order to drive the energy-requiring reactions in living systems. This emphasis might lead you to believe that the criterion for a reaction to occur is that it be exothermic. When stronger chemical bonds replace weaker ones, the products *are* more stable (lower potential energy), and it seems reasonable that a chemical reaction should be likely. Many observed reactions are exothermic, but many endothermic reactions also occur. You have seen several examples in this and in previous chapters.

We observe changes taking place around us all the time. We call many of these changes

> In everyday use, "spontaneous" usually means impulsive or arising without forethought and often carries the implication of happening quickly. In science, spontaneous simply indicates that a change is possible or feasible and can occur.

spontaneous, which in chemical terms means that the change is possible and can occur without apparent external cause. The process may be slow, but, given enough time, it does occur. *Speed of change is not a criterion for spontaneity.* Two examples of spontaneous change are dissolving ammonium chloride, $NH_4Cl(s)$, or calcium chloride, $CaCl_2(s)$, in water (Chapter 2, Investigate This 2.20). In these cases, if you just add the solute to some water, the dissolution occurs without any further intervention on your part. The enthalpy change for dissolving ammonium chloride in water is positive:

$$NH_4Cl(s) \rightarrow NH_4^+(aq) + Cl^-(aq) \qquad\qquad \Delta H° = +14.8 \text{ kJ} \qquad\qquad (7.53)$$

The enthalpy change for dissolving calcium chloride in water is negative:

$$CaCl_2(s) \rightarrow Ca^{2+}(aq) + 2Cl^-(aq) \qquad\qquad \Delta H° = -82.9 \text{ kJ} \qquad\qquad (7.54)$$

The sign of the enthalpy change is reversed when a reaction is written in reverse. An exothermic reaction going in reverse is endothermic; the initial and final states are reversed. Since both endothermic and exothermic reactions can be spontaneous, the sign of ΔH can't be used to predict the way reactions actually go; enthalpy alone does not enable us to understand or predict the direction of spontaneous chemical change. However, understanding the directionality of change is essential to explaining, for example, the driving forces in the organization of living cells. To address the problem of directionality of change, we have to account for changes in both the system *and* its surroundings. And we need another thermodynamic state function — *entropy* — the subject of the next chapter.

Section 7.12. Outcomes Review

Everything you do, including thinking, requires energy. In this chapter we examined various forms of energy and transfers of energy between chemical and physical systems. We learned that the first law of thermodynamics is a statement of the principle of conservation of energy. Changes in the internal energy of a system, ΔE, are combinations of thermal and mechanical energy (heat and work) transfers to or from the system. We found that the same change in internal energy could be brought about by different heat and work transfers. A change in internal energy is a function only of the initial and final states of the system, but heat and work transfers depend upon the path taken from the initial to the final state of the system.

We introduced a new state function called enthalpy, H, that accounts for the thermal energy component of an internal energy change when pressure-volume work is involved in the change. Calorimetric measurements at constant pressure provide a measure of ΔH for reactions. We can use these enthalpy data (among others) to get bond enthalpies and enthalpies of formation for compounds. Using these values, we can calculate the enthalpy change for essentially any reaction. We cannot, however, use energy to predict the direction of changes; we will pursue that goal in the next chapter.

Check your understanding of the ideas in the chapter by reviewing these expected outcomes of your study. You should be able to:

• identify the forms of energy transferred in physical and chemical changes and show how energy is conserved in the changes [Sections 7.1, 7.2, 7.3, 7.4, and 7.10].

• draw molecular level representations of thermal energy (undirected kinetic energy) and mechanical energy (directed kinetic energy) transfers [Sections 7.2, 7.3, and 7.10].

• identify whether a thermal energy transfer occurs by radiation, conduction, and/or convection [Section 7.3].

• define and identify the variables that are functions of state and those that are functions of the path for a given change [Sections 7.4 and 7.10].

• identify the system and the relevant surroundings for a given change [Section 7.5].

• define and identify open, closed, and isolated systems [Section 7.5].

• use the data from calorimetric measurements to calculate the thermal energy transferred to or from a reacting system [Section 7.6].

• use the data from constant pressure calorimetric measurements to calculate the enthalpy change for the reacting system [Section 7.6].

• define and give examples of homolytic bond cleavage reactions [Section 7.7].

• write equations for homolytic bond cleavage and homolytic bond formation for compounds, use bond enthalpies to calculate the enthalpy changes associated with these processes, and obtain the enthalpy change for gas phase reactions [Section 7.7].

• draw enthalpy level diagrams that show how bond enthalpies combine to give the enthalpy change for a reaction [Section 7.7].

• use bond enthalpies to predict whether, for given atoms, reactions will favor singly-bonded or multiply-bonded products [Section 7.7].

• define standard states for elements and compounds and write the equations whose enthalpy changes are the standard enthalpies of formation of the compounds [Section 7.8].

• use standard enthalpies of formation to calculate the standard enthalpy change for a reaction [Section 7.8].

• draw enthalpy level diagrams that show how standard enthalpies of formation combine to give the standard enthalpy change for a reaction [Section 7.8].

• define coupled reactions and identify examples based on the definition [Section 7.9].

• show whether coupling between two reactions would be an energetically favorable combination [Section 7.9].

• state the first law of thermodynamics in terms of internal energy, heat, and work and use it to analyze a change that occurs by different pathways [Section 7.10].

• explain the difference between ΔE and q_V and between ΔH and q_P [Section 7.10].

• calculate ΔE, given ΔH and appropriate pressure and volume or $P\text{-}V$ work data for a reaction, and *vice versa* [Section 7.10].

• determine whether a process is consistent with (allowed) by the first law of thermodynamics [Section 7.11].

• use the kinetic-molecular model of gases to explain the observed effects of changes in P, V, n, or T [Section 7.13].

• calculate the final value for P, V, n, or T, given their initial values and the changes in three of the variables [Section 7.13].

• use the ideal gas equation to calculate the $P\text{-}V$ work done on or by a chemical reaction system [Section 7.13].

Section 7.13. EXTENSION — Ideal Gases and Thermodynamics

7.63. Investigate This   *What happens when a gas is compressed or heated?*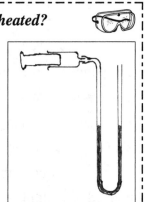

Do this as a class investigation and work in small groups to discuss and analyze the results. Draw about 40 mL of air into a 50-mL plastic syringe and connect the outlet of the syringe to a U-tube made of plastic tubing that is about half full of colored water (for visibility). Adjust the syringe plunger so the liquid levels are the same in both sides of the U-tube, as shown in the illustration.

(a) Push the syringe plunger in to decrease the volume of air in the syringe by several milliliters. Record your observations on the liquid levels in the U-tube.

(b) Readjust the syringe plunger until the U-tube liquid levels are the same again. While holding the volume constant, warm the syringe by holding it in your hand or directing a warm stream of air onto it. Record your observations on the liquid levels in the U-tube.

7.64. Consider This   *What causes changes when a gas is compressed or heated?*

(a) How, if at all, did the liquid levels in the U-tube change when you compressed the air (decreased the volume) in the syringe in Investigate This 7.63? What change(s) in the properties of the gas sample was(were) responsible for any change in the liquid levels? Explain the connection clearly.

(b) How, if at all, did the liquid levels in the U-tube change when you warmed the air in the syringe? What change(s) in the properties of the gas sample was(were) responsible for any change in the liquid levels? Explain the connection clearly.

In Investigate This 7.63, the liquid-containing U-tube is a **manometer**, a device for measuring gas pressures. When the liquid levels are the same in both arms of the manometer, the gas pressures pushing down on the liquid surfaces are the same, as shown in Figure 7.21(a). When the levels are unequal, a higher gas pressure is pushing down on the liquid in the arm with the lower level. This higher pressure is equal to the pressure of gas pushing down on the other arm plus the pressure of the column of liquid that is above the lower level, Figure 7.21(b).

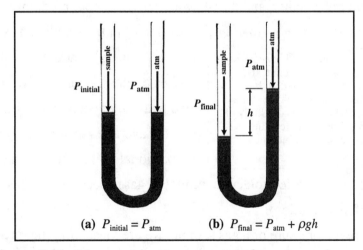

(a) $P_{initial} = P_{atm}$ (b) $P_{final} = P_{atm} + \rho gh$

Figure 7.21. Manometric measurement of a gas sample.

The manometer in Investigate This 7.63 is represented here. (a) shows the equal pressures of the sample and the atmosphere when the liquid levels are equal. (b) shows how the pressure of the sample is determined when the liquid levels are not equal. ρgh is the pressure in kPa due to the column of liquid with a density of ρ g·mL^{-3} and height of h meters acted on by gravitational acceleration, $g = 9.80$ m s^{-2}.

7.65. Consider This   *How are manometer readings interpreted and quantified?*

Suppose you have set up a syringe and manometer as shown in the illustration in Investigate This 7.63. After moving the syringe plunger you observe that the liquid level in the manometer arm attached to the syringe is 15.7 cm higher than the level in the open arm.

(a) What conclusion can you draw about the gas pressure in the syringe? What motion of the syringe plunger produced this pressure? Explain your responses.

(b) If the atmospheric pressure in the room is 100.4 kPa, what is the pressure of the gas in the syringe? *Hint:* The density of liquid (water) in the manometer is 1.00 g·mL^{-3}.

Kinetic-molecular model of gases

In Investigate This 7.63, you found that the level of liquid in the arm of the manometer connected to the syringe went down when the gas was compressed or warmed. Both compression and warming of a gas sample increase its pressure. Why? To answer this question we use a molecular-level model of gases. We have been using the model implicitly since Chapter 1, where we presented illustrations and animations (in the *Web Companion*) of molecules in the gas phase. In this section, we are going to use the model to explain the observed behavior of gases, which we will describe by the ideal gas equation. Then we will apply the equation to a chemical reaction occurring at either constant volume or constant pressure and see how it can account quantitatively for the results.

In our model for gases, we assume that molecules in the gas phase are far apart, so they do not often interact with one another, and that they are moving about randomly. We assume that the molecules act like hard spheres (marbles, for example) in their collisions with the walls of

> *Kinetic* and related terms (from Greek *kineein* = to move) also appears in many other contexts: kinetic energy is the energy associated with motion; kinematics is the branch of physics that deals with motion; kinesthesiology is the study of muscular movement in animals; and telekinesis is the yet-to-be-proved movement of objects by use of thought alone.

their container. In these the collisions a molecule pushes on the wall and the wall pushes back with an equal and opposite push. Finally, we assume that the average kinetic energy of the molecules is proportional to the temperature of the gas (in kelvin). This is called the **kinetic-molecular model** of gases because it is based on molecular motion ("kinetic" refers to motion).

When we measure the pressure of a gas sample, we are measuring the effect of the collisions of the moving gas molecules with the walls of their container, including our measuring device, such as the surface of the liquid in a manometer. Two factors can cause the pressure to change. First, the number of molecules colliding with the walls in a given time might change. The number of molecules hitting the walls per unit time is proportional to the number of molecules

---

per unit volume of the gas. Second the average speed with which the molecules strike the walls might change. If the average kinetic energy of the molecules changes, their average speed changes.

7.66. Consider This   *How does kinetic-molecular theory apply to Investigate This 7.63?*

(a) How does the number of molecules per unit volume of gas change in part (a) of Investigate This 7.63? Are your manometric observations consistent with this change? Use the kinetic-molecular model of gases to explain your response.

(b) How does the average kinetic energy of the gas molecules change in part (b) of Investigate This 7.63? Are your manometric observations consistent with this change? Use the kinetic-molecular model of gases to explain your response.

(c) Imagine that you remove some of the molecules from a flask containing a gaseous sample (keeping the temperature constant). If the flask is connected to a manometer, what would you expect to observe? Use the kinetic-molecular model of gases to explain your response.

The ideal-gas equation

Your results in Investigate This 7.63 and your analyses in Consider This 7.66 provide you the direction and the kinetic-molecular explanation for pressure changes that occur when you change the volume, temperature, and number of molecules in a gaseous sample. You found that the pressure increases when the volume decreases (an inverse relationship), that the pressure increases when the temperature increases (a direct relationship), and that the pressure will decrease if the number of molecules in the sample decreases (a direct relationship). We can show these relationships in a tentative equation that combines them all:

$$P \propto \frac{nT}{V} \tag{7.55}$$

Here, n is the number of moles of gas, a measure of the number of molecules in the sample.

7.67. Check This   *Relationships among the properties of a gas sample.*

Explain how equation (7.55) embodies the relationships in the preceding paragraph.

Equation (7.55) was derived from quantitative studies of gases begun in the 17th century and was a part of the evidence used to develop the kinetic-molecular model during the 19th century. Equation (7.55) implies that the effects of changes in the properties of a gas sample are independent of one another and, in fact, independent of the identity of the gas. The kinetic-molecular model for gases is consistent with this independence. The model treats all gases as hard spheres, so the identity of the gas makes no difference. The model assumes that the average kinetic energy is the same for all samples of gas molecules at the same temperature. The kinetic energy does not depend on the size of the sample or the number of molecules per unit volume.

The relationships among the variables in equation (7.55) are usually written in this form:

$$PV = nRT \qquad\qquad (7.56)$$

Equation (7.56) is called the **ideal-gas equation** and the proportionality constant, R, is called the **gas constant**. The value of the gas constant is 8.314 J·mol^{-1}·K^{-1}, when pressure is given in kPa

> Pressure measurements are still often given in atmospheres, atm, so another useful value for R is 0.08206 L·atm·mol^{-1}·K^{-1}.

and volume in liters. The ideal-gas equation is an approximation to the behavior of real gases.

Molecules are not hard spheres, but are somewhat squishy, so collisions between them and with surfaces are not ideal. Molecules in the gas phase are not entirely independent of one another: they take up space and attract one another by polar and dispersion forces. The attractions depend on the distance between molecules and the time they spend near one another, so they are less important when there are few molecules per unit volume (low pressure) or moving rapidly (high temperature). The ideal-gas equation provides a good approximation for the behavior of most gases near or below atmospheric pressure and at room temperature or above.

7.68. Worked Example *Using the ideal-gas equation*

The pressure in an automobile tire reads 38 psi (pounds per square inch) on a tire gauge at the start of a trip when the tire is at 21 °C. At the end of the trip, the tire has warmed to 42 °C. Assuming that the volume of the tire does not change and that no air leaves the tire, what is the air pressure in the tire at the end of the trip? What would the tire gauge read?

Necessary information: We need to know that a tire gauge reads the pressure of the tire *above* atmospheric pressure, which is about 15 psi. We need the ideal-gas equation and the temperature in kelvin, which we get by adding 273 degrees to the Celsius temperature.

Strategy: Convert the temperatures to kelvin, calculate the initial pressure of air in the tire, and apply the ideal-gas equation to the initial and final conditions to find the final tire pressure. Since the number of moles of gas in the tire and the volume of gas in the tire do not change during the trip, one way to do this problem is to rearrange the ideal-gas equation to put all the constant terms on one side. Since these terms do not change, we can equate the initial and final values of the variable side of the equation and solve for the unknown value.

Implementation:

     initial temperature: 21 °C + 273° = 294 K

---

final temperature: 42 °C + 273° = 315 K

initial pressure: 38 psi + 15 psi = 53 psi

Rearranging the ideal-gas equation to group the constant terms on the right, gives:

$$\frac{P}{T} = \frac{nR}{V} = \text{a constant}$$

Therefore:

$$\frac{P_i}{T_i} = \frac{P_f}{T_f} = \frac{53 \text{ psi}}{294 \text{ K}} = \frac{P_f}{315 \text{ K}}$$

$P_f = 57$ psi   The tire gauge will read 42 psi (= 57 psi − 15 psi)

Does the answer make sense? You know from Investigate This 7.63 that the pressure of a sample of gas increases when it is warmed up (at constant volume). This is also the result of this calculation, so it makes sense. The Kelvin temperature increased by about 7% ($\approx 21/304$) and the pressure also increased by about 7% ($\approx 4/55$), as we expect from their direct relationship. Note that we can express the pressure in any units that are convenient for the problem, as long as only initial and final ratios are involved.

7.69.   Check This   *Using the ideal-gas equation*

Imagine that you do an experiment like the one in Investigate This 7.63, but instead of a U-tube manometer, you use an electronic pressure sensor interfaced to a computer to measure the pressure. If you start with the syringe plunger at a volume of 44.7 mL at 100. kPa, to what volume will you have to move the plunger to get a gas pressure of 150. kPa.

7.70.   Worked Example   *Using the ideal-gas equation*

A 525 mL sample of gas initially at 167.4 kPa and 212 °C underwent a change to 101.3 kPa and 37 °C. What is the final volume of the gas?

Necessary information: We need the ideal-gas equation and the conversion from °C to K.

Strategy: Convert the temperatures to kelvin and use an approach similar to Worked Example 7.68. Rearrange the ideal-gas equation to put all the constant terms on one side, equate the initial and final values of the variable side of the equation, and solve for the unknown value.

Implementation: In this problem, the number of moles of gas does not change, so write the ideal gas equation as:

$$\frac{PV}{T} = nR = \text{a constant}$$

Therefore, $$\frac{P_iV_i}{T_i} = \frac{P_fV_f}{T_f}$$

and $$V_f = \left(\frac{P_i}{P_f}\right)\left(\frac{T_f}{T_i}\right)V_i = \left(\frac{167.4 \text{ kPa}}{101.3 \text{ kPa}}\right)\left(\frac{310 \text{ K}}{485 \text{ K}}\right)(525 \text{ mL}) = 555 \text{ mL}$$

Does the answer make sense? Consider the changes in pressure and temperature separately and then combine the effects. If the pressure on a gas is *decreased*, its volume will *increase* in inverse proportion. In this case, the pressure is reduced to about 3/5 of its initial value, which would result in an increase in volume by a factor of about 5/3. If the temperature of a volume of gas is *decreased*, its volume will *decrease* in direct proportion. In this case, the temperature is reduced to about 3/5 of its initial value, which would result in a decrease of the volume by a factor of 3/5. The pressure and temperature effects on the volume are in opposite directions and just about cancel one another. The volume should not change very much and this is the result we got, so it makes sense.

7.71. Check This   *Using the ideal-gas equation*

(a) To a sample of gas at 57.8 kPa and 289.2 K in a rigid (constant volume) container is added a second gas. The final pressure and temperature of the gas mixture are 95.8 kPa and 302.7 K, respectively. What is the *ratio* of the number of moles of added gas to the number of moles of gas originally present? *Hint:* All ideal gases act the same. A mixture of ideal gases can be treated as if all the molecules are the same.

(b) If the container in part (a) has a volume of 547 mL, how many moles of each gas are present in the final mixture? Explain how you get your answer.

Analysis of constant volume and constant pressure processes

The reason for introducing gas behavior and the ideal-gas equation is to allow us to use them to compare ΔE and ΔH quantitatively for reactions carried out at constant volume and constant pressure. These analyses are applicable only to reactions that involve gases as reactants and/or products. For many other reactions, as we showed in Section 7.10, $\Delta E \approx \Delta H$.

7.72. Investigate This   *Do reactions in open and capped containers differ?*

Do this as a class investigation and work in small groups to discuss and analyze the results. Use a plastic soft drink bottle, a balance, a watch or timer, a plastic syringe with a short needle, two ±0.1 °C thermometers that read from 0 to 100 °C, and a 24/40 ribbed rubber septum. Solid

---

sodium hydrogen carbonate (baking soda, NaHCO₃) and 6 M hydrochloric acid (HCl) solution
are the reactants in this system. The set up for the investigation is shown in the photograph.

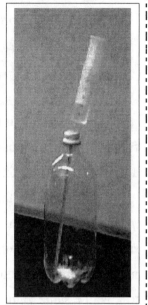

   (a) Add 6.8 g of NaHCO₃(s) to the plastic bottle. Place one
thermometer in the bottle, with the thermometer bulb at the bottom. Use
the other thermometer to measure the temperature of a solution of 6 M
HCl(aq) to the nearest 0.1 °C. Use the syringe to dispense 20.0 mL of the
HCl solution into the plastic bottle. Record temperature and time as you
swirl the mixture to be sure all the solid contacts the acid solution. While
continuing to mix, record the temperature of the mixture approximately
every ten seconds for about two minutes.

   (b) Repeat the experiment with these changes: After the NaHCO₃(s)
and thermometer are in the bottle, seal the bottle with a septum cap. Put
the syringe needle through the septum cap to inject the acid into the bottle
and then remove the needle. *IMPORTANT*: When the reaction is complete
and you have made all your observations, remove the needle from the syringe. Insert the needle
through the septum cap to release gas pressure from inside the bottle.

7.73.   Consider This   *Why do reactions in open and capped containers differ?*

   Hydrogen carbonate ion reacts with hydronium ion to produce carbon dioxide gas and water.
The reaction with solid sodium hydrogen carbonate (baking soda) is:

$$NaHCO_3(s) + H_3O^+(aq) \rightarrow CO_2(g) + Na^+(aq) + 2H_2O(l) \tag{7.57}$$

   (a) In Investigate This 7.72, what was the maximum change in temperature (from the starting
temperature of the acid) in the open bottle? in the capped bottle? Were these results surprising?
Why or why not?

   (b) Which one of the reactions in Investigate This 7.72, was carried out at constant volume?
at constant pressure?

   (c) Does your answer to part (b) help you interpret the results you reported in part (a)? Why
or why not?

   You found that the temperature change when reaction (7.57) is carried out in an uncapped
container is different from the temperature change for the reaction carried out in a capped
container; the thermal energy transfers are different in the uncapped and capped containers. The
reaction that occurs is the same in both containers, so the enthalpy change, ΔH, must be the same

---

in both cases. This is also true for ΔE. Since energy is conserved, some other form of energy must also be different in the two cases. What is different, as you saw in Section 7.10, is the work.

Reaction (7.57) produces carbon dioxide gas. In the capped container, the added gas

> **WEB Chap 7, Sect 7.11.1**
> Use an interactive animation of
> Investigate This 7.72 to help
> analyze the changes.

increases the pressure inside the container, but nothing moves, because the gas is trapped. This reaction is carried out at constant volume; no work is done on or by the reacting system. When the reaction is carried out in the open container, the carbon dioxide pushes back the atmosphere as it fills the bottle and pushes the air out. The reacting system does work pushing back the atmospheric gases. The reaction is carried out at constant atmospheric pressure.

You have found that reaction (7.57), the hydrogen carbonate-acid reaction, is endothermic; q_V and q_P are both positive quantities. The thermal energy transfer is larger when the reaction is run at constant pressure: $q_P > q_V$. Since the reaction produces gas, ΔV is positive and $w = -P\Delta V$ is negative. The heat and work terms in equation (7.49) must combine (work canceling some heat) to give the same value for ΔE as is obtained in equation (7.48) for the constant volume reaction:

$$\Delta E = q_V = q_P + w = q_P - P\Delta V \tag{7.58}$$

The analysis here is similar to what you did in Consider This 7.58. The directions (signs) of the quantities in equation (7.58) are shown in Figure 7.22. To find out whether the relationships among q_V, q_P, and w shown in the figure (and asserted in Consider This 7.58) are true, we need to determine their numeric values.

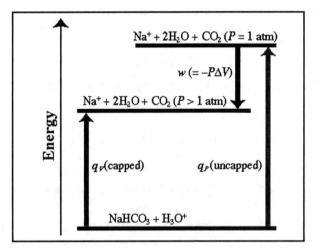

Figure 7.22. Relationships among q_V, q_P, and w for reaction (7.57).

7.74.   Consider This   *What are q_V and q_P for the hydrogen carbonate-acid reaction?*

Use your liquid volume and temperature data from Investigate This 7.72 (or the data from the

---

Web Companion) to estimate values of q_V and q_P for reaction (7.57). Assume that no thermal energy is transferred to or from the contents of the bottle during the reactions. Treat this setup just like the calorimeters discussed in Section 7.6 and use your data to do calculations like the ones there. In the absence of any other information, assume that the specific heat of the solution is the same as for water.

Pressure-volume work calculation

Your calculations in Consider This 7.74 provide estimates of the numeric values of q_V and q_P for reaction (7.57), but to test equation (7.58) we still need a value for the work term, $-P\Delta V$. The easiest approach is to use the ideal gas equation, $PV = nRT$, to describe the gas, because we can then express the pressure-volume work in terms of variables we already know, the temperature and the number of moles of reactants and products. Write the ideal gas equation for the initial conditions (number of moles of reactant gases, n_i) and the final conditions (number of moles of product gases, n_f):

$$\text{Initial:} \quad PV_i = n_i RT$$

$$\text{Final:} \quad PV_f = n_f RT$$

Subtracting the initial conditions from the final conditions yields:

$$P(V_f - V_i) = (n_f - n_i)RT$$

$$P\Delta V = (\Delta n)RT \tag{7.59}$$

Finally, substituting for $P\Delta V$ in equation (7.58) gives:

$$q_V = q_P - (\Delta n)RT \tag{7.60}$$

7.75. Worked Example   *Converting q_P (or ΔH) to q_V (or ΔE)*

In Check This 7.51, you calculated $\Delta H°_{reaction} = -126.3$ kJ for bromination of one mole of ethene at 298 K:

What is $\Delta E°_{reaction}$ at 298 K for this reaction?

Necessary Information: We need to keep in mind that q_V and q_P are, respectively, measures of ΔE and ΔH.

Strategy: Determine Δn for the reaction and substitute in equation (7.60).

Implementation: If one mole of gaseous ethene reacts, one mole of gaseous bromine also reacts and one mole of gaseous product is formed. Thus two moles of gaseous reactants go to one mole of gaseous product and $\Delta n = (1 \text{ mol}) - (2 \text{ mol}) = -1 \text{ mol}$.

$$q_V = q_P - (-1 \text{ mol})(8.314 \text{ J·mol}^{-1}\text{·K}^{-1})(298 \text{ K}) = q_P + 2.5 \text{ kJ}$$

Substitute $\Delta E^\circ_{\text{reaction}} = q_V$ and $\Delta H^\circ_{\text{reaction}} = q_P$ to get:

$$\Delta E^\circ_{\text{reaction}} = -126.3 \text{ kJ} + 2.5 \text{ kJ} = -123.8 \text{ kJ}$$

Does the Answer Make Sense? The reaction is exothermic, so thermal energy leaves the reacting system. Since the number of moles of gas decreases in the reaction, the volume change is negative, $\Delta V < 0$. If the change occurs at constant pressure, work is done on the reacting system and mechanical energy is added to the system. For a change at constant volume, no work is done on or by the system, but the change in internal energy is the same as for the constant pressure process. Since no work is added to the constant volume system, less thermal energy has to leave the system for the same change in internal energy. This is the result we got: $|q_V| < |q_P|$. Compare the numerical values here with those in Check This 7.62; be sure you can explain the correlations. Draw a diagram, modeled after Figure 7.22, to show the relationships among q_V, q_P, and w for this system.

7.76. Consider This *What is w for the hydrogen carbonate-acid reaction?*

(a) How many moles of $NaHCO_3$ did you use in the reactions in Investigate This 7.72?

(b) How many moles of H_3O^+ (HCl) did you use in the reactions?

(c) How many moles of CO_2 were formed in the reactions?

(d) What is $P\Delta V$ for the reaction carried out at constant pressure?

(e) Within the uncertainties of the experimental measurements, are your results from part (d) and from Consider This 7.74 consistent with equation (7.58). Explain why or why not.

7.77. Check This *$\Delta E_{reaction}$ and $\Delta H_{reaction}$ for the hydrogen carbonate-acid reaction*

Values of $\Delta E_{\text{reaction}}$ and $\Delta H_{\text{reaction}}$ are usually reported for a mole of reactant undergoing the reaction. Use your results from Consider This 7.74 and the number of moles reacting from Consider This 7.76 to get $\Delta E_{\text{reaction}}$ and $\Delta H_{\text{reaction}}$ for one mole of $NaHCO_3$ reacting according to reaction equation (7.57). Explain your procedure.

Index of Terms

Chapter 7 Problems

Section 7.1.   Energy and Change

7.1.   A slice of cheese pizza typically contains 180 Calories of food energy. The human heart requires about 1 J of energy for each beat.

(a)   Assuming that when you metabolize the pizza, all of the food energy will be used to keep your heart beating, how many heart beats can this slice of pizza sustain? About how many minutes would this keep your heart going? *Hint*: 1 Calorie = 1 kcal.

(b) Assume the efficiency of use of this food energy is about the same as that calculated for capturing the energy from glucose oxidation, Section 7.9. What are your answers for part (a) in this case? Explain your reasoning.

7.2.   Pasta is largely starch and a deep-fried potato chip is mostly starch and fat.

(a)   What do your observations from Investigate This 7.1 tell you about the energy value of these foods? Do they combine exothermically with oxygen? What is the evidence and reasoning for your response?

(b)   Starch is a polymer of glucose. Is this composition consistent with your answers in part (a) and other data from Investigate This 7.1? Explain your reasoning.

7.3.   Predict how the results of Investigate This 7.1 would differ if a baked, rather than a fried, potato chip were used. What will happen if a raw potato slice is used? What do you predict will be observed if a chip using Olestra® is used? In each case, explain the reasons for your prediction.

Section 7.2.   Thermal Energy (Heat) and Mechanical Energy (Work)

7.4.   Several different types of energy have been discussed in this chapter. These include kinetic energy, mechanical energy (work), potential energy, and thermal energy (heat). How are these terms related?

7.5.   Figure 7.2 illustrates a model for the movement of combustion gases.

(a)   Draw a similar diagram for a tethered helium-filled balloon. *Hint:* Unlike Figure 7.2, there will only be one type of particle, representing helium atoms.

(b)   Will the helium atoms in the tethered balloon exhibit any directed motion? Why or why not? What will happen to the motion of the helium atoms in the balloon when the tether is released? Explain briefly.

7.6. Consider a drop of water in a waterfall. At the very top of the fall, the molecules in the drop are moving randomly. While it is falling (pulled down by gravity), its molecules have an overall downward motion. When it hits the pool at the bottom, the molecular motion again becomes random.

(a) The water at the bottom of the waterfall is a little warmer than the water at the top. What can you say about the change in internal energy of your drop of water as it goes from the top of the waterfall to the bottom?

(b) Has work been done or has thermal energy been transferred to cause the change in the internal energy of the water drop? Explain where the work and/or thermal energy come from.

7.7. When heat is added to H_2O to do work, why does the vapor form of water generate more work than either ice or liquid water?

7.8. Why is it important to you, as a consumer, for automotive engineers to design engines that convert the highest possible percentage of energy to work?

Section 7.3.   Thermal Energy (Heat) Transfer

7.9. For each case below, is thermal energy being transferred by radiation or by a contact process of conduction and/or convection? Briefly justify your choice.

(a) A silver spoon at room temperature, when placed in a cup of hot water, becomes too warm to touch.

(b) An apple pie is baked in an electric oven.

(c) The water in an outdoor swimming pool cools from 25 °C to 20 °C as the summer season changes into autumn.

(d) You sunbathe with your back exposed to the sun, until your back feels very warm.

7.10. At the end of a skating season, an indoor ice rink was closed. Without any outside cooling, how will the roof and the ice rink floor reach a common temperature? Will it be a slow or fast process? Explain your answer.

7.11. After being immersed in cold water, the core body temperature of an individual may be abnormally low. Design three methods, one using convection, the second using conduction, and the third using radiation to transfer heat to warm up an individual as quickly and safely as possible. Which method will be the most efficient in this case?

Section 7.4.   State Functions and Path Functions

7.12.   Your monthly bank statement gives the opening balance for your account, details the transactions that have occurred, and reports the ending balance. Which aspects of this statement correspond to state functions and which to path-dependent functions?

7.13.   The quarterback starts a play on his own 35 yard line, but is pushed back by the defense to his own 25 yard line. He then successfully completes a pass to the opponent's 40 yard line. What is the number of yards gained in the play? Is the number of yards gained a state function or a path-dependent function?

7.14.   Today's most advanced fossil-fuel burning power plants producing electricity operate at an efficiency of about 42%. Suggest some reasons why a higher percentage of the energy from burning the fossil fuels is not converted to work. Do you think it is theoretically possible to convert 100% of the energy into work? Explain your reasoning.

7.15.   Consider two different cases in which a fully charged battery becomes totally discharged.
    Case 1: The battery is used to provide power to a flashlight.
    Case 2: The battery is used to provide power for a child's toy car.
(a) Will the change from the battery being fully charged to being totally discharged be a state function or a path-dependent function in each case? Explain your reasoning.
(b) Will there be any thermal energy transfer by radiation or by contact in either use of the battery? Explain your reasoning?

7.16.   Consider a C-172 airplane (small private plane), a helicopter, a parachute jumper, and a hawk, all presently at 1892 feet above the Clemson, South Carolina airport, which is at an altitude of 892 feet. Each plans a landing at the airport. Comment on the state functions and path functions for each of these airborne objects.

Section 7.5.   System and Surroundings

7.17.   Which ending(s) for the following sentence is(are) correct? Explain your reasoning in each case. The enthalpy change for a system open to the atmosphere is
    (a)   dependent on the identity of the reactants and products
    (b)   zero
    (c)   negative
    (d) q_P
    (e)   positive

7.18. Give "real life" examples of the following:

    (a) an open system

    (b) a closed system

    (c) an isolated system

    (d) an exothermic process

    (e) an endothermic process

7.19. A properly stoppered Thermos® bottle containing coffee is nearly an isolated system. The stopper prevents water vapor from escaping, while the vacuum construction keeps heat from being lost to the surroundings. How would you turn this isolated system into an open system? A closed system?

7.20. The earth is sometimes described as a closed system, particularly when considering regional or global environmental problems. For example, acid rain may fall at great distance from the original site of emissions responsible for the observed effect. Ozone depletion near the South Pole is linked to the emission of fluorocarbons from uses in industrial nations. While this view does convey a sense of the many connections that affect environmental issues, is it correct to think of the earth as a closed system from the standpoint of thermodynamics? Why or why not?

7.21. This diagram shows an experiment in which a hot silver bar is added to a beaker of water at room temperature.

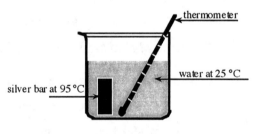

    (a) If the object of the experiment is to find the specific heat of the silver, define the system and the surroundings.

    (b) If you already know the specific heat of the silver, the same experimental set up might be used to determine if there is heat loss through the walls of the glass beaker or from the thermometer itself. Define the system and surroundings in that experiment.

    (c) Does this experimental diagram illustrate an open, closed, or isolated system? Give the reasoning behind your choice.

Section 7.6.   Calorimetry and Introduction to Enthalpy

7.22. In a calorimetric measurement, why might it be important to know the heat capacity of the calorimeter? The calorimeter heat capacity is the amount of thermal energy the calorimeter (container, thermometer, *etc.*) transfers per degree temperature change.

7.23. Choose the best response to complete the final sentence. Explain why you make this choice and what is wrong with each of the others. The equation for decomposition of gaseous ammonia, $NH_3(g)$, is:

$$NH_3(g) \rightarrow 1/2 N_2(g) + 3/2 H_2(g) \quad \Delta H = -45.9 \text{ kJ}$$

This reaction equation and enthalpy change indicate that the *formation* of gaseous ammonia

(a) evolves 45.9 kJ for each mole of ammonia formed.

(b) evolves 23 kJ for each mole of nitrogen used.

(c) absorbs 45.9 kJ for each mole of ammonia formed.

(d) absorbs 23 kJ for each mole of nitrogen used.

(e) is an exothermic process.

7.24. A 10.0 g sample of an unknown metal was heated to 100.0 °C and then added to 20.0 g of water at 23.0 °C in an insulated calorimeter. At thermal equilibrium, the temperature of the water in the calorimeter was 25.0 °C. Which of the metals in this list is most likely to be the unknown?

Specific Heat Data
$H_2O(l)$ 4.184 $J \cdot g^{-1} \cdot °C^{-1}$
$Au(s)$ 0.13 $J \cdot g^{-1} \cdot °C^{-1}$
$Ag(s)$ 0.22 $J \cdot g^{-1} \cdot °C^{-1}$
$Al(s)$ 0.90 $J \cdot g^{-1} \cdot °C^{-1}$

7.25. A student received 55.0 g of unknown metal from his laboratory instructor. To identify this metal, he decided to calculate its specific heat. He heated the metal sample to a temperature of 98.6 °C in a water bath. Then he transferred it slowly to a calorimeter containing 100. mL of water at 24.6 °C. The maximum temperature reached by the calorimeter system was 25.8 °C. His lab instructor, who was watching the experiment, explained to the student that he had introduced a large amount of error and asked him to repeat the experiment with the correct procedure. In the second trial, the temperature of the water bath was 98.4 °C and the initial temperature of water in the calorimeter was 24.3 °C. The maximum temperature after the sample was immersed in the calorimeter was 26.5 °C. What error had the student made? Explain how you know.

7.26. A student mixed 100. mL of 0.5M HCl and 100. mL of 0.5M of NaOH in a Styrofoam® cup calorimeter. The temperature of the solution increased from 19.0 °C to 22.2 °C. Is this process exothermic or endothermic? Assuming that the calorimeter absorbs only a negligible quantity of heat, that the density of the solution is 1.0 $g \cdot mL^{-1}$, and that its specific heat is 4.18 $J \cdot g^{-1} \cdot °C^{-1}$, calculate the molar enthalpy change for the reaction:

$$HCl(aq) + NaOH(aq) \rightarrow H_2O(l) + NaCl(aq)$$

7.27. Instant cold packs, such as those used to treat athletic injuries, contain ammonium nitrate and a separate pouch of water. When the pack is activated by squeezing to break the water pouch, the ammonium nitrate dissolves in water. This is an endothermic reaction and the change in enthalpy is 25.7 kJ·mol^{-1} of ammonium nitrate dissolved.

(a) Why does this endothermic reaction produce a cold sensation on your skin?

(b) The cold pack contains 125 g of water to dissolve 50.0 g of ammonium nitrate. What will be the final temperature of the activated cold pack, if the initial room temperature is 25 °C? Assume that the specific heat of the solution is the same as that for water, 4.184 J·g^{-1}·°C^{-1}.

(c) What other assumptions do you make in carrying out the calculation in part (b)?

(d) The final temperature is about 7 °C. Does this result justify the assumptions in parts (b) and (c)? If not, explain which assumption(s) might not be valid.

Section 7.7.   Bond Enthalpies

7.28. Ethanol, by itself or in blends, is an important alternative fuel for gasoline-powered engines. The molecular equation for the complete combustion of ethanol is:

$$C_2H_5OH(l) + 3O_2(g) \rightarrow 2CO_2(g) + 3H_2O(l)$$

(a) Rewrite the balanced equation using full Lewis structures for each reactant and product molecule.

(b) Use your model set and build a model of each reactant and product molecule in this reaction.

(c) Refer to the Lewis structures, the models, and Table 7.3 to determine the total enthalpy input that will be required to break all the bonds in the reactants.

(d) Refer to the Lewis structures, the models, and Table 7.3 to determine the total enthalpy that will be released in forming all bonds in the products.

(e) What is the net enthalpy change in this combustion reaction?

(f) Ethanol is also a fuel when consumed by humans. Will your result in part (e) also give an estimate for ethanol's fuel value in our bodies? Why or why not?

7.29. Consider this reaction in which two amino acids join to form a peptide bond, releasing a water molecule:

Use average bond enthalpies to estimate the enthalpy change for this reaction.

7.30. Ethanol, CH_3CH_2OH, and dimethyl ether, CH_3OCH_3, are isomers. Use average bond enthalpies to determine which of the isomers is the more stable. By how much? Clearly explain your reasoning and the calculations you do to get your result.

7.31. The glucose molecule can exist in a number of different forms, including the open-chain and cyclic forms represented below. By convention, the carbons in the chain are numbered 1 through 6 beginning at the top, as shown. Carbon-1 in the cyclic form is the one at the far right of the structure. To make the cyclic form, the oxygen bonded to carbon-5 in the linear form becomes the ring oxygen in the cyclic form where it is bonded to both carbon-1 and carbon-5. Which is the more stable, the open-chain or the cyclic structure? How much more stable? Show your reasoning and calculations clearly.

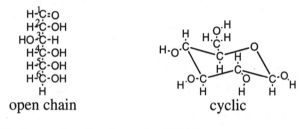

open chain   cyclic

7.32. Two isomers with the molecular formula CH_2N_2 are:

diazomethane   diazirine

Diazomethane and diazirine are both gases. Under appropriate conditions, each can react to give ethene and nitrogen:

$$2CH_2N_2(g) \rightarrow H_2C=CH_2(g) + 2N_2(g)$$

(a) Use average bond enthalpies to calculate $\Delta H°_{reaction}$ for diazomethane and for diazirine undergoing this reaction. Are the reactions exothermic or endothermic?

(b) Construct enthalpy-level diagrams (like Figure 7.11) for the two reactions. Which isomer is the more stable? Are there any factors that might complicate use of bond enthalpies for these molecules?

(c) Does our molecular bonding model help explain the relative stability? Explain why or why not.

7.33. The table at the right gives the molecular formulas and enthalpies of combustion for three common sugars found in living organisms. What conclusions can you draw from these data about the bonding in these molecules? Explain your response.

Sugar	Formula	$\dfrac{\Delta H_{combustion}}{kJ \cdot mol^{-1}}$
fructose	$C_6H_{12}O_6$	–2812
galactose	$C_6H_{12}O_6$	–2803
glucose	$C_6H_{12}O_6$	–2803

7.34. Disaccharides are sugars formed by the combination of two simpler sugars. A maltose molecule, for example, is a combination of two glucose molecules:

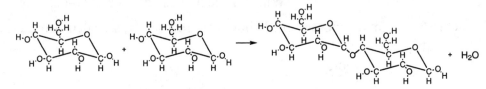

(a) Estimate $\Delta H°_{reaction}$, using bond enthalpies. Is it necessary to do any numerical calculations to make this estimate? Explain why or why not.

(b) Use your result from part (a) and the data from the Problem 7.33 to estimate the enthalpy of combustion of maltose. The experimental value is –5644 kJ·mol⁻¹. How well does your estimate compare to the experimental value? How do you explain the agreement or lack of agreement?

7.35. The reaction between hydrazine and hydrogen peroxide in rocket engines can be represented as:

$$H_2NNH_2(g) + 2HOOH(g) \rightarrow N_2(g) + 4H_2O(g)$$

(a) Use average bond enthalpies to estimate the standard enthalpy change (in kJ per mole of hydrazine) for the reaction.

(b) Compare the energy released in the hydrazine reaction to the energy that would be obtained if two moles of ammonia were oxidized by hydrogen peroxide.

$$2NH_3(g) + 3HOOH(g) \rightarrow N_2(g) + 6H_2O(g)$$

(c) Which provides the most energy per gram of fuel, hydrazine or ammonia?

7.36. WEB Chap 7, Sect 7.7.2-3. On *Web Companion* page 7.7.2 you select the bonds that are broken in the reactants and made in the products.

(a) Use the data on page 7.7.3 to calculate the enthalpy change for this process.

(b) How does your result in part (a) compare with the analysis on page 7.7.3 where all reactant bonds are broken and all product bonds formed?

(c) Explain why the comparison in part (b) comes out the way it does.

7.37. Carbon–hydrogen bond-dissociation enthalpies are shown for two different hydrogens in the reactions of propene and 2-butene shown here. The enthalpies for the two different bond dissociations are quite different. The dissociation of a carbon–hydrogen bond on a carbon that is next to a doubly bonded carbon, reactions (1) and (3), requires less energy.

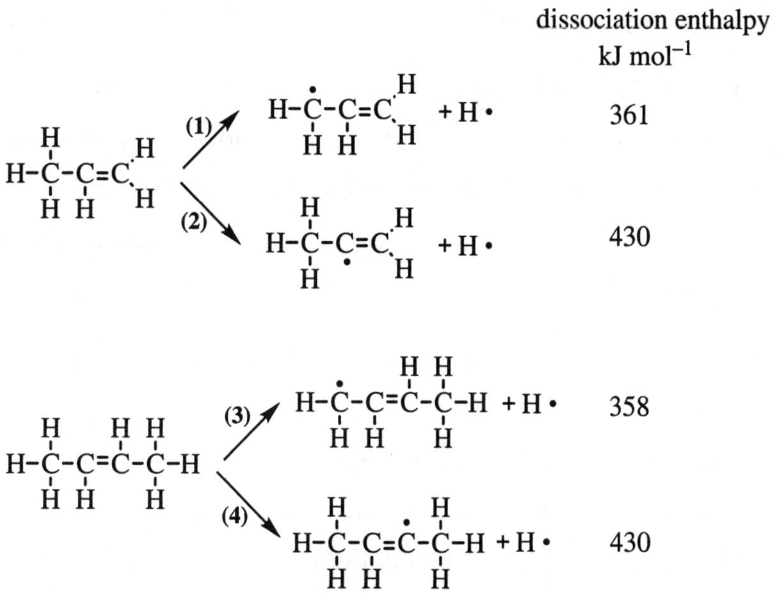

dissociation enthalpy
kJ mol^{-1}

(1) ... 361

(2) ... 430

(3) ... 358

(4) ... 430

(a) Are the propyl and butyl free radicals formed in reactions (1) and (3), respectively, more stable (lower enthalpy) or less stable (higher enthalpy) than those formed in reactions (2) and (4)? Clearly explain the reasoning for your response, using enthalpy level diagrams, if these aid your explanation.

(b) The unpaired electron in the propyl and butyl free radicals formed in reactions (1) and (3) is often shown as participating in delocalized π bonding (see Chapter 5, Section 5.7), with the two π electrons from the adjacent double bond. Do the bond-dissociation enthalpies and your interpretation in part (a) support this model? How would delocalization of the unpaired electron orbital affect the energy of the radical? Explain your reasoning clearly.

Section 7.8.   Standard Enthalpies of Formation

7.38. You usually see sulfur in the form of a light yellow powder, the rhombic crystalline allotrope of sulfur. If you melt this solid (m. p. 112.8 °C), pour the liquid into a filter paper cone, and then unfold the cone as the sample cools, you find that the liquid has solidified in dark yellow needles. This is the triclinic crystalline allotrope of sulfur. After several days, the needles become covered by a light yellow layer. Which allotrope of sulfur is probably its most stable form under standard conditions? Explain your answer.

7.39. Urea (Latin *urina* = urine), $H_2NC(O)NH_2$, is a water-soluble compound made by many organisms, including humans, to eliminate nitrogen. Use standard enthalpies of formation to find the enthalpy for the reaction producing urea from ammonia and carbon dioxide.

$$2NH_3(g) + CO_2(g) \rightarrow H_2NC(O)NH_2(s) + H_2O(l)$$

Show all your work so your procedure is clear.

7.40. In Problem 7.39, you used enthalpies of formation to calculate the enthalpy for making urea from ammonia and carbon dioxide. How could you estimate the enthalpy of vaporization of urea? Explain the procedure you would use and then carry it out. *Hint*: What other way could you estimate the enthalpy of the reaction forming urea? How do the two ways differ?

7.41. During fermentation of fruit and grains, glucose is converted to ethanol and carbon dioxide according to this reaction:

$$C_6H_{12}O_6(s) \rightarrow 2C_2H_5OH(l) + 2CO_2(g)$$

(a) Using the data in Appendix XX, calculate $\Delta H°_{reaction}$.

(b) Is this reaction exothermic or endothermic?

(c) Which has higher enthalpy, the reactant or the products?

(d) Calculate ΔH for the formation of 5.0 g of $C_2H_5OH(l)$.

(e) What quantity of heat is released when 95.0 g of $C_2H_5OH(l)$ is formed at constant pressure?

7.42. Consider this gas phase reaction in which methanoic (formic) acid reacts with ammonia. The product formamide contains a peptide bond and there is the release of a water molecule:

(a) Use average bond enthalpies to estimate the enthalpy change for this reaction. How does this $\Delta H°$ compare to the $\Delta H°$ you calculated in problem 7.29? Would you have expected this result? Why or why not?

(b) Use standard enthalpies of formation to calculate the change in enthalpy for this reaction, assuming reactants and products are gases in their standard states. The enthalpies of formation of gaseous methanoic acid and formamide are –379 and –186 kJ·mol^{-1}, respectively.

(c) Compare the results of your calculations in parts (a) and (b). Offer some reasons to help explain your comparison.

7.43. The combustion of ammonia is represented by this equation:

$$4NH_3(g) + 5O_2(g) \rightarrow 4NO(g) + 6H_2O(g)$$

Experimentally, we find $\Delta H^\circ_{reaction} = -905$ kJ for the reaction as written.

(a) Use the standard enthalpies of formation for $NO(g)$ and $H_2O(g)$ from Appendix XX and the standard enthalpy change for the reaction to calculate the standard enthalpy of formation for $NH_3(g)$.

(b) How does your result in part (a) compare with the standard enthalpy of formation for $NH_3(g)$ in Appendix XX? Explain why they are the same (or different).

7.44. In the process known as coal gasification, coal can be reacted with steam and oxygen to produce a mixture of hydrogen, carbon monoxide, and methane gases. These gases are desirable as fuels and they can also serve as the starting material for the synthesis of other organic substances such as methanol, used for the production of synthetic fibers and plastics. Consider these three reactions that can take place in the process of coal gasification.

> Reaction 1: $C(s) + H_2O(g) \rightarrow H_2(g) + CO(g)$
>
> Reaction 2: $C(s) + \frac{1}{2}O_2(g) \rightarrow CO(g)$
>
> Reaction 3: $C(s) + 2H_2O(g) \rightarrow CH_4(g) + O_2(g)$

Use standard enthalpies of formation to find the enthalpies of reaction for each of these three reactions. If it is possible to control the reaction conditions to favor one or more of these reactions, which is the most energetically favorable? the least energetically favorable? Explain your reasoning.

Section 7.9.   Harnessing Energy in Living Systems

7.45. What is a coupled reaction? What is its importance in biological reactions?

7.46. What is the role of ATP in biological reactions?

7.47. Coupled reactions take place in living systems, but coupled processes have many applications in industry as well. For example, waste heat from a power plant can be captured and used to do work, such as desalting seawater.

(a) Other than the availability of seawater, what do you think might limit the usefulness of this example of a coupled process?

(b) Can you think of any other examples of energetically coupled processes from your personal experience or other courses you have taken? Briefly describe such a process. You might find it useful to research this topic using web-based resources.

7.48. Many ATP^{4-}-coupled reactions require that Mg^{2+} be present as well. What do you think is the function of Mg^{2+} in these reactions? *Hint*: See Chapter 6, Section 6.6.

7.49. The aerobic reaction sequence of glycolysis involves the complete oxidation of pyruvic acid, $CH_3C(O)C(O)OH$ (= $C_3H_4O_3$). This reaction is coupled to the formation of ATP^{4-}, according to this overall reaction:

$$2C_3H_4O_3(l) + 5O_2(g) + 30ADP^{3-}(aq) + 30HOPO_3^{2-}(aq) + 30H^+(aq) \rightarrow$$
$$6CO_2(g) + 34H_2O(l) + 30ATP^{4-}(aq)$$

(a) Calculate the $\Delta H°$ for the uncoupled oxidation of pyruvic acid to CO_2 and H_2O.

(b) Calculate the $\Delta H°$ for the coupled reaction and determine the percentage of energy converted to ATP^{4-}.

Section 7.10. Pressure-Volume Work, Internal Energy, and Enthalpy

7.50. An inventor claims to have built a device that is able to do about 0.8 kJ of work for every 1 kJ of thermal energy put into it. He is looking for investors to buy shares in his company, so he can commercialize his invention. Would you invest in his company? Explain why or why not.

7.51. The volume of a gas is decreased from 10. L to 1.0 L at a constant pressure of 5.0 atmospheres. Calculate the work associated with the process. Is work done *on the gas* or *by the gas*? (1 L·atm = 101.3 J)

7.52. A hot air balloon is inflated by using a propane burner to heat the air in the balloon. If during this process 1.5×10^8 J of heat energy cause the volume of the balloon to change from 5.0×10^6 L to 5.5×10^6 L, what are the values of q, w, and ΔE for this process? Assume that the balloon expands against a constant pressure of 1.0 atmosphere. (1 L·atm = 101.3 J)

7.53. Photographic flash bulbs, once more common than they are now, produced light from the reaction of either zirconium or magnesium wire with oxygen in the bulb, for example:

$$2Mg(s) + O_2(g) \rightarrow 2MgO(s) + energy \text{ (light and heat)}$$

They became quite warm when used, but only rarely did they actually explode! Discuss the changes in ΔE, q, and w for the chemical reaction that takes place in a flash bulb.

7.54. A gaseous chemical reaction occurs in which the system loses heat and contracts during the process. What are the signs, "+" or "-", for $P\Delta V$, ΔE and ΔH? Explain your reasoning.

7.55. Consider two experiments in which a gas is compressed in a closed syringe by pushing the plunger against the trapped gas. Both experiments use the same size syringe.

Experiment A   Experiment B

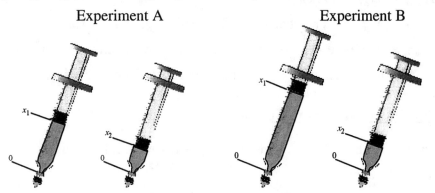

(a) How does the pressure-volume work performed in Experiment **A** compare to that performed in Experiment **B**? Assume the external pressure is the same in both experiments. Explain your reasoning.

(b) Assume that a third experiment is performed using a syringe with a wider diameter than the one used in the first two experiments. How would the pressure-volume work compare with that in Experiments **A** if Δx is the same? Assume the pressure is the same in both experiments.

7.56. Consider the changes that occur in this reaction:

$$3H_2(g) + N_2(g) \rightarrow 2NH_3(g)$$

Is work done by the system (gaseous reactants and product) or on the system? Explain your answer. Assume that this reaction occurs at constant temperature and pressure.

7.57. Most of the nitrogen used to inflate an automobile airbag is produced by the reaction of sodium azide:

$$2[Na^+][N_3^-](s) \longrightarrow 2Na(g) + 3N_2(g)$$

A driver's side airbag might typically contain about 95 g of sodium azide.

(a) A mole of nitrogen gas occupies about 25 L at 25 °C. How large is the airbag this nitrogen will inflate?

(b) How much work is done by the nitrogen as it inflates the airbag?

(c) If this exothermic reaction is investigated in a constant volume calorimeter, will the measured thermal energy release be greater or less than the thermal energy released at constant pressure? How large will the difference be? Clearly explain your reasoning.

7.58. For this reaction, as written, $\Delta H° = -484$ kJ:

$$2H_2(g) + O_2(g) \rightarrow 2H_2O(g)$$

When 0.5 mol of $H_2(g)$ was reacted with 0.3 mol of $O_2(g)$, at a constant pressure of 1.0 atm, the change in volume was –6.1 liters. Calculate how much work was done and the value of $\Delta E°$ for this reaction. Is work done on the system or by the system?

Section 7.11. What Enthalpy Doesn't Tell Us

7.59. Indicate whether each of the following statements is true or false. If a statement is false, write the correct statement.

(a) Consider a 10 mL sample of pure water at 25 °C and 1 atm pressure (state A). The sample is cooled to 1 °C and then the pressure is reduced to 0.5 atm (state B). It takes 5 hours to carry out the change from state A to state B. The sample is then heated and the pressure is raised to 1 atm. In one minute the water is back at 25 °C (the sample is back at state A). The internal energy change in going from state A to B is equal to, but opposite in sign to the internal energy change going from state B to A.

(b) The work done in changing from state A to B and the work done in changing from state B to A in part (a) are numerically the same but opposite in sign.

(c) The enthalpy change for a change of state of a system is independent of the exact state of the reactants or products.

(d) At constant pressure the amount of heat absorbed or evolved by a system is called the enthalpy change, ΔH.

(e) If volume does not change, the amount of heat released during a change of state of a system is equal to the decrease in internal energy of that system.

(f) If a reaction is spontaneous, it is always exothermic.

(g) If the enthalpy change for this reaction, $N_2(g) + O_2(g) \rightarrow 2NO(g)$, is 180.5 kJ, then the enthalpy change for this reaction, $1/2N_2(g) + 1/2O_2(g) \rightarrow NO(g)$, is 90.2 kJ.

7.60. When 1 mol of $NH_4Cl(s)$ is dissolved in water, forming $NH_4^+(aq)$ and $Cl^-(aq)$, the change in enthalpy is 14.8 kJ·mol⁻¹. When 1 mol of $CaCl_2(s)$ is dissolved in water, forming $Ca^{2+}(aq)$ and $Cl^-(aq)$, the change in enthalpy is –82.9 kJ·mol⁻¹.

(a) Which beaker felt cool to touch after the salt dissolved in water? Which one felt warm?

(b) What would you have to do to maintain the temperature of each beaker at 25°C?

(c) To determine $\Delta H°$ for dissolution of salts in water, what factors must you consider?

7.61. Given that ΔH°_f for $Cl^-(aq) = -167.4$ kJ·mol^{-1}, use the enthalpy of reaction data from Problem 7.60 to calculate ΔH°_f for $NH_4^+(aq)$ and for $Ca^{2+}(aq)$.

7.62. A nineteenth-century chemist, Marcellin Berthelot, suggested that all chemical processes that proceed spontaneously are exothermic. Is this correct? Give some examples that justify your answer.

Section 7.13. EXTENSION — Ideal Gases and Thermodynamics

7.63. Consider two samples of gas in identical size containers. Which of these statements is true? Explain your reasoning for each choice. Rewrite the false statements to make them true.

(a) If the temperature of the two samples is the same, then the pressure of gas in each container is the same.

(b) If the temperature and pressure of the two samples are the same, then the number of moles of gas in each container is the same.

(c) If the gas in each container is the same and the temperature and number of moles of gas in each container are the same, then the number of collisions with the wall per unit time is the same in both containers.

7.64. Molecules of all gases at the same absolute (Kelvin) temperature have the same average kinetic energy of translation, $KE = 1/2mu^2$, where m is the mass of a molecule and u is the average speed. If a gas is in a container that has one or more tiny holes, the molecules will be able to escape through the holes. Faster moving molecules escape more readily.

(a) A rubber balloon was inflated with helium gas and floated in the air at the end of its string. The next day, the balloon was somewhat smaller and no longer floated. Explain why. *Hint*: When a thin sheet of rubber is stretched, tiny holes form in the sheet.

(b) Two identical rubber balloons were blown up to the same size, one with helium gas and the other with air. The next day, both balloons were somewhat smaller, but not the same size. Which gas was in the larger balloon? Explain your answer.

7.65. A change involving gases is carried out at constant pressure in a cylinder with a piston, like the one illustrated in Figure 7.20. The final position of the piston is higher in the cylinder than it was initially. What possible change(s) might occur in the system enclosed in the cylinder to cause the observed change in the position of the piston? Clearly explain how the change(s) you suggest would account for the movement of the piston.

7.66. Problem 7.57(a) says a mole of nitrogen gas occupies about 25 L at 25 °C (and 1 atm pressure). Show that this is true. What is the volume for a mole of oxygen? Explain.

7.67. A sample of gas is put into a rigid (fixed volume) container at –3 °C and a pressure of 38.3 kPa. The container is then placed in an oven at 267 °C.

(a) What pressure would you expect to measure for the gas in the container at this higher temperature? Explain.

(b) The measured pressure of the gas in the container at 267 °C is 137.2 kPa. How does this experimental value compare with the pressure you calculated in part (a)? If they are different, what factor(s) could account for the difference? Clearly explain your reasoning.

7.68. Analysis of a liquid hydrocarbon sample shows that the ratio of carbon to hydrogen atoms in the compound is one-to-one. A 0.237 g sample of the liquid is placed in a 327-mL flask attached to a pressure sensor. The air is pumped out of the flask and then the flask is sealed and heated to 150 °C, at which temperature the liquid has all vaporized and the pressure in the flask is 33.2 kPa.

(a) How many moles of the hydrocarbon sample are in the flask? Explain how you get your answer.

(b) What is the molar mass of the hydrocarbon? What is the molecular formula of the hydrocarbon? Explain how you get your answers.

7.69. A mineral sample contains several different ionic compounds, including calcium carbonate, $CaCO_3(s)$. A 1.587 g sample of the mineral was heated to decompose the calcium carbonate to $CaO(s)$ and $CO_2(g)$, The gas was collected and measured and had a pressure of 83.35 kPa at 295.3 K in a 127.5 mL collection container.

(a) How many moles of gas were evolved by the sample? Explain how you get your answer.

(b) How many moles of $CaCO_3(s)$ were present in the mineral sample. Explain.

(c) What percentage of the mineral sample is $CaCO_3(s)$. Explain how you get your answer.

7.70. The fermentation of glucose produces ethanol and releases carbon dioxide. This equation represents the reaction.

$$C_6H_{12}O_6(s) \rightarrow 2C_2H_5OH(l) + 2CO_2(g)$$

(a) Do $\Delta E°$ and $\Delta H°$ have the same value for this reaction? Explain your reasoning.

(b) If $\Delta E°$ and $\Delta H°$ differ, which is larger? by how much? Explain clearly.

General Problems

7.71. A constant-pressure calorimeter similar to the one shown in Figure 7.7 can be used to study the reaction of magnesium with hydronium ion:

$$Mg(s) + H_3O^+(aq) \rightarrow Mg^{2+}(aq) + H_2(g)$$

100.0 mL of 0.5 M hydrochloric acid, HCl(aq), was put in the calorimeter; its temperature was 19.32 °C. Then, 0.1372 g of Mg metal was added to the acid and the temperature observed until it reached a maximum of 25.69 °C.

(a) What is $\Delta H_{reaction}$, in kJ per mole of Mg reacted?

(b) Calculate $\Delta E_{reaction}$, in kJ per mole of Mg reacted.

(c) How could you measure $\Delta E_{reaction}$ directly? Are data like those in this problem precise enough to measure any difference between $\Delta E_{reaction}$ and $\Delta H_{reaction}$? Explain why you answer as you do.

7.72. Several elegant experiments have been done to determine the amount of work that *molecular motor*s from living cells can do. The motor in the illustration consists of seven proteins, three α, three β, and a γ, that come together as shown. The motor is too small (about $10 \times 10 \times 8$ nm) to be observed, so a long actin filament is attached to the motor shaft, the γ protein, and movement of the filament is observed under a microscope. The rotation of the shaft is in 120° steps. Each step appears to require the 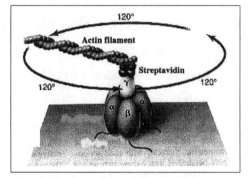 coupled hydrolysis of one ATP molecule, reaction (7.37). Calculations based on the observed motion indicate that the motor is able to produce a force of at least 100×10^{-12} N. In one complete revolution, the top of the motor shaft moves about 1 nm.

(a) How much work does the motor do in one complete revolution? (1 J = 1 N·m).

(b) How much energy (enthalpy) is provided by ATP during one complete revolution?

(c) What is the approximate efficiency of this motor? Explain how you arrive at your conclusion.

7.73. Calculate the $\Delta H°_{reaction}$ for the hydrolysis of a molecule of glycylglycine to two molecules of glycine using the $\Delta H°_{combustion}$ data for glycylglycine and glycine. Assume that the nitrogen atoms in both molecules are oxidized to NO_2. Explain your procedure.

$\Delta H°_{combustion}$ for glycine, $H_2NCH_2C(O)OH = -981$ kJ·mol^{-1}

$\Delta H°_{combustion}$ for glycylglycine, $H_2NCH_2C(O)NHCH_2C(O)OH = -1,996$ kJ·mol^{-1}

7.74.   The diagram shows a glass demonstration version of a fire syringe (also called a fire piston). A tiny amount of a flammable material, cloth or paper, is placed at the closed end of the syringe. The plunger is inserted and then thrust quickly into the piston, compressing the air inside to about $1/20^{th}$ of its initial volume. The temperature of the air rises almost instantly to about 1000 K and ignites the flammable material, which burns with a bright flash.

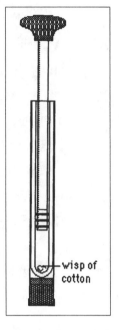

wisp of cotton

(a)   The internal energy of a gas depends on its temperature. When the temperature of the gas changes, its change in internal energy depends on its constant-volume molar heat capacity (C_V), the number of moles of gas, and the temperature change: $\Delta E = n \times C_V \times \Delta T$. C_V for air is 21 kJ·mol^{-1}. What is ΔE for the gas compressed in a fire syringe that is 15 cm long and has a cross-sectional area of 0.20 cm^2? One mole of gas at 300 K occupies about 25 L.

(b)   The compression in a fire syringe is so rapid that the gas has no time to transfer thermal energy to its surroundings during the compression. How much work is done on the gas during the compression? What pressure (in atm) on the plunger is required to obtain this much work? Explain clearly how you obtain your answer. (1 L·atm = 101.3 J)

(c)   To get a better "feel" for the effort required to operate a fire syringe, calculate the force in pounds required to produce the pressure you calculated in part (b). Use the fact that a pressure of 1 atm = 15 pound·in^{-2}. Show clearly how you do the necessary unit conversions. Is the force required a reasonable amount for a human to produce?

(d)   Make a molecular level drawing showing the motion of the molecules of air in the syringe before, during, and immediately after the compression. Clearly show how they obtain the increase in energy you calculated in part (a).

> Fire syringes made of hollow wood like bamboo have apparently been used to start fires for several centuries. A fire piston was patented in England in 1807. The diesel engine works on this same principle: the air in the engine cylinder is compressed quickly and the fuel is injected just before the compression stroke is complete.

(e)   Why does the amount of flammable material in the syringe have to be small? If the temperature of the gas reaches 1000 K, couldn't it set fire to a larger quantity? *Hint*: Consider the heat capacity (or specific heat) of a piece of solid cloth or paper.

7.75. Given these atomization reactions and their corresponding enthalpy changes, calculate $\Delta H°$ for the combustion of 1 mol of $PH_3(g)$ to yield $H_2O(g)$ and $P_2O_5(g)$. Show your reasoning clearly.

$PH_3(g) \rightarrow P(g) + 3H(g)$ $\qquad\qquad$ $\Delta H° = 965\ kJ \cdot mol^{-1}$

$O_2(g) \rightarrow 2O(g)$ $\qquad\qquad$ $\Delta H° = 490\ kJ \cdot mol^{-1}$

$H_2O(g) \rightarrow 2H(g) + O(g)$ $\qquad\qquad$ $\Delta H° = 930\ kJ \cdot mol^{-1}$

$P_2O_5(g) \rightarrow 2P(g) + 5O(g)$ $\qquad\qquad$ $\Delta H° = 3382\ kJ \cdot mol^{-1}$

Chapter 8. Entropy and Molecular Organization

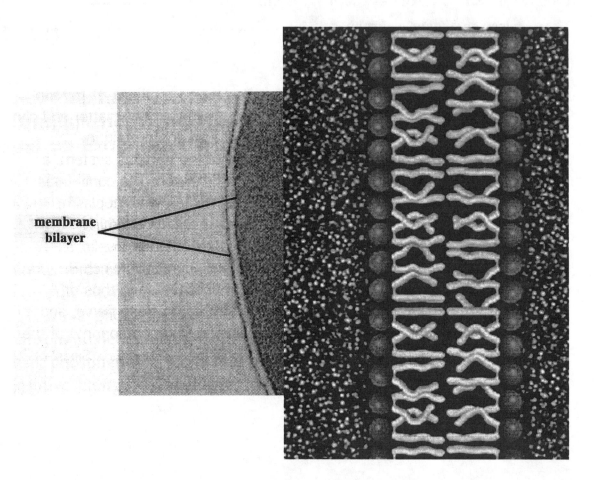

membrane
bilayer

The electron micrograph is a thin layer of a red blood cell that has been specially prepared to show the membrane that encloses the cell. The two parallel layers of the membrane are labeled. The drawing shows a simple model of a tiny portion of a phospholipid bilayer cell membrane with water molecules on both sides of the membrane as they would be in an intact cell. The yellow "fingers" in the central part of the bilayer represent the hydrocarbon chains of long-chain carboxylic acids connected by ester bonds to the polar "heads," of the phospholipid molecules.

Chapter 8. Entropy and Molecular Organization

"Things fall apart; the centre cannot hold;
Mere anarchy is loosed upon the world"

W. B. Yeats (1865-1939), The Second Coming

These lines from Yeats do seem to describe many aspects of the world as we experience it. Is this really the way things are? The notion of anarchy (disorder) is certainly supported by substantial evidence: apples rot, iron rusts, glass breaks, paper tears, batteries run down. The natural order of things *does* seem to favor *dis*order. On the other hand, consider something as familiar as snowflakes on a winter day. The complex crystals shown in Figure 8.1 originate from ordinary water freezing in the atmosphere and falling to earth. How is it that some things become so disorganized while others become so highly organized?

Figure 8.1. Snowflakes

Living systems represent a huge degree of organization. Complex molecules are arranged into complex arrays to form cells. Cells are organized into complex arrays to form organs. Organs are organized into complex arrays to form plants and animals. The illustration on the facing page represents a tiny portion of the molecules in a cell membrane. Long hydrocarbon chains of the membrane are stacked next to one another in the center of the figure. Water molecules are less organized than the membrane molecules, and can move about randomly on both sides of the membrane Water molecules can also pass through the membrane from one side to the other. The net direction of movement of water molecules through a cellular membrane depends upon the concentrations of the solutions on either side of the membrane. Water molecules can move through the membrane from the less concentrated solution to the more concentrated solution; this movement of the water molecules is called **osmosis**.

There is little or no enthalpy (or energy) change in the formation of cell membranes or when osmosis occurs. At the end of the previous chapter, we pointed out that the first law of thermodynamics (conservation of energy) cannot explain the direction of changes. Spontaneous processes can occur endothermically, exothermically, or without appreciable input or output of enthalpy. Enthalpy changes do not provide a way to predict the direction in which a process will occur. We have also seen that molecular organization increases in some changes — precipitation, for example — and decreases in others, such as dissolution. Therefore, molecular organization alone cannot be directing the changes. The central objective for this chapter is to identify and characterize the property of systems that is responsible for change. We will find that *entropy* is the thermodynamic quantity that tells us the direction processes will go spontaneously. Let's begin our discussion with the processes of mixing and osmosis.

Section 8.1.   Mixing and Osmosis

8.1.   Investigate This   *Does water move through a hollow carrot filled with syrup?*

Do this as a class investigation, but work in small groups to discuss the results. Bore a cylindrical hole about $3/4$ the length of a large carrot. Insert a 10-cm length of a clear, rigid plastic straw or glass tubing a short distance into the hole and seal it to the carrot. Add enough dark sugar syrup — such as dark corn syrup or pancake syrup — to the cavity in the carrot to bring the liquid level a centimeter or two above the carrot. Mark the level of the syrup on the tubing. Support the carrot and tubing upright in a container of water. The water level should be just about to the top of the carrot. Record your observations on the set up every five minutes or so.

8.2.   Consider This   *What direction does water move through a hollow carrot?*

(a) Did the column of liquid in the tubing rise or fall in Investigate This 8.1? Does the volume of liquid in the cavity of the carrot increase or decrease? Explain your reasoning.

(b)   WEB   Chap 8, Sect 8.4.1. Play the animation. How does the animation compare to what you observed in Investigate This 8.1? Explain.

(c) What is your explanation for your observations from the investigation and the animation from the *Web Companion*? Compare your interpretation with that of other students.

Mixing

Mixing is familiar to you in many changes you have observed, for example, milk mixing into coffee or sugar dissolving and mixing into tea. Figure 8.2 illustrates such a change. When a drop of concentrated dye solution is added to a container of water, the dye molecules begin to mix into the water and become diluted. If we wait long enough, the dye molecules become dispersed

> **WEB Chap 8, Sect 8.3.1-2**
> View a movie and molecular-level animation of the mixing process

uniformly throughout the entire container. Mixing to form a uniform solution is the spontaneous direction of this process. As the figure shows, no matter how long you wait, you never observe a homogeneous solution of dye molecules unmixing, that is, all coming together in a single drop somewhere in the container.

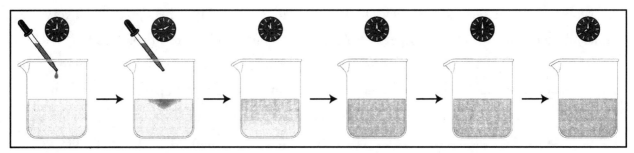

Figure 8.2. A drop of dye mixing into water.
By four o'clock the soluble dye has formed a uniform mixture with the water. The resulting uniform solution will not spontaneously unmix, no matter how long you wait.

Osmosis

Osmosis is fundamentally a mixing process: a concentrated solution becomes less concentrated by mixing with additional solvent. This happens when a barrier (usually a thin sheet of material called a **membrane**) allows solvent molecules to pass through, but prevents passage of solute ions or molecules. For example, picture, as in Figure 8.3(a), a concentrated solution on one side of a membrane and an equal level of a less concentrated solution (or pure solvent) on the other side. After some time, the level of the more concentrated solution will rise, and the level of the less concentrated solution will fall [Figure 7.3(b)]. The solvent has diffused through the membrane, mixing with and diluting the concentrated solution.

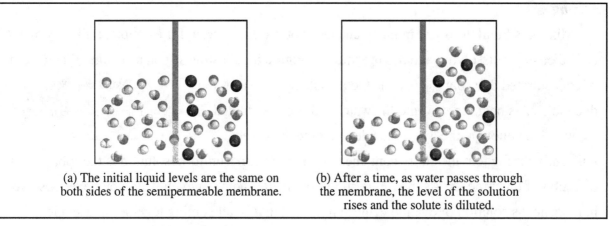

(a) The initial liquid levels are the same on both sides of the semipermeable membrane.

(b) After a time, as water passes through the membrane, the level of the solution rises and the solute is diluted.

Figure 8.3. Changes that occur in osmosis.

Membranes that allow some species to pass through readily while others go through very slowly or not at all are called **semipermeable**. Plant and animal cells are surrounded by semipermeable membranes — the model of a cell membrane in the chapter opening illustration is one example. (We will say more about biological membranes in Section 8.11.) If osmosis occurs in

> A permeable (*per* = through + *mea*re = to pass) barrier allows everything to pass through and an impermeable (*im* = not) barrier allows nothing to pass through. A semipermeable barrier is in between.

Activity 8.1, water must pass through the membranes of many layers of plant cells on its way to diluting the sugar solution in the center of the carrot. The rising column of solution is evidence that the solution volume is increasing due to the osmosis of water from the pure water outside the carrot.

Why is it that we observe mixing, but not unmixing of homogeneous solutions? Why is it that osmosis always takes place in a direction that moves solvent so that it dilutes a concentrated solution? The answer to these questions is that a mixed state or a more dilute state is more probable (more likely to occur) than an unmixed or concentrated state. In order to see why this is so, we need to know more about probability and how molecular arrangements are related to probability. Then we will return to see how these ideas are applied to understanding mixing and osmosis.

Section 8.2.   Probability and Change

We often use the terms chance and probability in a casual way, but probability has a precise mathematical meaning that we can use to understand change. Some changes will always occur,

> Lotteries and other games of chance are based on probabilities. The games are always designed so that the probability that the players will lose is greater than the probability that they will win.

and the probability of their occurrence is assigned a value of 1. Some changes will never occur, and the probability of their occurrence is assigned a value of 0. Everything else is in between and more care is required to define the probability.

8.3. **Consider This** *Probability of change*

What numerical value (0, 1, or somewhere in between) would you assign to the probability of each of these changes being observed? Explain each of your answers.

(a) A book pushed off a table falls to the floor.

(b) A solution of sugar in water separates into pure solid sugar and liquid water.

(c) A glass of liquid water sitting on your desk freezes.

(d) An ice cube sitting on your desk melts.

(e) A discarded piece of paper moves from your wastebasket to your desktop.

(f) A teaspoon of sugar dissolves in a teaspoon of water.

With a little help from mathematical probability, we can use our model of the atomic and molecular world to understand the direction of change. More importantly, the model enables us to predict changes that will occur in new situations where we have no previous observations to go on. The molecular-level explanatory and predictive model for physical and chemical changes is straightforward: *If a system can exist in more than one observable state* (**mixed or unmixed, for example**)*, any changes will be in the direction toward the state that is most probable.* The hard part is to figure out the probability of each possible observable state of the system, so that we can apply the probability model.

Probability

The **probability** that a particular state of a system will be observed is measured by the number of ways that the molecules and the energy in the system can be arranged to give this state. For the moment, we will focus on molecular arrangements and then take up energy arrangements in Section 8.5. Think about the drop of dye mixing with water in Figure 8.2. Before mixing, all the dye molecules are together in a drop, and any particular dye molecule must be found in this tiny volume. After mixing, that same dye molecule might be found anywhere in the total volume of the solution. In the mixture, the dye molecules are dispersed among the water molecules. The number of possible arrangements of the dye molecules increases, making the mixed state more probable than the state before mixing. The system always changes toward the mixed state, and the soluble dye molecules are never observed to separate from the solution.

If we start in Figure 8.2 with 300 mL of water, we have about 17 moles ($= \frac{300 \text{ g}}{18 \text{ g·mol}^{-1}}$) of water. Seventeen moles of water is about $(17 \text{ mol}) \times (6 \times 10^{23} \text{ molecule·mol}^{-1}) = 10^{25}$ molecules

---

of water. Imagine that the 300 mL of water is divided into 10^{25} tiny boxes, each containing a water molecule. When the drop of dye is added, the number of boxes increases slightly and a small fraction of them are now occupied by dye molecules. We assume that each of the tiny boxes can be occupied by either a water molecule or a dye molecule. Even if only one in a million of the molecules is a dye molecule, there are about 10^{19} dye molecules in the mixture.

To count the number of arrangements of this many molecules requires using statistical methods that obscure the simple, basic idea of counting. To get at the basic idea, we will consider simple model systems that contain only a few particles (molecules) and then generalize to more realistic systems. To use our counting results we make a ***fundamental assumption***: *each distinguishably different molecular arrangement of a system is equally probable.* This is a postulate that cannot be proved and is only accepted because it predicts results that we observe. A new arrangement is **distinguishable** from another if you turn your back, someone rearranges the molecules, and you can tell the new arrangement from the original when you look again at the system. Exchanging identical objects does not produce a new arrangement.

Section 8.3. Counting Molecular Arrangements in Mixtures

8.4.   **Investigate This**   *How are distinguishable arrangements counted?*

Try an example of the kind of counting we'll need to do for this simple model. Use five labeled boxes and three identical, unlabeled objects (such as the candies shown here) to place in them.

(a) How many different, *distinguishable* arrangements can you make of three candies among the five boxes, if each box can hold only one candy? As you try new arrangements, make a table to list and count the different ways of placing candies in the boxes. The arrangement above might be shown as (1,0,1,1,0). How many distinguishable arrangements are there? Can you tell that you have found all possibilities? Compare your method for finding the number of distinguishable arrangements with methods used by your classmates. Did you miss any arrangements or duplicate any?

(b) For a more challenging investigation, find the number of distinguishable arrangements for three objects in six boxes. Again, compare your results with those of other students.

Mixing model

One strategy you might have used for finding the distinguishable arrangements in Investigate This 8.4 is to start with the three objects in boxes 1, 2, and 3; this is one arrangement, (1,1,1,0,0). Moving the object in box 2 to box 4 creates a second arrangement, (1,0,1,1,0), which is the example shown in the investigation. Moving the object into box 5 creates a third arrangement, (1,0,1,0,1). If you start with the (1,1,1,0,0) arrangement and interchange the objects in boxes 1 and 2, you will get a (1,1,1,0,0) arrangement; this is not a new arrangement. Your result for the investigation should have been ten distinguishable arrangements.

In order to relate what you did in Investigate This 8.4 to a realistic system such as the mixing of a dye into water, we need to create an analogous model for the dye solution. Figure 8.4 shows an exploded two-dimensional view of fifteen of the 10^{25} boxes available for a water molecule or a dye molecule. We will use these fifteen boxes, as you used the five boxes in the investigation, to model a two-dimensional mixing process as the dye molecules spread throughout the solution.

> **WEB Chap 8, Sect 8.3.3-5**
> Practice counting arrangements in an interactive animation of mixing.

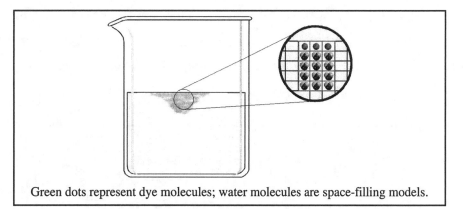

Green dots represent dye molecules; water molecules are space-filling models.

Figure 8.4. Model of a drop of dye molecules added to water.

Arrangements for the unmixed system

The dye-mixing model contains three dye molecules (the solute, indicated by green dots) and twelve water molecules (the solvent, indicated by the space-filling models) in the fifteen boxes. When the dye is first added, the molecules occupy only the top layer of boxes. Only one arrangement, the one shown in Figure 8.4, is possible for the three dye molecules. The number of distinguishable molecular arrangements associated with a particular observable state is usually given the symbol W. For the dye molecules (solute) in this

> Think of W as the number of Ways of arranging identical particles (molecules) or identical energy quanta, as we will show in Section 8.5.

model, there is only one way to arrange three molecules in the top three boxes, so $W_{solute} = 1$.

Similarly, there is only one way of arranging the twelve identical water (solvent) molecules in the remaining twelve boxes (each contains one molecule, as shown in Figure 8.4), so $W_{solvent} = 1$.

Any arrangement of the solute particles can be paired with any arrangement of the solvent particles. The total number of molecular arrangements associated with a given state of the system, W_{system}, is the *mathematical product* of the number of distinguishable arrangements of solute and solvent particles for that state:

$$W_{system} = (W_{solute}) \times (W_{solvent}) \tag{8.1}$$

For our fifteen-box system, in the state that has three solute particles in the top three boxes and twelve solvent molecules in the remaining twelve boxes, W_{system} is:

$$W_{system} = (W_{solute}) \times (W_{solvent}) = 1 \times 1 = 1 \tag{8.2}$$

This result makes sense; if every dye molecule looks like every other dye molecule, and every water molecule looks like every other water molecule, then the arrangement in Figure 8.4 is the only one possible.

Arrangements for the mixed system

As the dye molecules and water molecules begin to mix, some dye molecules move to the next layer of boxes, and those that leave are replaced in the first layer by water molecules. With the three dye molecules confined to the top two layers, there are *twenty* distinguishable arrangements, as you probably discovered if you completed the challenge activity, Investigate This 8.4(b). All twenty possible arrangements are shown in Figure 8.5.

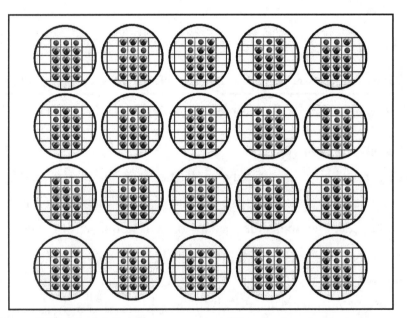

Figure 8.5. Model for arrangements of dye molecules in water
Dye molecules have mixed from Figure 8.4 into the top two layers.

When the three dye molecules have been arranged among the top six boxes, there are only twelve boxes left for the twelve water molecules, and there is only one distinguishable way to distribute these molecules among the boxes. Thus, as before, $W_{solvent} = 1$. For our fifteen-box system, in the state that has three solute particles distributed among the top six boxes and twelve solvent molecules in the remaining twelve boxes, W_{system} is:

$$W_{system} = (W_{solute}) \times (W_{solvent}) = 20 \times 1 = 20 \tag{8.3}$$

Our assumption is that the more probable state will be the one having the largest number of distinguishable arrangements. The state with the dye molecules occupying any of six boxes instead of only three is the more probable. Solute molecules spread out (giving more arrangements) rather than clump together (with fewer possible arrangements). The increased number of distinguishable arrangements makes the mixed state more probable and, therefore, more favored.

As mixing continues, the number of distinguishable arrangements (W) increases rapidly. The three dye molecules have 84 distinguishable arrangements in the top three layers, 220 in the top four, and 455 in all fifteen boxes. The plot in Figure 8.6 shows W as a function of the number of boxes available per molecule in our sample. For example, a value of "5" on the x-axis represents the case of three molecules in 15 boxes. The x-axis is proportional to the *volume* (expressed as the number of boxes) that is available to the solute particles in the system. As the molecules spread through the solution, the number of arrangements rapidly increases.

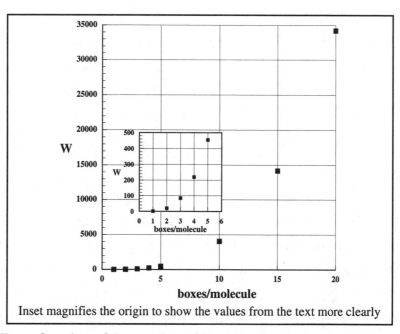

Inset magnifies the origin to show the values from the text more clearly

Figure 8.6. *W* as a function of the number of boxes/molecule for three dye molecules.
No continuous curve is shown because only integer numbers of boxes are possible for this system.

8.5. **Check This** *Unmixing*

WEB Chap 8, Sect 8.3.6. Play the animation. Explain what the animation represents and how you can conclude that unmixing is an unfavorable process.

Section 8.4. Implications for Mixing and Osmosis in Macroscopic Systems

Diffusion

The process modeled in the previous section corresponds to diffusion in a liquid or gas. **Diffusion** is the movement of molecules or ions from a region of higher concentration (many

> One important diffusion process is the movement of oxygen from your arterial blood stream (where its concentration is high) into your cells (where the oxygen concentration is low).

particles in a small volume) into regions of lower concentration. In the model system, Figures 8.4 and 8.5, the lower concentration region was

represented by boxes that initially contained only water molecules. Diffusion always occurs in the same direction, with solute particles moving from regions of higher concentration toward regions of lower concentration.

Figure 8.6 shows that the number of molecular arrangements increases continuously as the volume (number of boxes) available to solute molecules increases. For a given solution, the largest number of molecular arrangements occurs when the particles are spread throughout its entire volume. Diffusion ultimately spreads particles uniformly throughout the available volume. This is the most favored arrangement, as you have observed for actual mixtures, such as milk in coffee, sugar in tea, or a dye in water.

The result represented graphically in Figure 8.6 applies qualitatively to many analogous systems. Expansion of a gas from a small volume to a larger volume, for example, increases the volume (number of boxes in the model) available to each gas molecule. The expanded state of the system is favored, since the value for *W* increases as the volume increases. Consequently, gases always expand to fill any container, occupying the maximum volume available.

Molecular model for osmosis

In osmosis, the solute molecules are prevented from diffusing, because they cannot pass through the semipermeable membrane into the pure solvent. However, if the solvent molecules move into the solution, the volume of the solution increases and there is a larger volume for the

> WEB Chap 8, Sect. 8.4.2-4
> Reinforce this discussion with interactive, molecular-level animations of osmosis.

solute molecules to move about in. Our model from Section 8.3 for countable numbers of molecules can be used to explain osmosis as a result of mixing of solute and solvent molecules.

The change shown in Figure 8.3 is represented schematically in Figure 8.7 for a countable

---

number of solvent and solute molecules. We start with nine molecules on each side of the semipermeable membrane (shown in yellow between the two liquids). On the left are nine solvent molecules in nine boxes and on the right are three solute molecules mixed with six solvent molecules in nine boxes. The numbers of possible molecular arrangements given in the figure for the solvent (W_{solv}) and solution (W_{soln}) are from Figure 8.6.

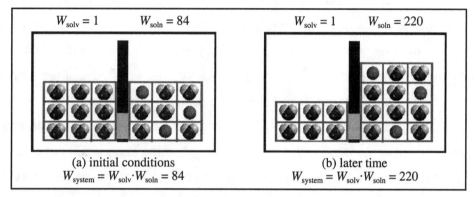

$$W_{solv} = 1 \qquad W_{soln} = 84 \qquad\qquad W_{solv} = 1 \qquad W_{soln} = 220$$

(a) initial conditions
$W_{system} = W_{solv} \cdot W_{soln} = 84$

(b) later time
$W_{system} = W_{solv} \cdot W_{soln} = 220$

Figure 8.7. Increase of molecular arrangements in osmosis.
Compare with Figure 8.3. The semipermeable membrane is represented in yellow here.

If three solvent molecules pass through the membrane from the solvent side to the solution side, there are now twelve boxes on the right occupied by three solute molecules and nine solvent molecules. The total number of molecular arrangements increases by almost a factor of three for this process, so the change is favorable and the process can occur as shown. Your results from Investigate This 8.1 show that the process does, in fact, occur in macroscopic systems. There, the water passed through the membranes of the carrot cells to dilute the solution, and raise its level by increasing its volume, in the center of the carrot. At some point, the net movement of water stops. We shall return in Section 8.13 to discuss osmosis in more quantitative terms to understand why the net movement of water stops. For now, however, we will go on to complete our more qualitative discussion of molecular and energy arrangements.

8.6.   **Check This**   *Transfer of solvent from solution to pure solvent*
    WEB   Chap 8, Sect 8.4.5-7. Starting from the same initial condition as in Figure 8.7(a), how does the number of molecular arrangements for the system change, if three solvent molecules pass from the solution to the pure solvent side of the membrane? Explain how the numeric values are obtained. Would you expect to observe this "reverse" osmosis? Why or why not?

Reflection and Projection

Everyone has experienced mixing processes and observed that homogeneous solutions do not "unmix." You may also have had some experience with osmosis, but might not have made a connection between mixing and osmosis. We have now made the connection through a molecular-level model that provides an explanation for the direction of these processes. Our approach is to consider an imaginary system of just a few molecules (or objects) and calculate how many distinguishable ways there are to arrange these molecules in a limited number of possible positions. Our assumption is that change is in the direction that produces an increase in distinguishable arrangements among the molecules of a system. We find that the number of arrangements increases as the volume available to the molecules increases.

At this point, our probability analysis is incomplete because it is based only on arrangements of particles and has not accounted for the role of enthalpy — the energy content of a substance. In Chapter 7, we saw that enthalpy changes, thermal energy transfers, were an outcome of chemical and physical changes. We will consider the roles of the two change factors — molecular arrangements and thermal energy transfers — together in Section 8.7 and following sections. To prepare for that discussion, we need to consider energy arrangements among the particles in a system.

Section 8.5.   Energy Arrangements Among Molecules

8.7.   **Investigate This**   *How is a different kind of distinguishable arrangement counted?*

Try this example of another kind of counting, using four soft candies of different colors and two toothpicks. How many distinguishable ways can you arrange zero, one, or two toothpicks stuck into the four candies if *any number of toothpicks can be placed in the same candy*? The pictures in this list show a few possible arrangements:

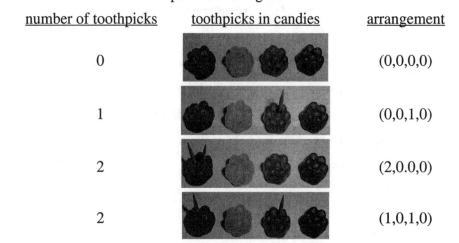

number of toothpicks	toothpicks in candies	arrangement
0		(0,0,0,0)
1		(0,0,1,0)
2		(2,0.0,0)
2		(1,0,1,0)

(a) How many arrangements are there for zero, one, and two toothpicks? How will you know that you have found all the possibilities? Compare your results with what other students get.

(b) For a more challenging activity, find the number of distinguishable arrangements for three toothpicks among four candies.

Model for energy arrangements

A solid made up of identical atoms in a regular array, such as a metallic crystal, is the simplest system we can use to count different arrangements of energy associated with individual atoms. The atoms are located at known positions in the crystal and are not free to move about.

> **WEB Chap 8, Sect 8.5.1-2**
> Use the interactive animations of a metallic crystal and check your counting results.

Consequently, they are distinguishable from one another by their position, just as the candies are distinguishable in Investigate This 8.7. Recall from the discussion in Chapter 4, Sections 4.4 and 4.5, that atomic systems can only absorb (or release) energy in certain amounts called **quanta**. For this discussion, we will assume the quanta are all the same size (that is, that they all are the same

> At the molecular level, all energy transfers are quantized. Macroscopic energy transfers *appear* to be continuous, because thermal energy quanta are so small. Many quanta must be exchanged before the transfer can be detected.

amount of energy). This is a good model for an atomic solid. In an atomic solid, any atom in the array can absorb as many quanta as are available and, just like the toothpicks in Investigate This 8.7, we can't tell one identical quantum from another.

Counting energy arrangements

Investigate This 8.7 is similar to Investigate This 8.4, but there is a fundamental difference. In Investigate This 8.4, the pattern of boxes was fixed, and the objects (molecules) could move among them to produce different arrangements. Only one object could reside in any particular box. For Investigate This 8.7, the candies are identifiable by color (and/or position) and variable numbers of toothpicks can be distributed to each one. Similarly, an atom in a crystal, identified by its location, can have different numbers of quanta. Some atoms have no quanta. Others have one. Still others have two, and so on. Just as moving molecules around produced distinguishable arrangements, quanta moving among atoms produce distinguishable arrangements. Figure 8.8 shows all possible arrangements for zero or one quantum distributed among four atoms in a solid, and two of the possible arrangements for two quanta. Compare these arrangements with those you found in Investigate This 8.7.

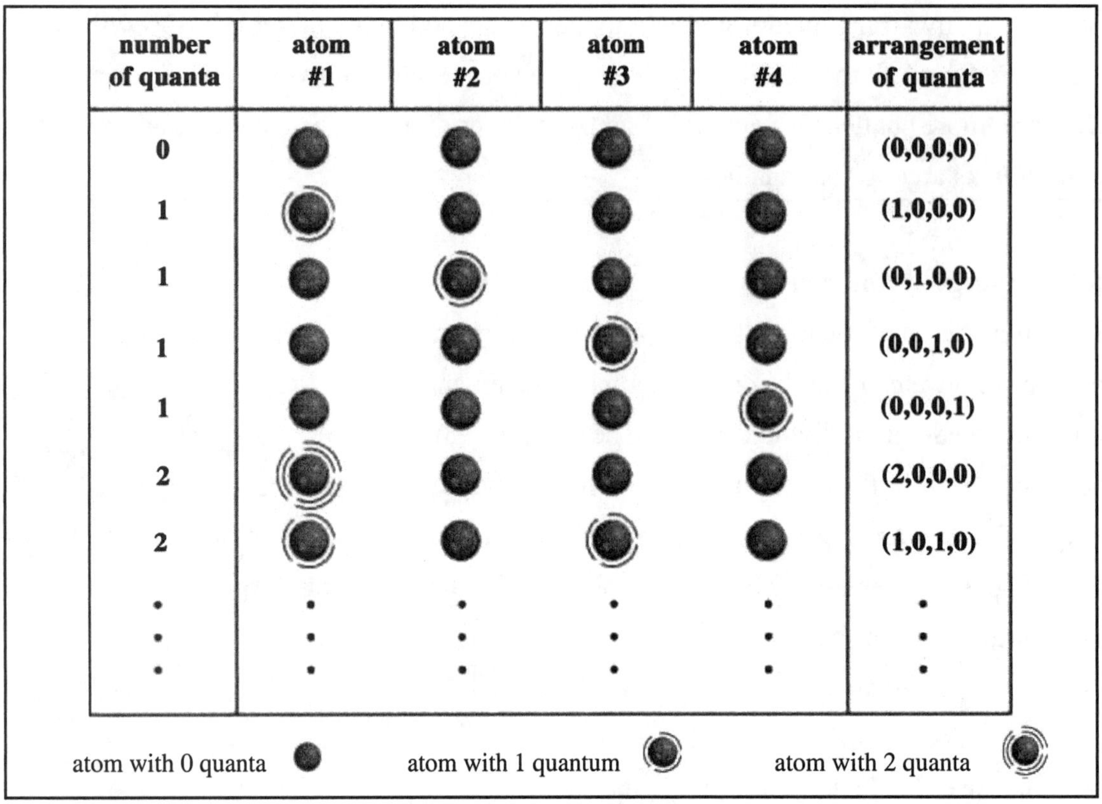

number of quanta	atom #1	atom #2	atom #3	atom #4	arrangement of quanta
0	●	●	●	●	(0,0,0,0)
1	◉	●	●	●	(1,0,0,0)
1	●	◉	●	●	(0,1,0,0)
1	●	●	◉	●	(0,0,1,0)
1	●	●	●	◉	(0,0,0,1)
2	◉	●	●	●	(2,0,0,0)
2	◉	●	◉	●	(1,0,1,0)
⋮	⋮	⋮	⋮	⋮	⋮

atom with 0 quanta ●   atom with 1 quantum ◉   atom with 2 quanta ◉

Figure 8.8. Arrangements of one, two, and three quanta in a four-atom solid.

Investigate This 8.7 demonstrates that the number of possible energy arrangements increases as the number of quanta increases. This is like the systems in Section 8.3, where we showed that the number of possible molecular arrangements increases as the volume of space available per particle increases. Using the approach of Investigate This 8.7, we can find the numbers of distinguishable energy arrangements, W, for increasing numbers of quanta. The results are shown in Figure 8.9, a plot of W as a function of energy (number of quanta) in a four-atom solid.

A four-atom system is too small actually to see, but we can use the results in Figure 8.9 to understand the direction of changes in observable systems. Let's begin by considering an energy

WEB Chap 8, Sect 8.5.3-7
Use interactive animations that model a process like the one in Worked Example 8.8.

transfer in four-atom systems that can be related to an observable result you already know: when a warm object and a cool object are placed in contact, thermal energy is transferred from the warm object to the cool object until they reach the same temperature.

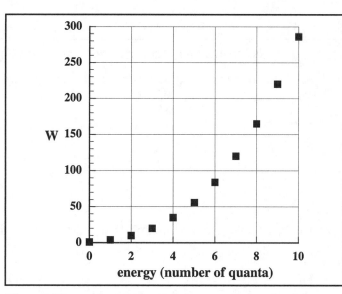

Figure 8.9. Plot of *W* as a function of energy content of a four-atom solid.
No continuous curve is shown because only integer numbers of quanta are physically possible.

8.8.   Worked Example   *Energy transfer between solids with different energies*

Imagine two identical four-atom solids, one having four quanta of energy, and the other having eight. **(a)** What observable property differs between these two solids? **(b)** Allow two quanta to be transferred from the solid with eight quanta to the solid with four quanta. Use the data from Figure 8.9 to determine whether this energy transfer should occur.

Necessary Information: We need Figure 8.9 and the method for obtaining total numbers of distinguishable arrangements that we used in Section 8.3. We need to recall that temperature is a measure of the energy content of a substance.

Strategy: **(a)** Relate the number of energy quanta in each solid to a measurable property of the solid. **(b)** An energy transfer will occur if the change increases the number of possible arrangements of energy quanta. Compare the total number of distinguishable arrangements of the energy quanta before and after the assumed energy transfer.

Implementation: **(a)** Adding energy to an object increases its energy content, producing an increase in temperature. Extracting energy from an object decreases its energy content, producing a decrease in temperature. Temperature is the measurable property that distinguishes the two solids. The four-atom solid with eight energy quanta has a higher temperature than the four-atom solid with four energy quanta. When two energy quanta are transferred, giving each four-atom solid six quanta of energy, the temperatures of the two solids will be the same.

(b) The total number of arrangements in a system, W_{tot}, is the product of the number of arrangements possible in each component of the system. In this case, there are two objects:

$$W_{tot} = W_{obj\ 1} \times W_{obj\ 2}$$

The total number of arrangements is the *product* of arrangements for each object because any energy arrangement in object 1 can occur in conjunction with any energy arrangement in object 2.

For the initial state,

$$W_i = W_{(8\ quanta)} \times W_{(4\ quanta)}$$

For the final state,

$$W_f = W_{(6\ quanta)} \times W_{(6\ quanta)}$$

From Figure 8.9, read values of W for four-atom solids with two, six, and four quanta.

$$W_{(4\ quanta)} = 35$$

$$W_{(6\ quanta)} = 84$$

$$W_{(8\ quanta)} = 165$$

Calculate the total number of initial and final arrangements:

$$W_i = W_{(4\ quanta)} \times W_{(8\ quanta)} = 35 \times 165 = 5775$$

$$W_f = W_{(6\ quanta)} \times W_{(6\ quanta)} = 84 \times 84 = 7056$$

There are more ways to arrange the energy quanta when both solids have six quanta compared to when one solid has four and the other has eight. The transfer of quanta from the warmer to the cooler solid increases the total number of arrangements of quanta, so the transfer should occur.

Does the Answer Make Sense? This problem represents energy transfer from a warm object (more energy quanta) to an identical cooler object (fewer energy quanta); both objects reach the same intermediate temperature (with each having the same number of energy quanta). This is the direction of observed changes in macroscopic systems.

8.9. **Check This** *Energy transfer from a cooler to a warmer solid*

Imagine two identical four-atom solids, one having two quanta of energy, and the other having six. Allow one quantum to be transferred from the solid with two quanta to the one with six. Use the data from Figure 8.9 to determine whether this energy transfer should occur. Explain how your result applies to observations on macroscopic systems.

The transfer of energy from a warm to a cool object, Worked Example 8.8, increases the number of ways that energy quanta can be distributed between the two objects. Since the transfer increases the number of arrangements of the quanta, our probability model predicts that the energy transfer will occur. The prediction agrees with our observations that energy is always transferred from a warm object to a cool object when they are placed in contact. In Check This 8.9, you found that transfer of energy from a cool to a warm object decreases the number of arrangements of the quanta. Our model predicts that this energy transfer will not occur; you never observe a warm object getting warmer when placed in contact with a cooler object.

Section 8.6.   Entropy

The ideas derived from the arrangement activities using atomic-scale systems in the previous sections also apply to systems that are large enough to study directly. For this purpose, we introduce a new thermodynamic quantity, **entropy**, S, which is a measure of the number of different ways that particles or quanta of energy can be arranged. (You can think of entropy as a measure of "Spreadedness.") Entropy is directly related to W, the number of possible arrangements of particles and energy: *the larger the number of possible arrangements, the larger the entropy of the system*. The entropy of a system is proportional to the logarithm (base e) of W:

$$S \propto \ln W \tag{8.4}$$

We have seen in the previous sections that the numbers of arrangements, W, for the components of a system combine multiplicatively to give the total number of arrangements for the system. But we want thermodynamic functions to combine additively, as you have seen for enthalpy in Chapter 7. The logarithms of W do combine additively; take equation (8.1) as an example:

$$\ln W_{system} = \ln [(W_{solute}) \times (W_{solvent})] = \ln W_{solute} + \ln W_{solvent} \tag{8.5}$$

Entropies, defined in terms of ln W, are additive. The proportionality constant between S and ln W is the **Boltzmann constant** and is given the symbol k. The full definition of S is:

$$S \equiv k \ln W \tag{8.6}$$

The Boltzmann constant has a value 1.68×10^{-23} J·K^{-1}. (The Boltzmann constant appears everywhere in models of molecular behavior as a proportionality constant relating the energy of molecules to their temperature in kelvin.) The ln W term has no units, so entropy has the same units as the Boltzmann constant, energy per degree. Equation (8.6) applies to individual molecules. For macroscopic systems, we use moles to specify the amount of substance and replace k by $N_A k$ (Avogadro's number times the Boltzmann constant). The new proportionality constant is the **gas constant, R**:

> Recall that we have met and used the gas constant before in Chapter 7, Section 7.13.

$$R = N_A k = (6.023 \times 10^{23} \text{ mol}^{-1})(1.68 \times 10^{-23} \text{ J·K}^{-1}) = 8.314 \text{ J·K}^{-1}\text{·mol}^{-1} \tag{8.7}$$

R has the units of J·K^{-1}·mol^{-1}, so molar entropy has units of J·K^{-1}·mol^{-1}.

Net entropy: the second law of thermodynamics

In the preceding sections, we have described two different origins for W. In the first, the different distinguishable arrangements in a system were a result of the distribution of molecules. We will call entropy calculated on that basis **positional entropy.** In the second, the distinguishable arrangements are a result of the distribution of energy quanta among molecules. We will call entropy calculated on that basis **thermal entropy**.

Since k is a constant, S depends only on the number of distinguishable molecular and/or energy arrangements of a system — the more arrangements, the higher the entropy. Higher entropy is associated with higher probability. **Net entropy change** is the sum of all the positional and thermal entropy changes *in the surroundings as well as in the system* being studied. *Observed changes always take place in a direction that increases net entropy.* This is the **second law of thermodynamics**. There is no way to "prove" the second law of thermodynamics. It is based on experience and observation. No change has ever been observed that fails to follow the second law, just as no change has ever been observed in which energy is not conserved (the first law).

Absolute entropy

Entropy, like enthalpy, is a state function. The entropy of a system in a particular state is determined entirely by the positional and thermal contributions to the number of distinguishable arrangements of the system. The entropy is independent of the pathway by which the system got into this state. Recall that enthalpy is a function of the internal potential energy of a system; we have no way of knowing an absolute value for enthalpy. Thermodynamic tables, Appendix XX, for example, give enthalpies of formation, ΔH_f°, the enthalpy *change* for the formation of the compound from its elements in their standard states. The elements are a *reference system*, which is *defined* to have an enthalpy of formation of zero. We *can* determine *absolute* values for entropies. Tabulated values for entropies of compounds and elements in their standard states are given as S° (no "Δ"). The reference point for

> Absolute zero has never been experimentally achieved, that is, $W = 1$ has not been attained for any real system. Temperatures of about 0.000001 K have been achieved.

entropies is **absolute zero** (K = 0°), the temperature at which $W = 1$, that is, where there is a single arrangement for particles (perfect order) and all particles are in their lowest energy state. When $W = 1$, for perfectly ordered systems, then $S_0^\circ = 0$.

Reflection and Projection

The discussions of distinguishable arrangements in the preceding sections leads to the definition of entropy as a measure of the number of different ways that particles or quanta of energy can be arranged. Change can occur if the change results in an increase in the possible arrangements of particles and/or quanta (both of which increase entropy). When the results are applied to the observable properties of substances, several important generalizations can be made:

- For soluble substances, diffusion of a component from a region of high concentration into one of lower concentration increases the entropy, ultimately producing a homogeneous mixture.

- There are more energy arrangements and greater entropy for systems in which two objects have the same temperature than when one object is warmer than the other. Transfer of energy from a warm body (more energy quanta per molecule) to a cool body (fewer energy quanta per molecule) increases the entropy of the overall system.

- The reference point for measuring entropy is absolute zero, at which temperature the entropy of a perfectly ordered substance is zero. Entropy values in thermodynamic tables are values for *absolute* entropy.

Changes such as those referred to in this list are often described as being "driven" by the entropy change, because the most probable condition is always the state of highest entropy. The notions that are embedded in the concept of entropy and the second law of thermodynamics become clearer when we study them in relation to real systems. We will do this in the next several sections, as we consider phase changes, other chemical changes, and colligative properties, including osmotic pressure.

Section 8.7.   Phase Changes and Net Entropy

8.10.   **Consider This**   *What are the relative probabilities of different phases?*

Consider the sequence of changes of state as a substance goes from solid to liquid to gas, as represented by the diagram. In the solid, a molecule occupies a fixed lattice position. It cannot exchange places with any of its neighbors. As a liquid or a gas, the molecules are not in fixed positions; they move. In the liquid, the collection of molecules occupies a definite volume. In the gas, the molecules move to fill the entire volume of their container.

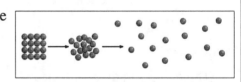

(a) Draw three squares enclosing the "volume" available to a molecule in the solid, liquid, and gas represented here. If the volume per molecule determines the number of molecular arrangements in a system, which phase has the largest number of arrangements available? What is the implication of your answer for the relative probabilities of the three phases?

(b)   WEB   Chap 8, Sect 8.7.1-3. Explain how the movies and interactive questions on these pages of the *Web Companion* are related to the diagram here and your responses in part (a).

As you have learned, in a solid, the molecules (or atoms) are not free to move about, except to vibrate in place. The volume available to each molecule is only slightly larger than the molecule itself. In a liquid, the molecules can move about anywhere within the liquid volume, so the volume available to each molecule is the entire volume of the liquid. Similarly, molecules in the gas phase can move about in the entire volume occupied by the gas. Since gases occupy larger volumes than liquids, gases always have more distinguishable arrangements than liquids, which, in turn, always have more distinguishable arrangements than solids. Therefore, for a given substance, the positional entropy of its gas is greater than the entropy of its liquid, which is greater than the entropy of its solid. When a gas expands, the molecules are farther apart, and the expanded gas has more distinguishable arrangements, because all the molecules have a larger volume to move in. Expansion of gases, as well as phase changes from solid to liquid, liquid to gas, and solid to gas, increase the positional entropy of the system.

8.11. **Check This**   *Direction of positional entropy changes*

For each of these changes, tell whether the positional entropy of the system increases, decreases, or stays the same, and explain why.

(a) Ice melts to form liquid water.

(b) Water vapor in the air condenses to form a cloud.

(c) Equal volumes of ethanol and water are mixed to form a homogenous solution.

(d) Water is poured from a graduated cylinder into a beaker.

(e) A piece of dry ice (solid carbon dioxide) sublimes to form carbon dioxide gas.

(f) Ethene (ethylene) molecules are polymerized to form polyethylene.

A dilemma

The preceding discussion leads to a dilemma. If the liquid phase is more probable (has more molecular arrangements) than the solid phase, and if the gaseous phase is more probable still, why are substances ever found in condensed phases? To answer this question, we have to consider the *net* entropy change for these changes (and all other changes as well). Recall that the net entropy change for a process is the sum of all the positional and thermal entropy changes for the process.

Melting and freezing water

To see how positional and thermal entropy combine to determine the direction of change, let's examine two familiar processes: the melting of ice and the freezing of water at a constant pressure of 1 bar, standard conditions. Our approach will be to consider these changes at

---

temperatures where we know they occur spontaneously and determine the relationship between the positional and thermal entropy changes that *must* be true, in order to give a positive net entropy change in each case. Then we will look at our models for positional and thermal entropy changes to see how they can be reconciled with the required relationships.

Melting changes a solid to a liquid, a change of the system from a small number of molecular arrangements (solid water) to a larger number (liquid water). The positional entropy of the water molecules increases in this change, $S^{\circ}_{liquid} > S^{\circ}_{solid}$, and favors the formation of the liquid:

$$\Delta S_{system\,(s \to l)} = \Delta S^{\circ}_{system\,(s \to l)} = S^{\circ}_{liquid} - S^{\circ}_{solid} > 0 \tag{8.8}$$

For ice melting, the standard entropy change, $\Delta S^{\circ}_{s \to l}$, is 22.0 J·K^{-1}·mol^{-1}. But melting does not occur in an isolated piece of ice. Melting occurs when energy is supplied from the surroundings to break some of the hydrogen bonds that hold the water molecules in position in the crystal.

8.12. Consider This   *What happens in an ice-water mixture at different temperatures?*

The melting point of ice is 273 K. The temperature of a mixture of ice and liquid water is 273 K. Figure 8.10 shows a container of this mixture sitting on a large block of metal which is at a temperature T.

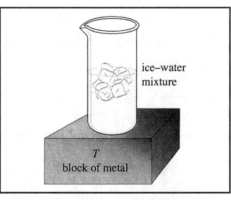

Figure 8.10. Ice-water mixture in contact with a large block of metal.

(a) If $T > 273$ K, what will you observe happening to the ice-water mixture? (This is equivalent to placing the container on a room-temperature countertop.) Will energy leave the block and enter the ice-water system or will energy transfer be from the ice-water to the block? Explain your response.

(b) If $T < 273$ K, what will you observe happening to the ice-water mixture? (This is equivalent to placing the container in the freezing compartment of a refrigerator.) Will energy leave the block and enter the ice-water system or will energy transfer be from the ice-water to the block? Explain your response.

You know that the ice in a glass of ice water will melt, if the glass is left on a counter. In order to melt the ice, energy must enter the mixture to break hydrogen bonds in the solid ice. The standard enthalpy change for ice melting is $\Delta H°_{s\to l} = 6.00 \text{ kJ·mol}^{-1}$. For the process in Consider This 8.12(a), this energy (enthalpy) has to come from the surroundings, which we take to be the block of metal at $T > 273$ K. For every mole of ice that melts, the enthalpy change in the surroundings is:

> We assume the ice-water mixture and metal block together are an isolated system: the only thermal transfers are between these components.

$$\Delta H_{surr\,(s\to l)} = -\Delta H°_{s\to l} = -6.00 \text{ kJ·mol}^{-1} \tag{8.9}$$

The surroundings lose energy, the thermal entropy of the surroundings decreases; $\Delta S_{surr\,(s\to l)} < 0$.

Net entropy change for ice melting. The net entropy change for the ice melting is the sum of the positional entropy change for the system (ice-water mixture) and the thermal entropy change for the surroundings (block of metal):

$$\Delta S_{net\,(s\to l)} = \Delta S_{system\,(s\to l)} + \Delta S_{surr\,(s\to l)} \tag{8.10}$$

Under the given conditions, $T > 273$ K, melting is the observed process, so this net entropy change must be positive. Figure 8.11 shows the relationship between the positional and thermal entropies which *must* be true for $\Delta S_{net\,(s\to l)} > 0$, that is, for the final entropy to be larger than the initial entropy. The arrows in the figure show that $|\Delta S_{system\,(s\to l)}| > |\Delta S_{surr\,(s\to l)}|$. The magnitude of $\Delta S_{system\,(s\to l)}$ must be greater than the magnitude of $\Delta S_{surr\,(s\to l)}$.

> **WEB Chap 8, Sect 8.7.5-6**
> Use this interactive activity to reinforce your understanding of these phase changes.

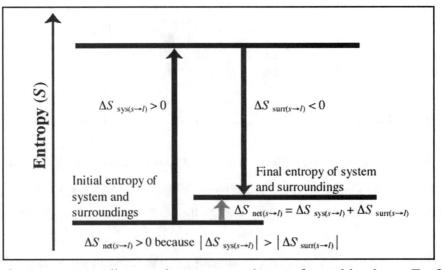

Figure 8.11. System, surroundings, and net entropy changes for melting ice at $T > 273$ K.
On these entropy diagrams, blue arrows represent positional entropy change in the system, red arrows represent thermal entropy changes in the thermal surroundings, and green arrows represent the net entropy change (the sum of the positional and thermal entropy changes).

8.13.   **Check This**   *Net entropy change for water freezing*

Figure 8.12 shows the relationship between the positional and thermal entropies which *must* be true, if $\Delta S_{net\,(l\rightarrow s)} > 0$, for water freezing at T < 273 K. This is the process in Consider This 8.12(b). Freezing water is the reverse of melting ice, so $\Delta S_{system\,(l\rightarrow s)} = -\Delta S_{system\,(s\rightarrow l)}$. Thus, the arrows representing ΔS_{system} in Figures 8.11 and 8.12 are the same length, but opposite in direction. Also, because freezing is the reverse of melting, $\Delta H_{surr\,(l\rightarrow s)} = -\Delta H_{surr\,(s\rightarrow l)}$.

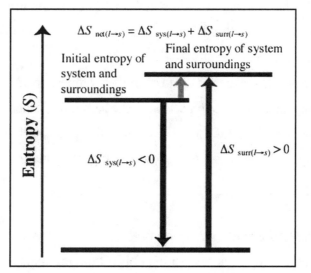

Figure 8.12. System, surroundings, and net entropy changes for freezing water at *T* < 273 K. The colors of the arrows have the same meaning as in Figure 8.11.

(a) Explain why the arrows representing $\Delta S_{system\,(l\rightarrow s)}$ and $\Delta S_{surr\,(l\rightarrow s)}$ go in the directions shown in Figure 8.12.

(b) Figure 8.11 showed the relationship between $|\Delta S_{system\,(s\rightarrow l)}|$ and $|\Delta S_{surr\,(s\rightarrow l)}|$, in order for $\Delta S_{net\,(s\rightarrow l)} > 0$. Write the appropriate relationship between $|\Delta S_{system\,(l\rightarrow s)}|$ and $|\Delta S_{surr\,(l\rightarrow s)}|$, in order for $\Delta S_{net\,(l\rightarrow s)} > 0$, that is, for the final entropy to be larger than the initial entropy, as shown in Figure 8.12.

Magnitude of thermal entropy changes

Figures 8.11 and 8.12 show that the positional (system) entropy changes for melting and freezing have opposite signs, as we would expect, since melting increases positional entropy and freezing decreases it. Similarly, the thermal (surroundings) entropy changes for melting and freezing have opposite signs. The thermal entropy of the surroundings decreases as energy is lost from the surroundings to melt ice. The thermal entropy of the surroundings increases as energy is gained by the surroundings when freezing water releases it. What may seem peculiar about these changes is that the magnitude of the thermal entropy change in the surroundings is different for the different temperatures of the surroundings, above and below 273 K. Since it is this difference that is responsible for ice melting when T > 0 and water freezing when $T < 0$, we need to examine these thermal entropy changes in more detail.

8.14. **Consider This**   *How does thermal entropy change depend upon temperature?*

   (a) Use the data from Figure 8.9 and the definition of entropy, equation (8.6), to calculate the change in entropy, ΔS, for the four-atom solid when it loses one quantum of energy to go from 10 to 9 quanta. Do the same for the loss of one quantum to go from 5 to 4 quanta. For which case is the *change* in entropy larger?

   (b) Which change in part (a) takes place at the higher temperature? Explain how you know.

   (c) The energy change in each case in part (a) is the same, one quantum, but the entropy changes and temperatures are different for the two cases. Propose a relationship that describes the relative sizes of entropy changes as equal energy changes are made at different temperatures. *Hint:* Recall the units of entropy.

Quantitative expression for thermal entropy change

   The transfer of one quantum of energy from an object having many quanta has less effect on the number of distinguishable arrangements than does the transfer of one quantum from an object having fewer quanta. The entropy decrease for loss of energy from an object at a higher temperature is smaller than the entropy decrease for the loss of the same amount of energy from the same object at a lower temperature. This result is general. *For a given energy change in a substance, the change in thermal entropy, ΔS, is always greater at a lower temperature.*

   You might find an analogy helpful in remembering this important result. Think of energy in terms of money. Suppose you have two friends, one of whom has $1000 and the other only $10. If you give each friend $5, which one will be affected more by getting the extra money? If, instead of giving each friend money, you ask each to contribute $5 to buy food for a small party, which one would be affected more by giving up the money? In both cases, the friend who has less money will be more affected by getting or giving the $5. The friend with less money (energy) corresponds to a system at lower temperature (less money per person). The greater monetary effect for your less wealthy friend is analogous to the greater change in entropy at a lower temperature.

   For constant pressure systems, we express thermal entropy change quantitatively as:

$$\Delta S = \frac{\Delta H}{T} \tag{8.11}$$

This is the *form* of the relationship you should have written for Consider This 8.14(c): for a given change in energy (enthalpy), the thermal entropy change is inversely proportional to the temperature. Equation (8.11) shows you why the units of entropy are $J \cdot K^{-1} \cdot mol^{-1}$. We could start from $S = k \ln W$, the definition of entropy, apply it to more realistic systems than you did in Consider This 8.14, and derive equation (8.11). In practice, we use equation (8.11) to express ΔS for energy transfers, because ΔH and T are easily measured quantities and finding all the ways to arrange thermal energy quanta in macroscopic systems is hard. We will not try to prove or derive equation (8.11), but will show examples of its application, We begin by continuing our present examples, melting ice and freezing water.

8.15. Worked Example   *Net entropy change for melting ice with surroundings at 283 K*

If the temperature of the block of metal in Figure 8.10 is 283 K, what is the net entropy change for melting one mole of ice in the ice-water mixture under standard conditions?

Necessary Information: We need to know that, for melting ice, $\Delta S^{\circ}_{s \to l} = 22.0 \ J \cdot K^{-1} \cdot mol^{-1}$ and $\Delta H^{\circ}_{s \to l} = 6.00 \ kJ \cdot mol^{-1}$.

Strategy: Calculate the thermal entropy change for the surroundings from equation (8.11) and add it to the positional entropy change for the system to get the net entropy change.

Implementation: Since the energy required to melt the ice leaves the surroundings, $\Delta H_{surr} = -\Delta H^{\circ}_{s \to l} = -6.00 \ kJ \cdot mol^{-1}$. The thermal entropy change for the surroundings is:

$$\Delta S_{surr} = \frac{\Delta H_{surr}}{T_{surr}} = \frac{-6.00 \times 10^3 \ J \cdot mol^{-1}}{283 \ K} = -21.2 \ J \cdot K^{-1} \cdot mol^{-1}$$

The positional entropy change for the system is:

$$\Delta S_{system} = \Delta S^{\circ}_{s \to l} = 22.0 \ J \cdot K^{-1} \cdot mol^{-1}$$

The net entropy change for melting ice with the surroundings at 283 K is:

$$\Delta S_{net} = \Delta S_{system} + \Delta S_{surr} = 22.0 \ J \cdot K^{-1} \cdot mol^{-1} + (-21.2 \ J \cdot K^{-1} \cdot mol^{-1}) = 0.8 \ J \cdot K^{-1} \cdot mol^{-1}$$

Does the Answer Make Sense? Our result shows that the net entropy change is positive for ice melting at 283 K, so the process should occur. Since the temperature of the surroundings is above 273 K, we know from experience that ice does melt. This problem is an example of the general case illustrated in Figure 8.11.

8.16. **Check This**   *Net entropy change for freezing water with surroundings at 263 K*

If the temperature of the block of metal in Figure 8.10 is 263 K, what is the net entropy change for freezing one mole of water in the ice-water mixture? Does your result predict that this process will be observed? Explain why or why not.

The dilemma resolved

Now we have the answer that resolves the dilemma raised by the results of Consider This 8.10 and posed at the beginning of the section: Why are substances ever found in condensed phases? An input of enthalpy *from* the surroundings is required to change a system from a low entropy condensed phase to a higher entropy phase; $\Delta H_{surr} < 0$. The familiar changes are solid to liquid (melting), liquid to gas (vaporization), and solid to gas (sublimation). Both the gain of positional entropy by the system ($\Delta S_{system} > 0$) and the loss of thermal entropy by the surroundings must be accounted for, as they are in this expression for the net entropy change:

$$\Delta S_{net} = \Delta S_{system} + \Delta S_{surr} = \Delta S_{system} + \frac{\Delta H_{surr}}{T} \tag{8.12}$$

The thermal entropy change for the surroundings is negative, but its numerical value decreases as the temperature increases. Thus, *at higher temperatures*, ΔS_{system} dominates equation (8.12), the net entropy change is positive, and *changes to phases of higher positional entropy are favored.*

From the preceding discussion, you can conclude that three factors influence the direction of a phase change.

- If the *positional entropy* of the system increases, the phase change is favored. Gases have greater positional entropy than liquids and liquids have greater positional entropy than solids.
- If *thermal energy* is transferred from the system to the surroundings, the phase change is favored. Exothermic changes (gases condensing to liquids or liquids freezing to solids) always increase the *thermal entropy* of the surroundings.
- Entropy change in the thermal surroundings is inversely proportional to the *temperature* of the surroundings.

8.17. **Consider This**   *What are net entropy changes for ice and water changes at 273 K?*

(a) Use equation (8.12) to determine the net entropy change for melting a mole of ice at 273 K and standard pressure, if it is in contact with surroundings that are also at 273 K. Would you expect any of the ice to melt? Explain why or why not.

(b) Do the same calculation for a mole of water freezing at 273 K when the surroundings are at 273 K. Would you expect any of the water to freeze? Why or why not?

Phase equilibrium

In Consider This 8.17, you were asked to predict what change, if any, would occur when ice or water at 273 K (the system) is placed in contact with surroundings at 273 K. You found in part (a) that, under these conditions, ΔS_{net}(ice melting) = 0.0 J·K^{-1}. And, in part (b), you found that ΔS_{net}(water freezing) = 0.0 J·K^{-1} for water at 273 K in contact with surroundings at 273 K. Neither the liquid nor the solid phase is favored at 273 K. There will be no *net* change in the amount of ice or water in the system and no *net* transfer of energy between the ice and/or water system and the surroundings. The system is said to be in **equilibrium**. There is no change in entropy for the change in either direction; $\Delta S_{net} = 0$. Under equilibrium conditions, the entropy changes in the system and the surroundings are exactly balanced for a change in either direction.

Section 8.8.   Gibbs Free Energy

Focusing on the system

Under standard conditions, the direction of a phase change can be determined using standard enthalpies and entropies in equation (8.12) to calculate the net entropy change for the system and surroundings. However, we are not usually interested in what is happening in the surroundings. In many cases, the surroundings are simply a convenient place to which to transfer thermal energy when the change is exothermic or a convenient source of thermal energy when the change is endothermic. The thermodynamic relationships that enable us to focus on changes in the system without explicitly considering the surroundings grow out of the work of J. Willard Gibbs, (American mathematical physicist, 1839-1903). The concepts Gibbs developed, enable us to use only system variables to determine which changes in the system can occur, and which changes are impossible. A new state function that resulted from Gibbs work is called the **Gibbs free energy**, *G*. Our task in this section is to discover what the Gibbs free energy is and how it is related to the ideas we have already developed.

The entropy change in the thermal surroundings depends only on the thermal energy transferred to or from the system at constant pressure and temperature. The actual process that occurs in the system to produce this energy transfer is irrelevant. Since thermal energy lost by the system is gained by the surroundings and *vice versa*, we can write:

$$\Delta H_{surr} = -\Delta H^{\circ}_{system} \tag{8.13}$$

We have already used this relationship (for $\Delta H_{system} = \Delta H^{\circ}_{phase\ change}$) in Worked Example 8.15 and, following that example, you used it in Check This 8.16 and Consider This 8.17. We do not have to restrict ourselves to standard conditions, but can also consider changes that occur at any constant pressure, that is, $\Delta H_{surr} = -\Delta H_{system}$. For these more general conditions, equation (8.12) for the net entropy change can be written as:

$$\Delta S_{net} = \Delta S_{system} - \frac{\Delta H_{system}}{T_{surr}} \tag{8.14}$$

ΔS_{net} as a function of system variables

If changes occur in systems at the same temperature as the thermal surroundings, we have $T_{system} = T_{surr} = T$ and equation (8.14) becomes:

$$\Delta S_{net} = \Delta S_{system} - \frac{\Delta H_{system}}{T} \tag{8.15}$$

Equation (8.15) provides us a powerful tool for understanding and predicting the direction of change in chemical systems, because it gives the net entropy change (whose sign tells us the direction of change) in terms of system variables only. As you use equation (8.15), keep in mind its restrictions: the system changes have to be measured at constant pressure and temperature and the temperatures of the system and thermal surroundings have to be the same.

8.18. Worked Example   *Formation of $N_2O_4(g)$ under standard conditions at 298 K*

The formation of $N_2O_4(g)$, dinitrogen tetroxide, from $NO_2(g)$, nitrogen dioxide, is represented by this equation:

$$2NO_2(g) \rightleftharpoons N_2O_4(g)$$

For the formation of one mole of $N_2O_4(g)$ at 298 K, the standard entropy and enthalpy changes are -175.8 J·K^{-1} and -57.2 kJ, respectively (from Appendix XX). Do these values seem to make sense in terms of positional entropies and bond formations? Can this reaction proceed to form $N_2O_4(g)$ under standard conditions?

Necessary Information: We need to recall that energy is released when bonds are formed.

Strategy: Use our models for entropy change to decide whether the entropy and enthalpy changes make sense. Then, substitute the numeric values for entropy, enthalpy, and temperature into equation (8.15) to get the net entropy change for $N_2O_4(g)$ formation and look at its sign to determine if the reaction is possible.

Reasoning: The reaction changes two moles of gaseous molecules to one mole. The formation of fewer product molecules means that the atoms in the products have less freedom to move about separately. For example, any nitrogen in the NO_2 reactant can be found anywhere in the system, but, after the reaction, pairs of nitrogens must be found together in half as many N_2O_4 molecules. Thus, a decrease in number of moles of gas in the reaction gives a negative value for the positional entropy change for the reaction, which is consistent with the experimental value.

The formation of the bond between two NO_2 molecules to give N_2O_4 releases energy, so the reaction must be exothermic, as the experimental value shows.

Implementation:

$$\Delta S_{net} = \Delta S_{system} - \frac{\Delta H_{system}}{T} = -175.8 \ J \cdot K^{-1} - \left(\frac{-57.2 \times 10^3 \ J}{298 \ K} \right) = -175.8 \ J \cdot K^{-1} + 191.9 \ J \cdot K^{-1}$$

$$\Delta S_{net} = 16.1 \ J \cdot K^{-1}$$

This positive value for ΔS_{net} means that the formation of N_2O_4 from NO_2 under standard conditions at 298 K is possible.

Does the Answer Make Sense? The large exothermicity of the reaction leads to a large positive change in the thermal entropy of the surroundings [accounted for as $(- \Delta H_{system}/T)$], which compensates for the negative change in the positional entropy of the reacting system. Since the thermal entropy change in the surroundings will decrease as the temperature increases, the reaction might not be favored at a higher temperature.

8.19.   **Check This**   *Formation of $N_2O_4(g)$ under standard conditions at 350 K*

Can the reaction in Worked Example 8.18, proceed under standard conditions at 350 K? Assume that the standard entropy and enthalpy changes for the reaction are the same at 350 K and 298 K. Is your conclusion the same as in Worked Example 8.18? Why or why not?

Shift to an energy perspective

Equation (8.15) can be rearranged by multiplying through by T to get:

$$T\Delta S_{net} = T\Delta S_{system} - \Delta H_{system} \tag{8.16}$$

All the terms in equation (8.16) have units of energy. Absolute temperatures are always positive, so the sign of $T\Delta S_{net}$ is determined by the sign of ΔS_{net} and is positive (> 0) for reactions that are possible and negative (< 0) for those that are impossible. However, the *conventional* way we deal with energies, is to *express changes that lead to a lower total energy* (the favorable direction) *as negative*, since the final energy is lower than the initial. Multiplying through equation (8.16) by -1 yields:

$$-T\Delta S_{net} = -T\Delta S_{system} + \Delta H_{system} = \Delta H_{system} - T\Delta S_{system} \tag{8.17}$$

8.20.   **Check This**   *Criterion for direction of change*

In terms of $-T\Delta S_{net}$, what are the criteria for reactions that are possible and impossible?

Gibbs free energy and the direction of change

For our purposes in this chapter and the remainder of the book, we could use the numeric values of $-T\Delta S_{net}$ as the criterion for spontaneity and equilibrium. However, for several reasons, some historical and some practical, a new thermodynamic function, the **Gibbs free energy**, G, (or simply the **free energy**) is usually used. The Gibbs free energy for a system is defined as:

$$G \equiv H - TS \tag{8.18}$$

In equation (8.18) all the variables refer to the system. *For a change in a system at constant pressure and temperature*, the change in Gibbs free energy, ΔG, is:

$$\Delta G = \Delta H_{system} - T\Delta S_{system} \tag{8.19}$$

If we compare equation (8.19) with equation (8.17), we see that:

$$\Delta G = - T\Delta S_{net} \tag{8.20}$$

Thus, in terms of ΔG, equation (8.19) gives the same criteria for changes that are possible and impossible as you found for $-T\Delta S_{net}$ in Check This 8.20:

$$\text{If } \Delta G \ (= - T\Delta S_{net} = \Delta H_{system} - T\Delta S_{system}) < 0, \text{ the change is possible.} \tag{8.21}$$

$$\text{If } \Delta G \ (= - T\Delta S_{net} = \Delta H_{system} - T\Delta S_{system}) > 0, \text{ the change is impossible.} \tag{8.22}$$

$$\text{If } \Delta G \ (= - T\Delta S_{net} = \Delta H_{system} - T\Delta S_{system}) = 0, \text{ the change is at equilibrium.} \tag{8.23}$$

8.21. **Worked Example** *Free energy change for urea formation at 298 K*

The formation of urea from ammonia and carbon dioxide is represented by this equation:

$$2NH_3(g) + CO_2(g) \rightleftharpoons H_2NC(O)NH_2(s) + H_2O(l)$$

For the formation of one mole of urea at 298 K, the standard entropy and enthalpy changes are -424 J·K^{-1} and -133.3 kJ, respectively. Calculate the standard Gibbs free energy change, $\Delta G°$ (as usual, "°" denotes the change under standard conditions), for formation of one mole of urea at 298 K. Does the free energy show that this change is possible?

Necessary information: We need the criteria represented by equations (8.21) – (8.23).

Strategy: Substitute the values for T, $\Delta H°_{system}$, and $\Delta S°_{system}$ into equation (8.19) and look at the sign of $\Delta G°$ to see if urea formation is possible under standard conditions at 298 K.

Implementation:

$$\Delta G° = \Delta H°_{system} - T\Delta S°_{system} = -133.3 \times 10^3 \text{ J} - (298 \text{ K})·(-424 \text{ J·K}^{-1})$$

$$\Delta G° = -133.3 \times 10^3 \text{ J} + 126 \times 10^3 \text{ J} = -7 \times 10^3 \text{ J} = -7 \text{ kJ}$$

A negative value for the free energy change, $\Delta G = \Delta G°$ in this example, means that the formation of urea under standard conditions at 298 K is possible.

Does the Answer Make Sense? We have no independent way to know whether this answer is correct, without knowing something about the reaction. Many organisms excrete nitrogenous waste by converting ammonia to urea by this reaction, so it is likely to be a possible reaction under these conditions, although perhaps not under others.

8.22.   Check This   *Free energy change for urea formation at 325 K*

Calculate the standard Gibbs free energy change, $\Delta G°$, for formation of one mole of urea at 325 K. Assume that the standard entropy and enthalpy changes for the reaction are the same at 325 K and 298 K. Is your conclusion the same as in Check This 8.18? Why or why not?

The relationships in equations (8.21) – (8.23) are illustrated graphically in Figure 8.13 for the solid-to-liquid phase change for water under standard conditions. ΔH and ΔS are assumed to be constant over the temperature range shown. You have made this same assumption in Check This 8.18 and 8.21. The assumption is not strictly true. Entropies and enthalpies do change with temperature, but the changes (for liquids and solids) are not large and can usually be neglected, if the range of temperatures is only a few tens of degrees.

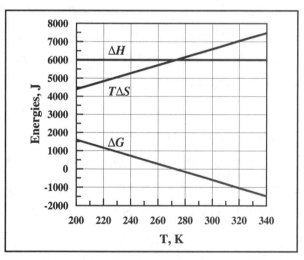

Figure 8.13. ΔH, $T\Delta S$, and ΔG vs. *T* for the change $H_2O(s) \rightarrow H_2O(l)$ at standard pressure.

8.23.   Worked Example   *Graphical relationship of $\Delta H°$, $T\Delta S°$, and $\Delta G°$ vs. T*

Show how the values for ΔG, at 240 K and 300 K, in Figure 8.13 are derived from the plotted values for ΔH and $T\Delta S$.

Necessary information: We need equation (8.19).

Strategy: Read the ΔH and $T\Delta S$ values at these temperatures from the graph and then calculate $\Delta H - T\Delta S$ to get ΔG.

Implementation:

At 240 K: $\Delta H = 6000$ J, $T\Delta S = 5300$ J, and $\Delta G = 6000 - 5300 = 700$ J

At 300 K: $\Delta H = 6000$ J, $T\Delta S = 6600$ J, and $\Delta G = 6000 - 6600 = -600$ J

Does the Answer Make Sense? The values we calculated are the values for ΔG on the plot.

8.24. Check This   *Significance of the graphical relationship of ΔH, $T\Delta S$, and ΔG vs. T*

What is the significance of the point where the ΔH and $T\Delta S$ lines cross in Figure 8.13? Under what condition(s) can the solid-to-liquid phase change occur? Under what condition(s) is the change not possible? Explain the reasoning for your answers.

Criterion for equilibrium

The criterion for equilibrium in equation (8.23) is a generalization of what we discussed at the end of Section 8.7 for phase equilibria. Under standard conditions, $\Delta S = \Delta S°_{net} = 0$ for a phase change at the temperature where the two phases are in equilibrium. More generally, *systems are at equilibrium when $\Delta G = -T\Delta S_{net} = 0$ for a change from the initial to final state under any constant pressure and temperature conditions.* For the solid-to-liquid phase change in water under standard conditions, Figure 8.13 shows that $\Delta G = \Delta G° = 0$ at 273 K. We know solid and liquid water are in equilibrium at this temperature.

8.25. Check This   *The melting point of benzene*

The standard enthalpy of formation and standard entropy of solid benzene, C_6H_6, are 38.4 kJ·mol^{-1} and 135 J·K^{-1}·mol^{-1}, respectively. The corresponding values for liquid benzene are 49.0 kJ·mol^{-1} and 173 J·K^{-1}·mol^{-1}. Use these data to calculate $\Delta H°$ and $\Delta S°$ for the reaction:

$$C_6H_6(l) \rightarrow C_6H_6(s)$$

Use equation (8.23) to estimate the freezing temperature of benzene under standard conditions, assuming that $\Delta H°$ and $\Delta S°$ do not change with temperature. See Table 8.2 to check your result.

Why a new thermodynamic variable?

In a more detailed treatment of thermodynamics, the Gibbs free energy plays a broader role than we are presenting here and it is convenient to have a single variable (that accounts for enthalpy and entropy) to make the mathematics easier. From a practical point of view, tables of thermodynamic properties (Appendix XX) always list standard free energies of formation (the difference in free energy between the compound in its standard state and its elements in their standard states) together with standard enthalpies of formation and standard entropies. Thus, it is easy to get ΔG° ($= -T\Delta S^\circ_{net}$) for a reaction. For many reactions in biological systems, enthalpies and entropies of reaction can't be measured, but the free energy of reaction can be determined from equilibrium measurements, as we will see in Chapter 9.

Why "free" energy?

Equation (8.19), $\Delta G = \Delta H_{system} - T\Delta S_{system}$, is a succinct combination of the first and second laws of thermodynamics. For any reaction, the internal energy change (exclusive of the pressure-volume work) is included in the enthalpy term. The entropy term takes account of the change in molecular arrangements that occurs during the reaction. What is "left over" is the *free* energy, that is, the energy that is *available* to do other forms of work in the surroundings. In Chapter 7, Section 7.11, we did a calculation to figure out the amount of work available from the thermal energy released by burning fuel. For that calculation, we equated the amount of work to the thermal energy change of the reaction, but noted that it is not possible to convert thermal energy completely into work. Equation (8.19) now shows that it is entropy, the second law, that limits this conversion and also gives the maximum amount of work that *is* available, the free energy. No real device can actually produce this much work in the surroundings, because there are always energy losses in its internal workings.

Reflection and Projection

The last three sections have taken the qualitative idea of changes proceeding in the direction that produces larger numbers of distinguishable molecular arrangements and expressed it in quantitative terms using entropy and enthalpy values for the ice–water phase change as concrete examples. The switch we made from a perspective that includes both the system and surroundings to a focus on the system alone requires close attention to the restrictions placed on the application of equation (8.15). The changes for which the equation can be used must take place at constant temperature and pressure and the temperature of the system and surroundings must be the same.

A new thermodynamic function, the Gibbs free energy, G, was introduced. Under constant temperature and pressure conditions, the Gibbs free energy change, ΔG, for a change in a system is $\Delta G = \Delta H_{system} - T\Delta S_{system}$. Within these constraints, we can predict whether a change is possible by determining the sign of its free energy change. If $\Delta G < 0$, the change is possible. If $\Delta G > 0$, the change is not possible. If $\Delta G = 0$, the initial and final states are in equilibrium.

We will use both free energy and entropy, whichever is more convenient in a particular case, to help understand physical and chemical changes. The next sections of this chapter deal with chemical reactions and then further examples of phenomena, some of which may be surprising, but many of which are familiar, that will provide practice using our criteria for the direction of change.

Section 8.9.   Thermodynamic Calculations for Chemical Reactions

Standard molar entropies of many compounds are given in Appendix XX. These values are derived from statistical calculations based on equation (8.6), from calorimetric experiments, and from equilibrium measurements. Appendix XX also provides values for the standard molar Gibbs free energies of formation of these compounds. The values for entropies and free energies can be combined, just as those for energies and enthalpies are combined, to calculate entropy and free energy changes for processes of interest. In Chapter 7, we combined tabulated values for standard enthalpies of formation to obtain the standard enthalpy changes for reactions. Here, we'll do the same for standard entropy and standard free energy changes.

General equation for calculating entropy changes

The overall standard entropy change for a reaction is the sum of all the standard entropies of the products minus the sum of all the standard entropies of the reactants. The mathematical expression for the difference of sums is:

$$\Delta S^{\circ}_{reaction} = \Sigma[n_j(S^{\circ})_j]_{products} - \Sigma[n_j(S^{\circ})_j]_{reactants} \tag{8.24}$$

In words, equation (8.24) tells us: "Multiply the number of moles of each product by the standard entropy for that product. Add all the product entropy terms together. Multiply the number of moles of each reactant by the standard entropy for that reactant. Add all the reactant energy terms together. Finally, subtract the reactant sum from the product sum." Equation (8.24) for the standard entropy change of a reaction is identical in form to equation (7.30), Chapter 7, Section 7.8, for the standard enthalpy change of a reaction.

Entropy change for glucose oxidation

An example we have used before is the oxidation of glucose to give carbon dioxide and water:

$$C_6H_{12}O_6(s) + 6O_2(g) \rightarrow 6CO_2(g) + 6H_2O(l) \tag{8.25}$$

In terms of molecules, equation (8.25) tells us that seven reactant molecules combine to produce twelve product molecules, as represented in Figure 8.14. Atoms that were bonded to one another in the reactants (the carbon atoms in glucose, for example) are less constrained in the products.

For example, the six carbon atoms, originally all in one glucose molecule, Figure 8.14(a), are freed to spread throughout the volume available in six separate carbon dioxide molecules, Figure 8.14(b). The number of molecular arrangements available to the products of the reaction is much greater than the number of molecular arrangements available to the reactants. Without doing any calculations, we can conclude that the collective entropy of the products is larger than the entropy of the reactants. The entropy change for the reaction described in equation (8.25) should be positive, and it may be quite large.

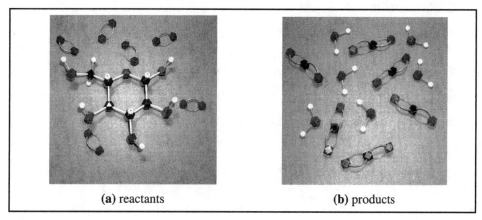

(a) reactants   (b) products

Figure 8.14. Molecular level representation of the oxidation of a glucose molecule.

8.26.   **Worked Example**   *Entropy change for oxidation of glucose*

Calculate the standard entropy change when one mole of glucose is oxidized by oxygen, reaction equation (8.25), at 298 K. Compare the result with the prediction that the entropy change should be positive and quite large.

Necessary Information: We need equations (8.24) and (8.25) and, from Appendix XX, the standard molar entropies, $S°$, at 298 K for each reactant and product.

compound	$S°$, $J \cdot K^{-1} \cdot mol^{-1}$
$C_6H_{12}O_6(s)$	212.1
$O_2(g)$	205.1
$CO_2(g)$	213.7
$H_2O(l)$	69.91

Strategy: Use equation (8.24) and the stoichiometry of reaction equation (8.25) to combine the standard molar entropies to get the standard entropy change for the reaction. Compare the result with our prediction that the entropy change should be large and positive.

Implementation:

$$\Delta S^{\circ}_{reaction} = \Sigma[n_j(S^{\circ})_j]_{products} - \Sigma[n_j(S^{\circ})_j]_{reactants}$$

$$= \{[(6 \text{ mol } CO_2)(213.7 \text{ J·K}^{-1}\text{·mol}^{-1})] + [(6 \text{ mol } H_2O)(69.91 \text{ J·K}^{-1}\text{·mol}^{-1})]\}$$

$$- \{[(1 \text{ mol } C_6H_{12}O_6)(212.1 \text{ J·K}^{-1}\text{·mol}^{-1})] + [(6 \text{ mol } O_2)(205.1 \text{ J·K}^{-1}\text{·mol}^{-1})]\}$$

$$= +259.0 \text{ J·K}^{-1}$$

Does the Answer Make Sense? The large, positive standard entropy change is consistent with our prediction that there are more ways to arrange the product molecules than there are to arrange the reactant molecules.

8.27. Check This *Predict the direction of the entropy change for glucose fermentation*

Humans have known how to produce ethanol by fermentation (Chapter 6, Section 6.11) for almost seven thousand years. The reaction for glucose fermentation is:

$$C_6H_{12}O_6(s) \rightarrow 2C_2H_5OH(l) + 2CO_2(g)$$

Do you expect the entropy change for this reaction to be positive or negative? Relatively large or relatively small? Explain your responses.

8.28. Check This *Entropy change for glucose fermentation*

Use standard entropies, S°, from Appendix XX to calculate $\Delta S^{\circ}_{reaction}$ for glucose fermentation. Do your calculated results agree with your prediction in Check This 8.27?

Free energy change for glucose oxidation

Both the entropy change in the system and the thermal entropy change for energy transfer between system and surroundings must be considered to determine whether a reaction can proceed in the direction written. We cannot be sure that glucose oxidation, reaction (8.25), can occur until we know the free energy change, $\Delta G_{reaction}$, for the reaction. We can use standard enthalpies of formation from Appendix XX and equation (7.30) to calculate $\Delta H^{\circ}_{reaction}$:

$$\Delta H^{\circ}_{reaction} = \Sigma[n_j(\Delta H_f^{\circ})_j]_{products} - \Sigma[n_j(\Delta H_f^{\circ})_j]_{reactants} \qquad (7.30)$$

We then substitute $\Delta H^{\circ}_{reaction}$ and the $\Delta S^{\circ}_{reaction}$, calculated in Worked Example 8.26, into equation (8.19) to calculate $\Delta G^{\circ}_{reaction}$ $(= \Delta H^{\circ}_{reaction} - T\Delta S^{\circ}_{reaction})$.

8.29. **Worked Example**   *Free energy change for oxidation of glucose*

Use the standard enthalpy and entropy of reaction to calculate the standard free energy change when one mole of glucose is oxidized by oxygen at 298 K.

Necessary Information: You calculated $\Delta H^{\circ}_{reaction} = -2801$ kJ in Chapter 7, Check This 7.50, and we calculated $\Delta S^{\circ}_{reaction} = 259.0$ J·K^{-1} in Worked Example 8.26.

Strategy: Substitute $\Delta H^{\circ}_{reaction}$ and $\Delta S^{\circ}_{reaction}$ into equation (8.19) to get $\Delta G^{\circ}_{reaction}$.

Implementation:

$$\Delta G^{\circ}_{reaction} = (-2801 \text{ kJ}) - (298 \text{ K})(+259.5 \text{ J·K}^{-1}) = (-2801 \text{ kJ}) - (+77.3 \text{ kJ}) = -2878 \text{ kJ}$$

Does the Answer Make Sense? The information we get from the enthalpy and entropy is that both $\Delta H^{\circ}_{reaction}$ and $\Delta S^{\circ}_{reaction}$ favor the oxidation of glucose. Thermal energy is transferred to the surroundings as the reaction progresses and the products of the reaction can be arranged in more ways than the reactants. The large, negative value for the standard free energy change confirms that there is a large driving force for the oxidation of glucose, which is an essential reaction in the metabolism of both plants and animals.

8.30. **Check This**   *Free energy change for fermentation of glucose*

Use the standard enthalpies of formation from Appendix XX to calculate $\Delta H^{\circ}_{reaction}$ for the fermentation of glucose (Check This 8.27). Combine this $\Delta H^{\circ}_{reaction}$ with your $\Delta S^{\circ}_{reaction}$ from Check This 8.28 to get $\Delta G^{\circ}_{reaction}$ for glucose fermentation. Do you expect this reaction to proceed under standard conditions? Why or why not?

Since Appendix XX lists standard free energies of formation, ΔG_f°, we can also use them to calculate $\Delta G^{\circ}_{reaction}$ directly:

$$\Delta G^{\circ}_{reaction} = \Sigma[n_j(\Delta G_f^{\circ})_j]_{products} - \Sigma[n_j(\Delta G_f^{\circ})_j]_{reactants} \qquad (8.26)$$

Equation (8.26) has the same form as equation (7.30) and equation (8.24).

8.31. **Worked Example**   *Free energy change for oxidation of glucose*

Use standard free energies of formation to calculate the standard free energy change when one mole of glucose is oxidized by oxygen at 298. K.

Necessary Information: We need equations (8.25) and (8.26) and, from Appendix XX, the standard free energies of formation, ΔG_f°, at 298 K for each reactant and product.

---

compound	ΔG°_f, kJ·mol^{-1}
$C_6H_{12}O_6(s)$	–910.52
$O_2(g)$	0
$CO_2(g)$	–394.36
$H_2O(l)$	–237.13

Strategy: Use equation (8.26) and the stoichiometry of reaction equation (8.25) to combine the standard molar free energies of formation, ΔG°_f, to get the standard free energy change for the reaction, $\Delta G^{\circ}_{reaction}$.

Implementation:

$$\Delta G^{\circ}_{reaction} = \Sigma[n_j(\Delta G^{\circ}_f)_j]_{products} - \Sigma[n_j(\Delta G^{\circ}_f)_j]_{reactants}$$

$$= \{ [(6 \text{ mol } CO_2)(-394.36 \text{ kJ·mol}^{-1})] + [(6 \text{ mol } H_2O)(-237.13 \text{ kJ·mol}^{-1})_{H_2O}] \}$$

$$- \{ [(1 \text{ mol } C_6H_{12}O_6)(-910.52 \text{ kJ·mol}^{-1})] + [(6 \text{ mol } O_2)(0 \text{ kJ·mol}^{-1})_{O_2}] \}$$

$$= -2878.4 \text{ kJ}$$

Does the Answer Make Sense? This direct calculation of the standard free energy of reaction and the combination of the standard enthalpy and entropy of reaction in Worked Example 8.29 give the same numeric result. The oxidation of glucose under standard conditions has a large driving force.

8.32. **Check This**   *Free energy change for fermentation of glucose*

Use the standard free energies of formation from Appendix XX to calculate $\Delta G^{\circ}_{reaction}$ for glucose fermentation (Check This 8.27). How does this result compare with your result in Check This 8.30?

Entropy change for dissolving ionic solids

We found, in Chapter 2, that dissolving ionic solids in water is an endothermic process for some compounds and exothermic for others. Entropy can help us understand the direction of these processes. For ions in solution, a concentration has to be specified for standard entropies and standard enthalpies and free energies of formation; the usual choice is 1 molal. **Molality** (*m*) is defined as the number of moles of solute dissolved in 1 kg of solvent. The entropy, enthalpy of formation, and free energy of formation of 1 m hydronium ion are the reference point and defined as zero. (For aqueous solutions, molality and molarity are almost the same: $1\ m \approx 1$ M.)

Your bones and teeth are partly made of solids containing calcium ions, Ca^{2+}, and phosphate ions, PO_4^{3-}. The simplest compound of these two ions is calcium phosphate, $Ca_3(PO_4)_2$. Although this compound has a low solubility in water, we can write the reaction for calcium phosphate dissolving and consider standard enthalpy and standard entropy changes for dissolving:

$$Ca_3(PO_4)_2(s) \rightleftharpoons 3Ca^{2+}(aq) + 2PO_4^{3-}(aq) \tag{8.27}$$

Equation (8.27) describes the reaction of a solid ionic compound, in which the ions are constrained to particular locations in the crystal, dissolving to yield ions that can move about in the entire volume of the solution; dissolution of an ionic solid is represented in Figure 8.15. More arrangements are possible for ions in aqueous solution than in crystalline solids, so we *expect* the standard entropy change for reaction (8.27), $\Delta S^{\circ}_{reaction}$, to be positive. Since calcium phosphate is quite insoluble at room temperature, reaction (8.27) does not proceed to an appreciable extent. Thus, we can conclude that $\Delta G^{\circ}_{reaction}$ ($= \Delta H^{\circ}_{reaction} - T\Delta S^{\circ}_{reaction}$) is probably positive. In order for $\Delta G^{\circ}_{reaction}$ to be positive, $\Delta H^{\circ}_{reaction}$ must be positive (an endothermic reaction) *and* numerically larger than $T\Delta S^{\circ}_{reaction}$. Otherwise, $\Delta G^{\circ}_{reaction}$ would be negative (and the solid would be soluble). Let's see what the thermodynamic data tell us.

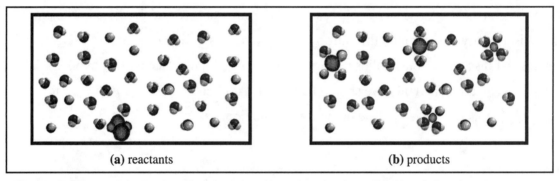

     (a) reactants   **(b)** products

Figure 8.15. Molecular level representation of an ionic solid dissolving in water.

8.33. **Worked Example** *Free energy change for dissolving calcium phosphate*

  Use standard enthalpies of formation and entropies to calculate the free energy change for dissolving one mole of calcium phosphate in water to form $Ca^{2+}(aq)$ and $PO_4^{3-}(aq)$ at 298 K.

  Necessary Information: We need equations (7.30), (8.19), (8.24), and (8.27) and, from Appendix XX, the standard molar enthalpies of formation, ΔH_f°, and standard molar entropies, S°, at 298 K for each reactant and product.

Species	ΔH°_f, kJ·mol^{-1}	S°_f, J·mol^{-1}K^{-1}
$Ca_3(PO_4)_2(s)$	–4121	+236
$Ca^{2+}(aq)$ (1 *m*)	–543	–56
$PO_4^{3-}(aq)$ (1 *m*)	–1277	–221

Strategy: Use equations (7.30) and (8.24) and the stoichiometry of reaction equation (8.27) to combine the standard molar enthalpies of formation and standard molar entropies to get the standard enthalpy and entropy changes for the reaction. Substitute these into equation (8.19) to get the standard free energy change for the reaction, $\Delta G^{\circ}_{reaction}$.

Implementation: The standard enthalpy of reaction is:

$$\Delta H^{\circ}_{reaction} = \{ [(3 \text{ mol Ca}^{2+})(-543 \text{ kJ·mol}^{-1})] + [(2 \text{ mol PO}_4{}^{3-})(-1277 \text{ kJ·mol}^{-1})] \}$$
$$- \{ [(1 \text{ mol Ca}_3(\text{PO}_4)_2)(-4121 \text{ kJ·mol}^{-1})] \} = -62 \text{ kJ}$$

The standard entropy of reaction is:

$$\Delta S^{\circ}_{reaction} = \{ [(3 \text{ mol Ca}^{2+})(-56 \text{ J·mol}^{-1}\text{·K}^{-1})] + [(2 \text{ mol PO}_4{}^{3-})(-221 \text{ J·mol}^{-1}\text{·K}^{-1})] \}$$
$$- \{ [(1 \text{ mol Ca}_3(\text{PO}_4)_2)(+236 \text{ J·mol}^{-1}\text{·K}^{-1})] \} = -846 \text{ J·K}^{-1}$$

The standard free energy change for the reaction is:

$$\Delta G^{\circ}_{reaction} = \Delta H^{\circ}_{reaction} - T\Delta S^{\circ}_{reaction} = -62 \text{ kJ} - (298 \text{ K})(-846 \text{ J·K}^{-1}) = +190 \text{ kJ}$$

Does the Answer Make Sense? $\Delta G^{\circ}_{reaction}$ is positive, as we expected for the standard free energy change for dissolving a compound that has a low solubility in water. The surprise here is that the signs of $\Delta S^{\circ}_{reaction}$ and $\Delta H^{\circ}_{reaction}$ are *opposite* from what we had predicted. The entropy for the solid is actually *greater* than the entropy for the ions in solution. In Chapter 2, Section 2.7, we found that dissolving ionic solids with multiply-charged ions is usually an exothermic process. If we had recalled this, we should have been skeptical about our *prediction* for the sign of $\Delta H^{\circ}_{reaction}$ in this case, even before doing the calculation.

8.34. Check This   *Free energy change for dissolving barium carbonate*

Barium carbonate is an insoluble solid:

$$\text{BaCO}_3(s) \rightleftharpoons \text{Ba}^{2+}(aq) + \text{CO}_3{}^{2-}(aq)$$

(a) Use the values for standard molar enthalpies of formation and standard molar entropies in Appendix XX to calculate the standard enthalpy and entropy changes for this reaction. Is the signs of the entropy change surprising? Why or why not?

(b) Calculate the standard free energy change for the reaction. Is the sign of the free energy change surprising? Why or why not?

Entropies of ions in solution

The discussion so far has treated ions dissolving into solution as if they were set free to move about independently of everything else in the solution. On the other hand, recall from Chapter 2 that the dissolving process involves ion–dipole attractions between the ions and the water

---

molecules of the solution. The ions exist in aqueous solution with a sheath of water molecules surrounding them in a more-or-less orderly array, as shown in Figure 8.15(b). The equation for dissolving calcium phosphate is more accurately written as:

$$Ca_3(PO_4)_2(s) + (m + n)H_2O(l) \rightleftharpoons 3[Ca(H_2O)_m]^{2+} + 2[PO_4(H_2O)_n]^{3-} \qquad (8.28)$$

Figure 8.15 is a schematic representation of the decrease in arrangements of water molecules that occurs when an ionic solid dissolves in water. Before dissolution, Figure 8.15(a), the water molecules are free to move about in the liquid. In the solution, Figure 8.15(b), water molecules around the hydrated ions are less free to move about because they are held next to one another by their attraction to the ions. The decrease in arrangements (decrease in entropy) is greater for multiply-charged ions, as you can see by comparing the entropies for various aqueous ions in Appendix XX.

Often, values for *m* and *n* in equation (8.28) are only approximately known, so the *(aq)* we usually write usefully represents the fact that the ions are solvated with some number of water molecules. However, writing the reaction as is done in equation (8.28) makes it clearer that there is organization on the product side as well as on the reactant side. In Chapter 2, we referred to this ordering as an unfavorable rearrangement of the water molecules. Now we see that the result of the ordering is a decrease in the entropy of the solution, compared to what we would predict based on mixing alone. Rather than assuming that dissolved ions in a mixture have greater net entropy than the pure ionic solid and water, always check tabulated entropy values.

Significance of ΔG and ΔG^o

Note that all of the above calculations give us the *standard* free energy change, $\Delta G^o_{reaction}$, for a reaction of the reactants in their standard states to give products in their standard states. Few reactions are actually carried out this way; $\Delta G^o_{reaction}$ is a *guide* to help us understand the direction of reaction. We really need $\Delta G_{reaction}$, the free energy change under the actual reaction conditions to tell for sure whether the reaction is possible or not. In Chapter 9, we will discuss how to get $\Delta G_{reaction}$ and the significance of $\Delta G^o_{reaction}$.

Reflection and Projection

By examining the number and physical states of the reactants and products of reactions, we made predictions about the sign of the entropy and enthalpy changes for the reactions. We tested these predictions by calculating standard entropy, enthalpy, and free energy changes for the reactions. In many cases, these predictions are valid. However, the qualitative idea that dissolved ions should always have greater entropy than the separated pure solid and the solvent proves to be untrue. In order to interpret entropies of solution, we need to use the solubility model developed in Chapter 2. The organization of water molecules about dissolved ions has to be taken into account and is especially important when the ions are multiply charged.

Positional entropy plays a fundamental role in controlling changes in these systems. It is not only ionic solutes that have an effect on the organization of water molecules. In the next sections, we will examine the consequences of the interactions of water with nonpolar molecules and molecules with polar and nonpolar regions.

Section 8.10. Why Oil and Water Don't Mix

8.35. Investigate This   *What happens when oil and water are mixed?*

For this investigation, use a small test tube and three plastic pipets containing these liquids: cooking oil (a clear, pale yellowish liquid), water, and a liquid dishwashing detergent.

(a) Place a few milliliters of water in the test tube and add about $1/2$ mL of oil. Cap the tube with your finger, mix by inverting the tube several times, and allow the tube to sit quietly for several seconds. Record your observations when no further change seems to be occurring.

(b) Add about $1/2$ mL of the dishwashing detergent to the test tube. Cap the tube with your finger and mix by inverting the tube several times. Try not to create a lot of suds. Allow the tube to sit quietly and observe. Record your observations when no further change seems to be occurring. We will return to this result in Consider This 8.40.

8.36. Consider This   *Why don't oil and water mix?*

In Investigate This 8.35(a) you investigated the solubility of oil in water. In Chapter 2, after you investigated the solubility of hexane in water (Investigate This 2.3), we explained the insolubility of nonpolar compounds in terms of unfavorable rearrangements of water molecules around any dissolved nonpolar molecule. How would you construct this explanation in terms of entropy?

As you know from Investigate This 8.35 and common experience, oil and water don't mix. Nonpolar compounds, such as cooking oil, are insoluble in water. Now that we have net entropy change (or free energy change) as a criterion for the directionality of processes, we can try to understand the factor(s) that make oil and water immiscible. We need to account for both thermal and positional entropy changes for the dissolution of nonpolar solutes in water. Let us look at thermal entropy first.

Dispersion forces are the attractive interactions that hold nonpolar molecules together in their solid and liquid states. Dispersion forces exist for all atoms and molecules, and, in hydrocarbons, they are the largest attractive forces. In polar molecules, attractions among the permanent dipoles are the strongest forces. Therefore, mixing hydrocarbon molecules and water molecules requires breaking some hydrogen bonds between the water molecules and disrupting dispersion interactions between hydrocarbon molecules. Disrupting attractions requires an input of energy. New attractions that occur among the mixed molecules release energy. When we measure the enthalpy change for dissolving hydrocarbons in water, $\Delta H_{solution}$ turns out to be very nearly zero in most cases. The disruptive and attractive interactions cancel one another. Since the enthalpy change for dissolving nonpolar molecules in water is about zero, the thermal entropy change (equation 8.11) is also about zero.

Oriented water molecules

If there is no change in thermal entropy for dissolving a hydrocarbon in water, it must be the positional entropy that is unfavorable and causes the insolubility. We know that mixtures of two substances always have higher positional entropy than the unmixed substances. Therefore, for hydrocarbons and water, there must be some other unfavorable entropy effect that is larger than the entropy of mixing. Figure 8.16 illustrates the origin of this unfavorable entropy effect.

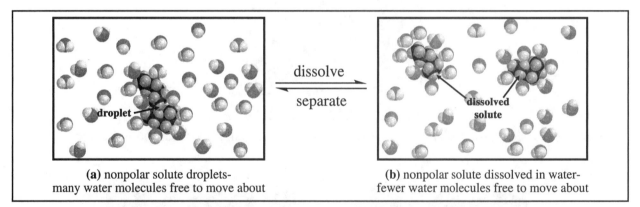

**(a) nonpolar solute droplets-
many water molecules free to move about**

**(b) nonpolar solute dissolved in water-
fewer water molecules free to move about**

Figure 8.16. Decrease in arrangements of water molecules when a nonpolar solute dissolves.
Compare the effects of ionic and nonpolar solutes on water molecules in Figure 8.15 and here.

There is experimental spectroscopic evidence that water molecules next to nonpolar molecules are oriented and lose some of their freedom of motion. When the nonpolar molecules are grouped together, not as many water molecules are restricted in their motion, Figure 8.16(a). If the nonpolar molecules mix into the water, more water molecules are needed to group around the individual nonpolar molecules and so more water molecules are restricted in their motion, Figure 8.16(b). Thus, fewer molecular arrangements of water molecules are possible. The loss of molecular arrangements reduces the entropy of the water molecules and makes the overall positional entropy change for mixing oil and water negative; oil and water don't mix.

Clathrate formation

Although hydrocarbons do not dissolve well in liquid water, the relatively ordered structure that water forms around them is much like ice. Thus, at lower temperatures, water can form ice-like structures around low-molecular-mass hydrocarbons. The cage structure shown in Figure 8.17 is typical for nonpolar molecules in water. Such mixtures, called **clathrates**, are

> "Clathrate" is derived from a Greek word meaning "lattice"; one substance is mixed into the crystal lattice of another.

characterized by molecules of the solute distributed throughout crystals of the solvent. Water, as a semi-solid host crystal, forms clathrates with many nonpolar molecules. One well-studied clathrate is formed between water and methane, CH_4. Methane produced by microorganisms is trapped as water clathrates at the bottom of seas and oceans and is a source of energy for other organisms as well as a potential source of fuel for humans. Figure 8.18 shows a methane clathrate mound with marine worms feeding on the methane.

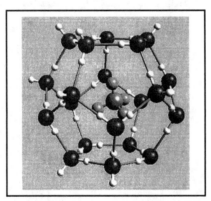

Figure 8.17. Ice clathrates are cages of water molecules around hydrocarbon molecules.

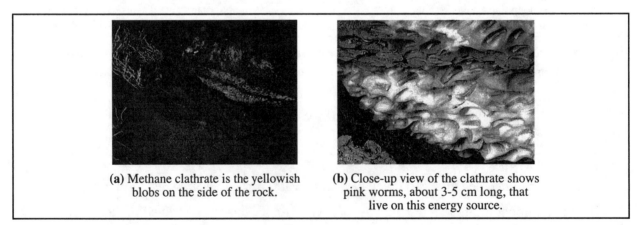

(a) Methane clathrate is the yellowish blobs on the side of the rock.

(b) Close-up view of the clathrate shows pink worms, about 3-5 cm long, that live on this energy source.

Figure 8.18. Organisms feeding on methane clathrate at the bottom of the sea.

Section 8.11. Ambiphilic Molecules: Micelles and Bilayer Membranes

8.37.   Investigate This   *How does light interact with solutions?*

Do this as a class investigation and work in small groups to discuss and analyze the results. Fill two transparent, colorless plastic or glass containers with water. To one of the containers add about one teaspoon of table sugar. To the other add 3-4 drops of liquid soap or detergent. Use separate stirrers to stir the contents of the containers until the solid(s) and/or liquid disappear and are thoroughly mixed with the water. Allow the containers to sit undisturbed for several minutes so the liquids are no longer swirling. Darken the room and direct a narrow beam of light through one of the solutions from the side while observing from the top. Repeat the observation with the other solution. Record your observations, especially noting any differences between the two solutions.

8.38.   Consider This   *What causes light to interact with a solution?*

Individual solute molecules in a solution are too small to affect the path of a beam of light. If there are "clumps" of solute molecules, the clumps may be large enough to interfere with a light beam and scatter it in all directions. How do you interpret any difference(s) between the observations you made on the two solutions in Investigate This 8.37?

In order to understand how cell membranes, like those represented in the chapter opening illustration, form, we will look at the properties of the molecules that form them. In particular, we'll examine how they interact with water and the part water plays in creating the membrane structure. We'll start with a similar but simpler model system, detergent molecules, which are less complicated than membrane molecules. In addition to helping us understand membranes, detergents have interesting and useful properties themselves, as you have observed in Investigate This 8.35(b) and 8.37.

Ambiphilic molecules

The structure of one kind of **detergent molecule**, $R-OSO_3^-$ (dodecyl sulfate), is shown in Figure 8.19. One end of the structure is highly polar, the ionized sulfate group, $-OSO_3^-$, and the rest of the molecule, $R-$, is a long nonpolar hydrocarbon "tail." Different detergents have different kinds of polar groups and different hydrocarbon chain lengths. Detergent molecules interact with water in two ways. The polar end of the detergent and the polar water molecules are attracted to one another. The polar end (head) of the detergent is called **hydrophilic** (*hydro* =

water + *philos* = loving) and is soluble in water. The nonpolar hydrocarbon tail acts like an oil; hydrocarbons are called **hydrophobic** (*hydro* = water + *phob*os = fearing or hating) because they don't mix with water to form solutions. Molecules like detergents that have both hydrophilic and hydrophobic regions are called **ambiphilic** (*ambi* = both + *philos* = loving).

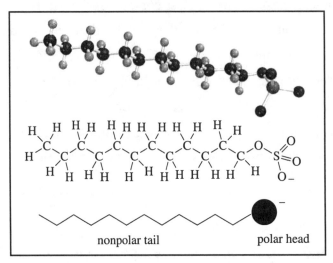

nonpolar tail   polar head

Figure 8.19. Different representations of a detergent molecule, dodecyl sulfate.

Micelle formation

Although we didn't use the term ambiphilic in Chapter 2, in Investigate This 2.3, we found

WEB Chap 8, Sect 8.13.1-3
Interactive animations reinforce
this explanation and illustration
of micelle formation

that such molecules have limited solubility in water. Their polar (hydrophilic) regions are soluble; their nonpolar (hydrophobic) regions are not. One way an ambiphile can satisfy both the hydrophilic and hydrophobic interactions is to form **micelles**, as shown in Figure 8.20.

 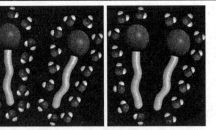

(a) Representations of an ambiphilic soap molecule (stearate).

(b) Ambiphilic molecular clustering frees water molecules and increases the entropy of the system.

(c) Micelles are formed as more of the cone-shaped ambiphilic molecules cluster.

Figure 8.20. Formation of a micelle from an ambiphilic soap molecule.

The diameter of the polar head of a detergent (or soap) molecule is relatively large compared to the diameter of its hydrocarbon tail, making the molecule shaped sort of like a thin cone, as shown in Figure 8.20(a). If the molecules cluster with their hydrophobic tails together, Figure 8.20(b), water molecules are freed from their interactions with the nonpolar tails. Freeing these water molecules increases the entropy of the solution which favors clustering. Notice that packing the cone-shaped detergent molecules makes the clusters form curved structures. Imagine this packing continuing in three dimensions to form a roughly spherical closed structure, Figure 8.20(c), which is the shape micelles take.

The polar heads in the micelle form an outer layer in contact with water and the hydrocarbon tails are tucked inside interacting with one another and freeing many water molecules. Micelle formation increases the net entropy of the solution by causing an increase in entropy of the water that is no longer organized around the hydrocarbon tails. A detergent micelle contains about 50-100 molecules and is about 3-5 nm in diameter. There can be a large number of micelles in the solution, so substantial amounts of detergent can mix with water this way. Although you can't see individual micelles, they are large enough to scatter light that shines through the mixture.

8.39.   Check This   *Evidence for micelle formation*

Are your observations for the solutions in Investigate This 8.37 consistent with this last statement about light scattering? Explain why or why not.

8.40.   Consider This   *How do ambiphilic molecules interact with nonpolar solutes?*

In Investigate This 8.35(b), you found that a mixture of water, oil, and detergent appears cloudy, but does not separate into a layer of oil on water. What interactions among the water, ambiphilic detergent, and nonpolar oil molecules could be responsible for this behavior? Draw a sketch showing your model of these interactions.

Detergent action

Consider what can happen if nonpolar molecules such as oil are present in an aqueous mixture that also contains ambiphilic detergent molecules. The hydrophobic tail of the detergent molecules and the nonpolar solute can interact to free some of the solvating water molecules, which is the kind of interaction we have illustrated in Figures 8.16 and 8.20(b). As micelles form, the nonpolar molecules associated with their hydrophobic tails become incorporated into the interior of the micelle, as illustrated in Figure 8.21. In Investigate This 8.35(b), the detergent added to the oil and water mixture formed micelles containing the oil molecules. Since the oil molecules could not get together to form droplets of oil, they could not form a separate layer.

When you use detergent (or soap) and water to wash your hands or soiled laundry, oils on your hands or on the soiled cloth become incorporated into micelles and are rinsed away with the water. The entropy increase in the solution, as micelles are formed, is responsible for your clean hands and clothes.

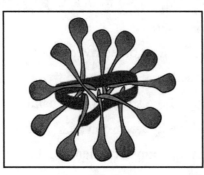

Figure 8.21. A nonpolar molecule in the hydrophobic interior of a detergent micelle.

Phospholipid bilayers

The ambiphilic molecules that make up cell membranes are **phospholipids**. **Lipids** [*lipos* = fat] are esters of long-chain carboxylic acids (fatty acids) with glycerol. Lipids are used by plants and animals for a variety of purposes including membrane structures and energy storage as fats and oils. Phospholipids, Figure 8.22, have two long hydrophobic tails and a polar head that contains charged phosphate and ammonium-like groups. The polar head is comparable in diameter to the two tails, so the overall shape of a phospholipid is approximately a cylinder, as shown in Figure 8.23(a). If the phospholipid molecules cluster together with their hydrophobic tails together, Figure 8.23(b), water molecules are freed from their interactions with the nonpolar tails. Freeing of the water molecules increases the entropy of the solution and favors clustering.

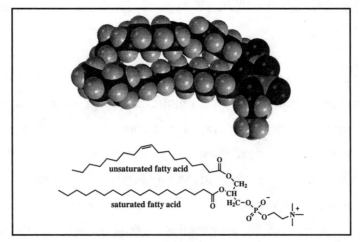

Figure 8.22. A phospholipid molecule, phosphatidyl choline.

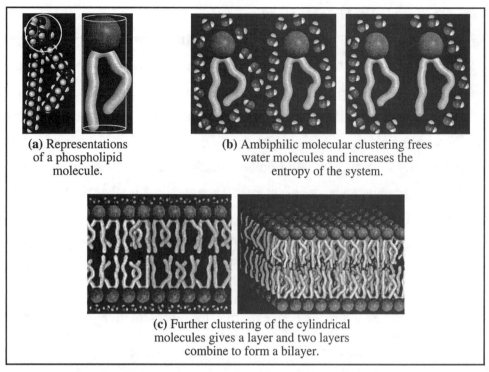

(a) Representations of a phospholipid molecule.

(b) Ambiphilic molecular clustering frees water molecules and increases the entropy of the system.

(c) Further clustering of the cylindrical molecules gives a layer and two layers combine to form a bilayer.

Figure 8.23. Formation of a phospholipid bilayer.

When cylindrical phospholipids pack together like cans on a supermarket shelf, a layer of molecules is formed, but this still leaves the ends of the hydrophobic tails available to interact with water. More water can be freed and entropy increased, if two layers come together with their hydrophobic sides facing to form a **phospholipid bilayer**, Figure 8.23(c) and the chapter opening illustration. Phospholipid bilayers are not rigid and can bend around and close on themselves, Figure 8.24. This closed structure is called a **liposome** or **lipid vesicle**. Liposomes can vary in size, but small ones are about 25 nm in diameter and contain about 2700 phospholipid molecules. Note that the liquid on the interior of a liposome is an aqueous solution.

> **WEB Chap 8, Sect 8.13.1**
> View an animation of bilayer formation illustrated in Figure 8.23.

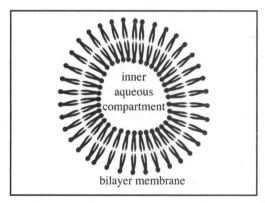

inner aqueous compartment

bilayer membrane

Figure 8.24. Formation of a liposome from a phospholipid bilayer.

Getting through a phospholipid bilayer membrane

All living cells are enclosed by **phospholipid bilayer membranes**, like those making up liposomes, and their membranes have the ambiphilic properties we have been describing. Liposomes are used to investigate some of the properties of cell membranes, such as how well various solutes can pass through them To do this, investigators make liposomes in an aqueous solution that contains the solute of interest. The solution on the inside of the liposome will contain the solute. After separating the liposomes from the solution, they are placed in a new solution that does not contain the solute of interest. Then the time it takes for the solute molecules to move out of the liposome into the surrounding solution is measured. From these experiments, we can determine the relative permeability of membranes for various molecules. A few results of such studies are given in Table 8.1.

Table 8.1. Relative membrane permeabilities for representative molecules and ions.

molecule	relative permeability
water, H_2O	10,000
glycerol, $C_3H_8O_3(aq)$	100
glucose, $C_6H_{12}O_6(aq)$	1 (assigned)
chloride ion, $Cl^-(aq)$	0.001
potassium ion, $K^+(aq)$	0.0001
sodium ion, $Na^+(aq)$	0.00002

A bilayer membrane is not a static structure; phospholipid molecules can shift about within their side of the bilayer. As the shifting and jostling of the phospholipid molecules occurs, small holes or pores form and disappear in the membrane. Each of these transient pores provides an opportunity for a molecule near the surface of the membrane to enter the bilayer or even cross it entirely. The molecules present in largest amount on both sides of a membrane are water, so a water molecule is the most likely to enter a membrane pore. As we've noted, interaction of a water molecule with the hydrocarbon tails is not very favorable; so any water molecules entering a membrane are likely to leave. They can either return through the side where they entered or pass all the way through the membrane and leave on the other side.

Both polarity and size are significant in whether a molecule can cross the membrane. Small polar molecules like glycerol cross less easily, but still quite readily. Larger polar molecules like sugars and amino acids cross even more slowly. The transient pores in the membrane are small, so these larger molecules simply have a hard time entering the membrane in the first place. Ions like $Na^+(aq)$, $K^+(aq)$, $Ca^{2+}(aq)$, and $HOCO_2^-(aq)$ don't easily cross phospholipid bilayer membranes at all because the water molecules associated with the ions have to cross the membrane with them and this cluster is larger than most of the pores. Thus size as well as their unfavorable interaction with the non-polar interior of the membrane keeps hydrated ions out.

8.41.   Consider This   *How do the properties of bilayer membranes help explain osmosis?*

In Investigate This 8.1, we used a carrot to set up an investigation of osmosis. Osmosis involves transfer of solvent molecules through a semipermeable membrane and we found that an increase in entropy as solvent molecules dilute a solution explains the observed direction of the process. Do the relative permeabilities of cell membranes in Table 8.1 help to explain why the carrot could be used for this investigation? Why or why not?

Reflection and projection

It seems that positional entropy should increase when mixtures form that provide larger volumes for one or both of the components to move in. We found, however, that ion hydration reduces the positional entropy of aqueous solutions and is the reason that ionic solids with multicharged ions are rather insoluble. Similarly, we find that the organization of water molecules around nonpolar solutes, also reduces the positional entropy of these aqueous solutions and is the reason that oil and water do not dissolve in one another, but form separate layers when mixed.

The outcomes are more complex when ambiphilic molecules are mixed with water, because their polar (hydrophilic) parts interact energetically favorably with water via polar interactions, while their nonpolar (hydrophobic) parts do not. The entropy of the mixture would be increased if the nonpolar parts could be separated from the water. Two ways to maintain the energetically favorable polar interactions and increase the entropy of the mixture are for the ambiphilic molecules to cluster into micelles or bilayers. Detergent molecules with a single nonpolar tail generally form micelles with the nonpolar tails on the inside and the polar heads on the outside of a roughly spherical cluster. Phospholipid molecules with double nonpolar tails generally form lipid bilayers, two flexible sheets of the molecules with their nonpolar tails facing one another on the inside of the bilayer and the polar heads on opposite sides of the bilayer in contact with the water.

Phospholipid bilayers can close on themselves and form the membranes that surround all living cells as well as the organelles (structures inside the cells) in eukaryotic cells. Molecules can cross from one side to the other of a phospholipid bilayer membrane, but they do so at different speeds, depending upon their size and polarity. Thus, bilayer membranes are semipermeable and osmosis takes place through them. We began the chapter discussing osmosis and now return to it and to a group of other properties, colligative properties, including osmotic pressure, all of which you can understand by analyzing their entropies.

Section 8.12. Colligative Properties of Solutions

8.42. **Investigate This**   *What is the freezing point of salt water?*

Do this as a class investigation and work in small groups to discuss and analyze the results. Fill a 250-mL beaker with crushed ice. Add about 50 mL of water and place a wooden stirrer and a thermometer in the beaker. Stir the mixture vigorously, and record the temperature of the ice–water mixture. Add about 90 g of salt, NaCl. Again stir vigorously for 2-3 minutes and record the temperature.

8.43. **Consider This**   *How does salt affect the freezing point of water?*

(a) In Investigate This 8.42, how do the temperatures compare before and after addition of the salt?

(b) How does the entropy of a salt solution compare to the entropy of pure water? Could the difference be responsible for the result you noted in part (a)? Explain.

Freezing point of a solution

In Section 8.7, we considered phase changes in pure substances. To explain the results from Investigate This 8.42, we need to consider the behavior of solutions of nonvolatile solutes. **Nonvolatile solutes** are solutes that cannot evaporate from the solution. Ionic solids (table salt) and polar covalent molecular solids (sugar) are examples. Our discussion of osmosis earlier in the chapter has already begun this consideration and now we will extend that discussion. The fundamental concept we build on is that mixing a soluble solute with a solvent makes the positional entropy of the solution greater than the positional entropy of the pure solvent. In Section 8.4, especially in Figure 8.7, you saw how this concept (expressed in terms of numbers of arrangements instead of entropy) explains the direction of osmosis. Since freezing and melting involve energy, the change in freezing temperature in Investigate This 8.42 must also involve energy. Thermal entropy also has to be accounted for in our explanation.

Entropy change for freezing from solution

Figure 8.25 shows the change that occurs when a solution of a nonvolatile solute begins to freeze. Note that the solid that freezes out of the solution is pure solvent. For example, when an aqueous solution of salt is frozen, the solid that forms is pure ice. We know that the entropy of a solution is greater than the entropy of pure liquid solvent, $S_{solution} > S_{solvent}$, and, also, that the entropy of a pure liquid is greater than that of a pure solid, $S_{solvent} > S_{solid}$. Thus, we can write (at the same temperature):

$$S_{solution} > S_{solvent} > S_{solid} \tag{8.29}$$

The solution has the greatest number of distinguishable arrangements of particles. The liquid has fewer arrangements, and the solid has the fewest of all. The entropy decrease to form solid solvent from a solution is always greater than the entropy decrease to form solid solvent from the pure liquid solvent. These relationships are shown graphically in Figure 8.26.

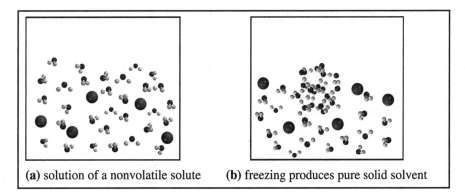

(a) solution of a nonvolatile solute   **(b)** freezing produces pure solid solvent

Figure 8.25. Changes that occur when a solution begins to freeze.

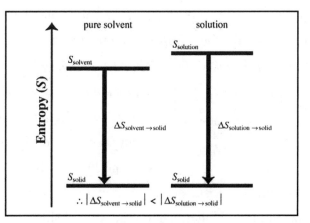

Figure 8.26. Positional entropy changes for freezing a pure solvent and a solution.

When the solution is in equilibrium with the solid solvent, $\Delta G_{\text{solution}\rightarrow\text{solid}}$ is zero:

$$\Delta G_{\text{solution}\rightarrow\text{solid}} = 0 = \Delta H_{\text{solution}\rightarrow\text{solid}} - (T_{\text{solution}\rightarrow\text{solid}})\Delta S_{\text{solution}\rightarrow\text{solid}} \tag{8.30}$$

Equation (8.30) can be rearranged to give:

$$\Delta H_{\text{solution}\rightarrow\text{solid}} = (T_{\text{solution}\rightarrow\text{solid}})\Delta S_{\text{solution}\rightarrow\text{solid}} \tag{8.31}$$

When pure liquid solvent is in equilibrium with the solid solvent, there is a similar equation:

$$\Delta H_{\text{solvent}\rightarrow\text{solid}} = (T_{\text{solvent}\rightarrow\text{solid}})\Delta S_{\text{solvent}\rightarrow\text{solid}} \tag{8.32}$$

The enthalpy change (ΔH) accompanying freezing is independent of whether the pure solid comes from the pure liquid solvent or from a solution, that is:

$$\Delta H_{\text{solvent}\rightarrow\text{solid}} = \Delta H_{\text{solution}\rightarrow\text{solid}} \tag{8.33}$$

Consequently, the right hand sides of equations (8.31) and (8.32) are also equal. Setting them equal and rearranging gives:

$$\frac{\Delta S_{\text{solvent}\rightarrow\text{solid}}}{\Delta S_{\text{solution}\rightarrow\text{solid}}} = \frac{T_{\text{solution}\rightarrow\text{solid}}}{T_{\text{solvent}\rightarrow\text{solid}}} \tag{8.34}$$

Figure 8.26 shows us that the ratio of entropy changes, $\dfrac{\Delta S_{\text{solvent}\rightarrow\text{solid}}}{\Delta S_{\text{solution}\rightarrow\text{solid}}}$, is less than one. Therefore, $\dfrac{T_{\text{solution}\rightarrow\text{solid}}}{T_{\text{solvent}\rightarrow\text{solid}}}$ is less than one, so $T_{\text{solution}\rightarrow\text{solid}}$ must be *less* than $T_{\text{solvent}\rightarrow\text{solid}}$. The freezing point *decreases* when a solute is added to a pure liquid, as you found in Investigate This 8.42. This phenomenon is known as **freezing point lowering** (also called freezing point depression).

Colligative properties

Note that none of the properties of the solute entered into this discussion of freezing point lowering of solutions. Likewise, the only property of the solute that is relevant to osmosis is that the solute not be able to pass through the semipermeable membrane. *Properties of solutions that do not depend on the identity of the nonvolatile solute (or solutes), but only on how many of them are present in the solution, are called* **colligative properties**. Colligative properties include freezing point lowering, **boiling point elevation** (solutions boil at a higher temperature than the

> Colligative is from Latin meaning to bind together or act together. Each solute molecule or ion in the solution acts together with all the others to produce the observed effect, freezing point lowering, for example.

pure solvent), and **osmotic pressure** (osmosis raises the level of the solution, so the pressure on the solution side of the membrane is higher than on the solvent side). Figure 8.27, which also repeats Figures 8.3 and 8.25, shows the change associated with each of these properties. The important thing to observe in each case is that the change involves a solution and its pure solvent.

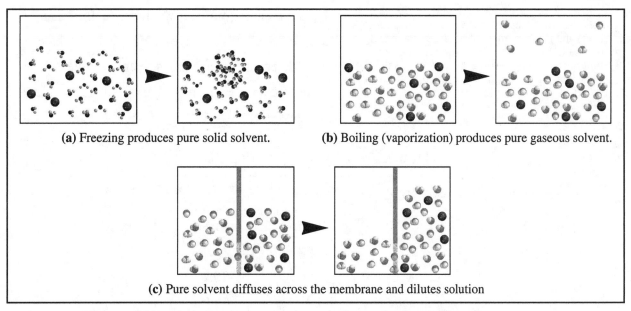

(a) Freezing produces pure solid solvent.   (b) Boiling (vaporization) produces pure gaseous solvent.

(c) Pure solvent diffuses across the membrane and dilutes solution

Figure 8.27. Changes that occur for solutions of nonvolatile solutes.

Every molecule or ion dissolved in a solution is included in the determination of the number of ways to arrange solute and solvent particles (W). Consequently, the entropy of a solution depends *only* on the relative numbers of solute and solvent particles, regardless of the identity of the solute particles. Equation (8.34) shows that the freezing point is inversely proportional to the entropy change for freezing solid solvent from the solution. Since the entropy change depends only on the numbers of particles, the freezing point also depends only on the number of particles.

Quantifying freezing point lowering

Relative numbers of solute and solvent particles (equivalent to relative numbers of moles) are easy to determine in the laboratory, as is the difference in freezing point between a solution and the pure solvent. For water, 1.0 mole of a molecular solute dissolved in 1 kg of water lowers the freezing point of the solution to –1.86 °C. If

> The "experimental" results given here are for an ideal solute that simply mixes among the water molecules without interacting with them. Values for k_{fp} are determined by extrapolating data for real solutions to the limit of high dilution. The simple equations we use here are valid only for dilute solutions.

the number of solute particles is increased to 2.0 mol (again dissolved in 1 kg of water), the freezing point is lowered to –3.72 °C. The relationship between freezing point lowering and solution concentration is:

$$\Delta T_{fp} = k_{fp} m \qquad (8.35)$$

Equation (8.35) shows that ΔT_{fp} ($= T_{fp\,(solution)} - T_{fp\,(solvent)}$), the change in freezing point of the solvent, is directly proportional to the molality of the solution. The freezing point of the solvent is always higher than that of the solution, so ΔT_{fp} is always negative. The proportionality factor,

> The value for k_{fp} can be derived from equations (8.31) and (8.32) combined with the relationship for the difference between the entropies of a solution and the pure solvent. The result is that k_{fp} depends upon the molar mass, enthalpy of fusion, and temperature of melting of the solvent.

k_{fp}, the **freezing point lowering constant**, is a characteristic of the solvent. Values for k_{fp} have been determined for many solvents; Table 8.2 gives you an idea of the range of values for a few solvents.

Table 8.2. Freezing point lowering constants.

Solvent	T_{fp}, °C	k_{fp}, °C·m^{-1} *
water	0.00	–1.86
benzene	5.50	–5.12
cyclohexane	6.5	–20.2
naphthalene	80.2	–6.9
camphor	179.8	–40.0

* k_{fp} is related to a temperature *difference*. The size of the difference is the same for both °C and K.

As an example, a 0.10 m aqueous solution of glucose in water will have a freezing point lowering of:

$$\Delta T_{fp} = (-1.86 \; °C·m^{-1}) \times (0.10 \; m) = -0.19 \; °C$$

The freezing point of the solution will be

$$T_{fp \, (solution)} = T_{fp \, (solvent)} + \Delta T_{fp} = 0.00 \; °C + (-0.19 \; °C) = -0.19 \; °C$$

As you can see, sensitive thermometers are required to measure freezing point lowering for aqueous solutions.

8.44. **Worked Example**   *Determining a freezing point lowering constant*

The freezing point of benzophenone, $C_{13}H_{10}O$, is 321.7 K (48.6 °C). In an experiment, 5.231 g of solid benzophenone was mixed with 0.127 g of biphenyl, $C_{12}H_{10}$, and the mixture was heated to melt the benzophenone and dissolve the biphenyl. The freezing point of the mixture was 320.2 K. Use these data to calculate the freezing point lowering constant for benzophenone.

Information Needed: We need equation (8.35), the definition of molality, and the molar mass of biphenyl.

Strategy: Use the masses of biphenyl and benzophenone and the molar mass of biphenyl to calculate the molality of the solution of biphenyl in benzophenone. Then substitute the molality and the freezing point lowering into equation (8.35) and solve for k_{fp}.

Implementation: The molar mass of biphenyl is 154.2 g·mol⁻¹, so the number of moles of biphenyl is:

$$0.127 \text{ g of biphenyl} = (0.127 \text{ g})\left(\frac{1 \text{ mol biphenyl}}{154.2 \text{ g}}\right) = 8.24 \times 10^{-4} \text{ mol}$$

The molality, m, of the solution of biphenyl in benzophenone is:

$$\frac{8.24 \times 10^{-4} \text{ mol biphenyl}}{5.231 \times 10^{-3} \text{ kg benzophenone}} = 1.574 \times 10^{-1} \text{ } m$$

The freezing point lowering is:

$$\Delta T_{fp} = T_{fp \text{ (solution)}} - T_{fp \text{ (solvent)}} = 320.2 \text{ K} - 321.7 \text{ K} = -1.5 \text{ K} = -1.5 \text{ °C}$$

The freezing point lowering constant is:

$$k_{fp} = \frac{\Delta T_{fp}}{m} = \frac{-1.5 \text{ °C}}{1.574 \times 10^{-1} \text{ } m} = -9.5 \text{ °C·}m^{-1}$$

Does the Answer Make Sense? What we have to go on are the values in Table 8.2 for the freezing point lowering constants of other compounds. The melting point and molar mass of benzophenone are similar to those of naphthalene and the freezing point lowering constant we calculated is also similar, so we are probably in the right ballpark. Use a handbook or web-based reference to check this freezing point lowering (or freezing point depression) constant for benzophenone.

8.45. Check This   *Determining molar mass of a solute from freezing point lowering*

Suppose you dissolve 0.243 g of a newly synthesized compound in 7.567 g of benzene and find that the freezing point of the solution is 5.12 °C.

(a) What is the molality of the solution? Give the reasoning for your answer.

(b) What is the molar mass of the new compound? Give the reasoning for your answer.

Effect of solute ionization on colligative properties

Since freezing point lowering is a colligative property and depends on the number of particles of solute, we need to know how many particles are produced per mole of the formula unit when a substance dissolves. For example, 1.0 mole of calcium chloride formula units, $CaCl_2$, dissolved in 1 kg of water produces 1.0 mole of $Ca^{2+}(aq)$ ions and 2.0 moles of $Cl^-(aq)$ ions. Each cation and anion has the same effect on the freezing point, so *one* mole of $CaCl_2$ depresses the freezing point as much as *three* moles of a dissolved molecular compound that does not ionize.

8.46. **Worked Example**   *Freezing point lowering by an ionic solute*

What is the freezing point of a solution prepared by dissolving 0.10 mol of magnesium sulfate, $MgSO_4$, in 1.00 kg of water?

Necessary Information: We need the k_{fp} for water from Table 8.2 and we need to know that magnesium sulfate dissolves to yield magnesium cations and sulfate anions.

Strategy: Calculate the total molality of solute particles. Then use equation (8.35) to get the freezing point lowering and the freezing point of the solution.

Implementation: 0.10 mol of magnesium sulfate dissolves to give 0.10 mol of magnesium cations and 0.10 mol of sulfate anions. The total number of moles of solute particles is, therefore, 0.20 mol and the solution is 0.20 m. The freezing point lowering is:

$$\Delta T_{fp} = (-1.86 \ °C \cdot m^{-1}) \times (0.20 \ m) = -0.37 \ °C$$

The freezing point of the solution is:

$$T_{fp \ (solution)} = T_{fp \ (water)} + \Delta T_{fp} = 0.00 \ °C + (-0.37 \ °C) = -0.37 \ °C$$

Does the Answer Make Sense? The freezing point lowering for the ionic solution is two times larger than the freezing point lowering we calculated for a 0.10 m glucose solution. Since this ionic solution has twice as many solute particles, this is the result we would expect.

8.47. **Check This**   *Freezing point lowering by a polar solute in a nonpolar solvent*

The freezing point of a 0.50 m solution of ethanoic acid, $CH_3C(O)OH$, in benzene is 3.96 °C.

(a) From the freezing point value, what is the molality of particles in the solution? Explain.

(b) How does your result in part (a) compare to the molality of the solution whose freezing point was measured? How can you explain any difference between these molalities? *Hint:* Consider possible hydrogen bonding interactions. See Problem 8.54.

Boiling point elevation

When we discussed freezing-point depression in solutions, we saw that the entropy of solutions was greater than the entropy of pure liquids, which was, in turn, greater than the entropy of solids. The explanation of higher boiling points for solutions of nonvolatile solutes follows the same line of reasoning applied to vaporization of pure solvents and solutions. In this case, the entropy of the solution is greater than the entropy of the pure liquid solvent and the entropy of the pure gaseous solvent is larger than either of the condensed phases:

$$S_{gas} > S_{solution} > S_{solvent} \tag{8.36}$$

8.48.   Consider This   *Is the entropy change larger for boiling a solution or pure solvent?*

(a) Use the data in equation (8.36) to draw a diagram showing the entropy changes for boiling a solution and its pure solvent. Model your reasoning on the logic used to construct Figure 8.26.

(b) What is the size of $|\Delta S_{solvent \to gas}|$ relative to $|\Delta S_{solution \to gas}|$? Does this relationship help you understand why a solution has a higher boiling point than its pure solvent? Explain. You might find it helpful to write a relationship analogous to equation (8.34).

Changes in boiling point are based on molality, just as were changes in freezing point:

$$\Delta T_{bp} = k_{bp}m \tag{8.37}$$

Table 8.3 lists **boiling point elevation constants, k_{bp}**, for the three liquid solvents included in Table 8.2. The boiling point elevation constant is positive, since ΔT_{bp} ($= T_{bp\ (solution)} - T_{bp\ (solvent)}$) and $T_{bp\ (solution)} > T_{bp\ (solvent)}$.

Table 8.3. Boiling point elevation constants.

Solvent	T_{fp}, °C	k_{fp}, °C·m^{-1} *
water	100.00	0.512
benzene	80.0	2.53
cyclohexane	80.7	2.79

8.49.   Check This   *Boiling point of a solution*

Calculate the boiling point of a 0.25 *m* biphenyl solution in benzene.

Section 8.13. Osmotic Pressure Calculations

Osmosis, was discussed in Sections 8.1 and 8.4. The explanation for osmosis in terms of molecular arrangements (Section 8.4), is parallel to the explanations for freezing point lowering and boiling point elevation, given in terms of entropies in the previous section. Now we will consider osmosis, in particular, osmotic pressure, more quantitatively, just as we did for those other colligative properties.

Our molecular model in Section 8.4 shows that the positional entropy of the solution continues to increase as the solution becomes more dilute, because the number of possible arrangements of the solute molecules increases. In Investigate This 8.1, we might expect water to continue to pass through the membrane (carrot) to the solution until the water is used up. That does not happen. Water finally does stop its *net* movement through the semipermeable

> *Osmose* is derived from a Greek verb meaning to impel or force.

membrane. The column of solution exerts pressure on the solution that is greater than the atmospheric pressure on the pure water. The column of solution stops rising when the pressure at the bottom of the column is large enough to stop the net movement of water into the solution. The pressure on the solution that is required to *just* stop net movement of water is called the **osmotic pressure**, symbolized by the uppercase Greek letter pi, Π.

When the solution level stops rising, the movement of water molecules through the membrane is the same in both directions. This means that the net entropy change for a solvent molecule passing from the solvent to the solution must be the same as that for the reverse process, a solvent molecule passing from the solution to the solvent. We know that entropy of mixing is gained when a solvent molecule enters the solution. There must be the same gain in entropy when a solvent molecule leaves the high pressure solution to enter the low pressure solvent.

The high pressure on the solution compresses the liquid. You can think of the compression as making the boxes occupied by the *solution* molecules in Figure 8.7(b) smaller. This means that the volume available to molecules in the solution is smaller. A smaller volume per molecule decreases the entropy of the solution. Net flow of solvent stops when the entropy decrease due to the compression by the osmotic pressure exactly compensates the entropy increase due to mixing. Liquids are not very compressible, so we would expect the pressure required to cause this decrease in entropy to be quite high. That is, we predict high osmotic pressures.

Quantifying osmotic pressure

Since it is a colligative property of solutions of nonvolatile solutes, the osmotic pressure of a solution depends only upon the number of molecules or ions present in the solution and not on

> Recall that molarity and molality are almost identical for dilute aqueous solutions.

their identity. When dealing with osmosis, *molarity* is used to express solute concentration rather than *molality*. The relationship between osmotic pressure and the molarity, M, of molecules or ions in a solution is:

$$\Pi = MRT \qquad\qquad (8.38)$$

> R has units of energy·mol^{-1}·K^{-1}. The energy units used for R determine the osmotic pressure units. Osmotic pressure is often given in atmospheres, for which $R = 0.08205$ atm·L·mol^{-1}·K^{-1}.

T is the temperature in kelvin and R is the gas constant. If the total concentration of solute particles in a solution is 0.10 M, the osmotic pressure at 25 °C (= 298 K) is:

$$\Pi = MRT = (0.10 \text{ mol·L}^{-1})(0.08205 \text{ atm·L·mol}^{-1}\text{·K}^{-1})(298 \text{ K}) = 2.4 \text{ atm} \qquad (8.39)$$

If you were to use water in a barometer to measure atmospheric pressure, you would have to use a barometer about 34 feet high to measure one atmosphere pressure. An osmotic pressure of 2.4 atm is equivalent to a column of water about 82 feet high! This is the height of a very tall tree or a seven-story building. Our prediction about the magnitude of osmotic pressures is correct.

Osmotic pressures are large and easy to measure, even for dilute solutions, which makes it possible to use osmotic pressure measurements to determine molar masses of compounds that are not very soluble in water. The technique is commonly used for large molecules encountered in polymer chemistry, biology, and biochemistry. An apparatus for measuring osmotic pressure of a solution is diagrammed in Figure 8.28. An aqueous solution to be measured is placed in the left-hand compartment and water is placed in the right compartment. The two compartments are separated by a membrane that is permeable only to water. Instead of letting the solution rise in the tube, an external pressure, P, is applied. The external pressure is increased until the net flow of water between the compartments stops. At this point $\Pi = P - P_{atm}$.

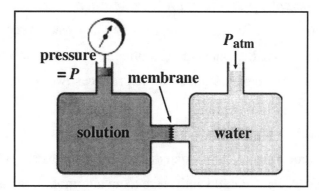

Figure 8.28. Device for measuring osmotic pressure.

8.50. Worked Example   *Molar mass determination by osmotic pressure measurement*

Suppose you have discovered a new protein and need to know its molar mass. Determine the molar mass of the protein using these data from an osmotic pressure measurement: the osmotic pressure is 121 Pa at 310 K for a solution of 5.6 mg of the protein in 11.0 mL of solution.

Necessary Information: We need to know equation (8.39) and the definition of molarity. The osmotic pressure is given in Pa (pascal), so we need to know R in units of pascals, liters, moles, and kelvin: $R = 8.314 \times 10^3 \text{ Pa·L·mol}^{-1}\text{·K}^{-1}$.

Strategy: We can calculate the molar mass, if we know the number of moles of protein in 5.6 mg of protein. Equation (8.39) shows that we can use the osmotic pressure to get the molarity of the solution used for its measurement. We know that molarity is mol·L^{-1}, so the molarity and volume of solution give us the number of moles of protein in the solution.

Implementation: The molarity of the solution is:

$$M_{soln} = \frac{\Pi}{RT} = \frac{121 \text{ Pa}}{(8.314 \times 10^3 \text{ Pa·L·mol}^{-1} \text{·K}^{-1})(310 \text{ K})} = 4.69 \times 10^{-5} \text{ M}$$

The moles of solute in V_{soln} liters of solution is:

$$\text{mol solute} = M_{soln} V_{soln} = (4.69 \times 10^{-5} \text{ M})(11.0 \times 10^{-3} \text{ L}) = 5.16 \times 10^{-7} \text{ mol}$$

The definition of moles of solute is rearranged to get the molar mass:

$$\text{molar mass} = \frac{\text{mass solute}}{\text{mole solute}} = \frac{5.6 \times 10^{-3} \text{ g}}{5.16 \times 10^{-7} \text{ mol}} = 1.08 \times 10^4 \text{ g·mol}^{-1}$$

Does the Answer Make Sense? A molar mass of about 10,800 g·mol^{-1}, is reasonable for a relatively small protein.

8.51.   Check This   *Osmotic pressure of an ionic solution*

Seawater contains many ions. For simplicity, assume that it can be represented as a 0.50 M solution of sodium chloride, NaCl. What is the osmotic pressure (in atmospheres) of seawater at 20 °C? Remember that each mole of sodium chloride produces two moles of ions in the solution.

Reflection and Projection:

Colligative properties of solutions have many applications in everyday life. Next time you use salt to melt ice from sidewalks or add antifreeze to an automobile radiator to prevent the coolant from freezing or boiling over, stop to consider that you are harnessing the power of entropy. The idea that mixtures are more probable than pure substances reflects our common experience, and the explanation in terms of ways of arranging particles and thermal quanta is consistent with that experience.

Sometimes, however, changes seem to deviate from our usual experiences or phenomena do not seem to be consistent with the sort of entropic ideas we have developed. To understand such cases, we have to examine the systems in greater detail and often discover that the examination deepens our concept of entropy and increases our ability to use the concept in new situations. We hope that the examples presented in this chapter will help as you begin to think through such new situations.

Section 8.14. The Cost of Molecular Organization

Recall that we started the chapter with a question: "How is it that some things become so disorganized while others become so highly organized?" It is *positional* organization and disorganization (water forming an intricate snowflake or sugar dissolving in iced tea) that usually catches our attention. However, we are not always aware of the subtle positional organization of molecules, as in water around ionic and molecular solutes, and we cannot see the organization and disorganization of energy quanta that must also be accounted for in answering the question.

In all spontaneous changes, the net entropy increases. This means that the number of arrangements of molecules and energy quanta at the end of the change is always greater than the number of arrangements before the change. There are fewer molecular arrangements for organized molecular systems than for disorganized systems. Therefore, when a change leads to greater positional organization of some molecules (water in a snowflake or phospholipids in a membrane), you have to look deeply into all the positional and energetic changes that occur elsewhere during the change to find the positional and/or energetic changes that increase the overall number of arrangements.

In many cases, energy released by a change is responsible for a large increase in the overall number of arrangements. That is, a large increase in thermal entropy of the surroundings produces a large increase in net entropy (or decrease in free energy of the system). One striking example of this situation is the oxidation of glucose, which we examined In Section 8.9 Worked Examples 8.29 and 8.31. In this case $\Delta G^{\circ}_{reaction} = -2878$ kJ per mole of glucose oxidized. The lion's share of this free energy change is the enthalpy released to the surroundings, 2801 kJ, which increases their thermal entropy. Recall that we found, in Chapter 7, Section 7.9, that the oxidation of glucose in biological systems is coupled to the production of ATP:

$C_6H_{12}O_6(s) + 6O_2(g) \rightarrow 6CO_2(g) + 6H_2O(l)$ $\Delta G^{\circ}_{reaction}$ (8.40)

$\underline{36ADP^{3-}(aq) + 36HOPO_3^{2-}(aq) + 36H_3O^+(aq) \rightarrow 36ATP^{4-}(aq) + 72H_2O(l)}$ $\Delta G^{\circ}_{\underline{ATP\ form}}$ (8.42)

$C_6H_{12}O_6(s) + 6O_2(g) + 36ADP^{3-}(aq) + 36HOPO_3^{2-}(aq)\ ) + 36H_3O^+(aq) \rightarrow$

$6CO_2(g) + 78H_2O(l) + 36ATP^{4-}(aq)$ $\Delta G^{\circ}_{coupled}$ (8.43)

The standard free energy change for the production of ATP under standard cellular conditions is positive, approximately 31 kJ per mole of ATP formed. Thus, $\Delta G^{\circ}_{ATP\ form} = 1100$ kJ ($= 36 \times 31$ kJ) for reaction (8.42). For the coupled reaction:

$$\Delta G^{\circ}_{coupled} = \Delta G^{\circ}_{reaction} + \Delta G^{\circ}_{ATP\ form} = -2878\ kJ + 1100\ kJ = -1778\ kJ \qquad (8.44)$$

This is an enormous driving force that is equivalent to a net entropy increase of almost 6000 J·K^{-1} for the coupled system. Some of the energy that would have gone into the

---

surroundings to increase their thermal entropy is captured in the coupled process and converted to chemical potential energy in the bonding of ATP.

In subsequent coupled processes, the chemical potential energy stored in ATP is used to drive other reactions as ATP is hydrolyzed to ADP and phosphate, the reverse of reaction (8.42). These coupled reactions also have negative free energy (positive net entropy) changes. Among these coupled reactions are those that synthesize the phospholipid molecules used to build the cellular membranes. As we have seen, the formation of the membranes is accompanied by an entropy increase when water is freed from its interactions with the phospholipids; it requires essentially no energy. Energy was, however, provided by the oxidation of glucose far back up the chain of reactions that formed the phospholipids.

Thus, the cost of all the organization and complex processes of life, whether driven by coupling with reactions like ATP hydrolysis or by more subtle positional entropy effects, is the large increase in thermal entropy when fuel (food) molecules are oxidized. As we go on in the book, we will further discuss ways the free energy released in these oxidations is captured.

But this analysis still leaves a question: Where does all the glucose come from in the first place? Green plants synthesize vast quantities of glucose every day from carbon dioxide and water. This is the reverse of reaction (8.40), so the standard free energy change for the synthesis reactions is 2878 kJ, a large *positive* free energy change. Plants must couple the synthesis with some other process(es) that has an even larger *negative* free energy change.

Synthesis of glucose in plants begins with the absorption of several quanta of visible light energy from the sun. This energy is ultimately released as free energy to drive the synthesis. The light energy itself is a result of the energy released in the sun's nuclear fusion reactions that we discussed in Chapter 3. Even though positional entropy decreases in a fusion reaction, the enormous release of energy increases the thermal entropy of the surroundings, so the net entropy change is very large and positive. Ultimately, therefore, the biological organization we see on Earth is "paid for" by the net entropy increase in the sun's nuclear fusion reactions. Electromagnetic radiation from the sun "couples" this increase to the synthesis of glucose on Earth.

Section 8.15. Outcomes Review

In this chapter, we introduced the thermodynamic function, entropy, that is responsible for the direction of all spontaneous changes. We found that net entropy is a measure of the number of distinguishable arrangements of atoms, molecules, and energy quanta in a system and its surroundings and that net entropy always increases in spontaneous changes. The net entropy is the sum of positional and thermal entropies of the system and surroundings. The examples we used to understand and analyze the role of net entropy were phase changes in pure compounds and in solutions, osmosis, oxidation of glucose, solubility of ionic and molecular solutes, and the formation of micelles, phospholipid bilayers, and cell membranes. We also introduced another thermodynamic function, the Gibbs free energy, that is directly related to net entropy changes, but is more convenient for some purposes. Net entropy change and free energy can be used interchangeably to understand and analyze actual or hypothetical changes.

Check your understanding of the ideas in the chapter by reviewing these expected outcomes of your study. You should be able to:

• relate the relative probability of two outcomes to the number of ways each outcome can be achieved [Sections 8.2, 8.3, and 8.4].

• calculate the number of distinguishable arrangements of a small number of identical particles (molecules) among a limited number of distinguishable locations [Section 8.3].

• relate mixing to osmosis and be able to explain both in terms of increasing number of distinguishable molecular arrangements [Sections 8.1 and 8.4].

• describe the conditions required for osmosis, predict the direction of osmosis, and calculate the osmotic pressure for aqueous systems [Sections 8.1, 8.4, and 8.13].

• calculate the number of distinguishable arrangements of a small number of identical energy quanta among a limited number of distinguishable atoms [Section 8.5].

• predict the direction and final outcome of energy transfer between two systems containing a countable number of energy quanta distributed among a countable number of distinguishable atoms [Section 8.5].

• relate the results for countable systems, both matter and energy distributions, to real systems [Sections 8.4 and 8.5].

• use the definition of entropy in terms of distinguishable arrangements to predict the relative entropies of different systems, for example, phases of matter or reactants and products of a reaction [Sections 8.7, 8.8, 8.9, 8.10, 8.11, and 8.12].

• distinguish between positional and thermal entropy changes for a process and combine them to determine net entropy change for the process [Sections 8.6, 8.7, 8.8, 8.9, 8.10, 8.11, and 8.12].

• use positional entropy changes and the relative thermal entropy changes for the same energy change at different temperatures to analyze and predict the direction of phase changes in pure compounds and solutions [Sections 8.7 and 8.12].

- state the criterion for equilibrium in chemical systems and relate the state of equilibrium to the positional and thermal entropy changes occurring [Sections 8.7 and 8.8].

- explain the relationship of Gibbs free energy change to net entropy change for a process and use the sign and/or magnitude of either one to predict whether the process is possible, not possible, or in equilibrium [Sections 8.8 and 8.9].

- calculate the free energy change for a process in a system using values for the enthalpy and entropy changes for the process [Sections 8.8 and 8.9].

- use standard enthalpies and standard free energies of formation, and standard entropies (from tabulated values) to calculate the standard enthalpy, free energy, and entropy changes for a reaction [Section 8.9].

- explain in words, equations, diagrams, and/or molecular-level sketches the origin and direction of positional and thermal entropy changes for dissolving ionic and molecular solutes in water and predict the observable outcomes [Sections 8.9 and 8.10].

- explain in words, equations, diagrams, and/or molecular-level sketches the origin and direction of positional and thermal entropy changes for formation of micelles and phospholipid bilayer membranes by ambiphilic molecules [Section 8.11].

- use words and/or molecular level sketches to describe the structure and properties of micelles, bilayer membranes, and liposomes [Section 8.11].

- calculate freezing point lowering, boiling point elevation, and osmotic pressure for solutions [Sections 8.12 and 8.13].

- use experimental values for colligative properties to determine the concentration of solutes in the solution and/or the molar mass of the solute [Sections 8.12 and 8.13].

- connect the formation of organized collections of molecules to increases in positional and/or thermal entropy in the system and surroundings that drive the organization [Sections 8.7, 8.10, 8.11, and 8.14].

- predict the direction of the positional entropy change(s) that must occur to produce the observed effects of heating or cooling a system, such as a rubber band [Section 8.16].

Section 8.16. EXTENSION — Thermodynamics of Rubber

8.52. Investigate This   *What happens when a stretched rubber band is heated?*

Do this as a class investigation and work in small groups to discuss and analyze the results. Use large paper clips or some other means to attach a 0.5-L bottle of water to one end of a rubber band that is about 0.8 cm wide and 10 cm long. Attach the other end of the rubber band to a sturdy support (such as the back of a chair or a coat hook) so that the rubber band and bottle hang freely without rubbing on anything. The full bottle of water should stretch the rubber band to about three times its relaxed length. Measure and record the length (in mm) of the rubber band from one end to the other. Use a hair dryer to warm the rubber band. Note whether its length changes, and how. Record the length when the rubber band is warm. Let the rubber band cool, and record its length again.

8.53. Consider This   *What is the direction of change when a rubber band is heated?*

Did the length of the rubber band in Investigate This 8.52 change in the direction you expected? Explain why or why not. Discuss the results, and possible entropy changes, with other members of your class.

Rubber bands are familiar objects, but some of their properties are not so familiar and are surprising. An entropy analysis of these properties leads to interesting insights about the behavior of rubber molecules. In Investigate This 8.52 you found that warming a stretched rubber band causes it to contract. The net entropy change for the contraction must be positive, $\Delta S_{net} > 0$, or equivalently, the free energy change must be negative, $\Delta G < 0$:

$$\Delta G = \Delta H_{rubber} - T\Delta S_{contraction} < 0 \tag{8.44}$$

The enthalpy change of the rubber is a result of a transfer of thermal energy *to* the rubber from the surroundings, so $\Delta H_{rubber} > 0$. Since ΔH_{rubber} is *positive*, ΔG can only be *negative* if $T\Delta S_{contraction}$ is *positive* and numerically *larger* than ΔH_{rubber}. In other words, the entropy of the rubber must *increase* as the rubber band contracts.

Molecular structure of rubber

To explain the increase in positional entropy of the rubber, we must take into account the structure of rubber molecules. Figure 8.29 illustrates rubber molecules in their relaxed state and in their stretched state. Rubber molecules contain between 4,000 and 20,000 carbon atoms linked together to form long chains of covalently-bonded hydrocarbons. In their relaxed state, these rubber molecules are tangled within themselves and with other rubber molecules, somewhat as diagrammed in Figure 8.29(a). The number of possible arrangements for random tangling is enormous.

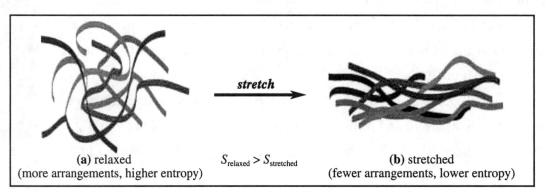

(a) relaxed $S_{relaxed} > S_{stretched}$ (b) stretched
(more arrangements, higher entropy)   (fewer arrangements, lower entropy)

Figure 8.29. Schematic illustrations of relaxed and stretched carbon chains in rubber.

When rubber is stretched a moderate amount, no chemical bonds are broken. Instead, the molecules are forced to begin untangling. An arrangement of long chains of molecules lying more-or-less parallel to each other, as in Figure 8.14(b), is lower in entropy than the tangled mass of molecules. Fewer arrangements are possible for the more orderly array:

$$\Delta S_{contraction} = S_{relaxed} - S_{stretched} > 0 \qquad (8.45)$$

The change in the organization of the molecules explains the positive value for the entropy change when a rubber band contracts. The entropy change is large enough that $T\Delta S_{contraction}$ is numerically larger than ΔH_{rubber}, producing a negative value for the free energy change, ΔG in equation (8.44). The surprising contraction of rubber when it is warmed is a result of the positional entropy of long carbon chains.

8.54. Investigate This   *What happens when a rubber band is stretched?*

Hold a rubber band, such as the one used in Investigate This 8.52, by its ends in its relaxed position. Place the rubber band against the sensitive | Some people find that their foreheads are more sensitive to temperature change than their lips.

part of your lip and allow it to come to the same temperature as your lip. Quickly stretch the rubber band to three-or-four times its relaxed length. Note any changes in the temperature of the rubber band. Now hold the stretched rubber band against your lip and allow the rubber band to come to the same temperature as your lip. Without letting go of the ends, quickly let the rubber band contract to its relaxed length. Note any changes in the temperature of the rubber band. Repeat the experiment to be sure of your observations.

8.55. **Consider This** *Does the model of rubber explain the rubber band results?*

Does the rubber warm, cool, or stay the same temperature when stretched? When it contracts? Do the relative magnitudes and signs of the enthalpy and entropy changes in equation (8.44) explain the observations? Explain why or why not.

The contraction of a stretched rubber band when it is heated is a bit surprising, since most solids expand when they are heated. The implications for the positional and thermal entropy changes that must occur lead to a better understanding of the molecular structure and properties of rubber. You will find that entropy analyses are often useful in cases like this.

Index of Terms

Chapter 8 Problems

Section 8.1.   Mixing and Osmosis

8.1.   Which processes involve an *increase* in organization for the system specified? Which involve a *decrease* in organization? Explain your reasoning briefly.

   (a) ice cream melting

   (b) students taking their places in a classroom with fixed seating

   (c) a shrub forming spring flowers

   (d) obtaining pure water from seawater

   (e) opening and shuffling a new deck of cards

   (f) raking leaves into a single pile

8.2.   Explain why a water soluble dye spreads out when it is mixed with water resulting in a uniformly colored solution. Will the reverse process ever happen?

8.3.   A few drops of green good coloring are added to some cookie dough.

   (a) Are these drops of food coloring likely to disperse on their own throughout the cookie dough? Explain your reasoning.

   (b) Compared to your answer in part (a), are drops of green food coloring more or less likely to disperse throughout a glass filled with water? Explain your reasoning.

8.4.   The membranes of red blood cells are semipermeable. Moving water out of the blood cell causes the cell to shrivel, a process known as *crenation*. Moving water into the blood cell causes the cell to swell and rupture, a process known as *hemolysis*. People needing body fluids or nutrients and who cannot be fed orally are given solutions by intravenous (or IV) infusion. An IV feeds nutrients directly into veins. Explain what would happen if an IV having a higher concentration than the solution within the red blood cell was given.

8.5.   Explain the role of osmosis in each of the following processes.

   (a) Swimming in a freshwater lake makes your body water-logged and increases your need to urinate.

   (b) Cucumbers wrinkle, lose water and shrink when placed in brine (concentrated salt water solution).

   (c) Prunes and raisins swell in water.

   (d You get thirsty and your skin wrinkles when you swim in the ocean.

   (e) Salt and sugar are used to preserve many foods.

   (f) Many saltwater fish die when placed in fresh water and *vice versa*.

Section 8.2.   Probability and Change

8.6.   Choose the best answer(s) and explain your choice(s).

When a dye solution mixes with water in an unstirred container, the mixing occurs because:

(i)   The mixed state has a higher number of distinguishable arrangements than the unmixed state.

(ii)   The mixed state has a lower heat content than the unmixed state.

(iii) The mixed state is more probable than the unmixed state.

8.7.   What numerical value (0, 1, or somewhere in between) would you assign to the chance of each of these changes being observed? Briefly discuss your reasoning, giving any conditions under which the probability could change.

(a)   A broken eggshell reforming into an unbroken eggshell.

(b)   A drop of food coloring dispersing throughout a cup of water at room temperature.

(c)   Drawing the ace of spades from a shuffled deck of cards.

(d)   Finding all of the pepperoni slices on one half of a pizza surface.

(e)   A large oak tree falling to the ground and then returning to the upright position.

(f)   Recovering six pairs of clean socks from the dryer after washing six pairs of dirty socks.

8.8.   Think about the examples from your daily experiences that would illustrate the statement: "If a system can exist in more than one observable state, any changes will be in direction toward the state that is most probable". Which state will be the most probable for:

(a)   water at –5 °C.

(b)   water at 130 °C.

(c)   sugar crystals in a glass of hot water.

(d)   a drop of perfume in a room.

(e)   iron filings in a magnetic field. (See Chapter 4, Section 4.3, Figure 4.11.)

Section 8.3.

8.9.   **(a)** Find the number of possible distinguishable arrangements for two identical objects in the five labeled boxes (one object per box) in Investigate This 8.4.

(b) How does the number of arrangements you found in part (a) compare with the number of possibilities found in Investigate This 8.4? Is this the result you expected? Why or why not? Explain why the results come out as they do.

8.10. **(a)** Refer to Figure 8.5. How many distinguishable arrangements are there if two, rather than three, soluble dye molecules are allowed to mix in the top two layers (6 boxes) of the system? Show your set of arrangements pictorially.

(b) Compare your result for two dye molecules mixing into six boxes with that shown in the figure for three dye molecules mixing into six boxes. Explain the relationship.

(c) Given your comparison in part (b), what conclusion can you draw about the affect of number of particles on the number of distinguishable arrangements for mixing into a fixed volume (number of boxes)? Does this result make sense? Explain.

8.11. In Figure 8.6, as the number of boxes per molecule increases ten-fold from 2 to 20, does the number of possible arrangements for three solute molecules in the cells also increase ten-fold? If it does not, by what factor does the number of arrangements increase? Explain your reasoning, using information from Figure 8.6.

8.12. For systems like those in Section 8.3, the number of distinguishable arrangements, W, of n identical objects among N labeled boxes is:

$$W = \frac{N!}{(N-n)!n!}$$

The exclamation point, "!", denotes a factorial. $N!$, pronounced "en factorial," is the product of all the integers from N to 1. For example, $5!$ (5 factorial) $= 5 \cdot 4 \cdot 3 \cdot 2 \cdot 1 = 120$. (Many scientific calculators have a factorial function key that gives you factorials of numbers up to 69 — limited by the maximum value the calculator can display.)

(a) Use the formula to calculate the number of distinguishable arrangements of 3 objects in 6 boxes. Is your result the same as the number of arrangements shown in Figure 8.5? Explain why or why not.

(b) Calculate the number of distinguishable arrangements for the other mixtures represented in Figure 8.6 and compare your results with the figure.

(c) Calculate the number of distinguishable arrangements for a mixture of 5 objects in 20 boxes and for a mixture of 10 objects in 20 boxes. Do your results reinforce the conclusion you drew in Problem 8.10(c)? Explain why or why not.

Section 8.4.   Implications for Mixing and Osmosis in Macroscopic Systems

8.13.   Imagine a living plant cell surrounded by a semipermeable membrane through which water can pass but sucrose cannot. The concentration of sucrose in this cell is 0.5%. By which of these processes can you increase the concentration of sucrose in the cell? Explain the reasoning for your choice and the reason(s) you reject the others.

     (i)   Place the cell in pure water.

     (ii)   Place the cell in a sucrose solution with a concentration greater than 0.5%.

     (iii) Place the cell in a sucrose solution with a concentration less than 0.5%.

     (iv) Any of the above will work.

8.14.   **(a)**   Starting with the initial arrangement of solute and solvent molecules shown in Figure 8.7(a), will the number of molecular arrangements increase or decrease after osmosis takes place? Explain.

     (b)   Draw a diagram modeled after Figure 8.7(b) to show what will happen if six of the nine solvent molecules pass from the solvent into the solution. How many distinguishable arrangements of the solution, the solvent, and the total system are possible after this change has occurred? Is the change likely to occur? Explain why or why not.

     (c)   Is the process in part (b) osmosis or reverse osmosis? Explain.

8.15.   The concentration of solute particles in a red blood cell is about 2%. Problem 8.4 states that red blood cells are susceptible to crenation or hemolysis. Red blood cells would probably shrivel the most when immersed in which of the following solutions? Explain the reasoning for your choice and the reason(s) you reject the others. (Sucrose cannot pass through the membrane of blood cells.)

     (i)   1% sucrose solution

     (ii)   2% sucrose solution

     (iii) 3% sucrose solution

     (iv) distilled water

Section 8.5.   Energy Arrangements Among Molecules

8.16.   **(a)**   If energy is transferred from a warmer object in contact with an identical, but cooler, object, how does the number of energy arrangements after the transfer compare to the number before the transfer? Explain.

     (b)   When net energy transfer stops, how will the final temperatures of the objects compare with the original two temperatures? Explain.

8.17. Consider two identical four-atom solids, one having four quanta of energy and the other having ten.

(a) What observable property differs between these two solids?

(b) Allow three quanta of energy to be transferred from the solid with ten quanta to the solid with four quanta of energy. Use the data from Figure 8.9 to determine whether this energy transfer is likely, explaining your reasoning.

(c) Compare the result from part (b) to the result obtained in Worked Example 8.8. In both cases, quanta were transferred from a warmer four-atom solid to a cooler four-atom solid. Are both transfers likely to take place? Explain the reasoning for your response.

8.18. Choose the best answer. Explain the reasoning for your choice and the reason(s) you reject the others. Which statement describes why thermal (heat) energy moves from a hot body to a cold body?

(i) There is more thermal energy in the final state.

(ii) There is more thermal energy in the initial state.

(iii) The number of distinguishable arrangements for the energy quanta is higher in the final state.

(iv) The number of distinguishable arrangements for the energy quanta is higher in the initial state.

8.19. For systems like those in Section 8.5, the number of distinguishable arrangements, W, of n identical quanta among N atoms in a solid is:

$$W = \frac{(N+n-1)!}{(N-1)!n!}$$

See Problem 8.12 for information about factorials.

(a) Use the formula to calculate the number of distinguishable arrangements of 2 quanta among 4 atoms. Is your result the same as the number of arrangements you found for two toothpicks in four candies in Investigate This 8.7? Explain why or why not.

(b) Calculate the number of distinguishable arrangements for the cases represented in Figure 8.9 and compare your results with the figure.

(c) Consider the transfer of four quanta of energy from an 8-atom solid with an initial 16 quanta to an identical 8-atom solid that initially has 10 quanta. Is this transfer likely to occur? Why or why not? Is there another transfer that is more likely to occur? If so what is it and why is it more likely?

Section 8.6.   Entropy

8.20. State the second law of thermodynamics in words and as an equation.

8.21. What is the difference between *positional* entropy and *thermal* entropy?

8.22. The second law of thermodynamics states that observed changes always take place in a direction that *increases net entropy.* Is this statement true if only thermal entropy is considered? If only positional entropy is considered? Give the reasons for your answers.

8.23. Why is it possible to tabulate absolute entropies, but not absolute enthalpies? That is, why does Appendix XX list $S°$ (not $\Delta S°$) and $\Delta H_f^{\,o}$, (not $H°$)?

8.24. Answer each of the following statements either true or false. If you decide that the statement is false, write the correct statement.

   (a) The entropy of a mole of solid mercury is higher than a mole of mercury vapor.

   (b) If the net entropy increases during a process, the process is spontaneous.

   (c) The entropy of a system depends on the pathway it took to its present state.

   (d) The more positive its net entropy change, the faster the process.

   (e) The entropy of one liter of a 0.1 M aqueous solution of NaCl is lower than the entropy of the undissolved 0.1 mole of NaCl and one liter of water.

   (f) The entropy is larger for a system of two identical blocks of copper in which the temperature of each block is 60 °C than for the same blocks of copper if one has a temperature of 20 °C and the other is at 120 °C.

   (g) There is no reference point for measuring entropy.

8.25. Predict whether the positional entropy change in each of these processes will be positive or negative. Explain your reasoning in each case.

   (a) precipitation of $BaSO_4$ upon mixing $Ba(NO_3)_2(aq)$ and $H_2SO_4(aq)$

   (b) cooling of the ammonia gas from room temperature to –50 °C, which results in: $NH_3(g) \rightarrow NH_3(l)$

   (c) formation of ammonia in the reaction: $N_2(g) + 3H_2(g) \rightarrow 2NH_3(g)$

   (d) decomposition of $CaCO_3$ in the reaction: $CaCO_3(s) \rightarrow CaO(s) + CO_2(g)$

   (e) formation of NO in the reaction: $N_2(g) + O_2(g) \rightarrow 2NO(g)$

8.26. Occasionally one hears that evolution violates the second law of thermodynamics. The claim is that the incredible precision with which the human body is arranged could not have evolved from less complex organisms without violating the second law of thermodynamics. What is the problem with this argument?

8.27. Do you think it is scientifically right to blame the second law of thermodynamics for clutter in a friend's workspace?

8.28. For each of these isomeric pairs of molecules, predict which one of the pair will have the higher entropy under the same pressure and temperature conditions. Explain the basis of each prediction.

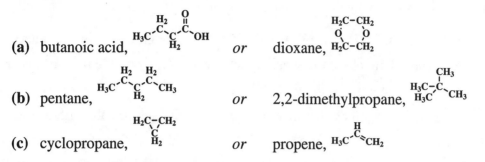

(a) butanoic acid,   *or*   dioxane,

(b) pentane,   *or*   2,2-dimethylpropane,

(c) cyclopropane,   *or*   propene,

Section 8.7.   Phase Changes and Net Entropy

8.29. Consider this statement: "Observed phase changes always take place in a direction that increases entropy." Is this statement true? Critically analyze this statement for scientific accuracy, explaining your reasoning.

8.30. Which member of each pair has the higher entropy? In each case, explain your reasoning.

   (a) $H_2O(s)$ *or* $H_2O(l)$

   (b) $CaCl_2(aq)$ *or* $CaCl_2(s)$

   (c) 5.0 g $H_2O(l)$ at 1 °C *or*   5.0 g $H_2O(l)$ at 60 °C

   (d) $H_2O(g)$ *or* $H_2O(l)$

   (e) $CO_2(g)$ *or* $CO_2(s)$, dry ice

8.31. The positional entropy change for water freezing is negative. Explain how it is possible for water to freeze.

8.32. For each of these changes, explain if the positional entropy of the system increases, decreases, or stays the same.

   (a) Solid CO_2 (dry ice) changes to carbon dioxide gas.

   (b) Liquid ethanol (C_2H_5OH) freezes.

   (c) Mothballs made of naphthalene ($C_{10}H_8$) sublime.

   (d) Plant materials burn to form carbon dioxide and water.

8.33. Consider an isolated system where a balloon containing steam, $H_2O(g)$, is immersed into a thermally isolated container of liquid water at 10.0 °C.

   (a) What will happen to the steam in the balloon?

   (b) Does the entropy of the water in the balloon increase or decrease? Explain.

   (c) Does the entropy of the water outside the balloon increase or decrease? Explain.

   (d) Is the net entropy change for this isolated system positive or negative? Explain.

8.34. **(a)** The melting point for H_2S is 187 K (–86 °C). Use a diagram similar to Figure 8.11 to explain the entropy changes for the system, the surroundings, and the net entropy change for melting H_2S at temperatures above 187 K.

   (b) How will the entropy diagram differ if liquid H_2S changes to solid H_2S at 187 K?

   (c) The phase change for water changing from solid to liquid takes place at 0 °C (273 K). Explain why the two substances have such different melting points. *Hint:* H_2S forms only weak hydrogen bonds.

8.35. Arrange the compounds ammonia, methane, and water in order of increasing entropy of vaporization, $\Delta S_{l \to g}$. Explain the reasoning for your ordering. *Hint:* Review Chapter 1, Section 1.7.

8.36. The complete combustion of ethanol, $C_2H_5OH(l)$, is represented by this equation:

$$C_2H_5OH(l) + 3O_2(g) \to 2CO_2(g) + 3H_2O(l)$$

For the combustion of one mole of ethanol under standard conditions, the standard entropy and standard enthalpy changes are –139 $J \cdot K^{-1}$ and –1367 $kJ \cdot mol^{-1}$, respectively.

   (a) Comment on these values in terms of positional entropies.

   (b) Comment on these values in terms of bond formation.

   (c) Considering positional entropy, is this reaction favored to produce carbon dioxide and water under standard conditions of temperature and pressure? Why or why not?

8.37. Consider the process: $Br_2(l) \to Br_2(g)$

   (a) Calculate $\Delta H°$ and $\Delta S°$ for this phase transition. Use the data in Appendix XX.

   (b) Calculate $\Delta S°_{net}$ for this phase transition at:

   (i) 0.0 °C **(ii)** 40 °C **(iii)** 70 °C.

   (c) At what temperature does the bromine begin to boil? Explain the reasoning you use to determine the boiling point.

8.38. Use the $\Delta H°$ and $\Delta S°$ for $H_2O(l) \to H_2O(g)$ to show how temperature influences whether the increase in the entropy of the system will favor the phase change.

8.39. **(a)** Use the standard entropies and standard enthalpies of formation for liquid and gaseous ethanal, CH_3CHO, from Appendix XX to find its standard entropy of vaporization, $\Delta S^{\circ}_{l \rightarrow g}$, and standard enthalpy of vaporization, $\Delta H^{\circ}_{l \rightarrow g}$.

(b) What is the boiling point of ethanal? Explain the reasoning you use to get your result, including any assumptions you make.

8.40. **(a)** Use the standard enthalpies of formation for liquid and gaseous methanol, CH_3OH, from Appendix XX to find its standard enthalpy of vaporization, $\Delta H^{\circ}_{l \rightarrow g}$.

(b) The boiling point of methanol is 65.0 °C. What is the standard entropy of vaporization, $\Delta S^{\circ}_{l \rightarrow g}$, for methanol? Explain the reasoning you use to get your result, including any assumptions you make.

(c) Use the data in Appendix XX to calculate $\Delta S^{\circ}_{l \rightarrow g}$ for methanol? How does your value from these data compare to your result in part (a)? How would you explain any difference?

Section 8.8.   Gibbs Free Energy

8.41. Consider the complete combustion of ethanol, $C_2H_5OH(l)$, given in Problem 8.36. For the combustion of one mole of ethanol at standard conditions, the standard change in Gibbs free energy is -1325 $kJ \cdot mol^{-1}$ at 298 K.

(a) Does the value of the Gibbs free energy change predict this reaction will take place at 298 K? Explain.

(b) At what temperature will this reaction be at equilibrium? Justify your response.

(c) Discuss why the numerical value you calculated in part (b) is likely to be in error.

8.42. Consider the reaction: $2H_2O_2(g) \rightarrow 2H_2O(g) + O_2(g)$. For this reaction at 298 K, $\Delta H = -106$ kJ and $\Delta S = 58$ $J \cdot K^{-1}$. Would you expect $H_2O_2(g)$ to be stable or likely to decompose at 298 K? Explain your reasoning.

8.43. A certain reaction is nonspontaneous at 300 K. The entropy change for the reaction is 130 $J \cdot K^{-1}$.

(a) Is the reaction endothermic or exothermic? Explain your response.

(b) What is the minimum value of ΔH for this reaction? Explain the reasoning for your answer.

8.44. The standard enthalpy of formation (ΔH°_f) and the standard free energy of formation (ΔG°_f) of elements are zero. But the standard absolute entropy (S°) is not zero. Explain why.

8.45. Changes can be characterized by the signs of ΔH and ΔS, for the system, as illustrated in this diagram. For example, processes with positive values for both ΔH and ΔS fall in the quadrant of the diagram labeled, "*B*".

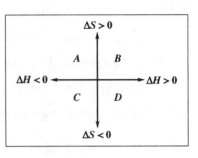

(a) Which quadrant(s), *A*, *B*, *C*, or, *D*, represents changes or processes that are always spontaneous? Explain your choice(s).

(b) Which quadrant(s) represents changes or processes that are never spontaneous? Explain your choice(s).

(c) Two of the quadrants represent processes that could *possibly* be spontaneous, depending upon the temperature at which the processes are carried out. Which quadrants are these? Which quadrant represents processes that could possibly be spontaneous at high temperatures? Which quadrant represents processes that could possibly be spontaneous at low temperatures? Clearly explain the reasoning for your choices.

(d) Which quadrant represents the freezing of a liquid? Explain.

(e) Which quadrant represents the melting of a solid? Explain.

8.46. At a certain temperature, ΔG for the reaction $CO_2(g) \rightleftharpoons C(s) + O_2(g)$, is found to be 42 kJ. Which of these statements is correct? Explain the reasoning for your choice and the reason(s) you reject the others.

(i) The system is at equilibrium.

(ii) The process is not possible.

(iii) CO_2 will be formed spontaneously.

(iv) CO_2 will decompose spontaneously.

8.47. Consider the oxidation of methane: $CH_4(g) + 2O_2(g) \rightarrow CO_2(g) + 2H_2O(g)$.

(a) By inspection, predict the change of entropy. Explain the basis of your prediction.

(b) Use the values of ΔH°_f and ΔG°_f from Appendix XX to calculate the change in standard enthalpy, free energy, and entropy for this reaction at 25 °C and 1 bar. Explain the procedures you use.

(c) Was your prediction in part (a) close to the calculated value in part (b)? Explain why or why not.

8.48. This reaction, $2H_2(g) + O_2(g) \rightarrow 2H_2O(l)$, is known to have a large negative free energy change at room temperature. Does this mean that if you mix hydrogen and oxygen at room temperature you should be able to see the reaction occur? Explain.

8.49. For this reaction, $C_{(graphite)} + 2H_2(g) \rightarrow CH_4(g)$, $\Delta H° = -74.8$ kJ and $\Delta S° = -80.8$ J·K^{-1}.

   (a) Plot, as was done in Figure 8.13, ΔH, $T\Delta S$, and ΔG as a function of temperature for the reaction under standard conditions. What assumptions must you make to construct your plot?

   (b) Use your plot to estimate the temperature at which this reaction is in equilibrium under standard conditions. Explain how you find the temperature.

   (c) Is the temperature you found in part (b) likely to be an accurate prediction? Explain why or why not.

8.50. What is the free energy change, ΔG, when one mole of water at 100 °C and 100 kPa pressure (standard pressure) is converted to gaseous water at 100 °C and 100 kPa pressure? The standard enthalpy change for this process is 40.6 kJ. Explain your answer.

8.51. We have shown that observed changes occur in the direction that increases the net entropy of the universe (or at least our tiny corner of it), ΔS_{net}, the sum of positional and thermal entropy changes for the process. You will find other writers who say that observed changes occur in the direction that minimizes the energy of the system undergoing change and maximizes its entropy. Explain clearly how these two viewpoints are actually the same. (We have used the first because it makes the connection between higher probability and direction of change more explicit.)

Section 8.9.   Thermodynamic Calculations for Chemical Reactions

8.52. For this combustion reaction, $CH_3COCH_3(g) + 4O_2(g) \rightarrow 3CO_2(g)) + 3H_2O(l)$, at 298 K, $\Delta G = -1784$ kJ and $\Delta H = -1854$ kJ.

   (a) Will the value of $T\Delta S$ for this reaction be about the same magnitude as for ΔG and ΔH? Make a qualitative estimate of this value.

   (b) Comment on what your estimated relative value for $T\Delta S$ tells you about the factor that contributes most to the net entropy change, ΔS_{net}, for this reaction.

   (c) Calculate the value of ΔS for this reaction, reporting your answer in J·K^{-1}.

8.53. For this reaction, $4Ag(s) + O_2(g) \rightarrow 2Ag_2O(s)$, at 298 K, $\Delta H° = -61.1$ kJ and $\Delta S° = -0.132$ kJ·K^{-1}.

   (a) Assuming that the reaction takes place at constant pressure, is it spontaneous at 25 °C? Explain.

   (b) Is it going to be spontaneous at 273 °C? Explain the reasoning for your response and clearly state any assumptions you make.

8.54. The thermodynamic data in this table have been obtained for the gas-phase dimerization reaction of methanoic acid (dotted lines represent hydrogen bonds):

	$\Delta H°_f$ kJ·mol^{-1}	$S°$ J·K^{-1}·mol^{-1}
HC(O)OH(g)	−362.63	251.0
[HC(O)OH]$_2$(g)	−785.34	347.7

(a) Calculate $\Delta H°$, $\Delta S°$, and $\Delta G°$ for the gas-phase dimerization at 298 K. Is the formation of the dimer from the monomers spontaneous under these conditions? Explain.

(b) Calculate the enthalpy change per hydrogen bond formed in the gas phase. Show your reasoning clearly.

(c) Why is it not useful to estimate the free energy or entropy of hydrogen bond formation in the same way you did the enthalpy in part (b)?

8.55. Use data from Appendix XX to calculate $\Delta H°$, $\Delta S°$, and $\Delta G°$ for each of these dissolution reactions:

$$KCl(s) \rightleftharpoons K^+(aq) + Cl^-(aq)$$
$$AgCl(s) \rightleftharpoons Ag^+(aq) + Cl^-(aq)$$

(a) Do your results support the solubilities we discussed in Chapter 2, Section 2.7? Explain why or why not.

(b) Do you predict that the solubility of KCl(s) should increase, decrease, or stay the same as the temperature is increased? Justify your answer thermodynamically.

(c) Do you predict that the solubility of AgCl(s) should increase, decrease, or stay the same as the temperature is increased? Justify your answer thermodynamically.

8.56. **(a)** For $3C_2H_2(g) \rightleftharpoons C_6H_6(l)$, use data from Appendix XX to find $\Delta G°$ at 298 K for the conversion of ethyne to benzene. In which direction is the reaction spontaneous under standard conditions? Explain.

(b) Should it be possible to find a temperature where this direction is reversed? Give a thermodynamic answer based on the reactants and check it with data from Appendix XX.

8.57. **(a)** Hydrazine, $N_2H_4(l)$, and its derivatives are used as rocket fuels. Several different oxidizers can be used with hydrazine. Use data from Appendix XX to calculate $\Delta H°$, $\Delta S°$, and $\Delta G°$ at 298 K for the reaction with $H_2O_2(l)$: $N_2H_4(l) + H_2O_2(l) \rightarrow N_2(g) + 2H_2O(l)$.

(b) Is the reaction spontaneous? Explain.

(c) In a rocket engine, the reactants will get hot. Are higher temperatures more favorable or less favorable for the reaction? Justify your answer thermodynamically.

8.58. Consider the reaction that converts pyruvic acid to ethanal and carbon dioxide:

$$CH_3COC(O)OH(l) \rightleftharpoons CH_3CHO(l) + CO_2(g)$$

(a) Use the data from Appendix XX to calculate $\Delta H°$, $\Delta S°$, and $\Delta G°$ for this reaction at 298 K. Are your results consistent with equation (8.19)? Explain why or why not.

(b) Is this reaction spontaneous under standard conditions? Explain.

(c) If the pressure of carbon dioxide is raised to 100 atm (about 10,000 kPa), how will the entropy of carbon dioxide be affected? Give the reasoning for your response. *Hint:* Recall what the ideal gas equation, Chapter 7, Section 7.13, equation (7.56), tells you about the relationship of the pressure to the volume of a gas and how the entropy of a system varies with the volume available to the molecules.

(d) How will the change in pressure of carbon dioxide affect the free energy change for the reaction? Will the reaction be more or less likely to proceed as written? Explain.

(e) Assuming that the reaction is in equilibrium before the carbon dioxide pressure is increased, what does Le Chatelier's principle predict about the effect on the equilibrium of increasing the pressure? Is this prediction consistent with your response in part (d)? Explain why or why not.

8.59. When water vapor is passed over hot carbon, this reaction occurs:

$$C(s) + H_2O(g) \rightarrow CO(g) + H_2(g)$$

The product mixture is called "water gas" or "synthesis gas," since it can be used as a starting material to synthesize many carbon-containing molecules.

(a) Use data from Appendix XX to calculate $\Delta H°$, $\Delta S°$, and $\Delta G°$ for this reaction at 298 K. Is the reaction spontaneous under standard conditions? Explain.

(b) Will increasing the temperature make the reaction more favorable or less favorable in the direction written? Justify your answer thermodynamically.

(c) Will the reaction be favored in the direction written by increasing or decreasing the pressure of gases in the system? Justify your answer thermodynamically.

(d) Given your answers in parts (b) and (c), what conditions would you recommend to optimize the formation of products.

    (i) low temperature and low pressure

    (ii) low temperature and high pressure

    (iii) high temperature and low pressure

    (iv) high temperature and high pressure

Section 8.10. Why Oil and Water Don't Mix

8.60.   The cellulose molecules (polymers of glucose) that make up paper are polar and hydrophilic. How does the structure of fatexplain how water beads up and does not get absorbed by "waxed" paper (paper impregnated with fatty substances)?

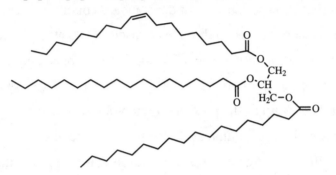

8.61.   **(a)** Do you think the positional entropy change for the formation of the methane-water clathrate from the pure compounds is positive or negative? Clearly explain the reasoning for your answer.

(b) What reason(s) can you think of for the clathrate formation being more favorable at lower temperatures? Clearly explain the reasoning for your answer.

(c) What reason(s) can you think of for the clathrate formation being more favorable at higher pressures of methane? Clearly explain the reasoning for your answer. *Hint*: The molar entropy of a gas decreases as its pressure increases at constant temperature. (Does this make sense to you? See Problems 8.58(c) and 8.93(e).)

8.62.   Benjamin Franklin was always interested in the world around him. He noted, for example, that a teaspoon of oil (about 5 mL) poured on the surface of a calm pond spread out to form an oil patch about $\frac{1}{2}$-acre (about 4000 m^2) in area. What do these data tell you about the size, in nm, of an oil particle (molecule)? Carefully note and explain any assumptions you make in answering this question.

Section 8.11. Ambiphilic Molecules: Micelles and Bilayer Membranes

8.63.   Suppose you have several ambiphilic compounds with the same polar head but non-polar tails of different lengths. How will the lengths of the tails affect the ability of the compounds to form micelles in water? Use drawings to help make your explanation clear.

8.64. The diagram here shows liquid water with a less dense non-polar liquid floating on it (oil on water).

(a) Suppose you have an ambiphilic compound that can't form micelles in water because its non-polar tail is too long. If you add just a little bit of this ambiphilic compound to the system above, where do you think the molecules of the compound will be in the system? As part of your answer draw a diagram to show where the ambiphilic molecules are and how they are interacting with the water and non-polar liquid.

(b) If you add more of the ambiphilic compound to the system, where will the molecules of the compound be in the system? As part of your answer draw a diagram to show where the ambiphilic molecules are and how they are interacting with the water and non-polar liquid. *Hint*: Consider a "reversed" micelle. What is reversed? How does this help solve the problem?

8.65. Suppose you have an ambiphilic compound that will form reversed micelles in the non-polar liquid in Problem 8.64. Do you think it would be possible to remove solutes from the water by getting them to dissolve in the reversed micelle in the non-polar liquid? (This would be sort of like the action of a detergent in water.) What kind of properties would the polar head of the ambiphilic compound need, in order to get a solute out of the water?

> Experiments are being done with reversed micelles and liquid CO_2 as the non-polar liquid to see if such extractions from water can be done without generating a lot of waste solvents that could pollute the environment.

8.66. Hepatocytes, the predominant cells in your liver, are roughly cubical with edge lengths of about 15×10^{-6} m (15 μm).

(a) About how many phospholipid molecules, Figure 8.22, are required to form the cell membrane? State all the assumptions you make to solve this problem and explain your solution method clearly.

(b) How many moles of phospholipid are required to form the membrane and what is the mass of the phospholipid in the membrane? State any further assumptions you make to solve this problem and explain your solution method clearly.

8.67. Mayonnaise has four main ingredients: oil, vinegar (mostly water), lemon juice (mostly water), and egg yolk. An egg yolk contains a lot of phospholipids, proteins, and fats to nourish the chick embryo in a fertilized egg. What is the purpose of the egg yolk in the mayonnaise? Clearly explain your reasoning.

8.68. To carry out their functions, membranes need to stay fluid, so that they retain their permeability and the proteins embedded in them can move about as necessary. One way that cold-blooded organisms use to keep their membranes fluid at different temperatures is to change the proportion of saturated and unsaturated fatty acids, Figure 8.22, in the phospholipid membrane molecules. Compared to a fish that lives in the tropics, would you expect a fish that lives in cold Arctic waters to have a higher or lower proportion of unsaturated fatty acids in its membranes? Clearly explain the reasoning for your response, including molecular level drawings, if that helps your explanation.

Section 8.12. Colligative Properties of Solutions

8.69. Oil-rich countries in the Middle East once considered towing icebergs from the Arctic Ocean to help solve their water shortages. Wouldn't the icebergs be made of salt water like the ocean water? Explain.

8.70. Explain why $S_{solution} > S_{solvent} > S_{solid}$.

8.71. Why does a colligative property depend only on the number of solute particles and not their chemical properties?

8.72. Which solution will have the lowest freezing point? Explain your answer.

   (i)   1 M NaCl

   (ii) 1 M $CaBr_2$

   (iii) 1 M $Al(NO_3)_3$

8.73. Explain why a mixture of antifreeze and water, instead of pure water, is typically placed in an automobile's radiator.

8.74. Which aqueous solution has the largest ΔT_{fp}? Explain your reasoning.

   (i)   1.5 m solution of glucose, $C_6H_{12}O_6$

   (ii) 1.5 m solution of sodium chloride

   (iii) 1.5 m solution of magnesium chloride

8.75. What is the freezing point of a solution made by dissolving 5.00 g of sucrose, $C_{12}H_{22}O_{11}$, in 100.0 g of water?

8.76.   A solution is made by mixing 10.00 g of n-decane, $CH_3(CH_2)_8CH_3$, with 100.00 g of benzene.

   (a) What is the molality of this solution?

   (b) Calculate the freezing point depression, ΔT_{fp}, for this solution.

   (c) What is the freezing point of this mixture?

8.77.   Solutions of a nonvolatile, nonionizing solute are prepared in benzene and in cyclohexane. Determine which solution will freeze at a lower temperature. Explain your procedure.

   (i)   0.30 m solution in benzene

   (ii) 0.10 m solution in cyclohexane

8.78.   **(a)** An aqueous solution of 1.0 m sodium chloride is predicted to have a freezing point of –3.72 °C. Explain why.

   (b) However, the actual freezing point is –3.53 °C. Why is the experimental freezing point lowering less than the predicted value?

8.79.   What is the freezing point of a solution made by dissolving 5.00 g of sodium sulfate, Na_2SO_4 in 100.0 g of water?

8.80.   Sorbitol is a sweet-tasting substance that is sometimes used as a substitute for sucrose. Its formula is $C_6H_{14}O_6$. What will be the freezing point of an aqueous solution containing 1.00 g sorbitol in 100.0 g of water?

8.81.   Calculate the change in the freezing point of water when 3.5 milligrams of hemoglobin (molar mass = 64,000 g) is dissolved in 5.0 mL of water. Would this method be accurate in determining molar masses of biomolecules of this size?

Section 8.13. Osmotic Pressure Calculations

8.82.   Both glucose solutions and physiological saline solutions ($NaCl(aq)$) are given to patients through intravenous injections. The average osmotic pressure of blood serum is 7.7 atm at 25 °C. To prevent osmosis through the semipermeable membrane of a red blood cell, the glucose solution or physiological saline solution given must have the same osmotic pressure as blood serum. (We say that these solutions are "isotonic" with blood serum). Calculate the concentration of glucose solution and saline solution that are isotonic with blood serum.

8.83. Suppose you dissolve 0.150 g of a newly isolated protein in 25.0 mL of water. The solution has an osmotic pressure of 0.00342 atm at 277 K. What is the molar mass of this protein?

8.84. Suppose you have a solution separated from pure water by a membrane that is only permeable to water, as shown in this diagram.

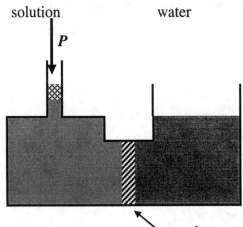

The solution will have a certain osmotic pressure, Π, determined by the solutes it contains. What would happen if a pressure, P, greater than this osmotic pressure is applied to the solution by a piston, as shown in the diagram? Use drawings to illustrate what is happening at the molecular level.

8.85. The process suggested in Problem 8.84 is called "reverse osmosis." What is it that is "reversed"? What practical uses can you think of for reverse osmosis?

8.86. Fresh water can be prepared from seawater by the process of reverse osmosis. (See Problem 8.84.) This method involves applying a pressure greater than the osmotic pressure to the seawater to force the flow of pure water from the seawater through the semipermeable membrane to the pure water. Calculate the minimum pressure that must be applied to a seawater sample to produce reverse osmosis. Assume that the concentrations of seawater's solutes are as follows: $[Cl^-]$ = 18,000 ppm; $[Na^+]$ = 10,500 ppm; $[Mg^{2+}]$ = 1200 ppm; $[SO_4^{2-}]$ = 870 ppm; $[K^+]$ = 379 ppm. The abbreviation "ppm" stands for "parts per million." One ppm is one milligram of the solute in 1000 g of solvent.

8.87. There is a brand of time release plant fertilizer called Osmocote®. The package says, "The unique resin coated granules are… easy to use. Within one week after application the soil moisture causes the granules to swell into plump capsules of liquefied plant food, which continuously release nutrients for approximately 9 months." What sort of properties does the resin coating have to have? Why do you think the company that makes this product decided on the name, "Osmocote"?

Section 8.14. The Cost of Molecular Organization

8.88. We can consider a growing plant (the system) to be an example of decreasing entropy. Small molecules, like CO_2 and H_2O, are built into complex, but orderly arrangements of macromolecules. Which of these statements applies to this example? Explain the reasoning for your choice and the reason(s) you reject the others.

(i) The second law of thermodynamics — that net entropy always increases in observed processes — is being violated.

(ii) The second law of thermodynamics is not being violated because of the entropy of the surroundings is increasing.

(iii) The second law of thermodynamics is not being violated because of the entropy of the surroundings is decreasing.

(iv) Plant growth is so complex that the laws of thermodynamics cannot be applied.

(v) None of these statements applies to this example.

8.89. Consider a fertilized chicken egg in an incubator, a constant temperature and pressure environment. The eggshell and its membrane are permeable to atmospheric gases. In about three weeks the egg will hatch into a chick.

(a) Take the egg as the system. Is this an open, closed, or isolated system? Explain. *Hint:* Review, Chapter 7, Section 7.5, if necessary.

(b) In the fertilized egg, proteins from the hen are formed into a highly ordered chick. Does the entropy of the system increase or decrease? Briefly explain why the development of the chick is or is not consistent with the second law of thermodynamics.

(c) What happens to the energy of the system as the chick develops? What forms of energy contribute to the change in energy (if any) of the system?

(d) Does the free energy of the system increase, decrease, or remain the same during development? How do you know?

Section 8.16. EXTENSION — Thermodynamics of Rubber

8.90. Suppose you are investigating a biological polymer that undergoes a reaction to give a product, polymer → product, and find that, at a given temperature, the reaction is spontaneous and endothermic.

(a) What can you tell about the signs of ΔH, ΔS, and ΔG for this reaction? Explain your reasoning.

(b) What do your results tell you about the structure of the product relative to the polymer? Explain your reasoning.

General Problems

8.91. The questions in this problem are based on the data in this table of values for the enthalpies of formation and entropies of several hydrocarbons in the gas phase. (Many of these compounds are liquids at 298 °C, so the gas-phase values here are calculated in standard ways from other experimental data.)

Enthalpies of formation and entropies for selected gas-phase hydrocarbons.

name	molecular formula	condensed structural formula	ΔH_f° kJ mol^{-1}	S° J K^{-1} mol^{-1}
ethene	C_2H_4	$H_2C=CH_2$	52.28	219.8
propene	C_3H_6	$H_2C=CHCH_3$	20.41	266.9
cyclopropane	C_3H_6	C's bonded in an equilateral triangle	53.30	237.4
1-butene	C_4H_8	$H_2C=CHCH_2CH_3$	1.17	307.4
cyclobutane	C_4H_8	C's bonded in a non-planar square	26.65	265.4
1-pentene	C_5H_{10}	$H_2C=CHCH_2CH_2CH_3$	−20.92	347.6
2-methyl-1-butene	C_5H_{10}	$H_2C=C(CH_3)CH_2CH_3$	−36.62	342.0
3-methyl-1-butene	C_5H_{10}	$H_2C=CHCH(CH_3)_2$	−28.95	333.5
2-methyl-2-butene	C_5H_{10}	$H_3CCH=C(CH_3)_2$	−42.55	338.5
trans-2-pentene	C_5H_{10}	$H_3CCH=CHCH_2CH_3$	−31.76	342.3
cyclopentane	C_5H_{10}	C's bonded in a non-planar pentagon	−77.24	292.9
1-hexene	C_6H_{12}	$H_2C=CH(CH_2)_3CH_3$	−41.7	386.0
cyclohexane	C_6H_{12}	C's bonded in a non-planar hexagon	−123.1	298.2
1-heptene	C_7H_{14}	$H_2C=CH(CH_2)_4CH_3$	−62.16	424.4
1-octene	C_8H_{16}	$H_2C=CH(CH_2)_5CH_3$	−82.93	462.8
1-nonene	C_9H_{18}	$H_2C=CH(CH_2)_6CH_3$	−103.5	501.2
1-decene	$C_{10}H_{20}$	$H_2C=CH(CH_2)_7CH_3$	−124.1	539.6
1-undecene	$C_{11}H_{22}$	$H_2C=CH(CH_2)_8CH_3$	−144.8	578.1
1-dodecene	$C_{12}H_{24}$	$H_2C=CH(CH_2)_9CH_3$	−165.4	616.5

(a) Use average bond enthalpies from Chapter 7, Section 7.7, Table 7.3 to calculate the difference you would expect between the enthalpies of formation of the C_3, C_4, C_5, and C_6 cylic compounds and their linear alkene isomers. Do you predict the cyclic or linear isomers to be more stable? How does your calculated difference compare with the experimental values from the table here? Is your prediction correct? Explain your reasoning. Use your molecular models to construct models of the cyclic and linear isomers. How, if at all, do the models affect your explanation and reasoning?

(b) Compare the entropies of the C_3, C_4, C_5, and C_6 cyclic compounds and their linear alkene isomers. Are the differences in the direction you would have predicted? Explain the basis for your prediction. Use the molecular models you constructed in part (a) to develop a rationale for the relative magnitudes as well as direction of the entropy differences. Clearly explain the basis for your reasoning.

(c) Are there trends in the data for the linear alkene isomers? Can you illustrate these graphically and draw more quantitative conclusions? Predict $\Delta H_f°$ and $S°$ for 1-tetradecene, $C_{14}H_{28}$. (What does a model of this compound look like?) Explain how you arrive at your predictions. How confident are you of these predictions?

(d) Can you use an analysis similar to that in part (c) to predict $\Delta H_f°$ and $S°$ for cyclooctane, C_8H_{16}? (What does a model of this compound look like?) Explain how you arrive at your predictions. How confident are you of these predictions?

(e) Write a discussion of the trends and correlations in the data for all the C_5H_{10} alkene isomers in the table. Use models as a basis for rationalizing and explaining what you find.

8.92. An inventor claims to have discovered a catalyst that will break water down to hydrogen and oxygen gases at room temperature without an input of energy. Use data from Appendix XX and thermodynamic arguments to counsel possible investors whether or not to invest their money to commercialize this catalyst.

8.93. Liquids evaporate, molecules leave the liquid and enter the gas phase. The pressure of the gas molecules in equilibrium with the liquid is the vapor pressure of the liquid at that temperature. Solutions of nonvolatile solutes have lower vapor pressures than the pure solvent; vapor pressure lowering is another colligative property.

(a) Which entropy is higher, that for pure solvent, $S_{solvent}$, or that for a solution, $S_{solution}$? Explain your response.

(b) The same input of enthalpy, ΔH_{vap}, is required to vaporize a given amount of liquid solvent from the pure solvent and from a solution. At a given temperature, how do the thermal entropy changes for vaporization of the same amount of liquid, $\Delta H_{vap}/T$, compare for pure solvent and a solution? Explain.

(c)   At equilibrium, the net entropy change, ΔS_{net}, must be zero for both vaporization of a pure solvent and vaporization of a solution. Given your answer in part (b), how must the positional entropy changes for (solvent $\rightarrow$ gas) and (solution $\rightarrow$ gas) compare? Explain.

(d)   Given the relative values in part (a) and your anwer in part (c), which gas must have the higher entropy, the one in equilibrium with pure solvent or the one in equilibrium with the solution? Explain your reasoning. *Hint:* A diagram similar to Figure 8.26 might be useful.

(e)   The entropy of a gas is inversely related to its pressure; higher pressure, more compressed gas has a lower entropy than the same gas at a lower pressure. Which gas in part (d) has the lower pressure? Is this result consistent with vapor pressure lowering by a nonvolatile solute. Explain the reasoning for your answers.

(f)   A beaker of water and a beaker of a 1 M aqueous glucose solution are placed side-by-side inside a sealed, transparent container held at constant temperature. If you observe this system for many days, what do you predict you will observe about the liquid levels in the two beakers? Explain the reasoning for your answer.

8.94.   The top diagram is a schematic representation of the binding of a substrate molecule (the squiggle) in the active site on an enzyme (the fat tilted "C"). In order to bind, the substrate has to fold into a shape that fits the site, as shown on the right of the diagram. The bottom diagram shows the interaction of the substrate with a metal ion (the little lozenge shape) in solution. Binding to the metal ion causes the substrate to fold as shown.

(a)   If you consider only the changes represented by the top diagram, what do you predict about the entropy change for this process? Clearly explain your reasoning.

(b)   How will the presence of the metal ion in the solution affect the entropy of binding of the substrate with the enzyme? What assumptions do you make? Explain your reasoning.

Chapter 9.   Chemical Equilibrium

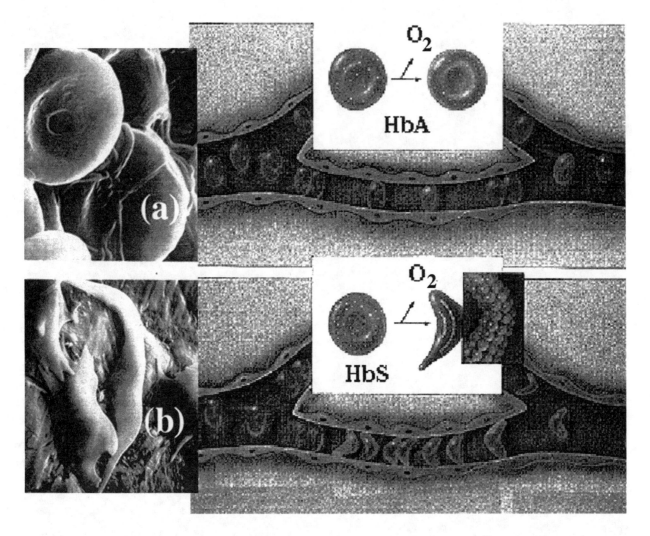

The electron micrographs on the left show red blood cells (a) from a normal adult with normal hemoglobin, HbA, and (b) from a person with sickle-cell disease, whose red blood cells contain sickle-cell hemoglobin, HbS. The illustrations on the right depict the fate of the cells as they pass through capillaries and release their oxygen to the surrounding cells. The uptake and release of oxygen in the lungs and capillaries, respectively, as well as the sickling of the cells are results of equilibrium processes. One of our tasks in this chapter will be to discover the factors that control equilibria and understand the conditions that favor one direction or the other for the systems illustrated here.

Chapter 9. Chemical Equilibria

> …the phenomena of chemical equilibrium play a capital role in all operations of industrial chemistry.
>
> *Henri Louis Le Chatelier, 1850-1936*

H. Le Chatelier was a French chemist who worked on problems in metallurgy, cements, explosives and other important industrial processes and products. His research led him to the above conclusion, but he could have expanded his comment on the central role of equilibrium to include all of chemistry: industrial reactions, laboratory studies, and living systems. Although few of the systems we see around us are at equilibrium, their changes toward equilibrium are what we observe. Thus, an understanding of equilibrium can help us understand and predict the direction of changes.

Examples of equilibria in a biological system are illustraed on the facing page. Our red blood cells are packed with the oxygen-carrying protein hemoglobin. Hemoglobin has a molar mass of about 66,000 g and each molecule can bind up to four oxygen molecules. In the lungs, where the oxygen pressure is high, oxygen binds to the heme groups on hemoglobin. When the blood reaches the capillaries, where the oxygen pressure is low (since it is used up by our cells), the process is reversed and hemoglobin releases oxygen. Binding of oxygen is an equilibrium process that favors bound oxygen when the oxygen concentration is high and release of oxygen when the oxygen concentration is low.

Normal blood cells are shaped like tiny candy mints, both before and after oxygen is released, as shown in part (a) of the illustration. Normal adult hemoglobin molecules, HbA, have little attraction for one another and shift about in the blood cell without distorting its shape as it passes through tiny capillaries. By contrast, the red blood cells of persons who suffer from sickle-cell disease have a mutant form of hemoglobin. The mutant, or HbS, molecules attract one another when the hemoglobin is deoxygenated and form long, stiff chains of HbS that distort the red blood cells into the sickle-like shape shown in part (b) of the illustration. The sickled cells get stuck in the capillaries and restrict the flow of blood and hence of oxygen to the cells, resulting in a great deal of pain and sometimes death. The mutation responsible for sickle-cell disease was discovered in experiments that depend on acid-base equilibria in proteins. Attractions between HbS molecules are a result of these equilibria.

In Chapter 8, we found that a system and its surroundings are in equilibrium, if the net entropy change for any change in the system and its surroundings is zero. In that chapter, this idea was developed for phase changes. In this chapter, our task is to extend it to equilibria in chemical reactions. Equilibria in all types of chemical reactions are fundamentally the same and can all be treated in exactly the same way. The change in Gibbs free energy, ΔG, is sometimes more convenient to use than net entropy change. Almost all arguments about directionality in biochemical reactions are based on free energies and we will look at some examples, including equilibria like those involved in oxygen transport by hemoglobin.

We will begin with a qualitative examination of the nature of chemical reactions at equilibrium to see what variables influence the equilibrium state. Then we will introduce an empirical approach to equilibrium constants, equilibrium constant expressions, and equilibrium system calculations. Finally, we will relate thermodynamics and equilibria and discover the power of this combination in a few chemical and biological systems.

Section 9.1.   The Nature of Equilibrium

9.1.   Investigate This   *How can we model a system at equilibrium?*

(a) Do this as a class investigation with two student investigators and work in small groups to discuss and analyze the results. Each investigator has a plastic basin about 10-cm deep. One investigator has a 100-mL beaker and the other a 250-mL beaker. Add water to fill one of the basins about two-thirds full; the other basin is empty at the start. In unison, the investigators dip their beakers into their basins and transfer any water in their beakers to the other investigator's basin. While continuing the water transfers for a few minutes, observe and record the water levels in the two basins.

(b) Repeat the procedure in part (a), except start with all the water in the other basin.

9.2.   Consider This   *What is dynamic equilibrium?*

(a) After several transfers of water in Investigate This 9.1, what do you observe about the water levels in the two basins? Does it make a difference which basin initially contains water? What would you predict the result to be if the water was initially distributed so some was in each basin? Explain your responses.

(b) Explain how the system of water transfer between basins can be considered a model for a chemical reaction system. Why is this called a dynamic equilibrium system?

Identifying systems at equilibrium

It is easy to be fooled about whether a chemical reaction system is at equilibrium. The *observable* properties of a system at equilibrium are unchanging, just as the liquid levels in Investigate This 9.1 are constant after the "reactions" have gone on for a time. The reactions in each direction are balanced and the net amounts of products and reactants are constant. The analogy in Investigate This 9.1 shows that the reactions continue, but the amount of product and reactant (water in the two basins) remains constant. This is a state of **dynamic equilibrium**: the observable state of the system is unchanging (amounts of reactants and products remain the same), but the reactions changing reactants to products and products back to reactants continue.

But a chemical system may only *appear* to be unchanging because the reactions are so slow that no change occurs during the time you make your observations. For example, consider wood, which is composed largely of cellulose, a polymer of glucose. You know that both the enthalpy and entropy of combustion of glucose greatly favor its oxidation, Chapter 8, Section 8.9. There is a strong driving force (large net entropy increase) to produce carbon dioxide and water from wood and oxygen. Yet wooden structures surrounded by a sea of atmospheric oxygen have existed for centuries. You can see the driving force in action if you get the reaction (combustion) started by heating the wood with a match. Wood and oxygen mixtures are not in equilibrium, but you can't tell this by just looking at the unchanging mixtures.

As in the wood-oxygen case, you can try disturbing an unchanging reaction system to see if it responds in a way that is consistent with being at equilibrium. Recall that **Le Chatelier's principle** states that a system at equilibrium responds to a disturbance in a way that minimizes the effect of the disturbance. We have been using this principle to predict how systems at equilibrium will adjust to changes in concentration or temperature and we will continue to use it in this chapter. For wood and oxygen, the disturbance (energy from the match flame) causes such an enormous change (transformation of the wood to new molecules) that you can be certain the system was *not* in equilibrium before the disturbance.

9.3. Investigate This   *Do solutions of Fe(NO₃)₃ and KSCN react when mixed?*

(a) Carry out this activity in small groups; discuss your observations and interpretations as you proceed. To each of three adjacent wells in a 24-well plate add about 1 mL (one-third of a well) of 0.002 M potassium thiocyanate, KSCN, solution. To each of the three wells, add *one drop* of 0.2 M iron(III) nitrate, Fe(NO₃)₃, solution. Record your observations on the appearance of the two reagent solutions and of their mixtures.

(b) To one of the mixtures add one more drop of the iron(III) nitrate solution. Record your observations on the appearance of this mixture compared to the others.

(c) To one of the remaining original mixtures add a *tiny* crystal of solid KSCN and observe what happens as the solid dissolves. Record your observations on the appearance of this mixture compared to the others.

9.4. **Consider This** *How do solutions of Fe(NO₃)₃ and KSCN react when mixed?*

(a) In Investigate This 9.3(a), do you have any evidence that aqueous solutions of $Fe(NO_3)_3$ and KSCN react? Explain.

(b) What did you observe when more $Fe(NO_3)_3$ solution was added to one of the wells in Investigate This 9.3(b)? How do you interpret any change that occurred?

(c) What did you observe when more KSCN was added to one of the wells in Investigate This 9.3(c)? How do you interpret any change that occurred?

9.5. **Check This** *Concentrations in the Fe(NO₃)₃–KSCN mixture*

A drop of solution from most droppers is about 0.05 mL of liquid. Thus, adding one drop of 0.2 M $Fe(NO_3)_3$ solution to 1 mL of water in Investigate This 9.3(a) dilutes the $Fe^{3+}(aq)$ by a factor of 20. What is the concentration of $Fe^{3+}(aq)$ in the $Fe(NO_3)_3$–KSCN mixtures you made in Investigate This 9.3(a) — assuming that no reaction occurs to use it up. What is the concentration of $SCN^-(aq)$ in this mixture? Which ion is present in higher concentration?

$Fe^{3+}(aq)$ reaction with $SCN^-(aq)$

Solutions of KSCN and $Fe(NO_3)_3$ are ionic. Mixtures of the two solutions contain $K^+(aq)$, $SCN^-(aq)$, $Fe^{3+}(aq)$, and $NO_3^-(aq)$ ions. All of these ions must be colorless, since the individual reagent solutions are colorless. However, as you found in Investigate This 9.3, their mixture is red or red-brown. Potassium nitrate, $KNO_3(s)$, is a white, crystalline solid that dissolves to give clear, colorless solutions, so reaction of the potassium, $K^+(aq)$, and nitrate, $NO_3^-(aq)$, ions is not responsible for the color. The other combination of ions that might be responsible for the red color is $Fe^{3+}(aq)$ with $SCN^-(aq)$. In fact, these ions react to form a red, one-to-one **metal ion complex** (a Lewis acid-base complex, Chapter 6, Section 6.6):

WEB Chap 9, Sect 9.1.1-2. View a movie of the reaction and animations of the separate ionic solutions.

$$Fe^{3+}(aq) + SCN^-(aq) \rightarrow Fe(SCN)^{2+}(aq) \tag{9.1}$$

| HF | HCl | HBr | HI |

Charge-density models of the hydrogen halides (to the same scale).

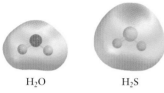

| H_2O | H_2S | H_2Se | H_2Te |

Consider This 6.28

Charge-density models of the hydrides of the first four members of the oxygen family of elements (to the same scale).

Figure 6.2

Charge-density model for the ethanoate ion.

(a)

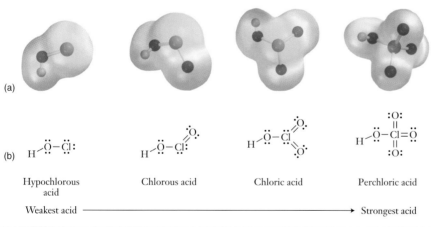

(b)

Hypochlorous acid   Chlorous acid   Chloric acid   Perchloric acid

Weakest acid ────────────────────────────────▶ Strongest acid

Figure 6.4

The oxyacids of chlorine. Both (a) charge-density models and (b) Lewis structures are shown for comparison.

P. 6-46

The reaction of carboxylic acid (ethanoic acid) with an alcohol (ethanol) to form an ester.

Ethanoic acid   Ethanol   Ethyl ethanoate   Water
                                    (an ester)

(a) Before the equivalence point

(b) At the equivalence point

Figure 6.12

An acid-base titration with phenolphthalein indicator. The flask contains ethanoic acid of unknown concentration and a few drops of phenolphthalein indicator, which is colorless in acid and red in base. The titrant is sodium hydroxide solution of known concentration. The titrant is added slowly while the flask is swirled. The addition is stopped just as the solution turns light pink.

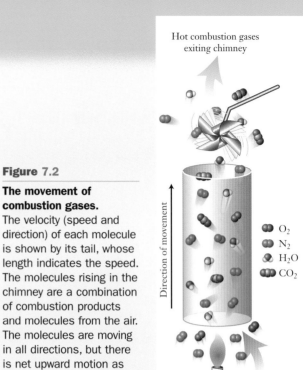

Figure 7.2

The movement of combustion gases.
The velocity (speed and direction) of each molecule is shown by its tail, whose length indicates the speed. The molecules rising in the chimney are a combination of combustion products and molecules from the air. The molecules are moving in all directions, but there is net upward motion as they are buoyed up by the denser gases that enter the chimney at the bottom.

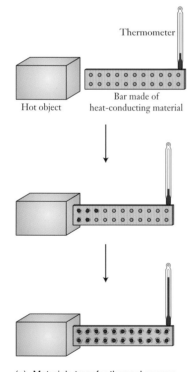

(a) Materials transfer thermal energy by conduction as vibrational energy (red) is transferred from one particle to the next

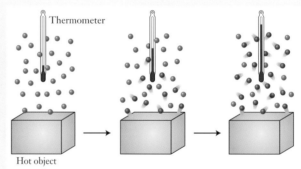

(b) Materials transfer thermal energy by convection as the more energetic particles (red) move from one place to another.

Figure 7.3

Conduction and convection of thermal energy.

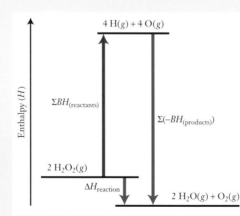

Figure 7.11

Enthalpy-level diagram for: 2 H₂O₂(g) → 2H₂O(g) + O₂(g). The reaction is assumed to be exothermic, as you found experimentally in Investigate This 7.32.

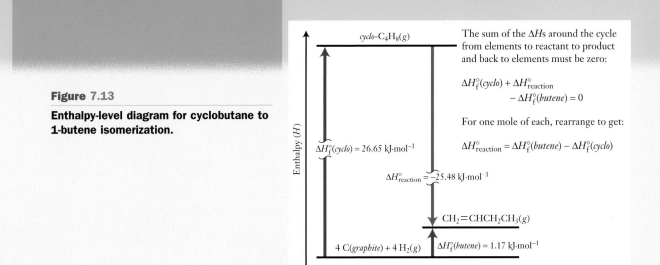

Figure 7.13

Enthalpy-level diagram for cyclobutane to 1-butene isomerization.

The sum of the ΔHs around the cycle from elements to reactant to product and back to elements must be zero:

$$\Delta H_f^\circ(cyclo) + \Delta H_{reaction}^\circ$$
$$- \Delta H_f^\circ(butene) = 0$$

For one mole of each, rearrange to get:

$$\Delta H_{reaction}^\circ = \Delta H_f^\circ(butene) - \Delta H_f^\circ(cyclo)$$

$$\Delta H_{reaction}^\circ = -25.48 \text{ kJ·mol}^{-1}$$

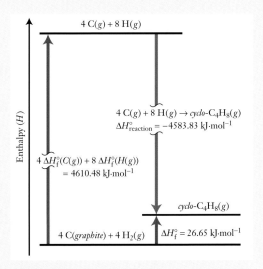

Figure 7.14

Comparison of formation of *cyclo*-C₄H₈ from its atoms and its elements. Energy is always released when a molecule is formed from its separated gas-phase atoms.

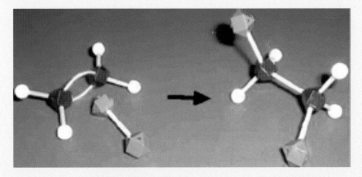

Worked Example 7.49

Molecular representation of the methanol oxidation.

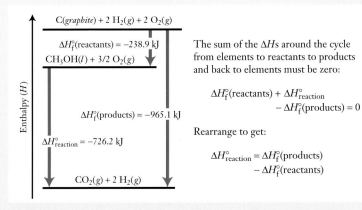

Check This 7.51

Molecular representation of ethene bromination.

The sum of the ΔHs around the cycle from elements to reactants to products and back to elements must be zero:

$$\Delta H_f^\circ(reactants) + \Delta H_{reaction}^\circ$$
$$- \Delta H_f^\circ(products) = 0$$

Rearrange to get:

$$\Delta H_{reaction}^\circ = \Delta H_f^\circ(products)$$
$$- \Delta H_f^\circ(reactants)$$

Figure 7.15

Enthalpy-level diagram for oxidation of methanol.

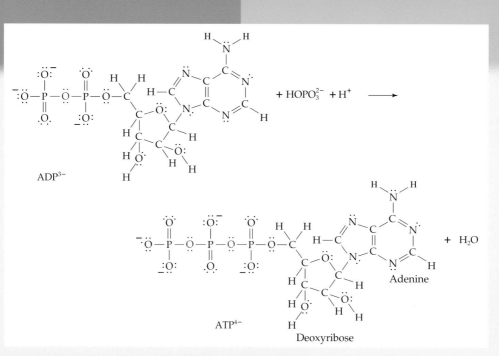

ADP^{3-} + HOPO$_3^{2-}$ + H$^+$ ⟶

ATP^{4-} + H$_2$O

Adenine

Deoxyribose

Figure 7.16

Reaction to form ATP^{4-} from ADP^{3-} and phosphate.

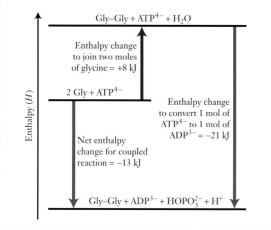

Gly–Gly + ATP^{4-} + H$_2$O

Enthalpy change to join two moles of glycine = +8 kJ

2 Gly + ATP^{4-}

Enthalpy change to convert 1 mol of ATP^{4-} to 1 mol of ADP^{3-} = −21 kJ

Net enthalpy change for coupled reaction = −13 kJ

Gly–Gly + ADP^{3-} + HOPO$_3^{2-}$ + H$^+$

Enthalpy (*H*)

Figure 7.18

Enthalpy coupling of diglycine synthesis to ATP^{4-} hydrolysis.

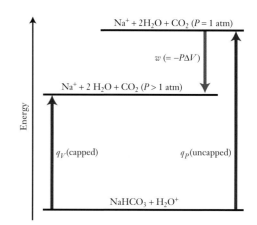

Na$^+$ + 2H$_2$O + CO$_2$ (*P* = 1 atm)

w (= −*P*Δ*V*)

Na$^+$ + 2 H$_2$O + CO$_2$ (*P* > 1 atm)

q_V (capped)

q_P (uncapped)

NaHCO$_3$ + H$_2$O$^+$

Energy

Figure 7.22

Relationships among q_v, q_p, and *w* for reaction (7.57).

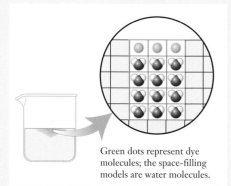

Green dots represent dye molecules; the space-filling models are water molecules.

Figure 8.4

Model of a drop of dye molecules added to water.

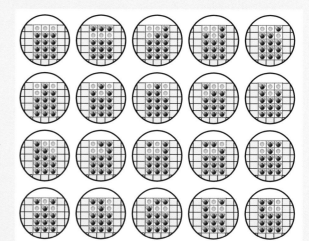

Figure 8.5

Model for arrangements of dye molecules in water. Dye molecules have mixed from Figure 8.4 into the top two layers.

Figure 8.7

Increase of molecular arrangements in osmosis. Compare with Figure 8.3. The semipermeable membrane is represented in yellow here.

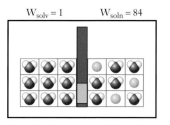

$W_{solv} = 1$ $W_{soln} = 84$

(a) Initial conditions
$W_{system} = W_{solv} \cdot W_{soln} = 84$

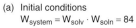

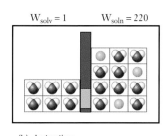

$W_{solv} = 1$ $W_{soln} = 220$

(b) Later time
$W_{system} = W_{solv} \cdot W_{soln} = 220$

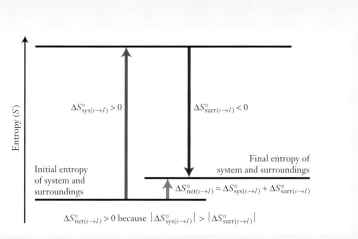

$\Delta S^\circ_{sys(s\to l)} > 0$

$\Delta S^\circ_{surr(s\to l)} < 0$

Entropy (S)

Initial entropy of system and surroundings

Final entropy of system and surroundings

$\Delta S^\circ_{net(s\to l)} = \Delta S^\circ_{sys(s\to l)} + \Delta S^\circ_{surr(s\to l)}$

$\Delta S^\circ_{net(s\to l)} > 0$ because $|\Delta S^\circ_{sys(s\to l)}| > |\Delta S^\circ_{surr(s\to l)}|$

Figure 8.11

System, surroundings, and net entropy changes for melting ice at *T* > 273 K. On these entropy diagrams, blue arrows represent positional entropy change in the system, red arrows represent thermal entropy changes in the thermal surroundings, and green arrows represent the net entropy change (the sum of the positional and thermal entropy changes).

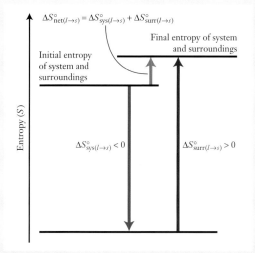

$\Delta S^\circ_{net(l\to s)} = \Delta S^\circ_{sys(l\to s)} + \Delta S^\circ_{surr(l\to s)}$

Final entropy of system and surroundings

Initial entropy of system and surroundings

Entropy (S)

$\Delta S^\circ_{sys(l\to s)} < 0$

$\Delta S^\circ_{surr(l\to s)} > 0$

Figure 8.12

System, surroundings, and net entropy changes for freezing water at *T* < 273 K. The colors of the arrows have the same meaning as in Figure 8.11.

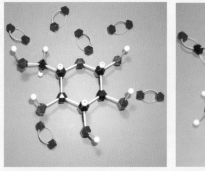

Reactants

Products

Figure 8.14

Molecular level representation of the oxidation of a glucose molecule.

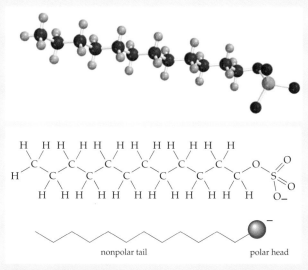

nonpolar tail   polar head

Figure 8.19

Different representations of a detergent molecule, dodecyl sulfate.

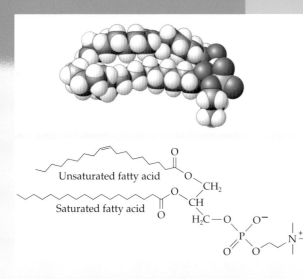

Figure 8.22

A phospholipid molecule, phosphatidyl choline.

Unsaturated fatty acid

Saturated fatty acid

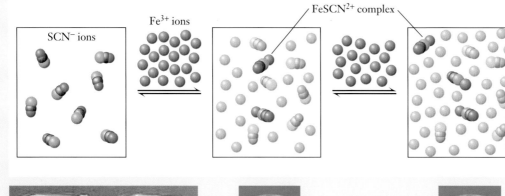

Figure 9.2

Equilibrium among Fe^{3+}(aq), SCN$^-$(aq), and Fe(SCN)$^{2+}$(aq) as Fe^{3+}(aq) is added.

SCN$^-$ ions

Fe^{3+} ions

FeSCN^{2+} complex

SCN$^-$

Fe^{3+}

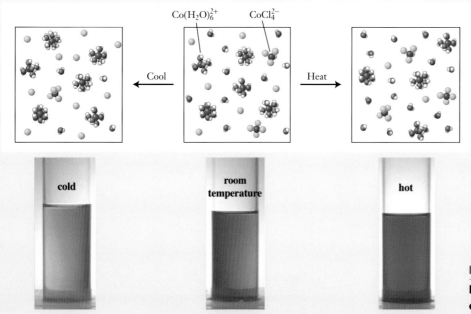

Co(H$_2$O)$_6^{2+}$ CoCl$_4^{2-}$

Cool

Heat

cold

room temperature

hot

Figure 9.3

Response of the Co(H$_2$O)$_6^{2+}$ CoCl$_4^{2-}$ equilibrium to heating and cooling.

Figure 9.9

Molecular-level representation of gravimetric analysis for calcium ion. (a) is the solution to be analyzed. In (b), a stoichiometric amount of sulfate ion is added and some $CaSO_4(s)$ precipitates. When more sulfate ion is added, in (c) the precipitation of $CaSO_4(s)$ is almost complete.

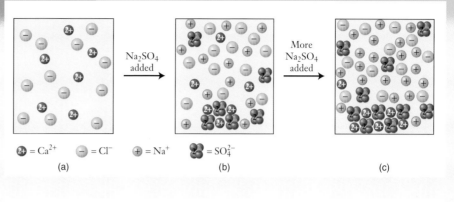

$2+ = Ca^{2+}$ $- = Cl^-$ $+ = Na^+$ $= SO_4^{2-}$

(a)   (b)   (c)

Figure 9.10

Common ion effect on solubility of $CaSO_4(s)$ in a solution containing $SO_4^{2-}(aq)$. (a) is a solution of $CaSO_4(s)$ dissolved in pure water. (b) is the same amount of $CaSO_4(s)$ added to a solution already containing $SO_4^{2-}(aq)$ ions; less solid dissolves.

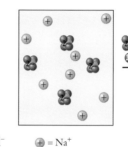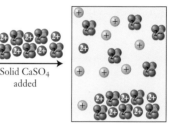

$2+ = Ca^{2+}$ $= SO_4^{2-}$ $+ = Na^+$

(a)   (b)

Figure 9.16

Effect of temperature on equilibrium mixtures of $NO_2(g)$ and $N_2O_4(g)$. In (a), both sample tubes are at the same temperature. In (b), one sample tube is in ice-water and the other in hot water.

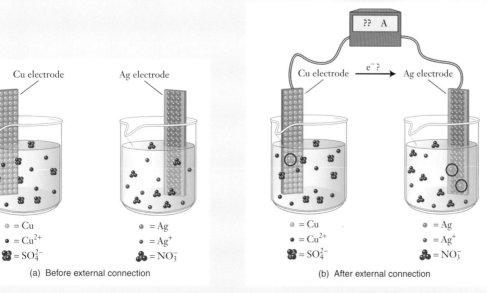

Figure 10.4

Does electron transfer occur between separate silver and copper wells? In (b), red rings highlight the loss of one copper atom and gain of two silver atoms at the respective electrodes.

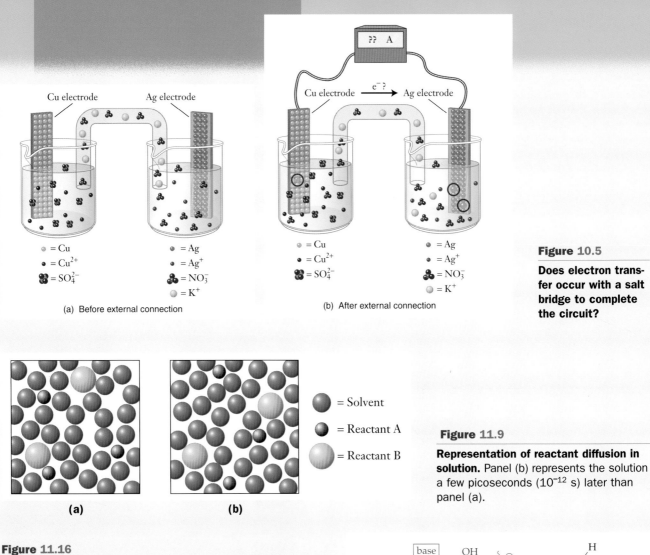

Figure 10.5

(a) Before external connection

(b) After external connection

⊚ = Cu
• = Cu²⁺
⬤ = SO₄²⁻

⊚ = Ag
• = Ag⁺
⬤ = NO₃⁻
○ = K⁺

Figure 10.5

Does electron transfer occur with a salt bridge to complete the circuit?

(a)

(b)

● = Solvent

◐ = Reactant A

○ = Reactant B

Figure 11.9

Representation of reactant diffusion in solution. Panel (b) represents the solution a few picoseconds (10⁻¹² s) later than panel (a).

Figure 11.16

Representation of substrate binding and active site interactions in ribonuclease. The solid blue represents the surface of the active site of the enzyme. Five of the amino acid side groups that are essential for binding and catalysis in the active site are shown. The abbreviations stand for histidine, lysine, serine, and threonine and the numbers are the positions of the amino acids along the 124 amino-acid protein chain. Folding of the protein chain brings amino acids from near both ends of the chain together as part of the active site. When pairs of electrons move as shown by the three curved arrows, a P–O bond is broken and the phosphodiester backbone of the ribonucleic acid molecule (in red) is broken. This is the function of the enzyme.

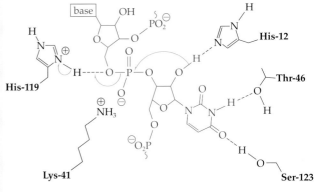

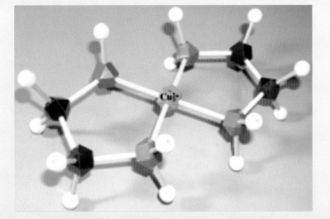

P. 10-75

Cu(en)₂²⁺, complex of Cu²⁺ with ethylenediamine (en).

The thiocyanate ion displaces one of the water molecules complexed with the iron(III) ion. This reaction (excluding the waters of hydration) is illustrated in Figure 9.1 and would explain your observations when the KSCN and $Fe(NO_3)_3$ solutions are mixed in Investigate This 9.3(a). You found in Check This 9.5 that the $Fe^{3+}(aq)$ in this mixture is present in excess over the $SCN^-(aq)$, as shown in the figure.

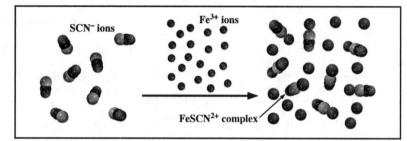

Figure 9.1. Stoichiometric reaction between $SCN^-(aq)$ and excess $Fe^{3+}(aq)$ to form $Fe(SCN)^{2+}(aq)$.

Equilibrium in the $Fe^{3+}(aq)$–$SCN^-(aq)$ system

In Investigate This 9.3(b), more $Fe^{3+}(aq)$ is added to an $Fe(NO_3)_3$–KSCN mixture and the light orange solution becomes a deeper orange-red. More of the red species, $Fe(SCN)^{2+}(aq)$, must have been formed. Since the addition of more $Fe^{3+}(aq)$ produces more $Fe(SCN)^{2+}(aq)$, some $SCN^-(aq)$ must have been left unreacted in the original mixture. The results suggest that reaction (9.1) reaches equilibrium with free $SCN^-(aq)$ ions remaining in the solution:

$$Fe^{3+}(aq) + SCN^-(aq) \rightleftharpoons Fe(SCN)^{2+}(aq) \tag{9.2}$$

This alternative view of the reaction and the two additions of $Fe^{3+}(aq)$ are illustrated at the molecular level in Figure 9.2(a). Figure 9.2(b) shows the observed colors in these solutions.

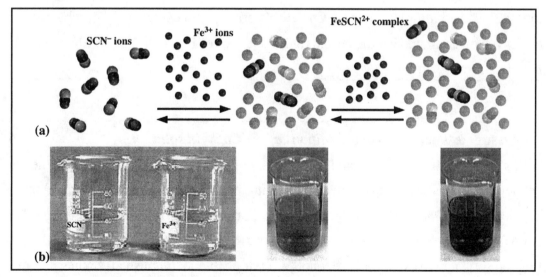

Figure 9.2. Equilibrium among $Fe^{3+}(aq)$, $SCN^-(aq)$, and $Fe(SCN)^{2+}(aq)$ as $Fe^{3+}(aq)$ is added.

The double **equilibrium arrows** in equation (9.2) and Figure 9.2 indicate that the reaction

> **WEB Chap 9, Sect 9.1.3.**
> View animations of the formation and dissociation of the $Fe(SCN)^{2+}(aq)$ complex

proceeds in both directions. We can explain our observations in Investigate This 9.3, if we assume that the reaction reaches an equilibrium state (balance of driving forces) in which there is still $SCN^-(aq)$ left in the mixture after the first addition of $Fe^{3+}(aq)$. The $Fe^{3+}(aq)$ in the second

> Where is thiocyanate found in nature? Collect about one milliliter of your saliva, dilute with an equal amount of water, mix, and filter the liquid. To the filtrate, add a few drops of the 0.2 M iron(III) solution you used in Investigate This 9.3. What do you observe? What can you conclude about thiocyanate in biological fluids?

addition can then react with some of this remaining $SCN^-(aq)$ to form more $Fe(SCN)^{2+}(aq)$. The second addition of $Fe^{3+}(aq)$ is a disturbance to the iron(III)-thiocyanate ion system that was in equilibrium. Le Chatelier's principle states that

the system will respond in a way that minimizes the affect of the disturbance, that is, to use up some of the extra $Fe^{3+}(aq)$ to form more $Fe(SCN)^{2+}(aq)$.

9.6.   **Check This**   *Dynamic equilibrium animation*

    (a)   WEB   Chap 9, Sect 9.1.4. Explain how you chose the phrase to complete the sentence on this page of the *Web Companion*.

    (b) On average, how many metal ion complexes are present in the tiny volume of solution represented in this animation? Explain how you get your answer and how it relates to part (a).

9.7.   **Check This**   *Adding SCN⁻(aq) to a Fe(NO₃)₃–KSCN solution*

    (a) In Investigate This 9.3(c), you added a tiny bit of $KSCN(s)$ to your $Fe(NO_3)_3$–KSCN solution. Is the response of the system to this disturbance consistent with the assumption that reaction (9.2) is in equilibrium? Use Le Chatelier's principle to explain why or why not.

    (b) Draw a diagram, modeled after the right-hand part of Figure 9.2(a), to illustrate your response in part (a).

9.8.   **Investigate This**   *What conditions affect CoCl₂ in solution?*

    Do this as a class investigation, and work in small groups to discuss and analyze the results. Place about 2 g of solid $CoCl_2 \cdot 6H_2O$ in a *dry* 125-mL erlenmeyer flask. Add about 50 mL of rubbing alcohol — 91% isopropyl alcohol (2-propanol) — to the flask and swirl to dissolve the solid. WARNING: Cobalt compounds are somewhat toxic; wear disposable gloves when handling the solid and the solutions. Alcohols are flammable; extinguish all flames before doing

---

this investigation. Divide the solution equally among four *dry*, 200-mm test tubes. Observe and record the appearance, especially the colors, of all the solids, liquids, and solutions.

(a) Add water a few drops at a time to the Co^{2+} ion solution in one of the test tubes. Mix thoroughly by swirling after each addition. Stop adding water when the solution becomes a lavender color, as in this photograph. Repeat this addition of water (to get a lavender color) with another of the solutions. Finally add this same amount of water to a third one of the solutions and then add about 2 mL more water. Record the colors of the solutions in all four test tubes.

(b) Place the test tube with one of the lavender solutions from part (a) in a beaker of ice-water. Place the test tube with the other lavender solution in beaker of 80–90 °C water. After two or three minutes, observe and record the colors of the solutions. Remove the test tubes from the water baths and allow them to return to room temperature. After two or three minutes, observe and record the colors of the solutions.

9.9.   **Consider This**   *How do solvents and temperature affect CoCl₂ in solution?*

(a) What color is the solution when solid $CoCl_2 \cdot 6H_2O$ is dissolved in alcohol in Investigate This 9.8? Is this the color you expected? Explain why or why not.

(b) What color is the solution to which you added the most water in Investigate This 9.8(a)? Is this the color you might have expected to see for a solution of $CoCl_2 \cdot 6H_2O$? Explain why or why not.

(c) In Investigate This 9.8(b), what changes do you observe when the lavender solutions are cooled and heated? What correlation is there between these changes and the observations you made on the solutions in part (a) of the investigation? Can you suggest a way to explain the correlation?

The Co^{2+} ion solutions in Activity 9.8 are pink or blue or lavender (a mixture of the pink and blue), depending upon the proportions of water and alcohol in the solvent. The pink and blue colors represent different species containing the Co^{2+} cation, so we conclude that different solvents cause changes in the equilibria involving this cation. The chemistry of this system is a little more complicated than the $Fe^{3+}(aq)$–$SCN^-(aq)$ system, because we have to be more careful to consider the interaction of water with Co^{2+} cations. To simplify our discussion, we'll ignore many intermediate reaction steps, since they do not affect the overall conclusions.

Co²⁺ complexes with water and chloride

Solid cobalt chloride is usually written as $CoCl_2 \cdot 6H_2O$. This solid is more accurately written as $(Co(H_2O)_6)Cl_2$. Six water molecules are bonded to each Co^{2+} cation (another Lewis acid-base metal ion complex) and it is this complex ion, $Co(H_2O)_6^{2+}$, that is responsible for the deep red-purple color of the solid. Water molecules in coordination complexes are usually rather loosely bonded to the metal ion and easily come off when the complex is in solution. In an aqueous solution, there are so many water molecules nearby that another quickly takes the place of the one that was lost, so the Co^{2+} cation in aqueous solution is present mainly as the pink (red) $Co(H_2O)_6^{2+}$ complex.

When the solid $Co(H_2O)_6)Cl_2$ is dissolved in 91% alcohol, only 9% of the solvent is water. A water molecule lost from the $Co(H_2O)_6^{2+}$ complex is not so rapidly replaced. Other reactions can occur, including reactions of the cobalt cations with the chloride anions:

$$Co(H_2O)_6^{2+}(aq) + 4Cl^-(aq) \rightleftharpoons CoCl_4^{2-}(aq) + 6H_2O(l) \qquad (9.3)$$

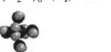

When chloride ions replace water molecules bound to the Co^{2+} cation, there is a change in the geometry that gives the new complex, $CoCl_4^{2-}$, a blue color. The blue complex is formed as the solid $(Co(H_2O)_6)Cl_2$ dissolves in alcohol, because the lack of water in the solvent favors the loss of water from $Co(H_2O)_6^{2+}$.

9.10.   Consider This   *What occurs when water is added to alcoholic Co²⁺ solutions?*

Addition of water to your alcoholic solutions of cobalt chloride in Activity 9.8(a) changed their color. Does equation (9.3) explain the changes you observed? Why or why not? Is the response of the system to this disturbance consistent with the assumption that reaction (9.3) is in equilibrium? Use Le Chatelier's principle to explain why or why not.

Temperature and the Co²⁺-water-chloride system

In Investigate This 9.8(b), you changed the temperatures of the Co^{2+} solutions and observed that this disturbance changes the colors of the solutions. When you removed energy from the lavender solution, by cooling it, the solution turned pink. When you added energy to the lavender solution, by warming it, the solution turned blue. When the solutions returned to their original energy states, back to room temperature, they once again became lavender. These observed color changes can be correlated with changes in the relative amounts of the $Co(H_2O)_6^{2+}$ and $CoCl_4^{2-}$ complexes in the solutions, Figure 9.3.

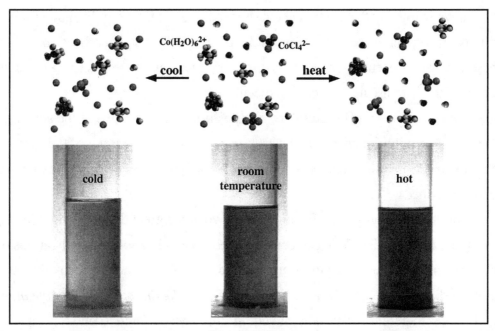

Figure 9.3. Response of the $Co(H_2O)_6^{2+}$–$CoCl_4^{2-}$ equilibrium to heating and cooling.

When the energy of the system decreases, more of the pink $Co(H_2O)_6^{2+}$ ion is formed. When the energy of the system increases, more of the blue $CoCl_4^{2-}$ ion is formed. The conclusion is that it requires an input of energy to the system to form the products in reaction (9.3). The reaction, as written, is endothermic, $\Delta H_{reaction} > 0$. These qualitative observations don't provide enough information to get a numerical value for $\Delta H_{reaction}$, but this simple experiment immediately gives you its sign.

9.11. Check This   *Temperature and Le Chatelier's principle*

Adding energy to the $Co(H_2O)_6^{2+}$–$CoCl_4^{2-}$ equilibrium system (raising the temperature) is a disturbance to the system. Use Le Chatelier's principle to explain the observed effect on the endothermic reaction (9.3).

Reflection and projection

In order to get some insight into chemical equilibrium, we began this section with an investigation of an analogous physical system that illustrates the dynamic nature of chemical equilibrium. At equilibrium, the chemical reactions that convert reactants to products and *vice versa* continue to occur, but both proceed at the same rate, so there is no net change in the concentrations in the system.

What we have learned in this section about the factors that affect equilibrium systems has been qualitative. From these specific examples, let us suggest three generalizations:

• Increasing the concentration of a reactant or product causes the reacting system to adjust in such a way as to try to use up the added molecules.

• Solution equilibria are affected by the composition and properties of the solvent.

• The way an equilibrium system adjusts to an increase in the energy of the system tells you whether the reaction, as written, is endothermic (more products formed) or exothermic (more reactants formed).

These important and useful generalizations can give you a great deal of insight into systems you haven't previously studied. You can consider these generalizations as applications of Le Chatelier's principle: a system at equilibrium responds to a disturbance in a way that minimizes the effect of the disturbance. *Le Chatelier's principle gives the direction but not the magnitude of the effects.* In order to get further insight into equilibrium systems and to make comparisons among them, we need to make our analyses more quantitative. As our treatments get more quantitative, keep in mind that the results have to be consistent with the qualitative, directional arguments you can make without the calculations.

Section 9.2.   Mathematical Expression for the Equilibrium Condition

9.12. Investigate This   *What is the pH of an acetic acid solution?*

Do this as a class investigation an work in small groups to discuss and analyze the results. Use a pH meter and pH electrode to measure the pH (±0.02 units) of a 0.10 M aqueous solution of acetic (ethanoic) acid. Record the pH of the sample.

9.13. Consider This   *Does the pH of the acetic acid solution make sense?*

What is the pH of the acetic acid solution in Investigate This 9.12? Is this a result you expected? Explain why or why not.

The investigations you did in Section 9.1 give you a feeling for how products and reactants are distributed in an equilibrium system and how the distribution changes when the system is disturbed. However, you did not determine the actual concentration of any species present in the solutions. In Investigate This 9.12, you have now determined the pH of an acetic (ethanaoic) acid solution and, from the pH, you can calculate the hydronium ion concentration, $[H_3O^+(aq)]$, in the solution. If all of the acetic acid had transferred its protons to water, the [$H_3O^+(aq)$] would have been about 10^{-1} M and the pH would have been about 1. However, the measured pH is nearer 3, so [$H_3O^+(aq)$] ≈ 10^{-3} M. Back in Chapter 2, we used this same result to argue that not all the acetic acid molecules transfer their protons to water; in other words, acetic acid is a weak acid. We can write a reversible, equilibrium reaction for the reaction of acetic acid with water:

$$CH_3C(O)OH(aq) + H_2O(aq) \rightleftharpoons H_3O^+(aq) + CH_3C(O)O^-(aq) \tag{9.4}$$

We have written $H_2O(aq)$ as a reminder that we are interested in a solution, not pure liquid water.

The equilibrium constant expression

Before we can further analyze this equilibrium system, we need to know more about how the concentrations in equilibrium systems are related to one another. To make our discussion more general, let's write a general equation for a reversible chemical reaction:

$$a\mathbf{A} + b\mathbf{B} \rightleftharpoons c\mathbf{C} + d\mathbf{D} \tag{9.5}$$

Equation (9.5) says that a fixed number (represented by a) of molecules or ions of **A** react with b molecules or ions of **B** to form c molecules or ions of **C** and d molecules or ions of **D**. The italicized lowercase letters represent the coefficients in the chemical equation, and the bold uppercase letters represent the formulas for reactants and products. At first, we will consider only reactions in which all species are in the same phase (in a solution or as a mixture of gases).

Experiments to determine equilibrium concentrations in solution were first carried out in the middle of the 19th century. The mathematical relationship of these concentrations to one another was determined empirically in 1863 by the Norwegian chemists C. M. Guldberg and P. Waage. Guldberg and Waage found that, as long as all species are in the same phase, the relationship of the equilibrium molar concentrations is:

$$K = \left(\frac{[\mathbf{C}]^c [\mathbf{D}]^d}{[\mathbf{A}]^a [\mathbf{B}]^b} \right)_{eq} \tag{9.6}$$

The ratio of equilibrium molar concentrations shown in equation (9.6) is the **equilibrium constant expression**. The subscript "eq" on the ratio reminds us that these are equilibrium concentrations. The numeric value of the

> Guldberg and Waage called the mathematical relationship in equation (9.6) the *law of mass action*, and you will still occasionally see that term used.

equilibrium constant expression is denoted by the **equilibrium constant, K**.

In the equilibrium constant expression, concentrations of products appear in the numerator, with each raised to the power equal to its coefficient in the chemical equation. Concentrations of reactants, written in the same way, appear in the denominator. If the system is at equilibrium and a concentration is changed, by adding more of one reactant, for example, all the concentrations will change as the reaction proceeds to use up some of the added reactant (Le Chatelier's principle) until equilibrium is reestablished and the ratio in equation (9.6) is again equal to K.

9.14. Worked Example *Equilibrium constant expression and equilibrium constant*

Write the equilibrium constant expression for the reaction of $Fe^{3+}(aq)$ with $SCN^-(aq)$ to form the $Fe(SCN)^{2+}(aq)$ complex, reaction (9.2). Assume that the number of symbols for each of the species in Figure 9.2 is directly proportional to the molarity of that species in the solution. If the solution shown after the first addition of $Fe^{3+}(aq)$ is at equilibrium, what is the numeric value of K, the equilibrium constant for the reaction?

Necessary information: We need reaction equation (9.2), equation (9.6), and the molarities of ions (proportional to numbers of symbols) in the solution: $[Fe^{3+}(aq)] = 18c$; $[SCN^-(aq)] = 6c$; and $[Fe(SCN)^{2+}(aq)] = 2c$. Here, "c" is the proportionality constant between molarity and numbers of symbols in the molecular level drawing.

Strategy: Write the equilibrium constant expression corresponding to reaction (9.2) and substitute the molarities to get a numeric value for K.

Implementation: For the reaction equation, $Fe^{3+}(aq) + SCN^-(aq) \rightleftharpoons Fe(SCN)^{2+}(aq)$, all the stoichiometric coefficients are unity, so the equilibrium constant expression and equilibrium constant are:

$$K = \frac{\left[Fe(SCN)^{2+}(aq)\right]}{\left[Fe^{3+}(aq)\right]\left[SCN^-(aq)\right]} = \frac{2c}{(18c)(6c)} = \frac{1}{54c}$$

Does the answer make sense? Without other information about the reaction system, you have no way to check this result. You can obtain further information in Check This 9.15.

9.15. Check This *Equilibrium constant expression and equilibrium constant*

(a) If the solution shown after the second addition of $Fe^{3+}(aq)$ in Figure 9.2 is at equilibrium, what is the numeric value of K, the equilibrium constant for the reaction of $Fe^{3+}(aq)$ with $SCN^-(aq)$ to form the $Fe(SCN)^{2+}(aq)$ complex? Make the same assumptions as in Worked Example 9.14.

(b) How does your result in part (a) compare to the result in Worked Example 9.14. Does the comparison suggest that the results "make sense?" Explain why or why not.

(c) The numeric value of the equilibrium constant for this reaction is about 1.3×10^2 M^{-1} at room temperature. What is the value of the proportionality constant, c, in our calculations? Explain how your get your value.

9.16. **Consider This**   *What can equilibrium constant expressions tell us?*

(a) Write the equilibrium constant expression for the reaction of acetic acid transferring protons to water, equation (9.4).

(b) Assume that you add more acetic acid to the solution in Investigate This 9.12. Use the equilibrium constant expression you wrote in part (a) to explain whether the pH of the solution would increase, decrease, or stay the same.

(c) Is your answer in part (b) consistent with Le Chatelier's principle? Explain why or why not.

Equation (9.6) is particularly useful for substances in solution, but *the ratio of molarities is a constant only in dilute solutions*. For accurate work, when concentrations are much above 0.1 M, activities rather than concentrations must be used in the equilibrium constant expression. You can think of the **activity** of a species as an "effective" concentration. In ionic solutions, for example, attractions between the negative and positive ions usually make the activity of the ions lower than their molar concentrations. Oppositely charged, neighboring ions attract one another and are not as free to interact with the other species in solution. In terms of reactivity, the ions appear to be present at lower concentration than they really are. Several solution models are available to calculate corrected ionic activities, but the calculations add a layer of complexity that is unnecessary for an understanding of equilibrium principles. In this text, we will stick mostly with dilute solutions, and use equilibrium constant expressions based on concentration, but with the warning that experimental and calculated results may be different.

Standard states in solution

Many calculations using equilibrium constants require using logarithms, but logarithms have meaning only when applied to dimensionless quantities, quantities without units. Immediately we have a problem with pH. We defined pH as the negative logarithm of the hydronium ion concentration:

$$pH \equiv -\log [H_3O^+(aq)] \tag{9.7}$$

But the hydronium ion concentration has units of molarity, so this equation is mathematically flawed.

Chemists avoid the dimensioned-number trap by defining a **standard state** for substances in whatever physical state they are found. This is the same thermodynamic standard state we introduced in Chapter 7, Section 7.8, and also used in Chapter 8. For substances in solution, the standard state is defined to be a 1 molal (*m*) concentration. In aqueous solutions, however, molality and molarity are almost the same, so we will take the standard state concentration in aqueous solutions to be 1 M. Now, **pH** can be defined as the *negative logarithm of the ratio of the molar concentration of hydronium ion to the standard state*:

$$pH \equiv -\log\left(\frac{\left[H_3O^+(aq)\right]}{1\ M}\right) \tag{9.8}$$

In equation (9.8), both the numerator and denominator of the argument for the logarithm have units of molarity, so we are now dealing with a dimensionless ratio that is *numerically equal* to the value of concentration expressed in moles per liter. From this point forward, we will use square brackets, [ ], *only* when dealing with concentrations including units. *Parentheses*, ( ), will be used in equilibrium constant expressions to denote that each term is the ratio of the molar solution concentration to the standard state concentration. Using this convention, equation (9.8) may be written as:

$$pH \equiv -\log (H_3O^+(aq)) \tag{9.9}$$

In Consider This 9.16, you wrote the equilibrium constant expression for acetic acid transferring protons to water in terms of molar concentrations. Let's abbreviate acetic acid as HOAc and acetate ion as OAc$^-$ and write the equilibrium constant expression in terms of ratios of molar concentrations of solutes to their 1 M standard state concentrations:

$$K = \left(\frac{\left(H_3O^+(aq)\right)\left(OAc^-(aq)\right)}{(HOAc(aq))\left[H_2O(aq)\right]}\right)_{eq} \tag{9.10}$$

We have a problem in equation (9.10), because the equilibrium constant ratio still contains the concentration of water, [H$_2$O(aq)], in the solution. The standard state for water is pure water, H$_2$O(l). Pure water is 55.5 M ($\approx$1000 g·L^{-1}/18.0 g·mol^{-1}). To be consistent with our molarity ratios for the solutes, we can write for the solvent, (H$_2$O(aq)) = [H$_2$O(aq)]/55.5 M. In dilute aqueous solutions, the water concentration is still very close to 55.5 M, so (H$_2$O(aq)) $\approx$ 1. The equilibrium constant expression then becomes:

$$K_a = \left(\frac{\left(H_3O^+(aq)\right)\left(OAc^-(aq)\right)}{(HOAc(aq))\left(H_2O(aq)\right)}\right)_{eq} = \left(\frac{\left(H_3O^+(aq)\right)\left(OAc^-(aq)\right)}{(HOAc(aq))}\right)_{eq} \tag{9.11}$$

This equilibrium constant is given the symbol K_a to remind us that it refers to an **acid** reacting to transfer a proton to water.

Not everyone is careful to make the distinction between concentration terms with units and terms that are ratios to the standard state. You will often see equilibrium constant expressions written using square brackets for the terms. When you see such expressions, check to see that concentration is expressed as molarity. If it is, simply strike the concentration units and deal with the pure number as the ratio of concentration to the standard state.

Other standard states

The standard state for gases is defined to be a pressure, P°, of 1 bar. One bar is defined as 10^5 $kg \cdot m^{-1} \cdot s^{-2}$. Thus, for a gas at a pressure P in bar, the ratio, P/P° = (gas), is the numeric value to be used in an equilibrium constant expression involving this gas. Pure solids or liquids are treated in a way that is analogous to water in dilute aqueous solutions. Because there is no way to vary the composition of a pure solid or liquid, the ratio to be used in equilibrium constant expressions is unity. All these standard state definitions and the resulting dimensionless concentration ratios to be used in equilibrium constant expressions are gathered in Table 9.1.

Table 9.1. Standard states and dimensionless concentration ratios.*

Physical state	Standard state	Dimensionless ratio
gas, P bar	gas at P° = 1 bar	P/P°
pure liquid	pure liquid	1
pure solid	pure solid	1
solute in aqueous solution, c M	c° = 1 M**	c/c°
solvent in dilute solution	pure liquid	1

   *   The external pressure is 1 bar for every standard state.

   **   Remember that the thermodynamic standard state unit for solutes in solution is *molal* concentration. Since molality and molarity are almost the same in dilute aqueous solutions and molarity is more familiar, we use molarity as the standard state unit.

Section 9.3.   Acid-Base Reactions and Equilibria

All equilibrium systems can be analyzed in essentially the same way. Equilibria in solutions of weak acids, weak bases, or both are particularly important, so we will begin our quantitative discussion of equilibria and equilibrium constant calculations with these systems. You have already seen examples of acid-base reactions in Chapters 2 and 6. In later sections, we will consider implications and extensions of acid–base equilibria, particularly in systems involving biologically important molecules, and extend the ideas to other equilibria.

The acetic acid-water reaction

To see how the concepts in Section 9.2 are applied, we'll continue our discussion of the reaction of acetic acid with water and the result from Investigate This 9.12. For all our analyses we will start with a balanced chemical reaction equation (or equations) and the equilibrium constant expression derived from it. For acetic acid in aqueous solution, we have already written equations (9.4) and (9.11):

$$CH_3C(O)OH(aq) + H_2O(aq) \rightleftharpoons H_3O^+(aq) + CH_3C(O)O^-(aq) \tag{9.4}$$

$$K_a = \left(\frac{\left(H_3O^+(aq)\right)\left(OAc^-(aq)\right)}{(HOAc(aq))\left(H_2O(aq)\right)} \right)_{eq} = \left(\frac{\left(H_3O^+(aq)\right)\left(OAc^-(aq)\right)}{(HOAc(aq))} \right)_{eq} \tag{9.11}$$

If we express all solute concentrations as molarities, all their standard states are 1 M. Thus, as we said in Section 9.2, we can substitute the numerical values of the equilibrium molarities into equation (9.11) to determine the equilibrium constant for the transfer of a proton from acetic acid to water.

9.17. **Worked Example**   *Acid equilibrium constant for acetic acid*

The pH of a 0.050 M solution of acetic acid was measured to be 3.02. What is the equilibrium constant, K_a, for reaction (9.4)?

Necessary information: We need equation (9.11) and the conversion, $(H_3O^+(aq)) = 10^{-pH}$.

Strategy: Substitute concentrations into the equilibrium constant expression (9.11), to get K_a. The pH of the solution gives us $(H_3O^+(aq))$. We know the concentration of acetic acid we started with and reaction (9.4) shows us that one mole of acetate ions forms for each mole of hydronium ions formed. To simplify the calculations, we will assume that we have 1 L of solution, so that the numeric value of the molarity, M, is equal to the number of moles of each species in the solution. We will also assume that we started with pure water, pH 7, to which we add enough acetic acid, HOAc, to make a 0.050 M solution.

Implementation:

in pure water at 298 K, pH 7.00: $(H_3O^+(aq)) = 10^{-7} = 1.0 \times 10^{-7}$

in the final solution, pH 3.02: $(H_3O^+(aq)) = 10^{-3.02} = 9.6 \times 10^{-4}$

A useful way to keep track of what is going on in chemical systems like this is to use a **change table**, a table you construct to show the initial number of moles of each species before reaction, the change in moles when the reaction occurs, and the final number of moles. The change table for this system is:

species	HOAc(aq)	OAc⁻(aq)	H₃O⁺(aq)
initial mol	0.050	0	1.0×10^{-7}
change in mol	9.6×10^{-4} reacts	9.6×10^{-4} formed	9.6×10^{-4} formed
final mol	$0.050 - 9.6 \times 10^{-4}$	9.6×10^{-4}	9.6×10^{-4}

The amount of acetic acid that reacts is only about 2% (1 part in 50). Usually, we neglect changes that are less than about 5%, unless we have data that are good enough to justify taking them into account. In this case, we neglect the amount of HOAc that reacts, so the concentration of HOAc(*aq*) is ≈ 0.050 M and (HOAc(*aq*)) ≈ 0.050 = 5.0×10^{-2}. Now we have all the data we need to substitute in equation (9.11) to get K_a:

$$K_a = \left(\frac{\left(H_3O^+(aq)\right)\left(OAc^-(aq)\right)}{(HOAc(aq))} \right)_{eq} = \frac{\left(9.6 \times 10^{-4}\right)\left(9.6 \times 10^{-4}\right)}{\left(5.0 \times 10^{-2}\right)} = 1.8 \times 10^{-5}$$

Does the answer make sense? We have known since Chapter 2, Section 2.13, that acetic acid is a weak acid that does not transfer all its protons to water in aqueous solution. Here we find that only about 2% of the acetic acid molecules transfer their protons. This means that reaction (9.4) does not proceed very far toward products before the system comes to equilibrium. Since not much product is formed, we expect the equilibrium constant for the reaction to be small, much less than 1. This is the answer we got; it makes sense.

9.18. Check This   *Acid equilibrium constant for acetic acid*

Use your experimental value for the pH of the 0.10 acetic acid in Investigate This 9.12 to calculate a value of K_a for acetic acid. If you find it helpful, use a change table to keep track of the chemical changes that take place. How does your value compare to the one we calculated in Worked Example 9.17 from data for a more dilute solution? If the two values differ significantly (more than a factor of two), suggest a possible explanation.

Solution pH for an acid of known K_a

In Worked Example 9.17 and Check This 9.18, we used measured values of pH and a knowledge of the composition of the solutions to calculate an acid equilibrium constant by substitution into an equilibrium constant expression. The problem you will meet more often is calculating the pH of a solution of a weak acid of known concentration and known acid equilibrium constant. This sort of problem might arise if you needed an acidic solution of a particular pH and wished to know what acid would give the appropriate pH. The approach is the same as above: write the balanced equilibrium reaction and corresponding equilibrium constant expression and then substitute known values and solve for the unknown pH.

9.19. Worked Example   *pH of an aqueous solution of lactic acid*

Calculate the pH of a 0.050 M solution of lactic acid, $CH_3CH(OH)C(O)OH$, which we will abbreviate as HOLac. The K_a for lactic acid is 1.4×10^{-4}.

Necessary information: We need the definition $pH = -\log(H_3O^+(aq))$, the balanced equation for the acid reaction (transfer of a proton from HOLac to H_2O), and the corresponding equilibrium constant expression:

$$HOLac(aq) + H_2O(aq) \rightleftharpoons H_3O^+(aq) + OLac^-(aq) \tag{9.12}$$

$$K_a = \left(\frac{\left(H_3O^+(aq)\right)\left(OLac^-(aq)\right)}{\left(HOLac(aq)\right)} \right)_{eq} = 1.4 \times 10^{-4} \tag{9.13}$$

Strategy: Let x stand for the number of moles per liter of $H_3O^+(aq)$ produced by reaction (9.12) when equilibrium is reached. We see from equation (9.12) that one $OLac^-(aq)$ ion is produced for every $H_3O^+(aq)$ formed in the solution; there are x moles per liter of $OLac^-(aq)$ in the equilibrium solution. Construct a change table, substitute into equation (9.13), and solve for x.

Implementation: The change table is:

species	HOLac(aq)	OLac⁻(aq)	H₃O⁺(aq)
initial mol	0.050	0	1.0×10^{-7}
change in mol	x reacts	x formed	x formed
final mol	$0.050 - x$	x	x

$$K_a = 1.4 \times 10^{-4} = \left(\frac{\left(H_3O^+(aq)\right)\left(OLac^-(aq)\right)}{\left(HOLac(aq)\right)} \right)_{eq} = \left(\frac{(x)\,(x)}{0.050 - x} \right)$$

Before trying to solve this equation for x, let's use a little chemical reasoning to simplify the arithmetic. The small equilibrium constant tells us that lactic acid is a weak acid, so the amount that reacts to form $H_3O^+(aq)$ and $OLac^-(aq)$ is small; that is, x is a small quantity. Let us *assume* that it is small enough to neglect relative to 0.050, so that we have $\left(HOLac(aq)\right) \approx 0.050$. We must *check this assumption* after we have found a value for x.

$$K_a = 1.4 \times 10^{-4} \approx \left(\frac{(x)\,(x)}{0.050} \right) = \frac{(x)^2}{0.050}$$

$$(x)^2 = 0.050\,(1.4 \times 10^{-4}) = 7.0 \times 10^{-6} \,;\, (x) = 2.6 \times 10^{-3} = \left(H_3O^+(aq)\right) = \left(OLac^-(aq)\right)$$

$$pH = -\log\left(H_3O^+(aq)\right) = -\log\,(2.6 \times 10^{-3}) = 2.59$$

Check the assumption: Our value for (x), 2.6×10^{-3}, is about 6% of 0.050 (3 parts in 50). This result is on the borderline of acceptability. In Worked Example 9.17, we said that about 5% is the "discrepancy" we are willing to accept between our simplifications and our calculated results (or measured values). In most cases, we settle for a result like we got here, because we are only interested in a ballpark estimate of the pH.

Does the answer make sense? Comparing the acid equilibrium constants, we see that lactic acid, $K_a = 1.4 \times 10^{-4}$, is a stronger acid (larger K_a) than acetic, $K_a = 1.8 \times 10^{-5}$ (Worked Example 9.17). We would, therefore, expect that a lactic acid solution would be somewhat more acidic (lower pH) than an acetic acid solution of the same concentration. The pH we calculated for the 0.050 M lactic acid solution, 2.59, is lower (more acidic) than that for a 0.050 M acetic acid solution, 3.02 (Worked Example 9.17). The answer makes sense.

9.20. Check This   *pH of an aqueous solution of benzoic acid*

Benzoic acid, $K_a = 6.4 \times 10^{-5}$, is sparingly soluble in water. A saturated solution of benzoic acid is 0.028 M. What is the pH of this solution? What assumption(s) do you make to get your answer? How good is(are) your assumption(s)?

benzoic acid

In Worked Example 9.19, we made the simplifying assumption that a negligible amount of the original weak acid reacted to transfer its protons to water. You probably made the same assumption in Check This 9.20. In both cases, the assumption leads to a result that is good to about 5 or 6%. If we wanted to be more mathematically correct, we could go back to the equilibrium constant expressions, rearrange them as quadratic equations and use the quadratic formula to solve them. For two reasons, we will not do more elaborate arithmetic for problems like these. First, the arithmetic tends to take us too far from the underlying chemistry of the systems. Second, and more fundamentally, equilibrium constants and the activities of ions in solution (which we only approximate by using molarities) are rarely known accurately enough to justify using more elaborate methods. If we really need to know the pH of a lactic acid or benzoic acid solution, we use a pH electrode and pH meter to measure it. The calculations give us a good idea what to expect, but we are not surprised when the experimental measurement is a bit different.

The autoionization of water

You might have noticed that, when we constructed the change tables in Worked Examples 9.17 and 9.19, we showed all of the hydronium ion, $H_3O^+(aq)$, in the final equilibrium solution coming from the added weak acid. This means that, in the equilibrium solution, we neglected the $H_3O^+(aq)$ from the reaction of water molecules with themselves:

$$H_2O(l) + H_2O(l) \rightleftharpoons H_3O^+(aq) + OH^-(aq) \tag{9.14}$$

Reaction (9.14) is often referred to as the **autoionization** of water, because it represents the reaction of one water molecule with another to produce a pair of ions. Water autoionization occurs in all aqueous solutions as well as in pure water. The equilibrium constant expression for this equilibrium reaction is:

$$K_w = \left(\frac{\left(H_3O^+(aq)\right)\left(OH^-(aq)\right)}{\left(H_2O(l)\right)\left(H_2O(l)\right)} \right)_{eq} \tag{9.15}$$

$$K_w = \left\{ \left(H_3O^+(aq)\right)\left(OH^-(aq)\right) \right\}_{eq} \tag{9.16}$$

> The curly brackets in equation (9.16) have no special significance except to enclose the concentration product. They are used to avoid possible confusion with parentheses and square brackets, which do have special meaning.

K_w is the **water autoionization constant**. We know that the pH of pure water is 7.00 at 298 K, so $(H_3O^+(aq))$ in the equilibrium reaction (9.14) is $10^{-7.00}$. Since one $OH^-(aq)$ is produced for each $H_3O^+(aq)$, we also have $(OH^-(aq)) = 10^{-7.00}$. Substitution into equation (9.16) gives:

$$K_w = \{10^{-7.00} \times 10^{-7.00}\} = 10^{-14.00} = 1.00 \times 10^{-14} \tag{9.17}$$

The equilibrium constant expression (9.16), with $K_w = 1.00 \times 10^{-14}$, *must* be true in all aqueous solutions at 298 K. If the solution contains other sources of $H_3O^+(aq)$ or $OH^-(aq)$ (such as added acids or bases), the concentrations of $H_3O^+(aq)$ or $OH^-(aq)$ from these sources must be accounted for in equilibrium constant expression (9.16).

9.21. Worked Example *Extent of water autoionization in an acetic acid solution*

What is the contribution of reaction (9.14) to the concentration of $H_3O^+(aq)$ in a 0.050 M acetic acid solution with a pH of 3.02 (as in Worked Example 9.17)?

Necessary information: We need to know the conversion: $(H_3O^+(aq)) = 10^{-pH}$ and the equilibrium constant expression (9.16), with $K_w = 1.00 \times 10^{-14}$.

Strategy: The only source of $OH^-(aq)$ in this solution is water autoionization, reaction (9.14). Thus, the concentration of $H_3O^+(aq)$ *produced by water autoionization* must be equal to the concentration of $OH^-(aq)$. Substitute the measured $(H_3O^+(aq))$ in this solution into equation (9.16) to get the concentration of $OH^-(aq)$ in the solution and, hence, the concentration of hydronium ion contributed by water autoionization.

Implementation:

$$(H_3O^+(aq)) = 10^{-pH} = 10^{-3.02} = 9.6 \times 10^{-4}$$

Substitute this result into equation (9.16) to get $(OH^-(aq))$:

$$K_w = 1.00 \times 10^{-14} = \left\{ \left(H_3O^+(aq)\right)\left(OH^-(aq)\right) \right\}_{eq} = \left\{ 9.6 \times 10^{-4} \left(OH^-(aq)\right) \right\}_{eq}$$

$$(OH^-(aq)) = \frac{1.00 \times 10^{-14}}{9.6 \times 10^{-4}} = 1.04 \times 10^{-11}$$

$$\therefore [OH^-(aq)] = 1.04 \times 10^{-11} \text{ M}$$

The concentration of $H_3O^+(aq)$ produced by autoionization of water is $[H_3O^+(aq)]_{autoionization} = [OH^-(aq)] = 1.04 \times 10^{-11}$ M.

Does the answer make sense? The concentration of $H_3O^+(aq)$ produced by autoionization of water is about eight orders of magnitude (10^8 times) smaller than the total concentration of $H_3O^+(aq)$ in the solution. Essentially all the $H_3O^+(aq)$ in the solution is produced by transfer of protons from acetic acid to water, as we assumed in Worked Example 9.17. Note that the amount of $H_3O^+(aq)$ produced by water autoionization in this solution is four orders of magnitude (10^4 times) smaller than in pure water. Le Chatelier's principle shows us that this makes sense. Addition of extra $H_3O^+(aq)$ from the acetic acid disturbs the equilibrium represented by equation (9.14). The system responds by "using up" a very tiny amount of the added $H_3O^+(aq)$, which also uses up some of the $OH^-(aq)$ and accounts for its low concentration in the solution.

9.22.   Check This   *Extent of water autoionization in an lactic acid solution*

(a) What is the contribution of reaction (9.14) to the concentration of $H_3O^+(aq)$ in a 0.050 M lactic acid solution with a pH of 2.59 (as in Worked Example 9.19)?

(b) Compare your answer in part (a) with the result for the acetic acid solution in Worked Example 9.21. Does the difference between the two results make sense? Explain your response.

9.23.   Investigate This   *What is the pH of an acetate ion solution?*

Do this as a class investigation and work in small groups to discuss and analyze the results. Use a pH meter and pH electrode to measure the pH (± 0.02 units) of a 0.10 M aqueous solution of sodium acetate, $CH_3C(O)ONa$ (or NaOAc). Record the pH of the sample.

9.24.   **Consider This**   *Does the pH of the acetate ion solution make sense?*

What is the pH of the acetate ion solution, OAc⁻(*aq*), in Investigate This 9.23? Is this a result you expected? Explain why or why not.

The acetate ion-water reaction

As you have seen above, most acetic acid molecules in aqueous solution do not transfer their protons to water. This must mean that acetate ions have an attraction for protons. When acetate ions are added to water, the water molecules will donate protons to the acetate:

$$H_2O(aq) + CH_3C(O)O^-(aq) \rightleftharpoons CH_3C(O)OH(aq) + OH^-(aq) \tag{9.18}$$

If reaction (9.18) occurs, aqueous solutions of acetate ion should be basic, that is, have a pH greater than 7. In Investigate This 9.23, your result for the 0.10 M sodium acetate solution (containing sodium cations and acetate anions), was probably a pH in the range 8.5 to 9.0. Reaction (9.18) proceeds to some extent, but it must not go far, because the solution is only weakly basic. Let's quantify the reaction by determining its equilibrium constant.

The equilibrium constant expression for reaction (9.18) is:

$$K_b = \left(\frac{\left(HOAc(aq)\right)\left(OH^-(aq)\right)}{\left(OAc^-(aq)\right)} \right)_{eq} \tag{9.19}$$

In equation (9.19), we have used our abbreviations for acetic acid and acetate ion and have given the equilibrium constant the symbol K_b to remind us that it refers to a **b**ase reacting to accept a proton from water.

9.25.   **Worked Example**   *Base equilibrium constant for acetate ion*

The pH of a 0.050 M solution of sodium acetate was measured to be 8.73. What is the equilibrium constant, K_b, for reaction (9.18)?

Necessary information: We need to know the conversion: $(H_3O^+(aq)) = 10^{-pH}$, the equilibrium constant expression (9.16), with $K_w = 1.00 \times 10^{-14}$, and equilibrium constant expression (9.19).

Strategy: Substitute concentrations into equilibrium constant expression (9.19), to get K_b. The pH of the solution gives us $(H_3O^+(aq))$ and we use equation (9.16) to find $(OH^-(aq))$. We know the concentration of acetate ion we started with and reaction (9.18) shows us that one mole of acetic acid forms for each mole of hydroxide ions formed. We assume that we start with pure water, pH 7, to which we add enough sodium acetate, NaOAc, to make a 0.050 M solution.

Implementation:

in pure water at 298 K, pH 7.00: $(H_3O^+(aq)) = (OH^-(aq)) = 10^{-7} = 1.0 \times 10^{-7}$

in the final solution, pH 8.73: $(H_3O^+(aq)) = 10^{-8.73} = 1.9 \times 10^{-9}$

$$(OH^-(aq)) = \frac{K_w}{(H_3O^+(aq))} = \frac{1.00 \times 10^{-14}}{1.9 \times 10^{-9}} = 5.3 \times 10^{-6}$$

The change table for this system is:

species	OAc⁻*(aq)*	HOAc*(aq)*	OH⁻*(aq)*
initial mol	0.050	0	1.0×10^{-7}
change in mol	5.3×10^{-6} reacts	5.3×10^{-6} formed	5.3×10^{-6} formed
final mol	$0.050 - 5.3 \times 10^{-6}$	5.3×10^{-6}	5.3×10^{-6}

The amount of acetate ion that reacts is only about 0.01%, which is negligible. The concentration of OAc⁻*(aq)* is 0.050 M, so $(OAc^-(aq)) = 0.050 = 5.0 \times 10^{-2}$.

$$K_b = \left(\frac{(HOAc(aq))\left(OH^-(aq)\right)}{\left(OAc^-(aq)\right)} \right)_{eq} = \left(\frac{(5.3 \times 10^{-6})(5.3 \times 10^{-6})}{0.050} \right) = 5.6 \times 10^{-10}$$

Does the answer make sense? Since the solution is only weakly basic, reaction (9.18) must not proceed far toward products before equilibrium is established. The small value of the equilibrium constant, K_b, is consistent with this experimental result.

9.26. Check This *Base equilibrium constant for acetate ion*

Use your experimental value for the pH of the 0.10 sodium acetate solution in Investigate This 9.23 to calculate a value of K_b for the acetate ion. How does your value compare to the one we calculated in Worked Example 9.25 for data from a more dilute solution? If the two values differ significantly (more than a factor of two), suggest a possible explanation.

Relationship between K_a and K_b for a conjugate acid-base pair

In any solution that contains acetic acid and acetate ion, the concentrations in the solution must satisfy *both* equations (9.11) and (9.19) simultaneously. To see the consequences of this requirement, let's calculate $K_a \cdot K_b$:

$$K_a \cdot K_b = \left(\frac{\left(H_3O^+(aq)\right)\left(OAc^-(aq)\right)}{(HOAc(aq))} \right)_{eq} \left(\frac{(HOAc(aq))\left(OH^-(aq)\right)}{\left(OAc^-(aq)\right)} \right)_{eq} \qquad (9.20)$$

Since we are considering a single solution, the concentration ratio of acetic acid in the first quotient cancels that in the second. The same is true for the acetate ion concentration ratio:

$$K_a \cdot K_b = \left\{ \left(H_3O^+(aq) \right) \left(OH^-(aq) \right) \right\}_{eq} = K_w \tag{9.21}$$

K_a and K_b for any conjugate acid-base pair are related to one another by equation (9.21); their product is always K_w. If you know either the conjugate acid or the conjugate base equilibrium constant, you can calculate the other.

9.27. Check This   *$K_a \cdot K_b$ for the acetic acid-acetate ion pair*

Use your experimental values for K_a and K_b from Check This 9.18 and 9.26 to calculate $K_a \cdot K_b$. Within the uncertainties of your values, is your result consistent with equation (9.21)?

---

pK

You will often find values for equilibrium constants given as **pK** instead of *K*. The "p" in this nomenclature is the same as the "p" in pH. The "p" means to take the negative logarithm (base 10) of the value represented by the other part of the symbol. For example, pH is the negative logarithm of the hydronium ion concentration ratio: $pH = -\log(H_3O^+(aq))$. pK is the negative logarithm of the equilibrium constant: $pK = -\log K$. Using "p" values makes some calculations easier. Equation (9.21), for example, can be transformed as follows:

$$-\log (K_a \cdot K_b) = -\log K_w \tag{9.22}$$

The right-hand and left-hand sides of equation (9.22) can be expressed as pK values:

$$-\log (K_a \cdot K_b) = -(\log K_a + \log K_b) = (-\log K_a) + (-\log K_b) = pK_a + pK_b$$

$$-\log K_w = pK_w = -\log (1.0 \times 10^{-14}) = -(-14.00) = 14.00$$

$$\therefore \ \ pK_a + pK_b = pK_w = 14.00 \tag{9.23}$$

The pK_a values for several common acids are given in Table 9.2. To get the pK_b value for the conjugate base of one of the acids, subtract pK_a from 14.00.

9.28. Check This   *Using the relationship of pK_a to pK_b*

When ammonia is dissolved in water, the water transfers protons to the ammonia to form a basic solution:

$$NH_3(aq) + H_2O(aq) \rightleftharpoons NH_4^+(aq) + OH^-(aq) \tag{9.24}$$

Write the K_b expression for reaction (9.24). Use the data in Table 9.2 to get pK_b and K_b for the reaction.

Table 9.2. pK_a values for several common acids at 298 K.

Many acids transfer more than one proton to water. Their successive pK values are labeled pK_{a1}, pK_{a2}, and so on.

Compound	Reaction	pK_a
sulfuric acid, pK_{a1}	$(HO)_2SO_2 + H_2O \rightarrow HOSO_3^- + H_3O^+$ transfer is "complete"	< 0
oxalic acid, pK_{a1}	$HO(O)CC(O)OH + H_2O \rightleftharpoons HO(O)CC(O)O^- + H_3O^+$	1.25
sulfuric acid, pK_{a2}	$HOSO_3^- + H_2O \rightleftharpoons SO_4^{2-} + H_3O^+$	1.99
phosphoric acid, pK_{a1}	$(HO)_3PO + H_2O \rightleftharpoons (HO)_2PO_2^- + H_3O^+$	2.15
glycine*, pK_{a1}	$^+H_3NCH_2C(O)OH + H_2O \rightleftharpoons {}^+H_3NCH_2C(O)O^- + H_3O^+$	2.35
aspartic acid**	$-CH_2C(O)OH + H_2O \rightleftharpoons -CH_2C(O)O^- + H_3O^+$	3.90
oxalic acid, pK_{a2}	$HO(O)CC(O)O^- + H_2O \rightleftharpoons {}^-O(O)CC(O)O^- + H_3O^+$	4.27
glutamic acid**	$-CH_2CH_2C(O)OH + H_2O \rightleftharpoons -CH_2CH_2C(O)O^- + H_3O^+$	4.42
acetic acid	$CH_3C(O)OH + H_2O \rightleftharpoons CH_3C(O)O^- + H_3O^+$	4.76
histidine**		6.00
carbonic acid, pK_{a1}	$(HO)_2CO + H_2O \rightleftharpoons HOCO_2^- + H_3O^+$	6.36
phosphoric acid, pK_{a2}	$(HO)_2PO_2^- + H_2O \rightleftharpoons (HO)PO_3^{2-} + H_3O^+$	7.20
"tris"	$(HOCH_2)_3CNH_3^+ + H_2O \rightleftharpoons (HOCH_2)_3CNH_2 + H_3O^+$	8.08
cysteine**	$-CH_2SH + H_2O \rightleftharpoons -CH_2S^- + H_3O^+$	8.36
ammonium ion	$NH_4^+ + H_2O \rightleftharpoons NH_3 + H_3O^+$	9.24
boric acid	$B(OH)_3 + 2H_2O \rightleftharpoons B(OH)_4^- + H_3O^+$	9.24
glycine*, pK_{a2}	$^+H_3NCH_2C(O)O^- + H_2O \rightleftharpoons H_2NCH_2C(O)O^- + H_3O^+$	9.78
phenol		9.98
carbonic acid, pK_{a2}	$HOCO_2^- + H_2O \rightleftharpoons CO_3^{2-} + H_3O^+$	10.33
tyrosine**		10.47
lysine**	$-(CH_2)_4NH_3^+ + H_2O \rightleftharpoons -(CH_2)_4NH_2 + H_3O^+$	10.69
phosphoric acid, pK_{a3}	$(HO)PO_3^{2-} + H_2O \rightleftharpoons PO_4^{3-} + H_3O^+$	12.15
arginine**		12.48
water	$H_2O + H_2O \rightleftharpoons H_3O^+ + OH^-$	14.00

* Typical pK_a values for the carboxylic acid and ammonium groups on the same carbon in amino acids are 1.5 to 2.5 and 9.5 to 10.5, respectively. These groups are used to make the amide bonds between amino acids in proteins. Only the amino acid acidic and basic side groups are left to give proteins their acid-base properties.

** Only the side groups are represented for these amino acids. The rest of the molecule is like glycine.

9.29. Consider This *What is the relationship of pH and pOH?*

(a) From equation (9.16), derive an equation that gives the sum of pH and pOH for any aqueous solution.

(b) What is the pOH of a solution with pH = 8.73? Use your pOH to determine $(OH^-(aq))$. Is your $(OH^-(aq))$ the same as that in Worked Example 9.25? Explain why or why not.

Reflection and projection

About 140 years ago, Guldberg and Waage showed that the results for several well-studied solution equilibria could be described in terms of equilibrium constants and equilibrium constant expressions that are ratios of the equilibrium concentrations of the products to the reactants. The form of the equilibrium constant expression for a reaction is related to its balanced chemical equation and the stoichiometric coefficients in the equation. Many further experimental studies have confirmed Guldberg and Waage's empirical observations and have extended equilibrium constant expressions to reactions involving gases and pure solids and liquids, as well as solutions. In order to make all the equilibrium constants consistent with one another, concentrations in equilibrium constant expressions are replaced by ratios of concentrations to a set of standard states, Table 9.1, for gases, pure solids and liquids, and solutes in solution.

Equilibrium constants can be calculated from measurements of concentrations in reactions at equilibrium in solution. Our first applications have been to aqueous acid-base reactions. We derived equilibrium constant expressions for K_a, K_b, and K_w and found that $K_a \cdot K_b = K_w$ (or $pK_a + pK_b = pK_w$) for conjugate acid-base pairs. Now we need to consider how to figure out what will happen in solutions that contain substantial concentrations of more than one acid and/or base. We will mainly focus on solutions of conjugate acid-base pairs (buffer solutions) and their effect on other species in solution, including biological molecules.

Section 9.4. Solutions of Conjugate Acid-Base Pairs: Buffer Solutions

9.30. Investigate This

Do this as a class investigation and work in small groups to discuss and analyze the results. Make up the three samples in this table in three, clean, dry, labeled sample vials or test tubes large enough to accommodate a pH electrode to measure the pH.

	#1	#2	#3
0.10 M HOAc	9.0 mL	5.0 mL	1.0 mL
0.10 M NaOAc	1.0 mL	5.0 mL	9.0 mL

Use a pH meter and pH electrode to measure the pH of each solution to ±0.02 pH unit and record the results.

9.31. Consider This   *Do the pHs of conjugate acid-base pair solutions make sense?*

As you do these analyses, consider your results from Investigate This 9.12 and 9.23, as well as Investigate This 9.30. Describe any correlations you find between the pH of the solutions and their composition. Do your correlations make sense in terms of the acid-base properties of acetic acid and acetate ion? Explain why or why not.

To understand the pH of solutions that contain several different acids and bases, biological solutions, for example, they must all be accounted for. In many cases, such solutions contain just a few predominant acid-base conjugate pairs that maintain the system at the appropriate pH. When we study the reactions of molecules, especially biomolecules, in the laboratory, we almost always use such conjugate acid-base pairs to control the pH of the solutions studied.

We can use the acid-base equilibrium systems introduced in Section 9.3 to prepare solutions of known and constant pH. These **buffer solutions** are used to control pH at a relatively constant value. Buffer solutions contain substantial concentrations of *both* the conjugate acid and conjugate base of a conjugate acid-base pair. If hydronium ion, $H_3O^+(aq)$, is added to a buffer solution, it will react with the conjugate base and be used up. The pH of the solution will not change as much as it would have without the buffer. A similar argument holds for the addition of hydroxide ion, $OH^-(aq)$, and its reaction with the conjugate acid of the buffer. The solutions whose pHs you measured in Investigate This 9.30 are buffer solutions, they contain substantial concentrations of both acetic acid and acetate ion, a conjugate acid-base pair. Before we examine quantitatively how a buffer solution works, let's analyze your results from the investigation.

We know the composition of the solutions and the pH of each in Investigate This 9.30. We can use these data to get K_a for acetic acid. You used the data from Investigate This 9.12 to determine this K_a, but analysis of the solutions in Investigate This 9.30 can give a more accurate value, because tiny amounts of impurities do not affect the pH as much.

9.32. Worked Example   *K_a for acetic acid*

The pH of a solution prepared by mixing 10.0 mL each of 0.050 M aqueous solutions of acetic acid and sodium acetate (similar to Sample #2 in Investigate This 9.30) is 4.77. Find K_a and pK_a for reaction (9.4), transfer of a proton from acetic acid to water.

Necessary information: We need the $(H_3O^+(aq)) = 10^{-pH}$ conversion. We also need to know that the number of moles of acetic acid (or acetate ion) in the stock solution that was added, $M_{stock} \cdot V_{stock}$, equals the number of moles of acetic acid in final mixture, $M_{mixt} \cdot V_{mixt}$:

$$M_{stock} \cdot V_{stock} = M_{mixt} \cdot V_{mixt} \qquad (9.25)$$

Strategy: As in previous examples, substitute concentrations into the equilibrium constant expression (9.11), to get K_a. The pH of the solution gives us $(H_3O^+(aq))$. Calculate the molarities of acetic acid and acetate ion in the mixture and construct a mole change table based on one liter of solution to get the values we need to substitute in equation (9.11). Since the solution is somewhat acidic, protons must have been transferred from HOAc to water.

Implementation:

$$(H_3O^+(aq)) = 10^{-4.77} = 1.7 \times 10^{-5}$$

$$\text{acetic acid: } M_{mixt} = \frac{M_{stock} \cdot V_{stock}}{V_{mixt}} = \frac{0.050 \text{ M} \cdot 0.010 \text{ L}}{0.020 \text{ L}} = 0.025 \text{ M}$$

$$\text{acetate ion: } M_{mixt} = \frac{M_{stock} \cdot V_{stock}}{V_{mixt}} = \frac{0.050 \text{ M} \cdot 0.010 \text{ L}}{0.020 \text{ L}} = 0.025 \text{ M}$$

species	HOAc(aq)	OAc⁻(aq)	H₃O⁺(aq)
initial mol	0.025	0.025	1.0×10^{-7}
change in mol	1.7×10^{-5} reacts	1.7×10^{-5} formed	1.7×10^{-5} formed
final mol	$0.025 - 1.7 \times 10^{-5}$	$0.025 + 1.7 \times 10^{-5}$	1.7×10^{-5}

The amounts of acetic acid that react and acetate ion that form are negligible compared to the amounts present initially, so we can neglect them when we calculate K_a:

$$K_a = \left(\frac{\left(H_3O^+(aq)\right)\left(OAc^-(aq)\right)}{(HOAc(aq))} \right)_{eq} = \frac{(1.7 \times 10^{-5})\,(0.025)}{0.025} = 1.7 \times 10^{-5}$$

$$pK_a = -\log(1.7 \times 10^{-5}) = 4.77$$

Does the answer make sense? This value for K_a agrees well with the value we calculated in Worked Example 9.17 and the pK_a agrees well with the value in Table 9.2.

9.33. Check This   *K_a for acetic acid*

(a) Use your experimental data for each of the three samples in Investigate This 9.30 to find three K_a and pK_a values for reaction (9.4), transfer of a proton from acetic acid to water. How do the three values compare? Is this the result you expected? Why or why not?

(b) How do these values compare with the one you got in Check This 9.18? Which do you think is the more reliable? Present your argument clearly.

We can analyze the data from Worked Examples 9.17 and 9.32 in terms of Le Chatelier's principle. With only acetic acid added to make the solution in Worked Example 9.17, the concentration of $H_3O^+(aq)$ formed by transfer of protons from acetic acid to water, reaction (9.4), is about 10^{-3} M. In Worked Example 9.32, acetate ion is added to the solution. In respose to this increase in concentration of one of the products of reaction (9.4), some of the added acetate ion reacts with $H_3O^+(aq)$ and lowers its concentration. We find that the concentration of $H_3O^+(aq)$ in this example is about 10^{-5} M, a 100-fold reduction from the solution without added acetate ion. Le Chatelier's principle explains the direction of the change in pH when acetate ion is added to an acetic acid solution.

Conjugate base-to-acid ratio

The calculations we made above in Worked Example 9.32 and Check This 9.33 are not difficult, but there is another way to treat these systems that can make the calculations even easier. We will do this for a generic acid, HA, and its conjugate base, A^-:

$$HA(aq) + H_2O(aq) \rightleftharpoons H_3O^+(aq) + A^-(aq) \tag{9.26}$$

The equilibrium constant expression for reaction (9.26) is:

$$K_a = \left(\frac{\left(H_3O^+(aq) \right) \left(A^-(aq) \right)}{\left(HA(aq) \right)} \right)_{eq} \tag{9.27}$$

Take the negative logarithm of both sides of equation (9.27) to get:

$$-\log K_a = pK_a = -\log(H_3O^+(aq))_{eq} - \log\left(\frac{\left(A^-(aq) \right)}{\left(HA(aq) \right)} \right)_{eq} \tag{9.28}$$

$$pK_a = pH - \log\left(\frac{\left(A^-(aq) \right)}{\left(HA(aq) \right)} \right)_{eq} \tag{9.29}$$

Equation (9.29) expresses the relationship among pK_a, pH, and the *ratio* of the conjugate base to conjugate acid concentrations in the solution. For some purposes, you need only the ratio of concentrations, not the actual concentrations, to characterize a solution. In Worked Example 9.32, we made the (good) approximation that changes in the concentrations of the acid and base were negligible compared to the concentrations initially present. This approximation is valid, if K_a is relatively small (not much reaction occurs) and the conjugate acid and base concentrations are relatively large. As a rule of thumb, if $100K_a$ is smaller than *both* the acid and the base concentration, you can use the approximation. In solutions that meet this condition, you can use the base-to-acid ratio rather than actual concentrations, in equation (9.29).

9.34. Check This   *Conjugate base-to-acid ratios and pK$_a$ for acetic acid*

(a) For each mixture in Investigate This 9.30, the base-to-acid ratio is the ratio of the volume of base to volume of acid used to make the mixture. Show why this is so. Use these volume ratios in equation (9.29) to get a value of pK_a for acetic acid from each mixture.

(b) How do your results in part (a) compare with those you got in Check This 9.33? Is this what you expect? Explain why or why not.

9.35. Investigate This   *How does a buffer respond to added H$_3$O$^+$(aq) or OH$^-$(aq)?*

Do this as a class investigation and work in small groups to discuss and analyze the results. Use four, clean, dry, labeled sample vials or test tubes large enough to accommodate a pH electrode. Put 10 mL of water in two of the vials. In each of the other two make a mixture of 5.0 mL 0.10 M aqueous acetic acid solution and 5.0 mL of 0.10 M aqueous sodium acetate solution.

(a) Measure and record the pH of one of the water samples. Add *one drop* of 1.0 M aqueous hydrochloric acid solution, mix the solution, and again measure and record the pH. Repeat this procedure with one of the acetic acid-acetate mixtures in place of the water.

(b) Repeat the procedure with the other water sample and the other acetic acid-acetate solution, but this time use *one drop* of 1.0 M aqueous sodium hydroxide solution instead of hydrochloric acid.

9.36. Consider This   *How does buffer pH respond to added H$_3$O$^+$(aq) or OH$^-$(aq)?*

(a) In Investigate This 9.35(a), how large is the *change* in pH when a drop of hydrochloric acid is added to 10 mL of water? to 10 mL of an acetic acid-acetate ion mixture? Do these results make sense? Explain why or why not.

(b) In Investigate This 9.35(b), how large is the *change* in pH when a drop of sodium hydroxide is added to 10 mL of water? to 10 mL of an acetic acid-acetate ion mixture? Do these results make sense? Explain why or why not.

How acid-base buffers work

   A buffer reduces the effect of some disturbance to a system. Your results from Investigate This 9.35 show why solutions of a weak acid and its conjugate base are called buffers. Addition of a drop of 1.0 M hydrochloric acid changes the pH of water from about 6 (the usual pH of water containing a little dissolved carbon dioxide from the air) to near 2, a pH change of about 4 units. The same amount of hydrochloric acid added to the acetic acid-acetate solution causes a pH drop of about 0.1 unit. Why is the affect on the pH of the buffer solution so much smaller? The molecular level drawings in Figure 9.4 can help you interpret these effects.

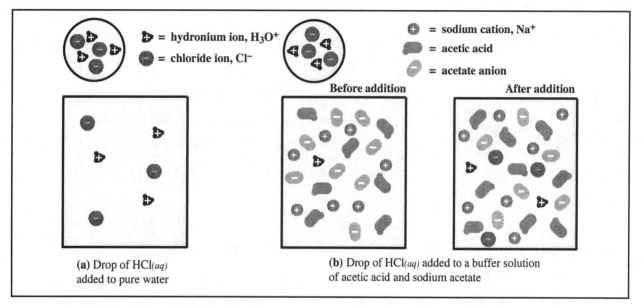

Figure 9.4.   Molecular-level representation of acid-base buffer action.

   If one drop (about 0.05 mL) of 1.0 M aqueous hydrochloric acid is added to 10 mL of water, the acid is diluted by a factor of about 200. In this solution, $[H_3O^+(aq)] \approx 0.005$ M and the pH will be about 2.3, as you observed in Investigate This 9.35(a), and as is represented in Figure 9.4(a). When the same amount of hydrochloric acid is added to the acetic acid-acetate solution, there is a base, acetate ion, to react with the hydronium ions from the hydrochloric acid. The pH doesn't change much, which means that not much hydronium ion is left unreacted. Acetate must be the predominant proton acceptor; acetate is a stronger base than water, as you can recall by reviewing Table 6.2 in Chapter 6, Section 6.4. The presence of a base stronger than water in the buffer solution makes the affect of adding $H_3O^+(aq)$ much smaller; there is only a small increase in pH. Figure 9.4(b) illustrates that most but not all of the added hydronium ion reacts with acetate anion to form acetic acid. At equilibrium, there is a little more hydronium ion in the solution than there was before the hydrochloric acid was added.

9.37. **Check This**   *Stoichiometry in buffer solutions*

(a) Before the hydrochloric acid is added in Figure 9.4(b), what are the positive and negative species in the solution? Is the solution electrically neutral? Explain.

(b) After the hydrochloric acid is added in Figure 9.4(b), what are the positive and negative species in the solution? Is the solution electrically neutral? Explain.

To analyze buffer action further, it is useful to rearrange equation (9.29) to a form that gives the pH of a solution as a function of known values for pK_a and the conjugate base-to-acid concentration ratio:

$$pH = pK_a + \log\left(\frac{(A^-(aq))}{(HA(aq))}\right)_{eq} = pK_a + \log\left(\frac{(\text{conjugate base})}{(\text{conjugate acid})}\right)_{eq} \tag{9.30}$$

> **WEB Chap 9, Sect 9.4.2-6.**
> Practice using equation (9.30) in an interactive format with graphical reinforcement.

We can use equation (9.30), often called the **Henderson-Hasselbalch equation,** to explain the buffering action when hydronium or hydroxide ions are added to a buffer solution. It can also help us figure out how to make buffer solutions with a desired pH.

Buffer pH

Before doing quantitative calculations, let's see what equation (9.30) tells us qualitatively. Consider a buffer solution in which the conjugate acid and conjugate base concentrations are the same, $[HA(aq)] = [A^-(aq)]$. In this solution,

$$\left(\frac{(A^-(aq))}{(HA(aq))}\right)_{eq} = 1, \text{ so}$$

$$pH = pK_a + \log\left(\frac{(A^-(aq))}{(HA(aq))}\right)_{eq} = pK_a + \log(1) = pK_a \tag{9.31}$$

If a solution contains equal concentrations of a conjugate acid and base pair, the pH of the solution is equal to the pK_a of the acid. If the conjugate base-to-acid ratio in a solution is not unity, Table 9.3 shows the *direction* that the pH of the solution varies from pK_a. Note from the table that when the relative amount of conjugate base increases, the solution becomes more basic ($[H_3O^+(aq)]$ decreases) and, when the relative amount of conjugate acid increases, the solution becomes more acidic ($[H_3O^+(aq)]$ increases). This correlation should help you remember the direction of change of the hydronium ion concentration, $[H_3O^+(aq)]$, and the pH when solution conditions are changed.

Table 9.3. Correlation of buffer pH with pK_a and conjugate base-to-acid ratio.

$\left(\dfrac{\text{(conjugate base)}}{\text{(conjugate acid)}}\right)_{eq}$	pH relative to pK_a	$[H_3O^+(aq)]$ relative to reference solution
= 1; equal concentrations	$pH = pK_a$	reference solution
> 1; base predominates	$pH > pK_a$	$[H_3O^+(aq)]$ decreases; more basic
< 1; acid predominates	$pH < pK_a$	$[H_3O^+(aq)]$ increases; more acidic

9.38. Check This   *pH and the conjugate acid-base ratio*

(a) Show how to get the results in Table 9.3 for $[HA(aq)] > [A^-(aq)]$ and $[HA(aq)] < [A^-(aq)]$. Give your reasoning clearly. *Note:* The logarithm of a number less than one is negative.

(b)   WEB   Chap 9, Sect 9.4.2-6. Practice the concepts in Table 9.3 by working through these interactive pages. After tracing the curve on page 6, write an explanation of how the curve relates to the graphic representation of the (conjugate base)/(conjugate acid) ratio shown below the graph. Also explain how the curve and graphic representation relate to your results in Investigate This 9.30 and 9.35.

9.39. Worked Example   *pH change when hydronium ion is added to a buffer solution*

An aqueous buffer solution that is 0.050 M in both acetic acid and acetate ion has a pH of 4.76 (= pK_a for acetic acid). If 0.05 mL of 1.0 M aqueous HCl solution is added to 10 mL of this buffer solution, what is the pH of the resulting solution?

Necessary information: We will need to use equation (9.30).

Strategy: Assume that all the added hydronium ion reacts with acetate ion to form acetic acid. Each mole of added hydronium reacts with one mole of acetate ion to form one mole of acetic acid. Using this stoichiometry, construct a change table to get the final concentrations of acetic acid and acetate ion and substitute these in equation (9.30) to get the final pH of the solution.

Implementation:

        initial mol HOAc(aq) = (0.050 M)·(0.010 L) = 5.0×10^{-4} mol

        initial mol OAc$^-$$(aq)$ = (0.050 M)·(0.010 L) = 5.0×10^{-4} mol

        mol H$_3$O$^+$$(aq)$ added = (1.0 M)·(0.00005 L) = 5.0×10^{-5} mol

species	HOAc(aq)	OAc⁻(aq)	H₃O⁺(aq)
initial mol	5.0×10^{-4}	5.0×10^{-4}	(pH = 4.67)
change in mol	5.0×10^{-5} formed	5.0×10^{-5} reacts	?
final mol	5.5×10^{-4}	4.5×10^{-4}	?

$$pH = pK_a + \log\left(\frac{(\text{conjugate base})}{(\text{conjugate acid})}\right)_{eq} = 4.76 + \log\left(\frac{4.5 \times 10^{-4}\,\text{mol}/0.010\text{L}}{5.5 \times 10^{-4}\,\text{mol}/0.010\text{L}}\right)$$

$$pH = = 4.76 - 0.09 = 4.67$$

Does the answer make sense? The reaction of hydronium ion with acetate ion formed some acetic acid and used up some acetate ion, so the conjugate acid and base concentrations are no longer equal; there is more acetic acid than acetate ion. Table 9.3 shows us that the pH should decrease and the calculation bears this out. However, even though 10% of the acetate has reacted, the pH changes by less than 0.1 unit. The buffer system resists change in the pH of the solution. How does this result compare with your observation in Investigate This 9.35?

9.40.   **Check This**   *pH change when hydroxide ion is added to a buffer solution*

An aqueous buffer solution that is 0.05 M in both acetic acid and acetate ion has a pH of 4.76. If 0.05 mL of 1.0 M aqueous sodium hydroxide solution is added to 10 mL of this buffer solution, what is the pH of the resulting solution? Show your work clearly. How does your result compare with your observation in Investigate This 9.35?

Preparing a buffer solution

To make a buffer with a desired pH, we choose a conjugate acid-base pair that has about the same pK_a as the desired pH. For this conjugate acid-base pair a base-to-acid ratio of unity will produce a solution close to the desired pH. Table 9.3 shows the *direction* you need to change the conjugate base-to-acid ratio to get the desired pH and equation (9.30) gives you a way to calculate the ratio that is required. For example, if you need a pH 7.85 buffer solution, Table 9.2

> "Tris" is an abbreviation for the compound, tris(hydroxymethyl)aminomethane, a commonly used buffer in biochemical studies.

shows that "tris," with a $pK_a = 8.08$, would be a good choice for the buffer. Since the desired pH is more acidic than the pK_a, the conjugate base-to-acid ratio has to be less than unity; that is, you need more conjugate acid than base. We find the required conjugate base-to-acid ratio by rearranging equation (9.30) and solving for the ratio:

$$\log\left(\frac{[\text{conjugate base}]}{[\text{conjugate acid}]}\right)_{eq} = pH - pK_a \qquad (9.32)$$

$$\log\left(\frac{[\text{conjugate base}]}{[\text{conjugate acid}]}\right)_{eq} = 7.85 - 8.08 = -0.23$$

$$\left(\frac{[\text{conjugate base}]}{[\text{conjugate acid}]}\right)_{eq} = 10^{-0.23} = 0.59$$

9.41. Consider This   *How do you prepare a buffer solution of known concentration?*

(a) Suppose you want to prepare a pH 7.85 "tris" buffer solution that contains 0.050 M *total* concentration of conjugate acid and conjugate base. What are the individual concentrations of the acid and base that will give this pH and a total concentration of 0.050 M? Clearly show how you get your answer.

(b) The acid form of "tris" is an ammonium-like ion that is available as its chloride salt; the molar mass of the salt is 157.6 g. The base, molar mass = 121.1 g, is also available as a pure solid. What masses of the acid salt and the base must be dissolved in a liter of solution to prepare the buffer in part (a)? Show all your work clearly.

Buffer solutions are probably the most widely used application of acid-base chemistry. All studies of biological systems are carried out in buffered solutions (or in living organisms whose internal solutions are buffered by several conjugate acid-base pairs). Many chemical reactions are studied in buffered solutions so that variations in the concentration of hydronium ion (and/or hydroxide ion) don't confuse interpretation. We use calculations based on equation (9.30) to find the pH of buffer solutions of a given composition and to determine the composition required to prepare a buffer solution of a desired pH. We can also use this equation to interpret the acid-base behavior of biomolecules, as we will see in the next section.

Section 9.5.   Acid-Base Properties of Proteins

9.42. Investigate This   *How is a protein solution affected by addition of $H_3O^+(aq)$?*

Do this as a class investigation and work in small groups to discuss and analyze the results. Place 250 mL of a basic 0.25% aqueous solution of casein (a protein found in milk) in a 400-mL beaker. Stir the solution at medium speed with a magnetic stirrer and a magnetic stir bar. Use a pH electrode and pH meter to monitor and read the pH of the solution. Add 3 M hydrochloric

---

acid solution several drops at a time to the stirred casein solution. After each addition, record your observations on the appearance of the solution and its pH. After the pH of the solution has reached about 2, reverse the pH change by adding 3 M sodium (or potassium) hydroxide solution a few drops at a time to the stirred casein solution. After each addition, record your observations on the appearance of the solution and its pH. Repeat to see if your observations are reproducible.

9.43.   Consider This   *How are protein solubility and pH related?*

(a) Did any of your observations in Investigate This 9.42 surprise you? Why or why not?

(b) How was the solubility of the protein (casein) related to the pH of the solution? At what pH was the most precipitate formed? Look at the data for amino acids in Table 9.2. Can you think of a way to relate your observations in Investigate This 9.42 to the properties of the amino acids that are combined to make the protein.

Recall from Chapter 2 that almost all biological molecules can act as proton donors and/or proton acceptors. We used this property to explain the solubility of these high molecular mass *ionic* molecules in aqueous solutions. The results from Investigate This 9.42 demonstrate that the solubility of casein is pH-dependent. Proteins all react somewhat differently to pH changes, but most show behavior similar to that of casein; their solubilities vary with pH. We have enough information from our discussion of acid-base equilibria to be able to understand this variation.

Acid-base side groups on proteins

We have seen previously (Chapter 1, Section 1.9 and Chapter 6, Section 6.7) that **proteins** are polymers of amino acids. The amine group and carboxylic acid group bonded to the same carbon on one amino acid form amide bonds with two neighboring amino acids to build the polymer chain. When this happens, these groups are no longer available to accept or donate

| Some of the acid-base side groups on proteins are buried inside the folded structure and do not contribute to interactions with the surrounding solution. |

protons. Acidic or basic **amino acid side groups** remain free to accept or donate protons. These groups give proteins their acid-base properties.

The seven amino acid side groups with acid-base properties are noted with a double asterisk (**) in Table 9.2.

Most real proteins have many acidic and basic side groups. Accounting simultaneously for

| WEB Chap 9, Sect 9.5.1-4. Practice using the concepts in this section with interactve animations of a small protein. |

the acid-base properties of all these side groups is complicated. To illustrate how the pH of the solution it is in affects the charge on a protein, we will use the greatly simplified protein model in

Figure 9.5. The two side groups shown are a carboxylic acid and an amine which have quite different pK_a values. We will use equation (9.32) to analyze the base-to-acid ratio for each side group as a function of pH and the applicable pK_a:

$$\log\left(\frac{[-COO^-]}{[-COOH]}\right)_{eq} = pH - pK_{a1} \tag{9.33}$$

$$\log\left(\frac{[-NH_2]}{[-NH_3^+]}\right)_{eq} = pH - pK_{a2} \tag{9.34}$$

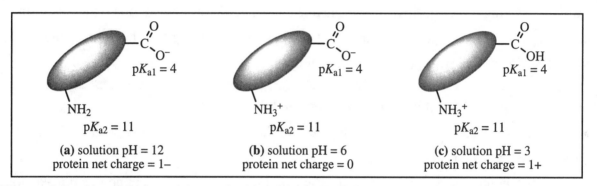

(a) solution pH = 12
protein net charge = 1–

(b) solution pH = 6
protein net charge = 0

(c) solution pH = 3
protein net charge = 1+

Figure 9.5. Simplified model protein (the ellipsoid) with only two acid-base side groups.

For the protein in a solution with a pH of 12, Figure 9.5(a), the right-hand side of both equations, (9.33) and (9.34), is positive because pH > pK_a. When the logarithm of the base-to-acid ratio is positive, the ratio is greater than unity, and the base form predominates. Thus, the protein is shown in Figure 9.5(a) with both groups in their base form. The amine is uncharged and the carboxylate ion has a negative charge, so the **net charge** (sum of the charges of the side groups) on the protein is 1–.

9.44.   Consider This   *How do side group charges on our model protein vary with pH?*

   (a) Why are the side groups on the model protein shown as they are in Figures 9.5(b) and 9.5(c) at solution pH 6 and 3, respectively, ? Clearly explain your answers.

   (b) Explain how we obtain the net charges shown for the protein in Figures 9.5(b) and 9.5(c).

Isoelectric pH for proteins

   In solutions with a high pH (basic solutions), proteins usually carry a net negative charge. Conversely, in solutions with a low pH (acidic solutions), proteins usually carry a net positive charge. At some intermediate pH, the net charge on a protein must be zero, as depicted for the model protein in Figure 9.5(b). The pH of the solution in which a protein has zero net charge is called the **isoelectric pH** (*iso* = same). At the isoelectric pH (sometimes called the **isoelectric point**), the net positive and net negative electric charges on the protein are the same, so the overall net charge is zero.

To see how these charges affect the properties of proteins, consider the solution of casein in Investigate This 9.42. At high pH all the casein molecules are negatively charged. At low pH they are all positively charged. In either case, the molecules repel one another and have no tendency to clump together and form a precipitate. At an intermediate pH, the isoelectric pH for casein, the molecules have no net charge and do not repel one another. The molecules do, however, have regions of positive and negative charge and the negative region on one molecule can attract the positive region on another. These attractions can bring together clumps of protein molecules and precipitate the protein from solution.

9.45.   **Consider This**   *What is the isoelectric pH for casein?*

(a) Are your observations from Investigate This 9.42 consistent with the picture of protein behavior just presented? Why or why not?

(b) What would you estimate the isoelectric pH of casein to be? Explain how you make your estimate.

9.46.   **Check This**   *Relating a* **Web Companion** *animation to the casein investigation*

(a) WEB Chap 9, Sect 9.4.3. Follow the directions on the page and observe the curve on the graph and the animation below the graph. Explain the correlation between the graphical plot and the animation.

(b) Explain how the graphical plot and the animation are related to your results in Investigate This 9.42. *Hint:* Work through the previous pages of this section of the *Web Companion*.

(c) What, if any, correlations do you see between the graphics on this page of the *Web Companion* and buffer solutions? Could these graphics be used to illustrate buffering action. Explain the reasoning for your responses.

Electrophoresis

A widely used technique for separating and analyzing biological molecules, **electrophoresis**,

Electrophoresis (pronounced ee-lek´-troe-fah-ree´-sis) is a combination of electric + *phorein* = to carry; carrying electricity or electric charge.

utilizes the acid-base principles we have been discussing. One type of electrophoresis apparatus is illustrated in Figure 9.6. Separation of molecules occurs within a thin slab of gel that is in contact with buffer solutions at both ends. Charged electrodes in the buffer solutions set up an electrical potential from one end of the gel to the other. Electrically charged sample molecules migrate through the gel under the influence of the electric field.

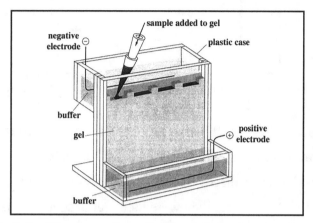

Figure 9.6. Diagram of a gel electrophoresis apparatus.

The apparatus shown in Figure 9.6 is designed to separate negatively charged molecules as they migrate from the negative (top) electrode toward the positive (bottom) electrode. Molecules with a higher negative charge migrate faster than those with a lower charge and move farther down the gel in a given time. Thus, molecules with different charges are separated from one another. We have seen above that the charge on a protein depends upon the pH of the solution in which it is dissolved. The buffer pH for electrophoresis separations of proteins is chosen to be high enough so that the proteins will have a negative charge.

Acids, bases, and sickle-cell hemoglobin

The chapter opening illustration shows what happens to the red blood cells (erythrocytes) of persons with sickle-cell disease when the hemoglobin in the cells is deoxygenated. Figure 9.7 shows the results of an electrophoretic analysis of sickle-cell hemoglobin, HbS, and normal adult hemoglobin, HbA. At the pH 8.6 chosen for the analysis both proteins have a net negative charge. However, HbA moves further toward the positive end of the gel; it migrates faster. HbA must therefore have a higher net negative charge than HbS.

9.47.   Consider This   *What is the difference between HbA and HbS?*

The negatively-charged side groups on hemoglobin at pH 8.6 are carboxylate anions. Does the electrophoresis result in Figure 9.7 suggest that HbS has more or fewer carboxylate side groups than HbA? Explain how you arrive at your conclusion.

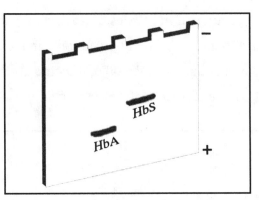

Figure 9.7. Results of electrophoresis of normal and sickle-cell hemoglobins at pH 8.6.

The hemoglobin molecule, Figure 9.8, consists of four separate protein chains (two α chains and two β chains) that are stacked together in a tetrahedral arrangement. After the difference in charge between HbA and HbS was discovered, further analysis of the separate chains showed

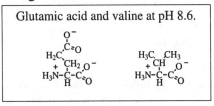

that they were identical except for one amino acid (out of 146) on each of the β chains. The sixth amino acid in the β chains of HbA is glutamic acid, but HbS has valine at this position. Glutamic acid, Table 9.2, has a carboxylic acid side group, but valine has a nonpolar, $(CH_3)_2CH-$, side group. The loss of the acid side group explains the results of the electrophoresis analysis, as you found in Consider This 9.47. In the folded protein chains, these side groups are on the outside of the molecule, as shown in Figure 9.8. HbS has a nonpolar, hydrophobic site where HbA has a polar, hydrophilic site.

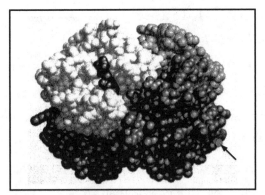

Figure 9.8. Molecular model of hemoglobin.
The α and β chains are at the top and bottom, respectively. The arrow pointing to the green dot locates where valine is substituted for a glutamic acid on one HbS β chain. This location on the other β chain is hidden from view.

We saw in Chapter 8, Section 8.11, that the entropy of aqueous solutions increases when nonpolar parts of molecules come together and release the water molecules that are "frozen" around them. The extra nonpolar site on one HbS molecule comes together with a nonpolar site on another and the process continues to form long insoluble chains of HbS molecules that are responsible for the shape and the rigidity of sickled red-blood cells.

Reflection and projection

Solutions that contain substantial concentrations of both members of a conjugate acid-base pair are buffer solutions that resist changes in pH when disturbed by the addition of hydronium ion or hydroxide ion. Buffers work because systems at equilibrium respond to a disturbance in a way that minimizes the affect on the equilibrium, Le Chatelier's principle. Addition of hydronium ion to a buffer solution, for example, disturbs the equilibrium by increasing the concentration of hydronium, so the conjugate acid-base-hydronium ion system is no longer in equilibrium. Some of the base reacts with some of the added hydronium to reduce the hydronium concentration, increase the conjugate acid concentration, and decrease the conjugate base concentration. The net effect of these changes is to restore the conjugate acid-base-hydronium ion equilibrium. The new equilibrium amount of hydronium ion is lower than it would have been for the same initial amount of hydronium ion added to pure water.

We applied these conjugate acid-base principles to the acid-base properties of proteins. The ratio of a protein's conjugate base-to-acid forms is controlled by the pH of the solution in which it is dissolved. We study a protein's acid-base properties by observing the effects of changing the solution pH. Among other properties, we can determine the isoelectric pH (net zero charge and minimum solubility) and can learn about the differences among proteins by electrophoresis.

Everything we have done with acids and bases has been directly based on the equilibrium concepts developed in the first part of the chapter. We will now see how these concepts are applied to another familiar chemical reaction, dissolution (and precipitation) of ionic solids in water and in solutions containing additional species.

Section 9.6.   Solubility Equilibria for Ionic Salts

9.48. Investigate This   *Is silver chromate insoluble?*

Do this as a class investigation and work in small groups to discuss and analyze the results. Use 10.0 mL of an aqueous 0.0015 M solution of silver nitrate, $AgNO_3$, in a 25-mL beaker on an overhead projector stage. Add one drop of an aqueous 0.0015 M solution of potassium chromate, K_2CrO_4, to the silver nitrate solution and swirl the mixture. WARNING: Cr(VI) compounds are suspect carcinogens. Wear disposable gloves when handling the solutions. Record your observations. Keep track of the amount you add as you continue to add potassium chromate solution, *one drop at a time*, until a precipitate forms and makes the solution cloudy.

9.49. Consider This   *How insoluble is silver chromate?*

In Investigate This 9.48, what volume of potassium chromate solution did you have to add in order for precipitation of silver chromate to begin? How do you explain this observation? *Note: One drop of aqueouos solution is about 0.05 mL.*

Solubility product

Mixing the two solutions in Investigate This 9.48 gives a solution containing $K^+(aq)$, $Ag^+(aq)$, $CrO_4^{2-}(aq)$, and $NO_3^-(aq)$ ions. The precipitate that forms must be $Ag_2CrO_4(s)$. The dissolution and precipitation of ionic salts are usually represented by dissolution reactions:

$$Ag_2CrO_4(s) \rightleftharpoons 2Ag^+(aq) + CrO_4^{2-}(aq) \tag{9.35}$$

The equilibrium reaction expression for this dissolution reaction is:

$$K = \left(\frac{\left(Ag^+(aq)\right)^2 \left(CrO_4^{2-}(aq)\right)}{\left(Ag_2CrO_4(s)\right)} \right)_{eq} \tag{9.36}$$

Recall, Table 9.1, that the concentration ratio a pure solid is unity, so $(Ag_2CrO_4(s)) = 1$. Thus, equation (9.36) becomes:

$$K_{sp} = \left\{ \left(Ag^+(aq)\right)^2 \left(CrO_4^{2-}(aq)\right) \right\}_{eq} \tag{9.37}$$

The equilibrium constant expression for dissolution of an ionic compound is just the product of the molar concentrations of the ions in the solution raised to their appropriate stoichiometric powers. The subscript "sp" reminds us that this equilibrium constant, K_{sp}, is the **solubility product** for the solid ionic compound. Keep in mind that *the solubility product is only valid when the ions in solution are in equilibrium with the solid.*

9.50. Worked Example   *Solubility product, K_{sp}, for silver chromate, $Ag_2CrO_4(s)$*

Suppose that, in Investigate This 9.48, a cloudy solution formed upon addition of the fifth drop of 0.0015 M potassium chromate solution. Estimate K_{sp}, the solubility product constant, for silver chromate, $Ag_2CrO_4(s)$. (Your results may be different; you can use the strategy here to get an estimate from your observation.)

Necessary information: We need to know the concentrations of the solutions used in Investigate This 9.48; the $Ag^+(aq)$ solution is 0.0015 M and the $CrO_4^{2-}(aq)$ solution is 0.0015 M. The volumes of solution used are 10.0 mL and 0.25 mL (the assumed five drops), for the $Ag^+(aq)$ and $CrO_4^{2-}(aq)$, respectively.

Strategy: Substitute $[Ag^+(aq)]$ and $[CrO_4^{2-}(aq)]$ (molar concentrations in the solution when solid is also present) into equation (9.37) to get K_{sp}. Assume that the solid *just* begins to form when the 0.25 mL of $CrO_4^{2-}(aq)$ has been added, so that a negligible number of moles of the $Ag^+(aq)$ and $CrO_4^{2-}(aq)$ will have reacted to form the precipitate we see. With this assumption, we can use the number of moles of $CrO_4^{2-}(aq)$ added and moles of $Ag^+(aq)$ originally present to calculate their concentrations in 10.25 mL of the mixture containing a trace of solid.

Implementation

mol $Ag^+(aq)$ originally present = $(10.0 \times 10^{-3}$ L$)\cdot(0.0015$ M$) = 1.5 \times 10^{-5}$ mol

mol $CrO_4^{2-}(aq)$ added = $(2.5 \times 10^{-4}$ L$)\cdot(0.0015$ M$) = 3.8 \times 10^{-7}$ mol

Only a trace amount of these ions has reacted to give precipitate, so these are the moles of these ions in equilibrium with the trace amount of solid. The concentrations of the ions are:

final molar concentration of $Ag^+(aq) = \left(1.5 \times 10^{-5} \text{ mol}\Big/0.01025 \text{ L}\right) = 1.5 \times 10^{-3}$ M

final molar concentration of $CrO_4^{2-}(aq) = \left(3.8 \times 10^{-7} \text{ mol}\Big/0.01025 \text{ L}\right) = 3.7 \times 10^{-5}$ M

$$K_{sp} = \left\{\left(Ag^+(aq)\right)^2 \left(CrO_4^{2-}(aq)\right)\right\}_{eq} = (1.5 \times 10^{-3})^2 (3.7 \times 10^{-5}) = 8 \times 10^{-11}$$

Does the answer make sense? The small value for the solubility product shows that silver chromate is not very soluble, as we have observed. But is our numeric value correct? The validity of the result depends upon whether our strategic assumption is any good. What if the concentrations were on the borderline of solubility when four drops of potassium chromate solution had been added, so that just a tiny amount more would have caused precipitation. For this case, we would have calculated $[Ag^+(aq)] = 1.5 \times 10^{-3}$ M, $[CrO_4^{2-}(aq)] = 3 \times 10^{-5}$ M, and $K_{sp} = 7 \times 10^{-11}$, which is almost the same as the value we calculated above. (What do you get from your results for Investigate This 9.48?) Although solubility products are often given with two (or even more) significant figures, these are usually not justified under actual solution conditions where the activities of ions are only approximated by their molarities and the existence of ion complexes that keep the ions in solution is not accounted for. Also, in experiments such as Investigate This 9.48, slow formation of solids is not uncommon.

9.51.   **Check This**   *Solubility product, K_{sp}, for silver chromate, $Ag_2CrO_4(s)$*

Assume that you carry out an experiment like that in Investigate This 9.48, but use a buret to add an aqueous 5.0×10^{-3} M solution of silver nitrate, $AgNO_3$, to 100 mL of a stirred, aqueous 2.0×10^{-4} M potassium chromate, K_2CrO_4, solution. You detect the first permanent cloudiness (deep red precipitate of Ag_2CrO_4) when you have added 9.8 mL of the silver solution. Estimate the numeric value for the solubility product constant, K_{sp}. Explain how you get your answer.

Solubility and solubility product

The **solubility** of an ionic salt in water is the number of moles of the solid that dissolve in one liter of solution to form a saturated solution. A **saturated solution** is one in which as much solid as possible has dissolved; the dissolved ions are in equilibrium with undissolved ionic solid. You can calculate the solubility product for an ionic salt from its solubility and *vice versa*. Such calculations do not agree exactly with values determined in other ways, because ionic interactions in solution are not accounted for by the simple solubility product relationships exemplified by equation (9.37). The calculations usually agree within one or two orders of magnitude, so the results are generally in the right ballpark. The solubility of "insoluble" ionic compounds varies a great deal. Table 9.4 gives the solubility products for several cation-anion combinations to give you an idea of the range of solubilities.

Table 9.4. Solubility products, K_{sp} and pK_{sp}, for several insoluble ionic compounds.

compound	K_{sp}	pK_{sp}	compound	K_{sp}	pK_{sp}
AgCl	1.8×10^{-10}	9.74	$Cu(IO_3)_2$	1.4×10^{-7}	6.85
AgBr	5.0×10^{-13}	12.30	CuC_2O_4	2.9×10^{-8}	7.5
AgI	8.3×10^{-17}	16.08	CuS	8×10^{-37}	36.1
Ag_2CrO_4	1.2×10^{-12}	11.92	$Fe(OH)_2$	7.9×10^{-16}	15.1
Ag_2S	8×10^{-51}	50.1	FeS	8×10^{-19}	18.1
$Al(OH)_3$	3×10^{-34}	33.5	$Fe(OH)_3$	1.6×10^{-39}	38.8
$BaCO_3$	5.0×10^{-9}	8.30	$MgCO_3$	3.5×10^{-8}	7.46
$BaSO_4$	1.1×10^{-10}	9.96	$Mg(OH)_2$	7.1×10^{-12}	11.15
BaC_2O_4	1×10^{-6}	6.0	$PbCl_2$	1.7×10^{-5}	4.78
$CaCO_3$	6.0×10^{-9}	8.22	$PbBr_2$	2.1×10^{-6}	5.68
$CaSO_4$	2.4×10^{-5}	4.62	PbI_2	7.9×10^{-9}	8.10
$Ca(OH)_2$	6.5×10^{-6}	5.19	$PbSO_4$	6.3×10^{-7}	6.20
CaC_2O_4	1.3×10^{-8}	7.9	PbS	3×10^{-28}	27.5

9.52. **Worked Example**   *Solubility product, K_{sp}, for magnesium hydroxide, $Mg(OH)_2(s)$*

A saturated solution of magnesium hydroxide (Milk of Magnesia®) contains 0.009 g of $Mg(OH)_2$ per liter of solution. What are the solubility and solubility product for $Mg(OH)_2$?

Necessary Information: We need the molar mass of $Mg(OH)_2$, 58 g·mol^{-1}. We'll use the stoichiometry of the dissolution reaction and solubility product constant:

$$Mg(OH)_2(s) \rightleftharpoons Mg^{2+}(aq) + 2OH^-(aq) \tag{9.38}$$

$$K_{sp} = \{(Mg^{2+}(aq))\cdot(OH^-(aq))^2\}_{eq} \tag{9.39}$$

Strategy: Use the mass of $Mg(OH)_2$ that dissolves in a liter of solution and the molar mass to calculate the solubility, s, in mol·L^{-1}. For every mole of $Mg(OH)_2(s)$ that dissolves the solution contains one mole of $Mg^{2+}(aq)$ and two moles of $OH^-(aq)$. Since s mol·L^{-1} dissolve, the concentration of $Mg^{2+}(aq)$ is s M and the concentration of $OH^-(aq)$ is $2s$ M. These values are substituted into the solubility product expression (9.39) to get K_{sp}.

Implementation:

$$\text{molar solubility of } Mg(OH)_2 = s = \frac{0.009 \text{ g} \cdot \text{L}^{-1}}{58 \text{ g} \cdot \text{mol}^{-1}} = 1.6 \times 10^{-4} \text{ M}$$

$$K_{sp} = (s)\cdot(2s)^2 = 4s^3 = 4(1.6 \times 10^{-4})^3 = 1.6 \times 10^{-11} \approx 2 \times 10^{-11}$$

Does the answer make sense? Using the stoichiometry of the solution reaction, we converted solubility to the solubility product. Our result is in the right range, about three times larger than the value in Table 9.4. This gives you an idea of the kind of agreement you can expect for ionic equilibria calculated without corrections for ionic interactions. When you do these kinds of calculations, be careful to use the stoichiometric coefficients in the reaction equation correctly to get the concentrations of each ion from the stoichiometry of the reaction, and then again as exponents in the solubility product expression.

9.53. Check This   *Solubility of silver phosphate, Ag$_3$PO$_4$*

The solubility product constant, K_{sp}, for silver phosphate, $Ag_3PO_4(s)$, is 2.8×10^{-18}.

(a) Write the dissolution reaction and the solubility product expression for $Ag_3PO_4(s)$.

(b) What is the molar solubility, s, of $Ag_3PO_4(s)$ in water? *Hint:* The strategy and calculations are the reverse of those in Worked Example 9.52. Use the stoichiometry of the reaction to write the concentrations of the ions in terms of s. Substitute these concentrations in the solubility product expression and solve for s.

(c) One handbook gives the solubility for $Ag_3PO_4(s)$ as 0.0071 g·L^{-1}. How does your value compare to this one?

Common ion effect

Gravimetric analysis is a method to analyze an ionic sample by precipitating the ion of interest as an insoluble salt with an appropriate counterion and determining the mass of the product. Suppose we wish to analyze a solution containing calcium ion by precipitating it as its sulfate, $CaSO_4(s)$. The dissolution reaction and solubility product expression for $CaSO_4(s)$ are:

$$CaSO_4(s) \rightleftharpoons Ca^{2+}(aq) + SO_4^{2-}(aq) \tag{9.40}$$

$$K_{sp} = \{(Ca^{2+}(aq)) \cdot (SO_4^{2-}(aq))\}_{eq} = 2.4 \times 10^{-5} \tag{9.41}$$

How are we going to make sure almost all the $Ca^{2+}(aq)$ gets precipitated? Let's think backwards for a moment. If we *dissolve* $CaSO_4(s)$ in water, its solubility is s $mol \cdot L^{-1}$ and the solubility product tells us that $s^2 = 2.4 \times 10^{-5}$, so $s = 5.0 \times 10^{-3}$ $mol \cdot L^{-1}$ (about 0.7 $g \cdot L^{-1}$). This is a moderate solubility. We have to find a way to reduce the solubility of the solid, so that it will precipitate and stay out of solution for our analysis. If we add extra $SO_4^{2-}(aq)$ (say by adding $Na_2SO_4(s)$ to the solution), the system should respond to this disturbance by using up some of the added anion. Reaction (9.40) will go in reverse to use up some of the added $SO_4^{2-}(aq)$, which also precipitates more of the $Ca^{2+}(aq)$. Figure 9.9 is a molecular-level representation of this procedure for gravimetric analysis. Let's see how the analysis works out quantitatively.

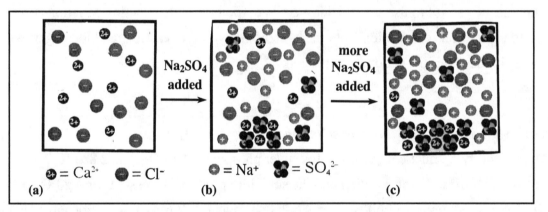

Figure 9.9. Molecular-level representation of gravimetric analysis for calcium ion.
(a) is the solution to be analyzed. In (b), a stoichiometric amount of sulfate ion is added and some $CaSO_4(s)$ precipitates. When more sulfate ion is added, in (c), the precipitation of $CaSO_4(s)$ is almost complete.

9.54. Worked Example *Solubility of $CaSO_4(s)$ in 0.50 M $SO_4^{2-}(aq)$ solution*

Find the molar solubility of $CaSO_4(s)$ in an aqueous 0.50 M solution of sodium sulfate, Na_2SO_4, that is, a solution that is 0.50 M in $SO_4^{2-}(aq)$ anion.

Necessary information: We need the solubility product for $CaSO_4(s)$, equation (9.41).

Strategy: Let s be the molar solubility of $CaSO_4(s)$ in the solution, construct a change table for the dissolution reaction, substitute the final molar concentrations into equation (9.41), and solve for s. We will assume, as we often have, that we have one liter of solution, so that numbers of moles and molarity have the same numeric value.

Implementation: The change table is:

species	$CaSO_4(s)$	$SO_4^{2-}(aq)$	$Ca^{2+}(aq)$
initial mol	0	0.50	0
change in mol	solid added; s dissolves	s goes into solution	s goes into solution
final mol	solid still present	$0.50 + s$	s

If the $CaSO_4(s)$ were dissolving in pure water, we know from above that s would be 5.0×10^{-3} M. This is the maximum s can be and it is only 1% of 0.50 M, so $0.50 + s \approx 0.50$:

$$K_{sp} = \{(Ca^{2+}(aq)) \cdot (SO_4^{2-}(aq))\}_{eq} = (s)(0.50)$$

$$s = \frac{K_{sp}}{0.50} = \frac{2.4 \times 10^{-5}}{0.50} = 4.8 \times 10^{-5} \text{ mol·L}^{-1}$$

Does the answer make sense? The solubility of $CaSO_4(s)$ in a solution that already contains $SO_4^{2-}(aq)$ is lower (one hundred-fold lower) than in pure water. This is exactly what we reasoned above on the basis of Le Chatelier's principle and the calculations bear out our prediction.

9.55.   **Check This**   *Solubility of $Cu(IO_3)_2(s)$ in water and 0.25 M $Cu^{2+}(aq)$ solution*

(a) What is the molar solubility of $Cu(IO_3)_2(s)$ in water? Use the K_{sp} from Table 9.4.

(b) What is the molar solubility of $Cu(IO_3)_2(s)$ in a 0.25 M aqueous solution of copper nitrate, $Cu(NO_3)_2$? How does your answer compare to your result from part (a)? Is this what you expected? Why or why not?

(c) Without doing any calculations, predict whether the solubility of $Cu(IO_3)_2(s)$ in a 0.25 M aqueous solution of sodium iodate, $NaIO_3$, will be larger, smaller, or the same as the solubility you calculated in part (b). Give your reasoning clearly.

In the systems we have just discussed, the ionic solid and the solution had an ion in common. The lowered solubility of the ionic solid in such a solution is an example of the **common ion effect**, the response of an ionic equilibrium system to an increase in concentration of one of the ions involved in the equilibrium. In Worked Example 9.54, since there is already $SO_4^{2-}(aq)$ in the solution, less $CaSO_4(s)$ has to dissolve to satisfy the solubility product. The common ion effect is another example of Le Chatelier's principle and the application of equilibrium principles, not a new concept. Figure 9.10 illustrates the common ion effect on the molecular level.

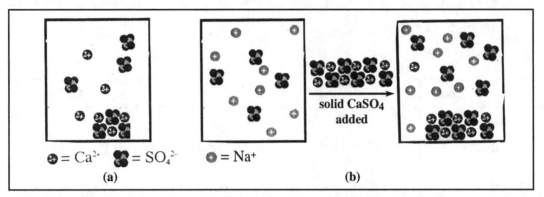

Figure 9.10. Common ion effect on solubility of $CaSO_4(s)$ in a solution containing $SO_4^{2-}(aq)$.
(a) is a solution of $CaSO_4(s)$ dissolved in pure water. (b) is the same amount of $CaSO_4(s)$ added to a solution already containing $SO_4^{2-}(aq)$ ions; less solid dissolves.

9.56.   **Check This**   *Compare Figures 9.9 and 9.10*

(a) What is the same and what is different about Figures 9.9(b) and 9.10(a)? Focus on the similarities and discuss why they are the same.

(b) How could you make the solution in Figure 9.10(a) identical to the one in Figure 9.9(b)? What does your answer imply about the solubility of an ionic compound in a solution that contains ions, but not a common ion? Explain clearly.

(c) What is the same and what is different about Figures 9.9(c) and the solution that results when the solid is added in Figure 9.10(b)? Focus on the similarities and discuss why they are the same.

(d) How could you make the resulting solution in Figure 9.10(b) identical to the one in Figure 9.9(c)? Does your answer reinforce the conclusion(s) you suggested in part (b)? Explain why or why not.

We do not often add a solid to a solution containing a common ion. However, in gravimetric analysis, we are interested in the solubility of a precipitate in a solution containing an excess of the counterion used to precipitate the ion of interest. This is exactly the same system we have been discussing, except that, in this case, the solid is formed when two solutions are mixed. We have to account for this reaction to determine how much of the common ion is present in the final solution. These cases are exemplified by Figures 9.9 and 9.10 and your analysis in Check This 9.56.

9.57.   Consider This   *How "complete" is the precipitation of $Pb^{2+}(aq)$ with $I^-(aq)$?*

In Investigate This 2.36, Chapter 2, Section 2.8, you mixed equal volumes of 0.34 M aqueous lead nitrate, $Pb(NO_3)_2$, solution and 0.74 M aqueous sodium iodide, NaI, solution.

(a) If no reaction had occurred in this mixture, what would be the concentrations of $Pb^{2+}(aq)$ and $I^-(aq)$?

(b) Assuming that the precipitation reaction between these ions goes to completion, which is the limiting reactant? What concentration of the other reactant remains in solution? Explain.

(c) What is the solubility of the solid in the solution containing the concentration of common ion you calculated in part (b)? Use the solubility product from Table 9.4 and clearly explain how you solve the problem and what assumptions you make.

(d) How good is our assumption that the precipitation reaction goes to completion? Has at least 99.9% of the possible solid remained undissolved in this solution? Show how you arrive at your conclusions.

You usually see the common ion effect associated with ionic solubilities. However, if you go back to our discussion of acid-base equilibria in Section 9.3, you will find several examples of the effect of changing the concentration of an ion that is common to both the reactants and products. One example would be the effect on the equilibrium concentrations when acetate ion is added to a solution that already contains acetate ion and acetic acid (a compound of acetate ion and H^+). Le Chatelier's principle makes it clear that the common ion effect is found everywhere in ionic equilibria.

Although solubility reactions look simple and we have treated them simply, they are not so simple. Many solute-solvent and solute-solute interactions occur in ionic solutions that affect these reactions. An example we did not discuss is the stepwise formation of intermediates, for example, $PbI^+(aq)$ and $PbI_2(aq)$ in the reaction that ultimately forms $PbI_2(s)$. A related example is formation of higher complexes, such as $PbI_3^-(aq)$ and $PbI_4^-(aq)$ in the lead-iodide system. Accounting for all these reactions and their equilibria is required for an accurate characterization of a solubility system. We will continue to neglect these complexities, but warn you that they exist and that they make the simple calculations only approximations (sometimes quite poor approximations).

Reflection and projection

We introduced the solubility of ionic compounds in water in Chapter 2 and used specific examples in subsequent chapters to illustrate various aspects of chemical reactivity. In Investigate This activities, you often produced precipitates by mixing solutions of ions that

---

reacted to form an insoluble solid. You could have attained the same end result (more slowly) by dissolving the solid in a solution of the counterions from the solutions you mixed to get a precipitate. It doesn't matter how the mixture of solid and solution are produced; at equilibrium, the product of the ionic concentrations in the solution (with appropriate stoichiometric exponents) is a constant, the solubility product, K_{sp}. You already knew that insoluble ionic compounds vary in their solubility. The solubility product quantifies the variability and Table 9.4 shows the great range of solubilities. The solubility product and solubility product expression also provide you the means to calculate the effects of added common ions on a system at solubility equilibrium.

In all the equilibrium calculations we have done, we have relied on the empirical relationship among the reactant and product concentrations, equation (9.6), that was determined by Guldberg and Waage in the 19th century. Together with the balanced chemical reaction, this equilibrium constant expression has been the starting point for all of our discussions. In Chapter 8, however, we stated the criterion for a system at equilibrium in terms of net entropy change, ΔS_{net}, or Gibbs free energy change, ΔG. For a system at equilibrium with respect to a change, we found that ΔS_{net}, or ΔG, is zero. How is the empirical equilibrium constant expression related to the thermodynamic criterion for equilibrium? That is the subject of the next section.

Section 9.7. Thermodynamics and the Equilibrium Constant

In Chapters 7 and 8, we learned how to use the data in Appendix XX, standard enthalpies of formation, ΔH°_f, standard entropies, S°, and standard Gibbs free energies of formation, ΔG°_f, to calculate standard enthalpy, entropy, and free energy changes for reactions. Also, in Chapter 8, we learned that, for a change in a reaction system, ΔS_{net}, or ΔG, is equal to zero for the change under equilibrum conditions. However, we have not learned how to calculate changes in these thermodynamic quantities under non-standard state conditions. What we need to learn is how to convert values we know, $\Delta G^{\circ}_{reaction}$, for example, to the corresponding values under non-standard state conditions, $\Delta G_{reaction}$.

Free energy of reaction and the reaction quotient

In Section 9.11, we show how to begin from the definition of entropy, Chapter 8, Section 8.6, and arrive at the relationship between $\Delta G_{reaction}$ and $\Delta G^{\circ}_{reaction}$ for a gas phase reaction. At the end of Section 9.11, we suggest that the relationship for the gas phase reaction can be applied to any reaction, including our general equation for a reversible reaction:

$$a\mathbf{A} + b\mathbf{B} \rightleftharpoons c\mathbf{C} + d\mathbf{D} \tag{9.5}$$

The relationship applied to reaction equation (9.5) is:

$$\Delta G_{\text{reaction}} = \Delta G^{\circ}_{\text{reaction}} + RT\ln\left\{\frac{(C)^{c}(D)^{d}}{(A)^{a}(B)^{b}}\right\} \qquad (9.42)$$

In equation (9.42), R and T are, respectively, the molar gas constant and the temperature in kelvin.

9.58. Check This   *Agreement of units in equation (9.42)*

Show that the units of the term on the far right of equation (9.42) are the same as those for the free energy change for a mole of reaction.

Equation (9.42) is the relationship we need to find the free energy change for a reaction under non-standard state conditions. The ratio in curly brackets in equation (9.42) is called the **reaction quotient** and is usually symbolized as Q:

$$Q = \left\{\frac{(C)^{c}(D)^{d}}{(A)^{a}(B)^{b}}\right\} \qquad (9.43)$$

$$\Delta G_{\text{reaction}} = \Delta G^{\circ}_{\text{reaction}} + RT\ln Q \qquad (9.44)$$

The reaction quotient should look familiar to you, since it has exactly the same *form* as the equilibrium constant expression for reaction (9.5). You write the reaction quotient for any reaction, just as you would write its equilibrium constant expression.

9.59. Worked Example   *Reaction quotient and free energy for a gas phase reaction*

Bromine, Br_2, and chlorine, Cl_2, gases can react to form the interhalogen compound (a diatomic species of two different halogen atoms), BrCl:

$$Br_2(g) + Cl_2(g) \rightleftharpoons 2BrCl(g) \qquad (9.45)$$

(a) Write the reaction quotient for reaction (9.45).

(b) The standard molar free energies of formation (at 298 K) of $Br_2(g)$, $Cl_2(g)$, and $BrCl(g)$ are 3.14, 0.00, and −0.88 kJ·mol⁻¹, respectively. What is the free energy change for reaction (9.45) at 298 K, if the pressures of $Br_2(g)$ and $Cl_2(g)$ are each 0.120 bar and the pressure of $BrCl(g)$ is 0.240 bar? Is the reaction spontaneous under these conditions? Why or why not?

Necessary information: In addition to the information in the problem statement, we need equation (9.42) [or, equivalently, equations (9.43) and (9.44)], the definition of the concentration ratio for gases from Table 9.1, and the value of R, 8.314 J·mol⁻¹·K⁻¹.

Strategy: Use the reaction equation (9.45) to write the reaction quotient (the same form as the equilibrium constant expression). Calculate $\Delta G^{\circ}_{\text{reaction}}$ and Q from the data in the problem,

---

combine them in equation (9.42) to determine $\Delta G_{reaction}$, and, from the sign of $\Delta G_{reaction}$, decide whether the reaction can occur under the specified conditions.

Implementation: (a) The reaction quotient is:

$$Q = \left\{ \frac{(BrCl_{(g)})^2}{(Br_{2(g)})(Cl_{2(g)})} \right\}$$

(b) $\Delta G^\circ_{reaction} = (2\ mol)(-0.88\ kJ\cdot mol^{-1}) - (1\ mol)(3.14\ kJ\cdot mol^{-1}) - (1\ mol)(0.00\ kJ\cdot mol^{-1})$

$\Delta G^\circ_{reaction} = -4.90\ kJ$

$$Q = \left\{ \frac{(0.240\ bar/1\ bar)^2}{(0.120\ bar/1\ bar)(0.120\ bar/1\ bar)} \right\} = 4.00$$

$\Delta G_{reaction} = \Delta G^\circ_{reaction} + RT\ln Q = -4.90\ kJ + (8.314\ J\cdot mol^{-1}\cdot K^{-1})(298\ K)\ln(4.00)$

$\Delta G_{reaction} = -4.90\ kJ + 3.43\ kJ = -1.47\ kJ$

Since $\Delta G_{reaction}$ is negative, the reaction is spontaneous (can occur) under the specified conditions.

Does the answer make sense? Without further information about this reaction (or other reactions for comparison) we cannot determine if the reaction is actually spontaneous under these conditions, although the existence of interhalogen compounds shows us that it *can* be.

9.60. Check This   *Reaction quotient and free energy for a gas phase reaction*

Consider reaction (9.45) with the same $Br_{2(g)}$ and $Cl_{2(g)}$ pressures as in Worked Example 9.59 and a $BrCl_{(g)}$ pressure of 0.500 bar. Is the reaction spontaneous under these conditions? Why or why not?

Standard free energy of reaction and the equilibrium constant

Recall that a reaction is in equilibrium, if $\Delta G_{reaction} = 0$ at the temperature and pressure in the system. Under these equilibrium conditions, we can rewrite equations (9.42) and (9.44) as:

$$\Delta G_{reaction} = 0 = \Delta G^\circ_{reaction} + RT\ln\left\{ \frac{(C)^c(D)^d}{(A)^a(B)^b} \right\}_{eq} = \Delta G^\circ_{reaction} + RT\ln Q_{eq} \qquad (9.46)$$

As you can see, the equilibrium value of the reaction quotient, Q_{eq}, is just the equilibrium constant expression:

$$Q_{eq} = \left\{ \frac{(C)^c(D)^d}{(A)^a(B)^b} \right\}_{eq} = K \qquad (9.47)$$

$$\Delta G^\circ_{reaction} = -RT\ln K \qquad (9.48)$$

Equation (9.48) provides an enormously useful relationship between the thermodynamics of a reaction and the observable composition of the chemical system at equilibrium. It is valid for any reaction, as long as we express the concentrations of the reactants and products as concentration ratios that are consistent with the standard states listed in Table 9.1, Section 9.2. If we know $\Delta G^{\circ}_{reaction}$ for a reaction, we can calculate the equilibrium constant for the reaction without doing any experimental measurements on the reaction system. One way to get $\Delta G^{\circ}_{reaction}$ is to calculate it from the standard free energies of formation in Appendix XX. Conversely, if we measure the equilibrium constant for a reaction, we can calculate $\Delta G^{\circ}_{reaction}$ without doing any calorimetric measurements. If we do calorimetric measurements and/or measure the temperature dependence of the equilibrium constant, we can also obtain $\Delta H^{\circ}_{reaction}$ and $\Delta S^{\circ}_{reaction}$. We will explore these possibilities in the rest of the chapter.

9.61. Investigate This   *How much urea will dissolve in water?*

Use a thin-stem plastic pipet and two capped, graduated conical plastic tubes containing, respectively, 4.0 g of urea, $H_2NCONH_2(s)$, and water. Read and record the volume of water to ± 0.1 mL. Use the pipet to transfer 4.0 mL of water from the water-containing tube to the tube containing urea, cap the tube, and invert several times to dissolve as much urea as possible. Add water one drop at a time to the urea solution, with thorough mixing between additions, until the urea is *just* dissolved. Return all unused water from the pipet to the tube containing pure water. Measure and record the volume of liquid in each tube to ± 0.1 mL.

9.62. Consider This   *What is the solubility of urea in water?*

(a) What is the volume of your just-saturated urea solution in Investigate This 9.61? What volume of water did you use to make this solution?

(b) How many moles of urea are dissolved in this just-saturated urea solution? What is the molarity of the urea, $[H_2NCONH_2(aq)]$, in the solution?

(c) How many moles of water are in this just-saturated urea solution? (Assume that the density of water is 1.0 $g \cdot mL^{-1}$.) What is the molarity of water, $[H_2O(aq)]$, in the solution?

The dissolution reaction for urea, Investigate This 9.61, is:

$$H_2NCONH_2(s) + H_2O(aq) \rightleftharpoons H_2NCONH_2(aq) \tag{9.49}$$

The equilibrium constant expression for the dissolution is:

$$K_{diss} = \left\{ \frac{(H_2NCONH_2(aq))}{(H_2NCONH_2(s))(H_2O(aq))} \right\}_{eq} \tag{9.50}$$

Since the urea is a pure solid, we know that $(H_2NCONH_2(s)) = 1$. In Consider This 9.62(b), you found the molar concentration of urea, $[H_2NCONH_2(aq)]$, and you know the standard state of a solute is 1 M, so you can calculate $(H_2NCONH_2(aq))$. A saturated solution of urea in water is far from dilute, so we have to take into account the molarity of water, $[H_2O(aq)]$, Consider This 9.62(c), in the solution to determine $(H_2O(aq))$:

$$(H_2O(aq)) = \left(\frac{[H_2O(aq)]}{55.5 \text{ M}} \right) \tag{9.51}$$

9.63. Consider This *What are $\Delta G°$, $\Delta H°$, and $\Delta S°$ for dissolution of urea in water?*

(a) Use your results from Consider This 9.62(b) and 9.62(c) to calculate K_{diss} for the dissolution of urea in water. Explain your approach clearly.

(b) What is $\Delta G°_{reaction}$ for the dissolution of urea in water? Explain your reasoning.

(c) In Chapter 7, Investigate This 7.22, you dissolved 6.0 g of urea (0.10 mol) in 100. mL of water in a Styrofoam® cup and analyzed the results in Consider This 7.23 and Check This 7.27 and 7.29. Is the dissolution reaction exothermic or endothermic? What is $\Delta H°_{reaction}$ for this dissolution reaction? Explain clearly how you get your answer.

(d) What is $\Delta S°_{reaction}$ for the dissolution of urea in water? Explain how you get your answer.

The reasoning and calculations you did in Consider This 9.63 are applications of two important thermodynamic equations:

$$\Delta G°_{reaction} = -RT\ln K \tag{9.48}$$

$$\Delta G°_{reastion} = \Delta H°_{reaction} - T\Delta S°_{reastion} \quad \text{(from Chapter 8)} \tag{9.52}$$

The example shows how an equilibrium constant measurement and a calorimetric measurement for a reaction provide enough data to obtain the standard state changes for the free energy, enthalpy, and entropy of the reaction. In the next section we will see how measurements of the equilibrium constant at two or more different temperatures can give the same information.

Section 9.8.   Temperature Dependence of the Equilibrium Constant

9.64.   **Investigate This**   *How does temperature affect the solubility of $PbI_2(s)$?*

Do this as a class investigation and work in small groups to discuss and analyze the results. Use a magnetic stir bar and a magnetic stirrer-hot plate combination to stir 100 mL of ice-cold water in a 250-mL beaker. Add 5 mL of a 1 M aqueous lead nitrate, $Pb(NO_3)_2$, solution and then 2 mL of ice-cold 0.05 M aqueous potassium iodide, KI, solution to the stirred water. WARNING: All lead compounds are toxic. Wear disposable gloves when handling the solutions. Watch for the appearance of a precipitate of lead iodide, $PbI_2(s)$. When a precipitate has formed, turn on the hot plate, heat the solution to about 70 °C, and record your observations. Remove the beaker from the heat, allow the solution to cool, and record your observations.

9.65.   **Consider This**   *Is the dissolution of $PbI_2(s)$ exothermic or endothermic?*

What did you observe in Investigate This 9.64 when you heated the mixture containing $PbI_2(s)$? when you allowed the heated solution to cool? Use your observations and Le Chatelier's principle to explain clearly whether $PbI_2(s)$ dissolution is exothermic or endothermic.

$$PbI_2(s) \rightleftharpoons Pb^{2+}(aq) + 2I^-(aq) \tag{9.53}$$

In Consider This 9.65, you used Le Chatelier's principle to analyze your experimental observations on the temperature dependence of the solubility of $PbI_2(s)$. We can make the analysis more quantitative by using the temperature dependence of the equilibrium constant. To get this temperature dependence, we combine equations (9.48) and (9.52), two equations for $\Delta G^\circ_{reaction}$, to give:

$$\Delta G^\circ_{reaction} = -RT\ln K = \Delta H^\circ_{reaction} - T\Delta S^\circ_{reaction} \tag{9.54}$$

We rearrange the right-hand equality of equation (9.54) to solve for $\ln K$:

$$\ln K = -\left(\frac{\Delta H^\circ_{reaction}}{R}\right)\frac{1}{T} + \frac{\Delta S^\circ_{reaction}}{R} \tag{9.55}$$

Although $\Delta H^\circ_{reaction}$ and $\Delta S^\circ_{reaction}$ vary with temperature, their variation is not large, so we will assume that they are constant. At two different temperatures T_1 and T_2, with equilibrium constants K_1 and K_2, respectively, we have:

$$\ln K_1 = -\left(\frac{\Delta H^{\circ}_{reaction}}{R}\right)\frac{1}{T_1} + \frac{\Delta S^{\circ}_{reaction}}{R} \qquad (9.56)$$

$$\ln K_2 = -\left(\frac{\Delta H^{\circ}_{reaction}}{R}\right)\frac{1}{T_2} + \frac{\Delta S^{\circ}_{reaction}}{R} \qquad (9.57)$$

Subtracting equation (9.56) from (9.57) gives:

$$\ln K_2 - \ln K_1 = \ln\left(\frac{K_2}{K_1}\right) = -\left(\frac{\Delta H^{\circ}_{reaction}}{R}\right)\frac{1}{T_2} + \left(\frac{\Delta H^{\circ}_{reaction}}{R}\right)\frac{1}{T_1} = -\left(\frac{\Delta H^{\circ}_{reaction}}{R}\right)\left(\frac{1}{T_2} - \frac{1}{T_1}\right) \qquad (9.58)$$

$$\ln\left(\frac{K_2}{K_1}\right) = \left(\frac{\Delta H^{\circ}_{reaction}}{R}\right)\left(\frac{T_2 - T_1}{T_2 \cdot T_1}\right) \qquad (9.59)$$

9.66.   Check This   *Rearranging equation (9.58)*

Show how equation (9.59) is obtained from equation (9.58).

9.67.   Worked Example   *$\Delta H^{\circ}_{reaction}$ from solubility temperature dependence*

One handbook gives the aqueous solubility of $PbI_2(s)$ as .44 g·L^{-1} at 0 °C and 4.1 g·L^{-1} at 100. °C. Use these data to find K_{sp} at both temperatures and then to determine $\Delta H^{\circ}_{reaction}$.

Necessary information: We need the molar mass of PbI_2, 461 g·mol^{-1}, and the relationship of the solubility product to the molar solubility, s, of $PbI_2(s)$: $K_{sp} = 4s^3$.

Strategy: Use the two values of s to get K_{sp} at 0 °C (273 K) and 100. °C (373 K) and then substitute into equation (9.59) and solve for $\Delta H^{\circ}_{reaction}$.

Implementation:

$$\text{at 273 K: } K_{sp}(273\ K) = 4\left(\frac{0.44\ \text{g·L}^{-1}\Big/461\ \text{g·mol}^{-1}}{1\ M}\right)^3 = 3.5 \times 10^{-9}$$

$$\text{at 373 K: } K_{sp}(373\ K) = 4\left(\frac{4.1\ \text{g·L}^{-1}\Big/461\ \text{g·mol}^{-1}}{1\ M}\right)^3 = 2.8 \times 10^{-6}$$

$$\ln\left(\frac{K_2}{K_1}\right) = \ln\left(\frac{2.8 \times 10^{-6}}{3.5 \times 10^{-9}}\right) = \left(\frac{\Delta H^{\circ}_{reaction}}{R}\right)\left(\frac{T_2 - T_1}{T_2 \cdot T_1}\right)$$

$$= \left(\frac{\Delta H^{\circ}_{reaction}}{8.314\ \text{J·mol}^{-1}\cdot\text{K}^{-1}}\right)\left(\frac{(373\ K) - (273\ K)}{(373\ K)(273\ K)}\right)$$

$$\Delta H^\circ_{\text{reaction}} = (8.314 \text{ J·mol}^{-1}\text{·K}^{-1})\left(\frac{(373 \text{ K})\cdot(273 \text{ K})}{(373 \text{ K}) - (273 \text{ K})}\right)\cdot\ln\left(\frac{2.8\times10^{-6}}{3.5\times10^{-9}}\right) = 56 \text{ kJ·mol}^{-1}$$

This is the standard enthalpy change for dissolving one mole of $PbI_2(s)$.

Does the answer make sense? The dissolution is endothermic, as you reasoned in Consider This 9.65 from the observed increasing solubility with increasing temperature. Note that the solubilities given in the problem statement confirm that the solubility of $PbI_2(s)$ increases with temperature. We have another check on our answer, the data for enthalpies of formation in Appendix XX. We calculate $\Delta H^\circ_{\text{reaction}}$ for one mole of $PbCl_2(s)$ dissolving:

$$\Delta H^\circ_{\text{reaction}} = (1 \text{ mol})\cdot\Delta H^\circ_{\text{f}}(Pb^{2+}(aq)) + (2 \text{ mol})\cdot\Delta H^\circ_{\text{f}}(I^-(aq)) - (1 \text{ mol})\cdot\Delta H^\circ_{\text{f}}(PbI_2(s))$$

$$\Delta H^\circ_{\text{reaction}} = (1.7 \text{ kJ}) + (-110.3 \text{ kJ}) - (-175.1 \text{ kJ}) = 67 \text{ kJ}$$

The $\Delta H^\circ_{\text{reaction}}$ we got from solubilities and the $\Delta H^\circ_{\text{reaction}}$ calculated from table values agree within about 20%. Considering the problems with solubilities that we have pointed out, this is reasonable agreement. Both values show that the reaction is endothermic. Note that all the reactants and products must be included in the calculation, even though they do not all appear in the solubility product expression. This is another reminder that the solubility product expression is only valid when the solution is in equilibrium with the solid.

9.68. Check This $\Delta H^\circ_{\text{reaction}}$ *from solubility temperature dependence*

(a) One handbook gives the solubility of $Ca(OH)_2(s)$ as 1.85 g·L^{-1} at 0 °C and 0.77 g·L^{-1} at 100. °C. Use these data and Le Chatelier's principle to predict the sign of $\Delta H^\circ_{\text{reaction}}$ for the dissolution of $Ca(OH)_2(s)$. Explain your reasoning clearly.

(b) Use the data in part (a) to find K_{sp} at both temperatures and then to determine $\Delta H^\circ_{\text{reaction}}$. Is the sign of $\Delta H^\circ_{\text{reaction}}$ consistent with your answer in part (a)? Explain why or why not.

(c) Check your result in part (b) using the values for standard enthalpies of formation in Appendix XX. Do your two values for $\Delta H^\circ_{\text{reaction}}$ agree reasonably well? Explain.

As you see, measurements of the equilibrium constant for a reaction at two different temperatures provide enough information to obtain $\Delta H^\circ_{\text{reaction}}$ for the reaction. This value can then be combined with one (or both) of the equilibrium constant values to get $\Delta S^\circ_{\text{reaction}}$, as well as $\Delta G^\circ_{\text{reaction}}$ and the equilibrium constant at other temperatures.

9.69. Worked Example $\Delta S^o_{reaction}$ *and* $\Delta G^o_{reaction}$ *from solubility temperature dependence*

Use the results from Worked Example 9.67 to get $\Delta S^o_{reaction}$ for the dissolution of $PbI_2(s)$ in water. What are $\Delta G^o_{reaction}$ and K_{sp} for this reaction at 298 K (25 °C)?

Necessary information: We'll need all the equalities in expression (9.54) and, from Worked Example 9.67, $K_{sp} = 3.5 \times 10^{-9}$ at 273 K and $\Delta H^o_{reaction} = 56 \times 10^3$ J·mol^{-1}. We also need to assume that $\Delta H^o_{reaction}$ and $\Delta S^o_{reaction}$ do not vary with temperature.

Strategy: Substitute the values for K_{sp} and $\Delta H^o_{reaction}$ into expression (9.54) and solve for $\Delta S^o_{reaction}$. Then substitute $\Delta H^o_{reaction}$ and $\Delta S^o_{reaction}$ into expression (9.54) to get $\Delta G^o_{reaction}$ at 298 K. Finally, substitute $\Delta G^o_{reaction}$ at 298 K into expression (9.54) to get K_{sp} at 298 K.

Implementation:

$$-RT\ln K_{sp} = \Delta H^o_{reaction} - T\Delta S^o_{reaction}$$

$$\Delta S^o_{reaction} = R\ln K_{sp} + \frac{\Delta H^o_{reaction}}{T} = (8.314 \text{ J·mol}^{-1}\text{·K}^{-1})\cdot\ln(3.5 \times 10^{-9}) + \frac{56 \times 10^3 \text{ J·mol}^{-1}}{273 \text{ K}}$$

$$\Delta S^o_{reaction} = -162 \text{ J·mol}^{-1}\text{·K}^{-1} + 205 \text{ J·mol}^{-1}\text{·K}^{-1} = 43 \text{ J·mol}^{-1}\text{·K}^{-1}$$

For $\Delta G^o_{reaction}$ and K_{sp} at 298 K, we have:

$$\Delta G^o_{reaction} = \Delta H^o_{reaction} - T\Delta S^o_{reaction} = (56 \times 10^3 \text{ J·mol}^{-1}) - (298 \text{ K})\cdot(43 \text{ J·mol}^{-1}\text{·K}^{-1})$$

$$\Delta G^o_{reaction} = 43 \times 10^3 \text{ J·mol}^{-1} = 43 \text{ kJ·mol}^{-1}$$

$$\Delta G^o_{reaction} = -RT\ln K_{sp}$$

$$\ln K_{sp} = -\frac{\Delta G^o_{reaction}}{RT} = -\left(\frac{43 \times 10^3 \text{ J·mol}^{-1}}{(8.314 \text{ J·mol}^{-1}\text{·K})\cdot(298 \text{ K})}\right) = -17.4$$

$$K_{sp} = 2.9 \times 10^{-8}$$

Does the answer make sense? The K_{sp} at 298 K is intermediate between the values at 273 K and 373 K (from Worked Example 9.67) and closer to the value at 273 K, as we would expect. Our calculated solubility product is about three times larger than the value in Table 9.4, which is good agreement. We can check the thermodynamic values by calculating them from the data in Appendix XX. We get 67 J·mol^{-1}·K^{-1} and 44 kJ·mol^{-1}, respectively, for $\Delta S^o_{reaction}$ and $\Delta G^o_{reaction}$. The standard entropy changes differ somewhat, but both show that the change is positive. The values for the standard free energy change are the same.

9.70. Check This $\Delta S^o_{reaction}$ *and* $\Delta G^o_{reaction}$ *from solubility temperature dependence*

(a) Use your results from Check This 9.68 to get $\Delta S^o_{reaction}$ for the dissolution of $Ca(OH)_2(s)$ in water. What are $\Delta G^o_{reaction}$ and K_{sp} for this reaction at 298 K (25 °C)?

(b) Check your result in part (a), using the values for standard entropies and free energies of formation in Appendix XX. Do the two values for $\Delta S^{\circ}_{reaction}$ and $\Delta G^{\circ}_{reaction}$ agree reasonably well? Explain. How does your value for K_{sp} compare with the value in Table 9.2?

Graphical determination of thermodynamic values

If you measure the equilibrium constant for a reaction at several temperatures, you can plot the data to obtain $\Delta H^{\circ}_{reaction}$ and $\Delta S^{\circ}_{reaction}$, and, hence, $\Delta G^{\circ}_{reaction}$. Equation (9.55) is the equation of a straight line if $\ln K$ is plotted as a function of $\frac{1}{T}$. The slope of the line is $-\left(\dfrac{\Delta H^{\circ}_{reaction}}{R}\right)$ and the intercept (at $\frac{1}{T} = 0$) is $\dfrac{\Delta S^{\circ}_{reaction}}{R}$. As an example, consider the temperature dependence of the equilibrium pressure of a gas, A, in equilibrium with liquid A. The vaporization reaction equation and equilibrium constant expression are:

$$A(l) \rightleftharpoons A(g) \tag{9.60}$$

$$K = \frac{\left(P_{A(g)}\right)}{\left(A(l)\right)} = \left(P_{A(g)}\right) \tag{9.61}$$

The equilibrium constant for vaporization is the numeric value of the equilibrium vapor pressure (in bar). Table 9.5 gives the vapor pressures as a function of temperature for two isomeric liquids.

Table 9.5. Vapor pressure variation with temperature for two $C_2H_2Cl_2$ isomers.

vapor pressure, torr	1.0	10.0	40.0	100.	400.	760.
	temperature, °C					
cis-1,2-dichloroethene	−58.4	−29.9	−7.9	9.5	41.0	59.0
trans-1,2-dichloroethene		−38.0	−17.0	−0.2	30.8	47.8

9.71. **Worked Example**   *Graphical determination of $\Delta H^{\circ}_{reaction}$ and $\Delta S^{\circ}_{reaction}$*

Plot the data in Table 9.5 and use the plot to determine $\Delta H^{\circ}_{reaction}$, $\Delta S^{\circ}_{reaction}$, and the vapor pressure at 298 K for the vaporization of *cis*-1,2-dichloroethene.

Necessary information: We need the equilibrium constant expression (9.61) and equation (9.55). We also need to know that 1 bar = 750 torr and temperature in K = °C + 273.1.

Strategy: Convert vapor pressures (v.p.) in torr to bar and temperatures in °C to K. Then calculate and plot $\ln K$ [$= \ln\{(v.p.)/1\ bar\}$] as a function of $\frac{1}{T}$, determine the slope and intercept of the line through the points, and use these to calculate $\Delta H^\circ_{reaction}$ and $\Delta S^\circ_{reaction}$.

Implementation: This table shows the converted data which are plotted in Figure 9.11.

v.p., torr	1.0	10.0	40.0	100.	400.	760.
v.p., bar	0.0013	0.0133	0.0533	0.133	0.533	1.013
$\ln K$	−6.65	−4.320	−2.932	−2.017	−0.629	0.0129
T, K	214.7	243.2	265.2	282.6	314.1	332.1
$\frac{1}{T}$, K^{-1}	0.00466	0.00411	0.00377	0.00354	0.00318	0.00301

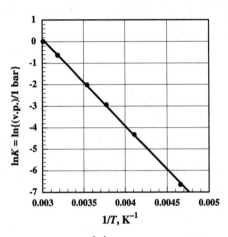

Figure 9.11. Plot of $\ln(v.p.)$ vs. $\frac{1}{T}$ for *cis*-1,2-dichloroethene.

The equation of the line (from the computer graphing software) in Figure 9.11 is:

$$\ln K = (-4.03 \times 10^3\ K)\left(\frac{1}{T}\right) + 12.22$$

Therefore, we get:

$$\text{slope} = -\left(\frac{\Delta H^\circ_{reaction}}{R}\right) = (-4.03 \times 10^3\ K)$$

$$\Delta H^\circ_{reaction} = (8.314\ J \cdot mol^{-1} \cdot K^{-1}) \cdot (4.03 \times 10^3\ K) = 33.5 \times 10^3\ J \cdot mol^{-1} = 33.5\ kJ \cdot mol^{-1}$$

$$\text{intercept} = \frac{\Delta S^\circ_{reaction}}{R} = 12.22$$

$$\Delta S^\circ_{reaction} = (8.314\ J \cdot mol^{-1} \cdot K^{-1}) \cdot (12.22) = 102\ J \cdot mol^{-1} \cdot K^{-1}$$

To get the vapor pressure at 298 K, we substitute in equation (9.55):

$$\ln K = \ln(v.p.) = -\left(\frac{\Delta H^\circ_{reaction}}{R}\right)\left(\frac{1}{T}\right) + \frac{\Delta S^\circ_{reaction}}{R} = (-4.03 \times 10^3\ K)\left(\frac{1}{298\ K}\right) + 12.22$$

$\ln(\text{v.p.}) = -1.303; \therefore$ vapor pressure at 298 K = 0.272 bar (= 204 torr)

Does the answer make sense? We have known since Chapter 1 that energy is required to vaporize a liquid and in Chapter 8 we learned that the entropy of a substance increases when it goes from the liquid to the gas phase. Therefore, the positive signs for $\Delta H^{\circ}_{reaction}$ and $\Delta S^{\circ}_{reaction}$ make sense. To judge whether the sizes of the changes make sense, you can compare them with other simple carbon compounds (such as methanol or ethanol) in Appendix XX and find that the values are fairly similar. Since 298 K (25 °C) lies between two of the temperatures in Table 9.5, we expect the vapor pressure to lie between the corresponding vapor pressures, as it does.

9.72.   **Check This**   *Graphical determination of $\Delta H^{\circ}_{reaction}$ and $\Delta S^{\circ}_{reaction}$*

(a) Plot the data in Table 9.5 and use the plot to determine $\Delta H^{\circ}_{reaction}$, $\Delta S^{\circ}_{reaction}$, and the vapor pressure at 298 K for the vaporization of *trans*-1,2-dichloroethene.

(b) How do your results in part (a) compare with the values for *cis*-1,2-dichloroethene from Worked Example 9.71? Are the *directions* of any differences consistent with the structures of the two isomers? Explain why or why not. *Hint:* Consider the polarites of the isomers.

Reflection and projection

The free energy change for a reaction can be expressed in terms of the standard free energy change and a reaction quotient that combines the concentration ratios of all the reactants and products. At equilibrium, where the free energy change is zero, the numeric value of the equilibrium reaction quotient is the equilibrium constant for the reaction. The equilibrium constant is directly related to the standard free energy change, which can be calculated from tabulated standard free energies of formation. We can apply these concepts (developed for gas phase equilibria in Section 9.11) to all reaction systems, if we are careful about how we define and use standard conditions.

Measuring the equilibrium constant for a reaction at two or more temperatures provides enough information to determine the standard enthalpy and entropy changes for the reaction. These values can be combined to find the standard free energy change and equilibrium constant for the reaction at other temperatures. The interrelationships of the equilibrium constant and thermodynamic variables provides us with numerous ways to analyze and understand equilibria, including those in biological systems, as we will discuss in the next section.

Section 9.9.   Thermodynamics in Living Systems

In a living cell, Figure 9.12, almost no reaction is at equilibrium. This is as it must be. A cell at equilibrium is unchanging, that is, dead. Only systems that are not in equilibrium can change. One reason why reactions in living cells are not in equilibrium is that many of the reactions are relatively slow, even though there are enzymes present to speed them up. Another reason is that so many competing reactions are going on that a molecule needed for one reaction might be used up by another before it can react by the first pathway.

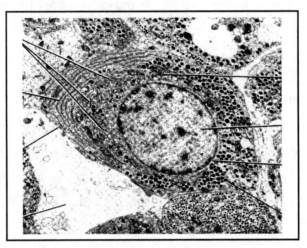

Figure 9.12. Electron micrograph of a thin section of a hormone-secreting cell.
A large number of chemical reactions go on continuously in all of the organelles, in an intact cell.

Free energy for cellular reactions

You might wonder why, if equilibria are not established for most reactions in living systems, we at all interested in the equilibria for these reactions. We wish to find the free energies of reactions in living cells, because these help us understand the directionality of these life processes. One approach is to carry out the reaction of interest in a test tube in the laboratory, allow it to come to equilibrium, determine its equilibrium constant, and thus get the standard free energy change for the reaction. We can then use this standard free energy, together with the reaction quotient under cellular conditions, to get the free energy change for the cellular reaction that is not at equilibrium.

Hydrolysis of ATP

As an example of this approach, consider a reaction we have discussed before, the hydrolysis of adenosine triphosphate, ATP:

$$ATP^{4-}(aq) + 2H_2O(l) \rightleftharpoons ADP^{3-}(aq) + HOPO_3^{2-}(aq) + H_3O^+(aq) \qquad (9.62)$$

The structures of ATP^{4-} and ADP^{3-} are shown in Chapter 7, Figure 7.16. For reaction (9.62), the reaction quotient expressed in terms of molar concentrations is:

$$Q = \frac{\left[ADP^{3-}(aq)\right]\left[HOPO_3^{2-}(aq)\right]\left[H_3O^+(aq)\right]}{\left[ATP^{4-}(aq)\right]} \tag{9.63}$$

We know that, at equilibrium, the numerical value of Q calculated from the concentrations in the solution is equal to the numeric value of K, the equilibrium constant. Unfortunately, these concentrations are difficult to obtain.

We encounter several problems trying to get the concentrations to substitute in equation (9.63). The pH in living cells is close to 7. Look at Table 9.2 and you will find that the conjugate acid-base pair, $(HO)_2PO_2^-(aq)$ and $HOPO_3^{2-}(aq)$, have a $pK_a = 7.20$. The conjugate base-to-acid ratio depends on the pH of the solution, so the concentration of $HOPO_3^{2-}(aq)$ depends on the pH of the solution. Exactly this same problem also occurs for both $ATP^{4-}(aq)$ and $ADP^{3-}(aq)$. The proton transfer to water from the end phosphate in each molecule also has a pK_a around 7. Therefore,

> Another complication in studying ATP and ADP is the dependence of their reactions, on the presence of Mg^{2+}. ATP and ADP form complexes with Mg^{2+} and it is these complexes that actually take part in enzyme-catalyzed reactions. Laboratory studies are usually carried out with added Mg^{2+} to mimic the conditions for ATP and ADP in a living cell.

the phosphate-containing compounds in reaction (9.62) are each present in at least two ionic forms and we don't know which ones take part in the reaction (or if they all do).

Equilibrium constant for hydrolysis of ATP

Since we lack information about the specific concentrations in these systems, one approach is simply to use the sum of all the forms present in the system. For example, suppose we add 2×10^{-6} moles of ATP to 1.00 milliliter of a reaction mixture. We say that the ATP concentration is 0.002 M and symbolize this total as [ATP], without regard to how much of each of its specific ionic forms is present. P_i is commonly used to symbolize the sum of the concentrations of the various forms of inorganic phosphate $((HO)_2PO_2^-(aq)$, $HOPO_3^{2-}(aq)$, and so on). Experiments (at 298 K) have shown that:

$$K = \left(\frac{[ADP(aq)][P_i(aq)][H_3O^+(aq)]}{[ATP(aq)]}\right)_{eq} \approx 10^{-2} \tag{9.64}$$

Biochemists usually take solutions at pH 7 as their reference standard (instead of pH 0). If we divide through equation (9.72) by the hydronium ion concentration ($= 10^{-7}$ M), we get:

$$\frac{K}{[H_3O^+(aq)]} = \frac{K}{10^{-7}M} = K' = \left(\frac{[ADP(aq)][P_i(aq)]}{[ATP(aq)]}\right)_{eq} \approx \frac{10^{-2}}{10^{-7}} = 10^5 \tag{9.65}$$

This nomenclature, K' and $\Delta G°'$, for systems with pH 7 as the standard state for hydronium ion will appear again in Chapter 10.

K' is the equilibrium constant for ATP hydrolysis *at pH 7*. The standard free energy change for hydrolysis of a mole of ATP *at pH 7* is:

$$\Delta G°'_{hydrolysis} = -RT \ln K' = -(8.314 \text{ J·mol}^{-1}\text{·K}^{-1})(298 \text{ K}) \ln(10^5) = -29 \text{ kJ·mol}^{-1} \quad (9.66)$$

9.73. Consider This   *What is the equilibrium cellular concentration of ATP?*

If the cellular concentrations of ADP and P_i are about 0.0002 M and 0.001 M, respectively, what ATP concentration is in equilibrium with these concentrations of hydrolysis products? Is your result consistent with the $\Delta G°'_{hydrolysis}$? Explain why or why not.

Free energy changes for cellular ATP reactions

The concentrations of ATP, ADP, and P_i vary from one part of a cell to another, but the concentration of ATP is usually about ten times higher than ADP and the P_i concentration is about 0.001 M. The free energy change for ATP hydrolysis under these conditions is:

$$\Delta G'_{hydrolysis} = \Delta G°'_{hydrolysis} + RT \ln Q' = \Delta G°'_{hydrolysis} + RT \ln \left\{ \left(\frac{[ADP(aq)]}{[ATP(aq)]} \right) [P_i(aq)] \right\}$$

$$\Delta G'_{hydrolysis} = -29 \text{ kJ·mol}^{-1} + (8.314 \text{ J·mol}^{-1}\text{·K}^{-1})(298 \text{ K}) \ln \left\{ \left(\frac{1}{10} \right)(0.001) \right\}$$

$$\Delta G'_{hydrolysis} = -29 \text{ kJ·mol}^{-1} + (-23 \text{ kJ·mol}^{-1}) = -52 \text{ kJ·mol}^{-1} \quad (9.67)$$

Under cellular conditions, the free energy change for ATP hydrolysis is almost twice as large as the standard free energy change. There is a strong driving force for ATP hydrolysis that can be *coupled* with other reactions (which would otherwise not be favored) to make life possible.

In Chapter 7, Section 7.9, we calculated the percentage of the *enthalpy* released in the complete oxidation of glucose that is captured by organisms to synthesize ATP. The percentage of the *free energy* release that is captured is a more appropriate measure of the efficiency of an organism. $\Delta G°_{reaction}$ is –2878.42 kJ for the oxidation of one mole of glucose:

$$C_6H_{12}O_6(s) + 6O_2(g) \rightarrow 6CO_2(g) + 6H_2O(l) \quad (9.68)$$

For the purposes of this approximate calculation, we will assume that $\Delta G_{reaction} \approx \Delta G°_{reaction}$ for the oxidation of glucose.

The synthesis of ATP from ADP and P_i is the reverse of hydrolysis. In the cell, the free energy change for ATP synthesis has the opposite sign but the same numeric value as the free energy change for hydrolysis: $\Delta G'_{synthesis} = 52$ kJ·mol^{-1}. Recall from Section 7.9 that about 36 moles of ATP can be formed for every mole of glucose oxidized. Synthesis of 36 moles of ATP requires about 1870 kJ. Under conditions in the cell, about 65% (= $\left[1870/2880\right] \times 100\%$) of the free energy released in the oxidation of glucose can be captured in the synthesis of ATP. The free energy available from glucose oxidation is used efficiently by our cells to maintain the ATP hydrolysis reaction far from equilibrium and, thus, provide the strong directional drive for life processes.

9.74. **Consider This**   *How do you analyze a coupled reaction?*

We can rewrite equation (9.62), *at pH 7*, as:

$$ATP(aq) + H_2O(l) \rightleftharpoons ADP(aq) + P_i(aq) \qquad \Delta G^{\circ}{}'_{(9.69)} = -RT\ln K' = -29 \text{ kJ·mol}^{-1} \qquad (9.69)$$

Many of the compounds used and formed in metabolic pathways are phosphate esters. An example is glycerol phosphate, $HOCH_2CHOHCH_2OPO_3^{2-}$ (abbreviated as R-OPO$_3^{2-}$). These esters hydrolyze to the alcohol and phosphate; R-OPO$_3^{2-}$ hydrolyzes to glycerol, R-OH:

$$R\text{-}OPO_3^{2-}(aq) + H_2O(l) \rightleftharpoons R\text{-}OH(aq) + P_i(aq) \qquad \Delta G^{\circ}{}'_{(9.70)} = -9 \text{ kJ·mol}^{-1} \qquad (9.70)$$

Our bodies need the phosphate ester, but the equilibrium favors the alcohol. ATP can transfer a phosphate group to the alcohol to make the phosphate ester:

$$R\text{-}OH(aq) + ATP(aq) \rightleftharpoons R\text{-}OPO_3^{2-}(aq) + ADP(aq) \qquad (9.71)$$

Equation (9.71) represents a direct coupling of two reactions to become one.

(a) What are $\Delta G^{\circ}{}'_{reaction}$ and K' for reaction (9.71)? Show clearly how you get your answers. *Hint:* If necessary, review Section 7.9 or 8.14 to recall how thermodynamic quantities are combined when reactions are coupled.

(b) If reaction (9.71) reaches equilibrium when $\left(\dfrac{[ATP(aq)]}{[ADP(aq)]}\right) = 10$ (conditions in a cell), what is the ratio of R-OPO$_3^{2-}$ to R-OH? Is this ratio advantageous to the cell? Explain your reasoning.

Your result in Consider This 9.74(b) demonstrates how the free energy released by ATP hydrolysis can be coupled to a reaction that requires an input of free energy to displace its equilibrium in a direction that is advantageous to the organism. This free energy is available because the ATP hydrolysis reaction is so far out of equilibrium. The observed concentration of ATP, about 2×10^{-3} M, in your cells is 10^9 times larger than the equilibrium concentration you calculated in Consider This 9.73.

How is the concentration of ATP maintained so far from equilibrium when it is continually hydrolyzed to drive such a large number of other reactions? The answer lies in the continuing supply of fuel molecules entering the cell and the speed of the reactions within the cell. The cell is an open system that exchanges matter, including the molecules that serve as fuels, with its surroundings. The synthesis reactions that use the fuel go fast enough to regenerate ATP as it is depleted, so its concentration is maintained approximately constant and far from equilibrium. The speed (rate) of reactions is the topic of Chapter 11.

Equilibria in oxygen transport

Oxygen is required by your cells to metabolize fuels like glucose, as in equation (9.68). The reaction shown in the chapter opening illustration is the release of the required oxygen from the oxygenated form of hemoglobin as red blood cells pass through your body's capillaries. We pointed out in the introductory discussion that the binding of oxygen to hemoglobin, Hb, is an equilibrium process with an equilibrium constant K_1:

$$Hb + O_{2(g)} \rightleftharpoons Hb(O_2) \qquad\qquad K_1 \qquad\qquad\qquad (9.72)$$

However, oxygen binding to hemoglobin is more complicated than this because four molecules of oxygen can bind to hemoglobin, one to each of the α and β subunits shown in Figure 9.8. Thus, there are three further successive oxygen-binding reactions, each adding an oxygen to the previous hemoglobin-oxygen complex and each with its own equilibrium constant:

$$Hb(O_2) + O_{2(g)} \rightleftharpoons Hb(O_2)_2 \qquad K_2 \qquad\qquad\qquad (9.73)$$

$$Hb(O_2)_2 + O_{2(g)} \rightleftharpoons Hb(O_2)_3 \qquad K_3 \qquad\qquad\qquad (9.74)$$

$$Hb(O_2)_3 + O_{2(g)} \rightleftharpoons Hb(O_2)_4 \qquad K_4 \qquad\qquad\qquad (9.75)$$

The subscripts on the equilibrium constants denote the number of bound oxygens in the product.

One way to characterize the binding of oxygen to hemoglobin is shown in Figure 9.13. This is a plot of the percent oxygen saturation as a function of the pressure of oxygen gas in contact with the protein. For example, an oxygen saturation of 25% means that one-fourth the maximum amount of oxygen is bound or that, on the average, each hemoglobin molecule has one bound oxygen molecule. This is an average, so some hemoglobins will have more than one bound oxygen and some will have none.

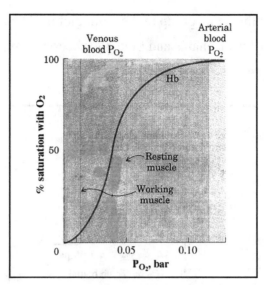

Figure 9.13. Oxygen-saturation curve for hemoglobin.

9.75. Check This   *Transport of oxygen by hemoglobin*

(a) What is the percent oxygen saturation of hemoglobin when it is in the arterial blood (blood that has recently been through the lungs)? in the venous blood (blood that has passed through oxygen-requiring tissues)?

(b) How do your answers in part (a) explain how hemoglobin is able to transport oxygen from the lungs to the tissues where it is required to oxidize the fuel molecules that provide energy for the cells in these tissues?

Note that the oxygen-saturation plot for hemoglobin, Figure 9.13, is shaped sort of like an "s." The oxygen-saturation curve starts up slowly and then the upward slope increases as more oxygen is bound and finally bends over as the system nears saturation. The shape is usually called *sigmoidal*, because the Greek letter sigma, σ, is equivalent to the Roman "s." Such curves appear often in studies of biological systems. Sigmoidal responses like the one for oxygen binding to hemoglobin indicate that there is some kind of cooperation among the reactions that cause the response. This usually means that a reaction and changes in one part of the protein make it easier for reactions to occur in other parts of the protein.

In the case of hemoglobin, the four protein chains in the deoxygenated form, Hb, are held together in a shape that is unfavorable for oxygen to bind to the protein. The chains are held together by both hydrophobic effects and by attractions between oppositely-charged amino acid side groups on the different chains. In order for the first oxygen molecule to bind, several of these interactions have to be disrupted, which requires an input of energy. Energy is released by the binding of the oxygen to the heme group in the hemoglobin. The *net* free energy change for disrupting the interactions between chains and binding successive oxygens is shown on a free energy diagram in Figure 9.14.

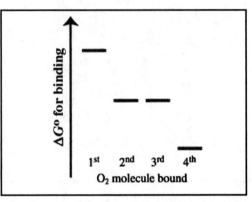

Figure 9.14. Free energy diagram for successive binding of oxygen to hemoglobin.

Figure 9.14 shows that the net free energy change for binding the first oxygen to hemoglobin is more positive (less favorable) than for binding the others. The release of energy by the binding is about the same for each oxygen, but the interactions that must be disrupted are fewer for the second and third oxygen which bind with about the same affinity. Binding of the fourth oxygen can be favorable, if the solution conditions are appropriate.

9.76. Check This   *Relative sizes of oxygen-binding equilibrium constants*

   (a) List the oxygen-binding equilibrium constants for reactions (9.72) through (9.75) in order of increasing size. Clearly explain how you decide on your order.

   (b) Hemoglobin transports oxygen inside blood vessels, but does not actually contact the cells that need the oxygen. The oxygen is transferred (through capillary and cell membranes) to myoglobin. Myoglobin, Mb, is a single protein chain that is similar to the individual chains in hemoglobin. Mb stores oxygen in your cells, especially muscles, and carries it to where the cell needs it. Use the information in Figure 9.15 to explain why oxygen can be transferred from hemoglobin to myoglobin at the low effective oxygen pressures, less than 0.05 bar, in the capillaries.

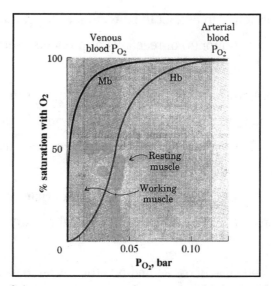

Figure 9.15. Comparison of the oxygen-saturation curves for hemoglobin and myoglobin.

In this chapter, we have limited our discussion to only a few equilibrium systems and the relationships between equilibria and thermodynamics. Even without further elaboration, you can see how important the concept of equilibrium is and can appreciate Le Chatelier's realization that it is central to all of chemistry. We will use equilibrium concepts from this chapter in the final two chapters of the book and you will meet equilibria often in all sciences.

Section 9.10. Outcomes Review

Many chemical systems that appear unchanging have reached a state of equilibrium where the forward and reverse reactions that characterize their chemistry are going at the same speed. We cannot detect any net change by observing these systems, but the reactions continue at the molecular level; equilibrium is dynamic. Le Chatelier's principle states that, if we disturb an equilibrium system by changing one or more concentrations or adding or subtracting energy, the system responds in a way that minimizes the disturbance. The equilibrium constant and equilibrium constant expression provide ways to quantify this qualitative (directional) principle and we applied them to acid-base, solubility, and biochemical reactions.

We connected the equilibrium constant to the thermodynamic variables introduced in the previous two chapters and found that this combination allows us to calculate equilibrium constants from thermodynamic data alone and conversely to use equilibrium measurements to obtain thermodynamic data. A particularly useful result of these connections is the ability to analyze the temperature dependence of equilibria quantitatively.

Check your understanding of the ideas in this chapter by reviewing these expected outcomes of your study. You should be able to:

- use Le Chatelier's principle to predict the direction of the response of an equilibrium system to changes in the concentration(s) of reactants or products [Sections 9.1, 9.3, 9.4, 9.5, 9.6, and 9.9].

- use Le Chatelier's principle and the response of an equilibrium system to heating or cooling to tell whether the reaction is endothermic or exothermic [Sections 9.1 and 9.8].

- explain how Le Chatelier's principle is a consequence of the dynamic nature of chemical equilibrium [Section 9.1].

- write the equilibrium constant expression using appropriate concentration ratios, hence defining the equilibrium constant, K, for a balanced chemical reaction [Sections 9.2, 9.3, 9.6, 9.7, and 9.9].

- find K_a and pK_a for a weak acid (or K_b and pK_b for a weak base) when you have the initial concentrations in the solution and the pH or pOH at equilibrium [Sections 9.3 and 9.4].

- find the pH and/or pOH of a solution of a weak acid or its conjugate base when you have the initial concentrations in the solution and the K_a or pK_a for the acid [Sections 9.3 and 9.4].

- find the concentrations of a weak acid and its conjugate base in a solution when you have the initial concentrations in the solution and the K_a or pK_a for the acid [Sections 9.3, 9.4, and 9.5].

- find the pH of a buffer solution when you have the initial concentrations of the weak acid and its conjugate base in the solution and the K_a or pK_a for the acid [Section 9.4].

- tell how to prepare a buffer solution of a specified pH and specified total concentration of an appropriately selected weak acid-base conjugate pair [Section 9.4].

- predict the direction of change of the net charge on a protein and its consequent behavior in electrophoresis if one amino acid is substituted for another in its structure [Section 9.5].

- find the solubility product, K_{sp} and pK_{sp}, for an ionic compound when you have data for the solubility of the compound and *vice versa* [Sections 9.6 and 9.8].

- find the solubility of an ionic compound of known K_{sp} or pK_{sp} in a solution containing a stoichiometric excess of one of the ions [Section 9.6].

- find $\Delta G^\circ_{reaction}$ for a reaction when you have the equilibrium constant, K, for the reaction [Section 9.7, 9.8, and 9.9].

- find the equilibrium constant, K, for a reaction when you have ΔG°_f for the reactants and products [Sections 9.7].

- find $\Delta H^\circ_{reaction}$, $\Delta G^\circ_{reaction}$, and $\Delta S^\circ_{reaction}$ for a reaction when you have calorimetric data for the reaction and its equilibrium constant, K [Section 9.7].

- find $\Delta H°_{reaction}$, $\Delta S°_{reaction}$, and $\Delta G°_{reaction}$ for a reaction when you have values of K for the reaction at two or more temperatures [Section 9.8].

- find $\Delta G°_{reaction}$ and K for a reaction at a temperature T when you have $\Delta H°_{reaction}$ and $\Delta S°_{reaction}$ for the reaction [Section 9.8].

- use the reaction quotient, Q, to find the free energy change, $\Delta G_{reaction}$, for a reaction that is not at equilibrium when you have the standard free energy change, $\Delta G°_{reaction}$, or K for the reaction and the concentrations in the non-equilibrium system [Sections 9.9 and 9.11].

- find $\Delta G°_{reaction}$ or $\Delta G_{reaction}$ for a coupled reaction when you have $\Delta G°_{reaction}$ or $\Delta G_{reaction}$ for the individual reactions that are coupled [Section 9.9].

- calculate the entropy of gases at non-standard pressures [Section 9.11].

- show how the entropy of gases at non-standard pressures is related to the free energy and reaction quotient fo a gas-phase reaction [Section 9.11].

Section 9.11. EXTENSION – Thermodynamic Basis for the Equilibrium Constant

---

9.77. Consider This   *What are some properties of nitrogen oxides?*

Nitrogen reacts with oxygen to form several oxides. One of the oxides of nitrogen is nitrogen dioxide, NO_2.

(a) Write a Lewis structure for NO_2; nitrogen is the central atom in the molecule. Can all three atoms have an octet of electrons in their valence orbitals? Why or why not?

(b) Do you expect NO_2 to be reactive or unreactive? Why or why not?

(c) Two molecules of NO_2 can react to form N_2O_4, dinitrogen tetroxide:

$$2NO_{2(g)} \rightleftharpoons N_2O_{4(g)} \qquad\qquad (9.76)$$

Write a Lewis structure for N_2O_4.

(d) Do you expect the entropy change for reaction (9.76) to be positive or negative? Give your reasoning.

(e) Do you expect reaction (9.76), formation of a bond between two $NO_{2(g)}$ molecules, to be exothermic or endothermic? Give your reasoning.

Most of the reactions we have discussed in this chapter occur in aqueous solutions. However, we begin our treatment of the relationship between equilibrium and thermodynamic variables with gases, because we can use our model for positional entropy from Chapter 8 to express the entropy of a gas as a function of its pressure. Except when they collide, molecules in the gas phase have little influence on one another. Thus, we do not have to be concerned about solvent effects, which are so important in solution reactions. Our example will be the reaction of $NO_2(g)$ molecules to form $N_2O_4(g)$, reaction (9.76) from Consider This 9.77. This simple reaction has the enormous advantage that one of the gases is a deep reddish-brown color, Figure 9.16, so we can tell a good deal about the reaction system, just by looking at it, as we did with the colored solution reactions in Section 9.1.

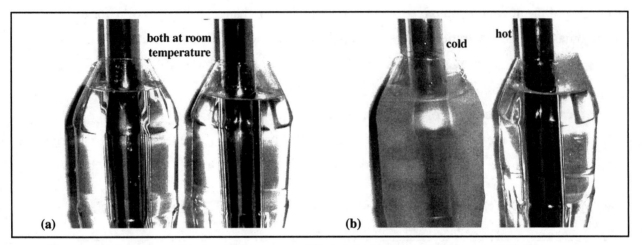

Figure 9.16. Effect of temperature on equilibrium mixtures of $NO_2(g)$ and $N_2O_4(g)$.
In (a), both sample tubes are at room temperature. In (b), one sample tube is in ice-water and the other in hot water.

9.78. Consider This *Which of the gases is colored: $NO_2(g)$ or $N_2O_4(g)$?*

(a) The two sealed tubes shown in Figure 9.16(a) contain identical, equilibrium mixtures of $NO_2(g)$ and $N_2O_4(g)$. When energy is added to this equilibrium system, Figure 9.16(b), is there more or less of the colored gas in the new equilibrium mixture? Give your reasoning clearly.

(b) *Assuming* that the colored gas is $NO_2(g)$, what do Le Chatelier's principle and your answer in part (a) tell you about the energetics of reaction (9.76)? Is the reaction exothermic or endothermic? Clearly explain your answer.

(c) *Assuming* that the colored gas is $N_2O_4(g)$, what do Le Chatelier's principle and your answer in part (a) tell you about energetics of reaction (9.76)? Clearly explain your answer.

(d) Which of your answers, part (b) or part (c), agrees with your answer in Consider This 9.77(e)? Is the colored gas $NO_2(g)$ or $N_2O_4(g)$? Explain your choice.

Changes under non-standard conditions

In Chapter 8, we found that the net entropy change for a reaction under standard conditions (one bar pressure) can be written in terms of the standard entropy and enthalpy changes for the reaction:

$$\Delta S^{\circ}{}_{net} = \Delta S^{\circ}{}_{reaction} - \frac{\Delta H^{\circ}{}_{reaction}}{T} \qquad (9.77)$$

Most reactions do not involve reactants and products in their standard states. We must use $\Delta H_{reaction}$ and $\Delta S_{reaction}$, enthalpy and entropy changes under non-standard conditions, to get ΔS_{net} for the reaction under the actual conditions:

$$\Delta S_{net} = \Delta S_{reaction} - \frac{\Delta H_{reaction}}{T} \qquad (9.78)$$

To analyze a gas phase reaction like $NO_2(g)$ forming $N_2O_4(g)$, we need to know how the enthalpy and entropy of each gas varies with its pressure at constant temperature, that is, $\Delta H_{P\,change}$ and $\Delta S_{P\,change}$ for:

$$gas\ (P^{\circ},\ T) \rightarrow gas\ (P,\ T) \qquad (9.79)$$

We will assume that the gases are ideal, Chapter 7, Section 7.13. The enthalpy of an ideal gas, like its total energy, is a function only of temperature, so $\Delta H_{P\,change} = 0$ for the constant temperature process in equation (9.79). Thus,

> The enthalpy of *ideal* gases is a function only of temperature; real gases can usually be treated as ideal, if they are not at high pressure.

$\Delta H_{reaction} = \Delta H^{\circ}{}_{reaction}$. Since the reaction is formation of a chemical bond, you know that the reaction must be exothermic; the product is at a lower energy than the reactants.

9.79. Check This $\Delta H^{\circ}{}_{reaction}$ for $2NO_2(g) \rightleftharpoons N_2O_4(g)$

Use the standard enthalpies of formation, $\Delta H^{\circ}{}_{f}$, in Appendix XX to calculate $\Delta H^{\circ}{}_{reaction}$ for reaction (9.76). Does your answer confirm your reasoning in Consider This 9.77(e) and 9.78(d)? Explain why or why not.

The entropy of a gas does depend upon its pressure. In Chapter 8, Section 8.3, we found that the number of arrangements of molecules in space, W, increases as the volume per molecule (or total volume, V) increases. The relationships among W, V, and P $(= nRT/V)$ for a gas are:

$$W \propto V \propto \frac{1}{P} \qquad (9.80)$$

The positional entropy change, $\Delta S_{P\,change}$, for one mole of gas going from standard pressure, $P^{\circ} = 1$ bar with a volume V°, to a pressure P with a volume V, reaction (9.79), is:

$$\Delta S_{P\,change} = R\ln W_V - R\ln W_{V^{\circ}} = R\ln\left(\frac{W_V}{W_{V^{\circ}}}\right) = R\ln\left(\frac{V}{V^{\circ}}\right) = R\ln\left(\frac{P^{\circ}}{P}\right) \qquad (9.81)$$

At a pressure P, the entropy of a gas, S_{gas}, is its standard entropy, $S^{\circ}{}_{gas}$, plus the entropy change for reaction (9.79), $\Delta S_{P\,change}$:

$$S_{gas} = S^o_{gas} + \Delta S_{P\,change} = S^o_{gas} + R\ln\left(\frac{P^o}{P}\right) = S^o_{gas} - R\ln\left(\frac{P}{P^o}\right) \qquad (9.82)$$

Recall from Section 9.2, Table 9.1, that (P/P^o) is the dimensionless ratio of pressures (in bar) we use in thermodynamics, so that the logarithmic terms we meet, as here, are mathematically correct. Since $P^o = 1$ bar, the numeric value of (P/P^o) is (P). We write the pressure in parentheses to remind ourselves that it represents the ratio. Using this nomenclature equation (9.82) becomes:

$$S_{gas} = S^o_{gas} - R\ln(P) \qquad (9.83)$$

9.80. Consider This   *What is the entropy of a compressed gas?*

(a) The standard entropy of carbon dioxide gas (at 298 K) is 213.7 $J\cdot K^{-1}\cdot mol^{-1}$. If carbon dioxide is compressed to a pressure of 2.8 bar, what is the entropy of one mole of the gas? Does the direction of the change from the standard entropy make sense? Explain why or why not.

(b) If the gas is expanded to a pressure less than one bar, does its entropy increase or decrease relative to its standard entropy? Explain the reasoning for your answer.

Entropy change for a gas phase reaction

Let's determine the entropy change for the reaction of two moles of $NO_2(g)$ at a pressure P_{NO_2} to give one mole of $N_2O_4(g)$ at a pressure $P_{N_2O_4}$. We will use equation (9.83) to find the entropy for each gas at these pressures and then combine the entropies to get, $\Delta S_{reaction}$ for reaction (9.76):

$$\Delta S_{reaction} = S_{N_2O_4} - 2\,S_{NO_2} = \{S^o_{N_2O_4} - R\ln(P_{N_2O_4})\} - 2\{S^o_{N_2O_4} - R\ln(P_{NO_2})\} \qquad (9.84)$$

$$\Delta S_{reaction} = \Delta S^o_{reaction} - R\{\ln(P_{N_2O_4}) - \ln(P_{NO_2})^2\}$$

$$\Delta S_{reaction} = \Delta S^o_{reaction} - R\ln\left\{\frac{(P_{N2O4})}{(P_{NO2})^2}\right\} \qquad (9.85)$$

9.81. Consider This   *What is $\Delta S_{reaction}$ for $2NO_2(g) \rightleftharpoons N_2O_4(g)$?*

(a) Use the standard entropies in Appendix XX to find $\Delta S^o_{reaction}$ for reaction (9.76) at 298 K. Is this the result you predicted in Consider This 9.77(d)? Explain why or why not.

(b) What is $\Delta S_{reaction}$ for a system in which the $NO_2(g)$ pressure is 0.243 bar and the $N_2O_4(g)$ pressure is 0.437 bar. Is the difference between $\Delta S^o_{reaction}$ and $\Delta S_{reaction}$ in the direction you expect? Explain why or why not. (Remember that standard pressure is 1 bar.)

Free energy change for a gas reaction

The change in Gibbs free energy, $\Delta G_{\text{reaction}}$, for a reaction at constant temperature and pressure is:

$$\Delta G_{\text{reaction}} = -T\Delta S_{\text{net}} = \Delta H_{\text{reaction}} - T\Delta S_{\text{reaction}} \tag{9.86}$$

We can account for pressure dependence of the free energy in gas reactions by substituting for $\Delta S_{\text{reaction}}$ from equation (9.85). Also, recall that $\Delta H_{\text{reaction}} = \Delta H^{\circ}_{\text{reaction}}$. These substitutions give:

$$\Delta G_{\text{reaction}} = \Delta H^{\circ}_{\text{reaction}} - T\Delta S^{\circ}_{\text{reaction}} + RT\ln\left\{\frac{(P_{N2O4})}{(P_{NO2})^2}\right\} \tag{9.87}$$

The first two terms on the right-hand side of equation (9.87) are equal to the standard free energy change for the reaction, $\Delta G^{\circ}_{\text{reaction}}$, so we have:

$$\Delta G_{\text{reaction}} = \Delta G^{\circ}_{\text{reaction}} + RT\ln\left\{\frac{(P_{N2O4})}{(P_{NO2})^2}\right\} \tag{9.88}$$

9.82. **Consider This**   *What is $\Delta G_{\text{reaction}}$ for $2NO_2(g) \rightleftharpoons N_2O_4(g)$?*

(a) Use the standard enthalpy of reaction from Check This 9.79 and standard entropy of reaction from Consider This 9.81 to calculate $\Delta G^{\circ}_{\text{reaction}}$ for reaction (9.76).

(b) Use your answer in part (a) to calculate $\Delta G_{\text{reaction}}$ under the conditions stated in Consider This 9.81(b). Is reaction (9.76) spontaneous under these conditions? Explain why or why not.

The reaction quotient

The ratio in curly brackets in equation (9.88) is the reaction quotient, Q:

$$\Delta G_{\text{reaction}} = \Delta G^{\circ}_{\text{reaction}} + RT\ln\left\{\frac{(P_{N2O4})}{(P_{NO2})^2}\right\} = \Delta G^{\circ}_{\text{reaction}} + RT\ln Q \tag{9.89}$$

The reaction quotient has the same *form* as the equilibrium constant expression for reaction (9.76). At this point we make a great logical leap and suggest that we might generalize the reasoning we have gone through for gases and apply equation (9.89) to all reactions, including our generalized reaction:

$$a\mathbf{A} + b\mathbf{B} \rightleftharpoons c\mathbf{C} + d\mathbf{D} \tag{9.5}$$

The result is:

$$\Delta G_{\text{reaction}} = \Delta G^{\circ}_{\text{reaction}} + RT\ln\left\{\frac{(C)^c(D)^d}{(A)^a(B)^b}\right\} \tag{9.42}$$

Equation (9.42) is the equation from which we began our discussion of the relationship

---

between thermodynamics and the equilibrium constant in Section 9.7. Our justification for accepting the application of equation (9.89) to all reactions is its success in analyzing and explaining equilibrium systems, a few of which we discussed in Sections 9.7, 9.8, and 9.9.

Index of Terms

Chapter 7 Problems

Section 9.1.   The Nature of Equilibrium

9.1. If $NH_3(aq)$ is slowly added to a light blue solution of $Cu^{2+}(aq)$, a pale blue precipitate, $Cu(OH)_2(s)$, forms. When additional $NH_3(aq)$ is added, the precipitate dissolves and a deep blue-violet solution results. The color results from the formation of $[Cu(NH_3)_4]^{2+}(aq)$.

(a) Write balanced chemical equations for these two equilibrium reactions, labeling the colors of the species containing copper ion.

(b) Will acidic or basic conditions favor the formation of $Cu(OH)_2(s)$ from $Cu^{2+}(aq)$? Use the first equation you wrote in part (a) to help explain your reasoning.

(c) Will mildly acidic conditions favor the formation of $[Cu(NH_3)_4]^{2+}(aq)$ from $Cu(OH)_2(s)$? Use the second equation you wrote in part (a) to help explain your reasoning.

(d) Are these equilibrium reactions also redox reactions? Explain your reasoning.

9.2. Explain how Figure 9.1 would be different if $Fe^{3+}(aq)$ and $SCN^-(aq)$ formed a one-to-two metal ion complex. Make a sketch of the new figure.

9.3. Consider this reaction at equilibrium:

$$Cd^{2+}(aq) + 6CN^-(aq) \rightleftharpoons Cd(CN)_6^{2-}(aq)$$
$$\text{colorless} \qquad\qquad\qquad \text{colorless}$$

(a) If the concentration of $CN^-(aq)$ is increased, how do you expect the concentrations of the other species to change? Explain your reasoning.

(b) In Investigate This 9.8(b), you were able to determine how temperature influenced the equilibrium between two cobalt(II) complexes. The same approach would not be effective with the equilibrium considered here. Offer a possible reason for this difference.

9.4. Consider this equilibrium reaction: $H_2(g) + Cl_2(g) \rightleftharpoons 2HCl(g)$. Discuss the validity of each of these statements.

(a) When this reaction has reached a state of equilibrium, no further reaction occurs.

(b) When equilibrium is established, the number of moles of reactants equals the number of moles of products for this reaction.

(c) The concentration of each substance in the system will be constant at equilibrium.

9.5. Consider this equilibrium reaction: $CO(g) + 2O_2(g) \rightleftharpoons 2CO_2(g)$. For this reaction at equilibrium, predict the effect of raising its temperature. Explain the reasoning for your prediction. *Hint:* Consider the bond enthalpies in Table 7.3, Chapter 7, Section 7.7.

9.6. Consider this equlibrium reaction: $CO(g) + 2H_2(g) \rightleftharpoons CH_3OH(g)$. For this reaction at equilibrium, predict the effect of each of these changes on the equilibrium. In each case explain the reasoning for your prediction.

(a) Some of the $CH_3OH(g)$ is removed while the volume is held constant.

(b) Some $CO(g)$ is added to the system while the volume is held constant.

(c) The system is compressed to a smaller volume.

(d) Some $N_2(g)$ is added to the system while the volume is held constant.

9.7. Consider this Lewis acid-base reaction between boric acid and water (Lewis acid-base reactions were discussed in Chapter 6.):

$$B(OH)_3(s) + 2H_2O(aq) \rightleftharpoons [B(OH)_4]^-(aq) + H_3O^+(aq)$$

Assume that the reaction is in equilibrium at a given temperature.

(a) How do you predict this equilibrium system will change, if more $B(OH)_3(s)$ is added to the mixture? Explain your reasoning.

(b) How do you predict this equilibrium system will change, if more $H_3O^+(aq)$ is added to the mixture? Explain your reasoning.

(c) How do you predict this equilibrium system will change, if $OH^-(aq)$ is added to the mixture? Explain your reasoning.

9.8. Consider the reaction equation for a weak acid, $HA(aq)$, transferring its proton to a water molecule: $HA(aq) + H_2O(aq) \rightleftharpoons H_3O^+(aq) + A^-(aq)$.

(a) Describe how you interpret this reaction equation at the molecular level. Be as complete as possible in describing the fate of each of the species represented.

(b) WEB Chap 9, Sect 9.4.1. Play the animation that represents the reaction of a weak acid in water. Is this representation consistent with the description that you wrote for part (a)? If not, what is(are) the difference(s)? Does the animation provide any insight into the reaction that is missing in the equation? Explain.

Section 9.2. Mathematical Expression for the Equilibrium Condition

9.9. About one percent of the world's energy resources is used to produce hydrogen and then ammonia from hydrogen and nitrogen. The reaction equation for the synthesis of ammonia is $N_2(g) + 3H_2(g) \rightleftharpoons 2NH_3(g)$. Write the equilibrium constant expression for the ammonia synthesis. If you want to calculate a numeric value for the equilibrium constant that is consistent with thermodynamic standard states, what units should be used for the dimensionless concentration ratio for each species?

9.10. An instructor used the following demonstration to show the difference between the dissociation of a strong acid and weak acid in water. She put the probes of a conductivity detector in 10.0 mL of distilled water. The detector was a simple type in which a small bulb lights up if the solution conducts electricity. In distilled water the bulb did not light up. She added 1 drop of 1 M HCl to the water and the bulb lit brightly. She rinsed the probe and put it into another 10.0 mL sample of distilled water. To this sample she added a 1 M solution of weak acid one drop at a time. When 50 drops of this acid had been added, the bulb lit as brightly as with the HCl. Estimate the % dissociation and the dissociation constant of the weak acid.

9.11. **(a)** Write the equilibrium constant expression for $CaCO_3(s) \rightleftharpoons CaO(s) + CO_2(g)$. What units must be used for each of the species, in order to make the equilibrium constant, K, dimensionless and consistent with the standard state definitions in Table 9.1? Explain.
(b) If the pressure of $CO_2(g)$ is 0.37 bar in a mixture of these species, what is the numeric value of K? Explain the reasoning for your answer.

9.12. This problem refers to the equilibrium systems represented by the molecular level drawings in Figure 9.2.
(a) The result of adding four more thiocyanate anions, $SCN(aq)^-$, to the equilibrium solution in the center of the figure is the formation of a total of three iron-thiocyanate complexes at equilibrium in the solution. Using the same assumptions as in Worked Example 9.14, calculate the equilibrium constant for this system. Explain your procedure.
(b) Is your equilibrium constant from part (a) consistent with those from Worked Example 9.14 and Check This 9.15? Explain why or why not.
(c) Is the result of adding the thiocyanate anion that is described in part (a) consistent with your observations in Investigate This 9.3(c)? Explain why or why not.

9.13. A gaseous mixture of HI, I_2, and H_2 in a 1.00 L container was held at a high temperature, 500 K, long enough for the reaction, $H_2(g) + I_2(g) \rightleftharpoons 2HI(g)$, to come to equilibrium and then rapidly cooled to "freeze out" the equilibrium mixture. Analysis of the cooled mixture gave these results: 2.21×10^{-3} mol HI, 1.46×10^{-3} mol I_2, and 2.09×10^{-5} mol H_2.
(a) What was the pressure, in bar, of each gas in the equilibrium mixture at 500 K? *Hint:* Recall the ideal gas equation, $PV = nRT$, with $R = 8.314 \times 10^{-2}$ L·bar·K^{-1}·mol$^-$1.
(b) Write the equilibrium constant expression for the reaction: $H_2(g) + I_2(g) \rightleftharpoons 2HI(g)$.
(c) What is the numeric value of the equilibrium constant for the reaction in part (b)? Explain your approach.

9.14. Acid-base indicators are weak acids: $HIn(aq) + H_2O(aq) \rightleftharpoons H_3O^+(aq) + In^-(aq)$. They are also dye molecules for which the colors of $HIn(aq)$ and $In^-(aq)$ are different. You can often analyze an acid-base indicator solution spectrophotometrically to determine the molar concentrations of the conjugate acid and base. For one such indicator, the concentrations were $[HIn(aq)] = 6.3 \times 10^{-5}$ M and $[HIn(aq)] = 8.7 \times 10^{-5}$ M in a solution with pH = 7.41.

(a) Write the equilibrium constant expression for the above proton transfer reaction.

(b) What is the numeric value of the equilibrium constant for the proton transfer reaction? Explain clearly how you get your answer.

Section 9.3.   Acid-Base Reactions and Equilibria

9.15. As you work through this problem and the next, use the information in this table.

Acid name	Formula	K_a
acetic acid	$CH_3C(O)OH$	1.8×10^{-5}
chloroacetic acid	$ClCH_2C(O)OH$	1.4×10^{-3}

(a) Write an equation for the acid equilibrium (proton transfer to water) for each acid.

(b) Write the equilibrium constant expression for K_a for each reaction in part (a).

(c) How will the pH of a 0.10 M solution of chloroacetic acid compare with that for a 0.10 M solution of acetic acid? Explain your prediction.

9.16. Use the K_a values given in the previous problem.

(a) Calculate the pH of a 0.050 M solution of acetic acid.

(b) Calculate the pH of a 0.050 M solution of chloroacetic acid.

(c) Discuss whether or not your results in parts (a) and (b) confirm the predictions made in part (c) of the previous problem.

9.17. The pK_w for water at 25 °C is 14.0 and is used as the basis for the pH scale. At 60 °C, the pK_w for water is 11.5.

(a) The pK_w for water refers to this reaction: $H_2O(l) + H_2O(l) \rightleftharpoons H_3O^+(aq) + OH^-(aq)$. Based on the two values for pK_w, is the autoionization reaction exothermic or endothermic? Explain your reasoning.

(b) If an aqueous solution at 25 °C has $[H_3O^+] = 1.0 \times 10^{-7}$ M, is the solution acidic, neutral, or basic? What is the pH of the solution?

(c) If an aqueous solution at 60 °C has $[H_3O^+] = 1.0 \times 10^{-7}$ M, is the solution acidic, neutral, or basic? What is the pH of the solution? How do your answers compare to those in part (b)? How do you explain the differences and similarities?

9.18. The position of equilibrium lies to the right (products are favored over reactants) in each of these reactions:

(i) $N_2H_5^+(aq) + NH_3(aq) \rightleftharpoons NH_4^+(aq) + N_2H_4(aq)$

(ii) $NH_3(aq) + HBr(aq) \rightleftharpoons NH_4^+(aq) + Br^-(aq)$

(iii) $N_2H_4(aq) + HBr(aq) \rightleftharpoons N_2H_5^+(aq) + Br^-(aq)$

Based on this information, what is the order of acid strength for species acting as acids in these reactions? Explain the reasoning for your order.

9.19. Calculate the pH of 0.1 M solutions of each of these acids and bases:

(a) lactic acid, $K_a = 1.4 \times 10^{-4}$ (c) aniline, $K_b = 4.3 \times 10^{-10}$

(b) benzoic acid, $K_a = 6.5 \times 10^{-5}$ (d) hydrazine, $K_b = 8.9 \times 10^{-7}$

9.20. Calculate the pH of a 0.250 M solution of aqueous ammonia. *Hint:* See Table 9.2 for the pK_a of the ammonium ion, the conjugate acid.

9.21. Estimate the percent ionization (the percent of the acid molecules that transfer their proton to water) of an aqueous 0.350 M hydrogen fluoride, HF, solution. Does your result justify using the approximation that a negligible amount, less than 5%, of the acid reacts? Explain. The K_a of HF is 6.46×10^{-4}.

9.22. The pH of lemon juice is about 3.4. Assume that all the acid in the juice is citric acid, H_3Cit. For the reaction, $H_3Cit(aq) + H_2O(aq) \rightleftharpoons H_3O^+(aq) + H_2Cit^-(aq)$, $K_a = 7.4 \times 10^{-4}$. Assume that none of the other protons in $H_2Cit^-(aq)$ is transferred to water in the lemon juice. What is the concentration of citric acid in lemon juice? Show clearly how you get your answer and state any further assumptions you make.

9.23. When simple salts of small, highly charged cations are dissolved in water, the resulting solutions are acidic. For example, a 0.1 M solution of aluminum chloride, $AlCl_3(s)$, in water has a pH of 2.9. To understand what is happening in these solutions, consider this reaction: $Al(H_2O)_6^{3+}(aq) + H_2O(aq) \rightleftharpoons H_3O^+(aq) + Al(OH)(H_2O)_5^{2+}(aq)$.

(a) Explain why the hydrated aluminum cation, $Al^{3+}(aq) = Al(H_2O)_6^{3+}(aq)$, is a weak acid. That is, explain what attractions and repulsions make the transfer of the proton more favorable in this case compared to $Na^+(aq)$, which shows no detectable acidity.

(b) What is the equilibrium constant expression and the numeric value of the equilibrium constant, K_a, for the proton transfer reaction.

(c) The pK_a for $Fe^{3+}(aq)$ is approximately 2.5. Which will be more acidic, a 0.1 M aqueous solution of Fe(III) or Al(III)? Explain your reasoning.

9.24. **(a)** Calculate K_a for an acid, HA, whose 0.215 M solution has a pH of 4.66. Does your result justify using the approximation that a negligible amount, less than 5%, of the acid reacts? Explain.

(b) What would be the pH of a 0.175 M aqueous solution of the salt NaA, the sodium salt of the conjugate base of HA, assuming the salt dissociates completely to $Na^+(aq)$ and $A^-(aq)$ ions in solution. Explain the reasoning for your answer.

9.25. **(a)** What is pK_b for the oxalate dianion, $^-O(O)CC(O)O^-(aq)$. Explain how you get your answer. *Hint:* See Table 9.2.

(b) About 35 g of sodium oxalate, the disodium salt of the oxalate dianion, dissolve in one liter of water. What is the pH of this solution? Explain.

9.26. **(a)** What are the pH and the percent ionization (the percent of the acid molecules that transfer their proton to water) of an aqueous 0.100 M methanoic acid, $HC(O)OH(aq)$, solution? Explain your reasoning. *Note:* $K_a = 1.8 \times 10^{-4}$ for methanoic acid.

(b) Predict, using Le Chatelier's principle, whether the percent ionization of 0.050 M methanoic acid will be larger, smaller, or the same as for the 0.100 M solution in part (a). Explain your reasoning.

(c) Calculate the percent ionization for the 0.050 M methanoic acid solution. Was your prediction in part (b) correct? Explain why or why not.

9.27. Explain why we almost always can disregard the hydronium ion contributed by the autoionization of water when we calculate the pH of solutions that contain an acid. Under what conditions would we have to take into account the autoionization of water?

Section 9.4. Solutions of Conjugate Acid-Base Pairs: Buffer Solutions

9.28. Which of these solutions contain buffer systems? Explain your reasoning.

   (i) a solution of $HClO_4$ and $NaClO_4$

   (ii) a solution of Na_2CO_3 and $NaHCO_3$

9.29. How do the concentrations of acetate ion, $[OAc^-(aq)]$, and hydronium ion, $[H_3O^+(aq)]$, change when sodium acetate, $NaOAc(s)$, is added to an aqueous solution of acetic acid? Explain your reasoning.

9.30. How do the concentrations of ammonium ion, $[NH_4^+(aq)]$ and hydroxide ion $[OH^-(aq)]$ change when ammonium bromide, $NH_4Br(s)$, is added to a solution of aqueous ammonia? Explain your reasoning.

9.31. Calculate the pH of a 0.250 M HOAc solution to which sufficient solid NaOAc is added to make the final sodium ion concentration, $[Na^+(aq)]$, 1.50 M. Assume that the volume on the solution does not change. Explain your reasoning.

9.32. Calculate the concentrations of HOAc and OAc^- in a 0.25 M acetate buffer solution at pH = 5.36. Explain your reasoning. *Hint:* Recall that the *sum* of the concentrations of acetate ion and acetic acid in this solution is 0.25 M.

9.33. **(a)** What is the pH of a buffer mixture composed of 0.15 M lactic acid ($HC_3H_5O_3$, $K_a = 1.4 \times 10^{-4}$) and 0.10 M sodium lactate?

(b) What is the pH half way to the equivalence point in a titration of 0.15 M lactic acid with 0.15 M sodium hydroxide? How does this pH compare with the pH you calculated in part (a)? Which pH is higher? Does this result make sense? Explain your reasoning.

9.34. **(a)** What is the pH of an aqueous solution that is 0.551 M phenol, HOC_6H_5, and 0.377 M sodium phenolate, $NaOC_6H_5$? Explain. *Hint:* See Table 9.2.

(b) What will be the pH of the mixture in part (a) after 0.100 moles of HCl gas are added to 1.00 L of the solution? Assume that the volume of the solution does not change. Explain your reasoning.

(c) What will be the pH of the mixture in part (a) after 0.125 moles of solid KOH are added to 1.00 L of the solution? Assume that the volume of the solution does not change. Explain your reasoning.

9.35. Many liquid household bleaches are about 5% by weight aqueous solutions of sodium hypochlorite, NaOCl, which completely dissociates to its ions in the solution. Hypochlorous acid, HOCl(*aq*), has a pK_a of 7.5.

(a) What is the molarity of sodium hypochlorite in bleach? Show how you get your answer. *Hint:* Assume that liquid bleach has the same density as water.

(b) Write the reaction that hypochlorite anion, $OCl^-(aq)$, undergoes with water. Write the equilibrium constant expression for this reaction. What is the usual symbol we use to designate the equilibrium constant for a reaction like this? What is the numeric value of this equilibrium constant? Explain your reasoning.

(c) What is the pH of a liquid bleach solution? Explain.

(d) If you wish to change the pH of bleach solution to 6.5, should you add sodium hydroxide or hydrochloric acid? Explain your reasoning.

(e) What is the ratio of the conjugate base to the conjugate acid in a bleach solution whose pH has been adjusted to 6.5? Explain.

9.36. Calculate the $H_2CO_3(aq)$ concentration in human arterial blood, given that the pH of blood is 7.41 and the $HCO_3^-(aq)$ concentration is 26×10^{-3} M. Explain the method you use to get your answer. *Note:* K_{a1} for $H_2CO_3(aq)$ is 8.0×10^{-7} at 37 °C.

9.37. **(a)** What volume of 6.0 M hydrochloric acid do you have to add to 250. mL of 0.10 M aqueous sodium acetate solution to give a solution with a pH of 4.30? Explain your reasoning.

(b) What are the concentrations of acetic acid and acetate ion in the pH 4.30 solution you prepared in part (a)? Is this a buffer solution? Explain your reasoning.

9.38. Combinations of phosphoric acid, $(HO)_3PO$, and hydrogen phosphate and phosphate salts can be used to prepare a variety of buffer solutions. Assume that you have available the acid, sodium dihydrogen phosphate, $Na(HO)_2PO_2(s)$, potassium hydrogen phosphate, $K_2(HO)PO_3(s)$, and sodium phosphate, $Na_3PO_4(s)$. Which of these compounds would you choose to make buffers with pH values of 2.00, 7.00, and 12.00? In each case, explain your choice.

Section 9.5.   Acid-Base Properties of Proteins

9.39. What is the isoelectric point of a protein? What are the properties of the protein at the isoelectric point? How are the solution conditions related to the isoelectric point?

9.40. **(a)** Write the equilibrium constant expressions for K_{a1} and K_{a2} for the amino acid glycine. *Hint:* See Table 9.2 for the reaction equations for these equilibria.

(b) Using the data in Table 9.2, calculate the isoelectric pH of glycine. Explain the procedure you use. *Hint*: Use an appropriate combination of the equilibrium constant expressions from part (a) and the condition that, at the isoelectric pH, the concentrations of the net positive and net negative forms of the amino acid are the same.

(c)   WEB   Chap 9, Sect 9.5.4. Show clearly how the calculation of the isoelectric pH for the dipeptide on this *Web Companion* page is related to the the procedure you used to answer part (b).

9.41. As we have seen, sickle-shaped blood cells result from just two minor changes in the total of 574 amino acids in hemoglobin. In both cases, glutamate is replaced by valine. Make sketches of a protein (see Figure 9.5) representing hemoglobin and show how the change of the two amino acids brings about the clumping together of hemoglobin that causes the change in shape of the blood cells. *Hint:* Consider formation of micelles and membanes represented in Figures 8.20 and 8.23 and the *Web Companion*, Chap 8, Sect 8.13.

9.42. You have isolated two small proteins, A and B, that you believe differ in only one amino acid. You think that one protein has a cysteine at position where the other has a histidine. When you analyze the proteins by electrophoresis (see Figures 9.6 and 9.7) at pH of 8.6, you find that protein B moves further toward the positive end of the gel.

(a) Is this result consistent with your hypothesis about the difference between the two proteins? If so, which protein has the cysteine and which the histidine? Explain the reasoning for your choice.

(b) If you cannot decide which protein is which from electrophoresis at this pH, what further experiment could you do to differentiate the two? Even if you can differentiate the two proteins, is there a further experiment you could do that would help to confirm your identification?

9.43. Explain how the observations you made in Investigate This 9.42 are related to sickle-cell disease.

9.44. Electrophoretic analysis is not limited to biomolecules such as proteins and nucleic acids. Other species that differ in charge, metal ion complexes, for example, can also be analyzed and separated by various forms of electrophoresis. In one experiment, a mixture of Fe(III) and Co(II) ions in an 8 M solution of hydrochloric acid was analyzed by paper electrophesis (a strip of filter paper substitutes for the gel in Figure 9.6). Fe(III) forms a light yellow chloride complex and Co(II) forms a blue chloride complex (as you observed in Investigate This 9.8). After several minutes of electrophoresis, a yellow spot on the paper was closer to the positive end of the paper than a blue spot.

(a) Based on this observation, what conclusions can you draw about the stoichiometry of the Fe(III) and Co(II) chloride ion complexes? Explain your reasoning.

(b) What would you hypothesize about the structures of the Fe(III) and Co(II) chloride ion complexes? Explain your reasoning. *Hint:* Consider what you learned in Chapter 6, Section 6.6, as well as in this chapter.

Section 9.6.   Solubility Equilibria for Ionic Salts

9.45. Use K_{sp} values for these ionic compounds to calculate the concentration of each ionic species in a saturated solution of the compound. Explain your reasoning for each.

(a) $BaSO_4$ (c) $Mg(OH)_2$

(b) Ag_2S (d) $PbBr_2$

9.46. How much solid calcium carbonate, $CaCO_3(s)$, will be obtained if 100. mL of a solution saturated with calcium carbonate is evaporated to dryness? Explain.

9.47. Calculate the mass of calcium fluoride, $CaF_2(s)$, that will dissolve in 100. mL of water. Explain your method. Assume that there is no volume change. The pK_{sp} of CaF_2 is 10.57.

9.48. Find the molar solubility and K_{sp} for $PbCrO_4(s)$, if 5.8×10^{-5} g dissolve in exactly 1 L of water. Explain your reasoning.

9.49. Calculate the pH and the concentrations of $Fe^{2+}(aq)$ and $OH^-(aq)$ when as much $Fe(OH)_2(s)$ as possible is dissolved in these solutions. Explain your reasoning in each case.

  (a) pure water

  (b) 0.25 M NaOH

  (c) 0.25 M $Fe(NO_3)_2$

9.50. Determine if a precipitate will form when 0.0025 mol of $Ba^{2+}(aq)$ (from $Ba(NO_3)_2$) and 0.00030 mol of CO_3^{2-} (from Na_2CO_3) are mixed in 250. mL of distilled water. Explain the reasoning for your response.

9.51. A solution of $Na_2SO_4(aq)$ is added dropwise to another solution containing 2.5×10^{-3} M $Pb^{2+}(aq)$ and 0.05 M $Ca^{2+}(aq)$. What is the composition of the precipitate that will form first? Explain. *Hint:* What is the minimum concentration of sulfate anion, $[SO_4^{2-}(aq)]$, required to begin the precipitation of each of the cations?

9.52. **(a)** Because of its low solubility and opaqueness to x-rays, a suspension of $BaSO_4(s)$ is used routinely in x-ray analyses of the intestinal tract. How much $Ba^{2+}(aq)$ is in 200 mL of a saturated barium sulfate solution given to a patient to drink?

  (b) The barium cation, $Ba^{2+}(aq)$, is rather toxic. How do your results in part (a) relate to the safety of the barium sulfate used in x-ray analysis?

9.53. If a patient is slightly allergic to $Ba^{2+}(aq)$, which one of the following methods would you choose to decrease the $Ba^{2+}(aq)$ concentration in order to minimize allergic discomfort when ingesting the $BaSO_4$ solution in the previous problem? Explain your choice and why you reject the others. *Note:* The dissolution reaction, $BaSO_4(s) \rightleftharpoons Ba^{2+}(aq) + SO_4^{2-}(aq)$, is endothermic.

  (i) Heat the saturated solution.

  (ii) Add sodium sulfate, Na_2SO_4.

  (iii) Add additional $BaSO_4$.

9.54. Determine the K_{sp} for a metal salt with formula $M(OH)_2$, if its saturated solution has a pH of 8.11. Explain your reasoning.

9.55. Determine the pK_{sp} for Ag_3AsO_4, if 0.000562 g dissolves in 1 liter of water.

9.56. Silver chloride, $AgCl(s)$, is white and silver chromate, $Ag_2CrO_4(s)$, is red. When chemists wish to analyze a solution for chloride ion they often titrate it with a standard solution of silver ion. A few drops of potassium chromate solution are added to the unknown chloride solution to act as an indicator. Explain how the chromate ion can act as an indicator. *Hint:* Examine Table 9.4 and consider the stoichiometry of the two solids.

9.57. The pK_{sp} for $Ni(OH)_2(s)$ is 17.2 and the pK_{sp} for $Fe(OH)_2(s)$ is 13.8. If you slowly add an aqueous solution of sodium hydroxide to a stirred aqueous solution containing 0.001 M $Ni^{2+}(aq)$ and 0.06 M $Fe^{2+}(aq)$, at what pH will a precipitate first begin to form? What will the precipitate be? Explain your reasoning. *Hint:* What is the minimum concentration of hydroxide, $[OH^-(aq)]$, required to begin the precipitation of each of the cations?

Section 9.7. Thermodynamics and the Equilibrium Constant

9.58. Consider this reaction, $H_2(g) + I_2(g) \rightleftharpoons 2HI(g)$, at equilibrium in a rigid (constant volume) reaction vessel.

(a) If the pressure of $I_2(g)$ is increased (by adding more I_2 to the vessel), how will the pressures of $H_2(g)$ and $HI(g)$ be affected as a new equilibrium is established? Explain your reasoning.

(b) If the pressure of $HI(g)$ is reduced (by removing some HI from the vessel), how will the rate of the reverse reaction compare to the forward reaction before a new equilibrium is established? Which direction will the equilibrium shift? Explain your responses.

9.59. Use free energies of formation, ΔG°_f, from Appendix XX to determine which of these two reactions has the larger equilibrium constant? Explain your choice.

(i) $H_2(g) + 1/2 O_2(g) \rightleftharpoons H_2O(l)$

(ii) $H_2(g) + 1/2 O_2(g) \rightleftharpoons H_2O(g)$

9.60. Urea, $(NH_2)_2CO(s)$, is an ingredient in many fertilizers and is occasionally used instead of salt to hasten the melting of ice on highways. The reaction for the industrial synthesis of urea is: $CO_2(g) + 2NH_3(g) \rightleftharpoons (NH_2)_2CO(s) + H_2O(l)$. What are the standard free energy change and equilibrium constant for this reaction at 298 K. Is the reaction spontaneous under standard conditions? Explain.

9.61. Consider the reaction $H_2(g) + I_2(g) \rightleftharpoons 2HI(g)$ at 500 K, for which Problem 9.13 provides the data you require to calculate the equilibrium constant, K. What is the standard free energy change for this reaction at 500 K? Explain.

9.62. Much of the ammonia, $NH_3(g)$, we synthesize, Problem 9.9, is used to make ammonium nitrate for fertilizer. The first step in making the nitrate (as nitric acid) is the oxidation of ammonia, $4NH_3(g) + 5O_2(g) \rightleftharpoons 4NO(g) + 6H_2O(l)$.

(a) Use the data in Appendix XX to determine the standard enthalpy change, $\Delta H^\circ_{reaction}$, for the oxidation of ammonia at 298 K. Is your result consistent with what you would get from average bond enthalpies (Chapter 7, Table 7.3)? Explain.

(b) Use the data in Appendix XX to determine the standard entropy change, $\Delta S^\circ_{reaction}$, for the oxidation of ammonia at 298 K. Is this the result you would predict from looking at the changes that occur in this reaction? Explain why or why not.

(c) Use the $\Delta H^\circ_{reaction}$ and $\Delta S^\circ_{reaction}$ values you calculated in parts (a) and (b) to determine the standard free energy change, $\Delta G^\circ_{reaction}$, for this reaction at 298 K. Use the data in Appenix XX to determine the $\Delta G^\circ_{reaction}$ from standard free energies of formation, ΔG°_f. Are the two values for $\Delta G^\circ_{reaction}$ the same? Explain why or why not.

(d) What is the equilibrium constant for the ammonia oxidation reaction at 298 K. Are products or reactants favored? Explain.

9.63. When calcium carbide, $CaC_2(s)$, a grayish solid, is dropped into water, the solution gets quite hot, gas bubbles out of the solution, and a white precipitate forms. The reaction that occurs, $CaC_2(s) + 2H_2O(l) \rightleftharpoons Ca(OH)_2(s) + C_2H_2(g)$, forms ethyne (acetylene), a flammable gas that coal miners once used in "carbide lamps" to light their way in the mines.

(a) What is the sign of $\Delta H_{reaction}$ for the ethyne-forming reaction? Explain how you know.

(b) What is the sign of $\Delta S_{reaction}$ for the ethyne-forming reaction? Explain how you know.

(c) What is the sign of $\Delta G_{reaction}$ for the ethyne-forming reaction? Explain how you know.

(d) Use the data in Appendix XX to calculate $\Delta H^\circ_{reaction}$, $\Delta S^\circ_{reaction}$, and $\Delta G^\circ_{reaction}$, for the ethyne formation reaction at 298 K. Are your results consistent with your answers in parts (a), (b), and (c)? Explain why or why not.

(e) Review the calculations in part (d). What is the major factor responsible for the size of the free energy change for this reaction?

9.64. Sketch a graph of $\Delta G_{reaction}$ as a function of the composition of a reaction mixture from a reactant-rich composition through equilibrium to a product-rich composition. Explain the reasoning you use to draw your graph.

9.65. At 298 K, the equilibrium constant, K, for the decomposition of calcite (chalk), $CaCO_3(s)$, to give calcium oxide, $CaO(s)$, and carbon dioxide, $CO_2(g)$, is 1.6×10^{-23}.

(a) Based on the equilibrium constant for this reaction, is there likely to be a higher concentration of $CO_2(g)$ near the chalkboard in a classroom compared to that in the rest of the room? Explain why or why not.

(b) Data for $CaO(s)$ are not given in Appendix XX. Use the data that are available there and, K, the equilibrium constant above, to estimate ΔG°_f for $CaO(s)$. Explain your procedure.

9.66. Compounds **A**, **B**, and **C** were dissolved in water in a reaction vessel and allowed to come to equilibrium. The graph shows the variation of concentration with time for this reaction: $3A(aq) \rightleftharpoons B(aq) + 2C(aq)$

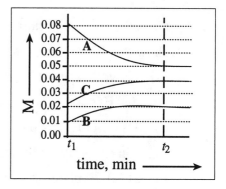

(a) Estimate the value of the reaction quotient, Q, at time t_1, the time the compounds were mixed.

(b) Is the reaction shifting in the forward or reverse direction to attain equilibrium? How can you tell?

(c) Is it reasonable to assume that equilibrium is reached at t_2? Explain why or why not.

(d) Estimate the value of the equilibrium constant, K, at t_2. Explain your procedure.

(e) Estimate the value of the standard free energy change for the reaction. Explain your reasoning.

9.67. The equilibrium constant, K, for the reaction $H_2(g) + I_2(g) \rightleftharpoons 2HI(g)$ at 700 K is 54.

(a) What is the standard free energy change for this reaction at 700 K? Explain.

(b) What is the reaction quotient, Q, in a mixture in which the pressures of $H_2(g)$, $I_2(g)$, and $HI(g)$ are, respectively, 0.040 bar, 0.35 bar, and 1.05 bar. Explain.

(c) What is the free energy of reaction under the conditions in part (b)? Is the sign of the free energy consistent with the relative values of Q and K? Explain why or why not.

(d) Is the mixture in part (b) likely to react to form more products, more reactants, or is it at equilibrium? Explain how you know.

9.68. Is methanol decomposition, $CH_3OH(l) \rightleftharpoons CO(g) + 2H_2(g)$, likely to occur spontaneously at 298 K and one bar pressure? Use thermodynamic reasoning to show why or why not.

9.69. Is the reaction of ethene with water, $CH_2CH_2(g) + H_2O(l) \rightleftharpoons CH_3CH_2OH(l)$, likely to occur spontaneously at 298 K and one bar pressure? Use thermodynamic reasoning to show why or why not.

Section 9.8.   Temperature Dependence of the Equilibrium Constant

9.70. Explain how temperature, pressure, and concentration affect a system at equilibrium.

9.71. Explain how the position of equilibrium is affected when the following changes occur for this exothermic reaction, $2SO_2(g) + O_2(g) \rightleftharpoons SO_3(g)$, initially at equilibrium.

(a) Some $SO_3(g)$ is removed.

(b) The pressure is decreased.

(c) The temperature is decreased.

(d) Some $O_2(g)$ is added.

9.72. Consider the equilibrium reaction $N_2(g) + O_2(g) \rightleftharpoons 2NO(g)$. Experimental measurements show that higher temperatures favor the formation of $NO(g)$.

(a) Do you expect this reaction to be exothermic or endothermic in the direction written? Explain your choice.

(b) Use average bond enthalpies from Table 7.2, Chapter 7, Section 7.7, to calculate the reaction enthalpy for this reaction. Does your result agree with the prediction you made in part (a). Explain why or why not. *Note:* The bond enthalpy for the bond in NO is 630. $kJ \cdot mol^{-1}$.

(c) Does the equilibrium constant for this reaction increase or decrease with increasing temperature? Justify your choice.

(d) How does the standard free energy of this reaction change as the temperature increases. How do you know? If the standard enthalpy of reaction is constant, how do you account for the direction of change in the standard free energy? Do the data in Appendix XX support your explanation? Show why or why not.

9.73. (a) Use the results from Worked Example 9.67 to calculate the solubility product (K_{sp}), molar solubility, and solubility in $g \cdot L^{-1}$, for $PbI_2(s)$ at 20. °C.

(b) One handbook gives a $PbI_2(s)$ solubility of 0.63 $g \cdot L^{-1}$ at 20. °C. How does your value from part (a) agree with this datum?

9.74. The solubility of $CaSO_4(s)$ at 303 K and 373 K is 2.09 $g \cdot L^{-1}$ and 1.62 $g \cdot L^{-1}$, respectively.

(a) What are the numeric values of K_{sp} for $CaSO_4(s)$ at 303 K and 373 K. Explain.

(b) What are $\Delta H^{\circ}_{reaction}$ and $\Delta S^{\circ}_{reaction}$ for the dissolution of $CaSO_4(s)$. Explain your procedure and any assumptions you make to get your answers.

(c) What are $\Delta G^{\circ}_{reaction}$ and K_{sp} for the dissolution of $CaSO_4(s)$ at 298 K? Discuss the agreement between your results and the data in Appendix XX and Table 9.4.

9.75. The coal gas (or water gas) reaction, $C(s) + H_2O(g) \rightleftharpoons CO(g) + H_2(g)$, produces a gaseous fuel mixture from a solid fuel. The product mixture is also a useful starting material for synthesis of carbon-containing compounds, such as methanol.

(a) Calculate $\Delta H°_{reaction}$, $\Delta S°_{reaction}$, and $\Delta G°_{reaction}$ for the coal gas reaction at 25 °C.

(b) At what temperature would $\Delta G°_{reaction}$ be zero? Explain your reasoning and state any assumptions you make.

(c) What is the equilibrium constant for the coal gas reaction at the temperature you calculated in part (b)? Explain.

9.76. (a) How is $\Delta G_{reaction}$ for a phase change, such as liquid $\rightleftharpoons$ gas, at equilibrium at one bar pressure related to $\Delta G°_{reaction}$ at the equilibrium temperature? Explain your reasoning.

(b) Use the data in Appendix XX to estimate the temperature at which rhombic and monoclinic sulfur are in equilibrium, $S(s, rhombic) \rightleftharpoons S(s, monoclinic)$, at one bar. Explain your reasoning and state any assumptions you make.

9.77. Consider the reaction $H_2(g) + I_2(g) \rightleftharpoons 2HI(g)$ at 500 K, for which Problem 9.13 provides the data required to calculate the equilibrium constant, K, and Problem 9.61 asks for the standard free energy change, $\Delta G°_{reaction}$, for the reaction at 500 K.

(a) If the equilibrium constant for this reaction is 794 at 298 K, what are $\Delta H°_{reaction}$ and $\Delta S°_{reaction}$? Explain your reasoning and state any assumptions you make.

(b) Based on your results from part (a), what is $\Delta G°_{reaction}$ at 298 K? How does this value for $\Delta G°_{reaction}$ compare with the one you calculate from the data in Appendix XX? Discuss your comparison.

9.78. For the vaporization of ethanol, $C_2H_5OH(l) \rightleftharpoons C_2H_5OH(g)$, the standard entropy change, $\Delta S°_{reaction}$, is 122 $J \cdot K^{-1} \cdot mol^{-1}$. The normal boiling point of ethanol at one bar pressure is 78.5 °C (351.6 K).

(a) Estimate $\Delta H°_{reaction}$ for ethanol vaporization. Explain your reasoning and state any assumptions you make.

(b) Write the equilibrium constant expression for ethanol vaporization. Explain why the numeric value of the equilibrium constant, K, is one at 78.5 °C. What is $\Delta G°_{reaction}$ for ethanol vaporization at 78.5 °C? Explain.

(c) What are the numeric values of K and $\Delta G°_{reaction}$ for ethanol vaporization at 25 °C? Explain your reasoning and state any assumptions you make.

(d) What is the vapor pressure (the pressure of the gas in equilibrium with the liquid) of ethanol at 25 °C? Explain your reasoning.

9.79. The variation of equilibirum vapor pressure for the isomeric alcohols, 1-propanol and 2-propanol, is given in this table.

vapor pressure, torr	1.0	10.0	40.0	100.	400.	760.
	temperature, °C					
1-propanol	−15.0	14.7	36.4	52.8	82.0	97.8
2-propanol	−26.1	2.4	23.8	39.5	67.8	82.5

(a) Plot these data and use the plots to determine $\Delta H^{\circ}_{reaction}$ and $\Delta S^{\circ}_{reaction}$ for the vaporization of each of these alcohols. *Note:* 1 bar = 750 torr and K = °C + 273.1.

(b) How, if at all, are the relative values of $\Delta H^{\circ}_{reaction}$ related to the structures of the two alcohols? Explain.

(c) How, if at all, are the relative values of $\Delta S^{\circ}_{reaction}$ related to the structures of the two alcohols? Explain.

(d) Which alcohol is the more volatile? Why is this isomer more volatile?

9.80. The pK_w for water is 14.0 at 25 °C and 11.5 at 60 °C. Use these data to determine $\Delta H^{\circ}_{reaction}$, $\Delta S^{\circ}_{reaction}$, and $\Delta G^{\circ}_{reaction}$ for water ionization. Explain your reasoning and state any assumptions you make. Discuss the agreement (or lack thereof) between these values and those you calculate from the data Appendix XX.

9.81. Phosphorus pentachloride, $PCl_5(s)$, sublimes readily, $PCl_5(s) \rightleftharpoons PCl_5(g)$. The equilibrium pressure of the gas is 40 torr at 102.5 °C and 400 torr at 147.2 °C.

(a) Use these data to determine $\Delta H^{\circ}_{reaction}$ and $\Delta S^{\circ}_{reaction}$ for the sublimation of $PCl_5(s)$. Explain your reasoning and state any assumptions you make. *Note:* 1 bar = 750 torr.

(b) Assuming that the solid does not melt, at what temperature is the sublimation pressure 760 torr (1 atm)?

9.82. (a) Consider a spontaneous endothermic reaction during which the entropy of the system increases. What is the sign of the free energy change for this reaction? Explain.

(b) Consider a spontaneous exothermic reaction during which the entropy of the system decreases. What is the sign of the free energy change for this reaction? Explain.

(c) If the temperature is changed could the reaction in part (a) become non-spontaneous? the reaction in part (b)? Explain why or why not for both cases.

9.83. Might this reaction, $CH_3OH(g) \rightleftharpoons CO(g) + 2H_2(g)$, occur spontaneously at about 1000 K and one bar pressure? Use thermodynamic reasoning to explain why or why not.

9.84. **(a)** Assume that rubbing alcohol is pure 2-propanol and use your results from Problem 9.79 to find the vapor pressure of 2-propanol at 35 °C, skin temperature. How is this result related to the use of rubbing alcohol to cool your skin? Explain.

(b) If 12 g of this rubbing alcohol evaporates from your skin, what is the change in the thermal energy of your body? Explain.

Section 9.9.   Thermodynamics in Living Systems

The data in this table may be needed in Problems 9.85 through 9.88. The free energies in the table are for the hydrolysis at pH 7 of the phosphorylated compounds to give the non-phosphorylated compound plus phosphate:

Free energy of hydrolysis at pH 7 and 25 °C

Compound	$\Delta G°'$, kJ·mol^{-1}
phosphoenolpyruvate, PEP	–62
creatine phosphate	–43
ATP (gives ADP)	–30
glucose-1-phosphate, Glu-1-P	–21
glucose-6-phosphate, Glu-6-P	–14
glycerol-3-phosphate, G3P	–9

$$\text{compound-P} + H_2O \rightleftharpoons \text{compound} + P_i$$

9.85. Consider this isomerizaton reaction: glu-6-P $\rightleftharpoons$ glu-1-P, at pH 7 and 25 °C.

(a) Write the hydrolysis reactions for glu-6-P and glu-1-P. Show how to combine these reactions to give the isomerization reaction.

(b) Calculate $\Delta G°'_{reaction}$ for the isomerization reaction. Explain your procedure.

(c) If this isomerization reaches equilibrium, what is the molar *ratio* of glu-6-P to glu-1-P in the reaction mixture? Explain your reasoning. *Hint:* Consider the equilibrium constant and equilibrium constant expression.

9.86. Consider this reaction: ATP + Pyr $\rightleftharpoons$ ADP + PEP (where Pyr = pyruvate).

(a) Combine hydrolysis reactions from the table to give the reaction of interest and calculate $\Delta G°'_{reaction}$ and the equiliibrium constant, K, for the the reaction at 25 °C.

(b) In many living cells, the molar *ratio* of ATP to ADP is about 10. If this reaction is at equilibrium in such a cell, what is the equilibrium molar ratio of pyruvate to phosphoenolpyruvate? Explain.

9.87. For each of these reactions, imagine that you start with a mixture of all the reactants and products at the same concentration in a pH 7 solution at 25 °C. For each case, tell which direction the reaction will go in order to approach equilibrium and give the reasoning for your choice.

(i) ATP + creatine phosphate $\rightleftharpoons$ ADP + creatine

(ii) ATP + Glu $\rightleftharpoons$ ADP + Glu-6-P

9.88. The stoichiometry of the glycolysis ("sugar breaking") pathway in organisms is:

$$Glu + 2ADP + 2P_i + 2NAD^+ \rightleftharpoons 2Pyr + 2ATP + 2NADH + 2H^+$$

The overall standard free energy change, $\Delta G^{\circ\prime}_{reaction}$, for glycolysis is about –84 kJ at 25 °C. Organisms also have a gluconeogenesis ("making new sugar") pathway to produce glucose from pyruvate with a stoichiometry we can represent as:

$$2Pyr + 6ATP + 2NADH + 2H_2O + 2H^+ \rightleftharpoons Glu + 6ADP + 6P_i + 2NAD^+$$

The overall free energy change for gluconeogenesis, $\Delta G^{\circ\prime}_{reaction}$, is about –38 kJ at 25 °C.

(a) Why isn't the gluconeogenesis pathway simply the reverse of the glycolysis pathway? Would there be a thermodynamic problem reversing glycolysis? Explain.

(b) Are the free energy changes for glycolysis and gluconeogenesis consistent with the differences between the overall reactions? Explain why or why not.

(c) Look again at Figure 7.17, Chapter 7, Section 7.9. The coupling shown in the figure was related to enthalpies of reaction, but the coupling is more accurately a representation of free energies of reaction. Discuss where glycolysis and gluconeogenesis would fit into the diagram if the arrows represent free energy changes.

9.89. In cells that need free energy faster than oxygen can get to them for complete glucose oxidation, such as our muscles during rapid strenuous exercise, pyruvate, $CH_3COC(O)O^-$, is reduced to lactate, $CH_3CHOHC(O)O^-$ (Lac): $Pyr + NADH + H^+ \rightleftharpoons Lac + NAD^+$.

(a) The reduction of pyruvate to lactate is a fermentation reaction and is required to enable glycolysis to proceed in the absence of oxygen to complete the glucose oxidation pathway. What is required for glycolysis (see preceding problem) that is provided by this fermentation reaction? Explain. *Hint:* See Chapter 6, Section 6.11.

(b) Combine the overall glycolysis reaction with this fermentation reaction to give the net reaction equation for the conversion of a glucose molecule to two lactates.

(c) If the standard free energy change, $\Delta G^{\circ\prime}_{reaction}$, for the fermentation reaction written above is about –25 kJ at 25 °C, what is the overall $\Delta G^{\circ\prime}_{reaction}$ for the glucose to lactate reaction you wrote in part (b)? Explain.

(d) What is the equilibrium constant expression for the conversion of glucose to lactate that you wrote in part (b). Explain.

(e) What is the free energy of reaction, $\Delta G^{\prime}_{reaction}$, (not the standard free energy) for the glucose to lactate reaction if the concentrations in a cell are: $[Glu] = 5 \times 10^{-3}$ M, $[Lac] = 5 \times 10^{-5}$ M, $[ATP] = 2 \times 10^{-3}$ M, $[ADP] = 2 \times 10^{-4}$ M, and $[P_i] = 1 \times 10^{-3}$ M. Is the reaction spontaneous under these conditions? Explain.

Section 9.11. EXTENSION – Thermodynamic Basis for the Equilibrium Constant

9.90. **(a)** If the pressure of a gas is changed from pressure P_1 to pressure P_2 at constant temperature, show that the change in entropy of the gas is: $\Delta S = R\ln(P_1/P_2)$. *Hint:* Consider equation 9.83.

(b) What is the sign of ΔS, if a gas is compressed? Show that the result you get by applying the equation from part (a) is the same as you get by reasoning from the entropy discussions in Chapter 8.

9.91. Consider this reaction, $2NO_2(g) \rightleftharpoons N_2O_4(g)$, at equilibrium at 298 K in a reaction vessel whose size can be changed. Initially, the pressures of $NO_2(g)$ and $N_2O_4(g)$ are 0.225 bar and 0.438 bar, respectively.

(a) Write the equilibrium constant expression for this reaction and calculate the equilibrium constant, K, at 298 K.

(b) Imagine that the volume of the reaction vessel is doubled while holding the temperature of the contents constant. If no reaction occurs, what is the pressure of each gas in the mixture? What is the numeric value of the reaction quotient, Q, in this mixture? Explain.

(c) How must the reaction mixture in part (b) change in order make Q move toward K and reattain equilibrium? Is this the same direction of pressure adjustments you would predict from Le Chatelier's principle. Explain why or why not.

(d) Calculate the pressures of the two gases at equilibrium in the new volume. *Hint:* You will need the quadratic equation or some estimation method, perhaps a spreadsheet.

General Problems

9.92. Consider the reaction $H_2(g) + I_2(g) \rightleftharpoons 2HI(g)$, for which you determined K at 500 K in Problem 9.13. If 4.00×10^{-3} mol of HI are placed in a 450. mL reaction vessel heated to 500 K, how many moles of I_2 will be present at equilibrium? Explain clearly the procedure you use to obtain your answer. *Hint:* If necessary, recall the ideal gas equation, $PV = nRT$, with $R = 8.314 \times 10^{-2}$ L·bar·K^{-1}·mol·1. Take into account the amount of $HI(g)$ that reacts and use the quadratic formula or an estimation procedure to solve the resulting equation.

9.93. How do the concentrations of acetate ion, $OAc^-(aq)$, and $H_3O^+(aq)$ change when hydrochloric acid, $HCl(aq)$, is added to an aqueous solution of acetic acid? How would you explain the change in terms of the common ion effect?

9.94. Fluorobenzene, C_6H_5F, and related fluorinated compounds, are important starting materials for the synthesis of many fungicides, drugs, and agricultural chemicals. Recently, this efficient process for making fluorobenzene has been reported:

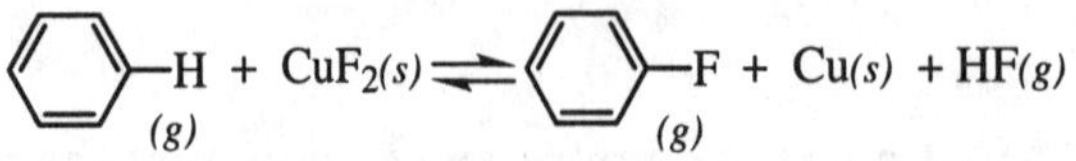

(The $HF(g)$ product is later recycled with oxygen to regenerate the $CuF_2(s)$ with only water as a by-product.) At 450 °C, the conversion of benzene to fluorobenzene is 5% and at 550 °C, the conversion is 30%.

(a) What is the equilibrium constant expression for this reaction? Assuming that the reaction is stoichiometric and comes to equilibrium in the reactor at about 0.1 bar total pressure of reactants plus products, what are the equilibrium constants for the reaction at 350 °C and 450 °C? Explain clearly the procedure you use to obtain your answers.

(b) What is the standard enthalpy of reaction, $\Delta H°_{reaction}$, for this process? Explain and state the assumptions you make.

(c) What is the standard entropy of reaction, $\Delta S°_{reaction}$, for this process? Explain and state the assumptions you make.

(d) If this process could be carried out at even higher temperatures, how high do you predict the temperature would have to be to convert 50% of the benzene to fluorobenzene? Explain your reasoning. *Hint:* What is the equilibrium constant for 50% conversion? How is the equilibrium constant related to the quantities in parts (b) and (c)?

9.95. A system at equilibrium responds to a disturbance in a way that minimizes the effect(s) of the disturbance (Le Chatelier's Principle). The disturbances that affect an equilibrium are changing the concentrations of the species in the reaction or changing the temperature of the system. Explain how qualitative, directional predictions based on Le Chatelier's Principle are related to the equilibrium constant expression, equation (9.47), and to the thermodynamics of equilibria embodied in equation (9.55).

Chapter 10. Reduction-Oxidation: Electrochemistry

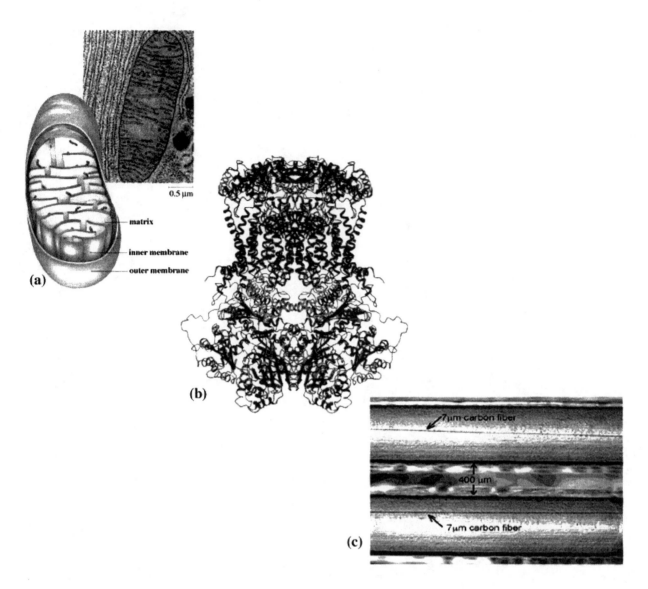

(a) A cutaway drawing and a transmission electron micrograph of a mitochondrion, the organelle where energy for eukaryotic cells is produced by coupling the oxidation of fuel molecules to the production of ATP. (b) Computer-generated model of the cytochrome bc_1 protein structure, which consists of 11 separate proteins and iron-ion complexes. A large number of these protein structures are located in the inner membranes of mitochondria and are responsible for part of the coupling pathway for energy conversion. (c) A tiny experimental electrochemical cell that uses the partial oxidation of glucose by oxygen to produce an electric current. The goal is to make these energy sources small enough to power therapeutic devices implanted in our bodies.

Chapter 10. Reduction-Oxidation: Electrochemistry

> …one must separate things in order to unite them. One must put
> them into their places as carefully as one handles fire and water…
>
> *I Ching (Wilhelm Baynes translation)*

Combustion, the *oxidation* of fuels by molecular oxygen, supplies the majority of the energy that supports the economy of the world. All multi-celled and many single-celled organisms also depend upon the energy from *oxidation* of biological fuel molecules by molecular oxygen. The redox (reduction-oxidation) processes in living cells take place quietly and in many steps. They do not create the large amounts of light and heat we get when we burn a fuel. Nevertheless, the *overall* thermodynamics of combustion and cellular oxidation are the same; we discussed them in Chapters 7 and 8. One goal of this chapter is to find out how organisms harness redox reactions to provide their energy.

The illustration on the facing page shows the structure of **mitochondria**, the powerhouses of living cells. Within the inner mitochondrial membrane, the protein complex cytochrome bc_1 (also illustrated), plays a central role in **respiration**, the process of biological oxidation of fuel molecules. The chemistry that occurs in this protein complex can be represented by this reaction involving the oxidation of reduced ubiquinone (ubiquinol), UQH_2, to ubiquinone, UQ, by Fe(III):

$$UQH_2 \quad + 2H_2O + 2Fe^{3+} \rightleftharpoons UQ \quad + 2H_3O^+ + 2Fe^{2+} \tag{10.1}$$

Reaction (10.1) in living cells is easier to analyze and understand, if we study it first in a different kind of cell. An **electrochemical cell** is a device that uses the free energy change of a redox reaction to move electrons through an external electrical circuit to do work. The batteries you use to power flashlights and electronic devices like cell phones and portable CD players are electrochemical cells. The facing-page illustration includes a photomicrograph of a miniature electrochemical cell that is based on the partial oxidation of glucose:

$$+ \tfrac{1}{2}O_{2(g)} \rightleftharpoons \tag{10.2}$$

10.1. **Consider This**   *What atoms are oxidized and reduced in reactions (10.1) and (10.2)?*

(a) In reaction (10.1), which reactant atom (or ion) is oxidized? Which reactant atom (or ion) is reduced? Explain how you know. If necessary, review Sections 6.9--6.11.

(b) Answer the same questions for reaction (10.2).

Before we analyze reactions (10.1) and (10.2) in more detail, we will examine several other simpler systems that will provide the concepts and practice we need in order to make sense of these two. In order to understand how an electrochemical or living cell works, we often need to consider the reduction and oxidation reactions separately. In many cases these reactions can actually be separated in space; so we separate the cell into its half reactions in order to learn more about it, as the opening quotation from the *I Ching* suggests. We will begin our discussion by reexamining the kind of investigations we carried out in Chapter 2, where we tested solutions for the presence of ions by finding out whether the solutions conducted an electric current. This discussion will help us relate electrical measurements to the chemistry occurring in reduction-oxidation reactions and will introduce important uses of electrochemistry.

Section 10.1. Electrolysis

10.2. **Investigate This**   *What happens when electric current passes through water?* 👓

(a) Do this as a class investigation and work in small groups to discuss and analyze the results. Half fill one well of a 6-well plate with distilled water and add one or two crystals of magnesium sulfate, $MgSO_4 \cdot 7H_2O$ (Epsom salt) and several drops of universal acid-base indicator. The color of the indicator ranges from red in acidic solutions through green in neutral solutions to blue in basic solutions. Use an overhead projector to show the well on a screen and note and record the color of the solution. Use thin carbon rods (pencil lead) as **electrodes**, electrical conductors used to make electrical contact between the solution and the external electrical components. Using wires with alligator clips, connect one end of each electrode to a 9-volt battery. Record which electrode is connected to the positive and which to the negative terminal of the battery.

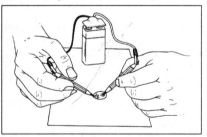

Immerse the other ends of the electrodes in the solution and hold them about two centimeters apart in the solution for 30-45 seconds and then remove them. Note and record any evidence for chemical reactions that you see occurring.

(b) After you remove the electrodes from the solution, stir the solution gently and record any evidence for chemical reactions that you see occurring.

10.3. Consider This   *What chemical reactions does an electric current cause?*

(a) What evidence of chemical reaction(s) did you observe in Investigate This 10.2(a) when the carbon electrodes were in the solution? Was a reaction associated with only one or with both of the electrodes? What do you think the product(s) of the reaction(s) is(are)? Explain the reasoning for your answers.

(b) What evidence of chemical reaction(s) did you observe in part (b) of the investigation? What do you think the product(s) of the reaction(s) is(are)? Explain your reasoning.

In Chapter 2, Section 2.3, when you tested solutions for electrical conductivity, you probably noticed that bubbles of gas were produced at the wires dipped in conducting solutions. In Investigate This 10.2, you also observed bubbles of gas produced at the carbon electrodes in the solution. In addition, the solution near each electrode changed color, indicating that the pH in these parts of the solution was changing. These pieces of evidence suggest that the electric current from the battery is causing chemical reactions at or near the surface of the electrodes as the ions in the solution conduct the current through the solution.

The acid-base indicator color changes show that the solution became more acidic near the electrode connected to the positive terminal of the battery and more basic near the electrode connected to the negative terminal of the battery. If you carry out this same investigation using sodium chloride (or many other ionic compounds) instead of magnesium sulfate in the solution, the results are the same: gas is produced at each electrode and the same pH changes occur. This information tells us that it is not the ionic compounds in the solution that are reacting at the electrodes. Water is the only other species in these solutions, so it must be the water whose atoms are being reduced and oxidized as a result of the electric current flow through the solution.

Electrolysis of water

Water molecules react with electrons from the battery at the negative electrode in the solution. The hydrogen atoms are reduced from an oxidation number of +1 in $H_2O(l)$ to zero in $H_2(g)$, hydrogen gas:

$$2H_2O(l) + 2e^- \text{ (from electrode)} \rightarrow H_2(g) + 2OH^-(aq) \tag{10.3}$$

Reaction (10.3) explains both the production of gas and formation of a basic solution at the negative electrode in Investigate This 10.2. Recall that, in Chapter 6, Section 6.10, we introduced

---

half reactions as a method for balancing reduction-oxidation reaction equations. A **half reaction** represents either the reduction or the oxidation taking place in a reduction-oxidation reaction. Reaction equation (10.3) is a reduction half reaction.

At the positive electrode electrons leave the reacting solution. Oxygen atoms are oxidized from a –2 oxidation number in water to zero in $O_2(g)$, oxygen gas:

$$2H_2O(l) \rightarrow O_2(g) + 4H^+(aq) + 4e^- \text{ (to electrode)} \tag{10.4}$$

In equation (10.4), as in Chapter 6, Sections 6.9 - 6.11, we have used $H^+(aq)$ instead of $H_3O^+(aq)$. We will continue to use $H^+(aq)$ in this chapter, in order to simplify reduction-oxidation equations.

This oxidation half reaction, reaction (10.4), explains both the production of gas and formation of an acidic solution at the positive electrode in Investigate This 10.2.

10.4. Consider This   *What happens to the $Mg^{2+}(aq)$ and $SO_4^{2-}(aq)$?*

(a) One possible result of reactions (10.3) and (10.4) would be to create a negatively-charged solution of $OH^-(aq)$ ions around the negative electrode and a positively-charged solution of $H^+(aq)$ around the positive electrode in Investigate This 10.2. Charge separation is an unfavorable process requiring a great deal of energy. Another possibility is movement of the $Mg^{2+}(aq)$ and $SO_4^{2-}(aq)$ ions to compensate for the production of the $OH^-(aq)$ and $H^+(aq)$ ions. Does the movement of the positive and negative ions shown in Figure 10.1 (a copy of Figure 2.8(c) in Chapter 2, Section 2.3) explain how charge separation is avoided? Why or why not?

(b) Sketch a new version of Figure 10.1 that explains, at the molecular level, the conductivity of ionic solutions, the flow of electrons between the electrodes outside the solution, and the observed production of bubbles at the electrodes.

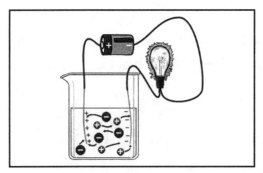

Figure 10.1. Representation of electrical conduction by an ionic solution.

In Investigate This 10.2, the overall reaction caused by the flow of electric charges in the circuit is the sum of half reactions (10.3) and (10.4). Before we add the half reactions, however, we have to account for the fact that an **electric circuit** is a closed pathway for the movement of *electric charge*. The same number of electrons must leave one electrode and enter the other electrode in an electric circuit. The flow of electric charges (ions and electrons) represented in Figure 10.1 and which you have shown in your sketch for Consider This 10.4(b) represents an electric circuit. Note carefully that *the charge carriers are not the same in all parts of an electric circuit*. Ions are the charge carriers in the solution and electrons are the charge carriers in the electrodes and wires. Inside the battery, ions are again the charge carriers, as we will find in Sections 10.2 and 10.3. To equalize the number of electrons leaving and entering the electrodes, we multiply equation (10.3) by two before adding to equation (10.4). The electrons will thus cancel out:

$$2\{2H_2O(l) + 2e^- \text{ (from electrode)} \rightarrow H_2(g) + 2OH^-(aq)\}$$
$$2H_2O(l) \rightarrow O_2(g) + 4H^+(aq) + 4e^- \text{ (to electrode)}$$
$$\overline{6H_2O(l) \rightarrow 2H_2(g) + O_2(g) + 4H^+(aq) + 4OH^-(aq)} \qquad (10.5)$$

When you removed the electrodes and stirred the solution in Investigate This 10.2(b), you mixed the hydronium and hydroxide ions together, they reacted to form water, and the entire solution returned to its original color:

$$4H^+(aq) + 4OH^-(aq) \rightarrow 4H_2O(l) \qquad (10.6)$$

Combining reactions (10.5) and (10.6) gives us the net reaction caused by the electric current:

$$2H_2O(l) \rightarrow 2H_2(g) + O_2(g) \qquad (10.7)$$

We often use the term **redox reaction** to refer to a net **red**uction-**ox**idation reaction. Redox reaction (10.7) represents the **electrolysis**, the breaking up (*lysis*) of a molecule (water in this case) by the action of the electric current flowing between the electrodes. An **electrolytic cell** is a solution (or ionic liquid) containing electrodes at which electrolysis reactions occur when an electric current from an external source flows between the electrodes.

10.5. Consider This   *How do you analyze the gases from water electrolysis?*

This diagram shows a simple electrolytic cell you could use to trap the gases produced by electrolysis of water. Before electrolysis began, the entire apparatus was filled with electrolyte solution. At which electrode is oxidation occurring? At which is reduction occurring? Clearly explain the reasoning for your answers. Which electrode (right or left) is attached to the negative terminal of the current source? Explain your choice.

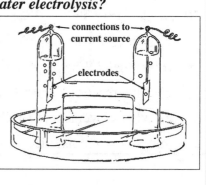

10.6. **Check This**   *Reactions in an electrophoresis apparatus*

Look back at the diagram of an electrophoresis apparatus in Figure 9.6, Chapter 9, Section 9.5. The gel, which is made using an ionic buffer solution, is in contact at each end with an ionic buffer solution. When the electrodes are connected to a source of electric current, gas bubbles are observed at the electrodes in the buffer wells. What gas(es) is(are) formed at each of the electrodes? Explain clearly and completely.

Anode and cathode

We call the electrodes in electrochemical cells the anode and cathode. The **anode** is the electrode at which oxidation occurs. The *anode acts as an electron acceptor* from the chemical reaction at the electrode. The **cathode** is the electrode at which reduction occurs. The *cathode acts as an electron donor* to the chemical reaction at the electrode. In the external circuit, electrons flow from the anode to the cathode. This nomenclature applies both to electrolytic cells in which an electric current causes chemical reactions and to galvanic or voltaic cells in which chemical reactions are used to produce an electric current. Our discussion of galvanic cells will begin in Section 10.2.

10.7. **Investigate This**   *What other reactions can occur in an electrolytic cell?*

Do this as a class investigation and work in small groups to discuss and analyze the results. Fill one well of a 6-well plate about one-third full of 0.1 M aqueous copper sulfate, $CuSO_4$, solution. Use thin carbon rods as electrodes and wires with alligator clips to connect them to a 6-V battery. Record which electrode is attached to the positive and which to the negative terminal of the battery. Immerse the electrodes in the solution and hold them about 2 cm apart for about 45 seconds and then remove them. Note and record any evidence for chemical reactions that you observe.

10.8. **Consider This**   *What are the products of electrolysis of a $CuSO_4$ solution?*

(a) In Investigate This 10.7, what reaction do you think occurs at the cathode in the electrolysis cell? What is your evidence?

(b) What reaction do you think occurs at the anode? What is your evidence?

(c) What movement of species do you think occurs in the electrolytic solution? Explain the reasoning for your answer.

Stoichiometry of electrolysis

When we discussed the stoichiometry of the chemical reactions that occurred in the electrolysis of water in Investigate This 10.2, we assumed that the number of electrons leaving the cathode (to reduce water) was equal to the number that entered the anode (from oxidation of water). This assumption is consistent with the chemical evidence that the moles of $H^+(aq)$ and $OH^-(aq)$ produced are the same. To better understand an electrochemical cell we need to be able to relate the number of moles of reaction products formed to the number of moles of electrons that enter (and leave) the cell.

We can get the number of moles of reaction products from water electrolysis by titrating the acid and/or base formed or by measuring the volume, pressure, and temperature of the hydrogen and/or oxygen formed. The reaction at the cathode in Investigate This 10.7 suggests a simpler technique. The deposit of solid metal on the cathode from a solution containing the $Cu^{2+}(aq)$ ion indicates that the reduction reaction is:

$$Cu^{2+}(aq) + 2e^- \rightarrow Cu(s) \tag{10.8}$$

The reduction process in reaction (10.8) is called **electrodeposition** because the metal is deposited by the action of the electric current through the cell. We can determine the number of moles of $Cu(s)$ formed (and $Cu^{2+}(aq)$ reduced) by simply weighing the cathode before and after reaction to get the mass of the electrodeposited metal. Under carefully controlled **electroplating** conditions many metals, including copper, nickel, silver, and gold, deposit in a thin layer (plate) on the cathode.

10.9. Check This   *The oxidation reaction in Investigate This 10.7*

Bubbles of gas are produced at the anode in Investigate This 10.7. What is the gas? What is the oxidation reaction that produces the gas? What, if any, are the other products of this reaction? What experimental tests could you do to find out whether your responses are correct? Explain.

To determine the number of moles of electrons that enter a cell, we need to make electrical measurements that "count" electrons. Recall from Table 3.1, Chapter 3, that the unit of electrical charge is the **coulomb, C**. The charge on an electron is 1.60218×10^{-19} C, so the charge on a mole (Avogadro's number) of electrons is:

charge per mole of $e^- = (6.0221 \times 10^{23}$ electron·mol^{-1})·$(1.60218 \times 10^{-19}$ C·electron^{-1})

charge per mole of $e^- = 96{,}485$ C·mol$^{-1} = 9.6485 \times 10^4$ C·mol^{-1} \qquad (10.9)

The charge per mole of electrons, 9.6485×10^4 C·mol^{-1}, is called the **Faraday**, symbolized *F*, to honor Michael Faraday (British scientist, 1791-1867), who developed the laws describing electrochemical stoichiometry. If we know the amount of charge that flows in the electric circuit connected to our electrolysis cell, we can calculate the number of moles of electrons that enter the cell. One way to measure the amount of charge that flows in an electrical circuit is to measure the current, in **amperes**, **A**, caused by the flow. The relationship between coulombs and amperes is $1\ A = 1\ C \cdot s^{-1}$, so the amount of charge that flows in a circuit is the amperage of the current times the length of time the current flows:

$$\text{(amount of charge, C)} = \text{(current, A)} \cdot \text{(time, s)} \qquad (10.10)$$

10.10. Worked Example   *Stoichiometry of nickel electrodeposition*

Suppose we do a nickel electroplating experiment and find that 0.272 g of nickel are deposited on the cathode when a current of 0.246 A is passed through the electrodeposition cell for 3640. s (about an hour). Are these data consistent with the stoichiometry of this reaction?

$$Ni^{2+}(aq) + 2e^- \rightarrow Ni(s) \qquad (10.11)$$

Necessary information: We need the molar mass of nickel, 58.69 $g \cdot mol^{-1}$, and the value of the Faraday, $9.6485 \times 10^4\ C \cdot mol^{-1}$.

Strategy: Compare the number of moles of electrons that entered the cell to the number of moles of nickel deposited to see if the ratio of electrons to nickel is two, as equation (10.11) predicts.

Implementation: The number of coulombs of charge that enter the cell is:

number of coulombs that enter = (0.246 A)·(3640 s) = 895 C

The number of moles of electrons represented by this charge is:

$$\text{moles of electrons} = (895\ C) \cdot \left(\frac{1\ mol}{96,485\ C} \right) = 0.00928\ mol$$

The number of moles of nickel deposited is:

$$\text{moles of nickel} = (0.272\ g) \cdot \left(\frac{1\ mol}{58.69\ g} \right) = 0.00463\ mol$$

$$\text{The stoichiometric ratio} = \frac{\text{mol electrons}}{\text{mol nickel}} = \frac{0.00928\ mol}{0.00463\ mol} = 2.00$$

Does the answer make sense? The ratio of moles of electrons that enter the cell to the moles of nickel deposited is two, as predicted by the reaction equation. The answer makes sense and helps to confirm that half reactions like equations (10.8) and (10.11) do represent the electrochemical stoichiometry of reactions that occur in an electrochemical cell.

10.11. **Check This**   *Electrochemical stoichiometry*

Electrolytic cells can be connected in series so that the same current passes through more than one cell and the reactions in them can be directly compared. This is the way Faraday carried out many of his quantitative experiments. He often used electrodeposition of silver as his comparison; the half reaction for the reduction of $Ag^+(aq)$ is:

$$Ag^+(aq) + e^- \rightarrow Ag(s) \tag{10.12}$$

(a) What do you predict will be the *ratio* of the mass of silver to mass of nickel deposited in an experimental set up with nickel and silver electrodeposition cells in series? Clearly explain your reasoning.

(b) If in the set up in part (a) a current of 0.106 A is passed for 7200. s (two hours), what mass of silver and what mass of nickel will be deposited? Is the *ratio* of these masses what you predicted in part (a)? Explain why or why not.

Applications of electrolysis

Electroplating of metals is an important application of electrolysis. Often, for example, the metal trim of motor vehicles is plated with chromium both to make it shiny and to make it resistant to corrosion. Jewelry made of less expensive metals can be gold plated to make it look like gold, but be more affordable than items made with a great deal more gold. Flatware (eating utensils — knives, forks, and spoons) is often called "silverware," because expensive flatware used to be made mostly of silver (compounded with other metals for strength). Nowadays, most silverware is silver plate, that is, a thin layer of silver electroplated onto a utensil made mostly of steel. The utensil looks like silver (and tarnishes like silver) but is less expensive than an old-fashioned silver utensil.

Far more important, however, are electrolytic processes for obtaining pure metals, such as aluminum and copper. Pure aluminum is a reactive metal and reacts rapidly with oxygen in the air to form aluminum oxide:

$$4Al(s) + 3O_2(g) \rightarrow 2Al_2O_3(s) \qquad \Delta G° = -3164.6 \text{ kJ} \tag{10.13}$$

This is an enormously favorable reaction that forms a very thin, unreactive protective coating on aluminum metal. The coating prevents oxygen and many other corrosive substances, including strong acids, from reacting with the metal beneath it. Objects made of aluminum resist corrosion and retain their metallic luster for a long time. On the negative side, the great stability of $Al_2O_3(s)$ makes it very difficult to obtain pure $Al(s)$ from its ores where it is usually present as its oxide.

Therefore, although aluminum is the most abundant metal in the Earth's crust (Chapter 3, Figure 3.26), it was rare and expensive until 1886 when two young inventors, Charles M. Hall

(American, 1863-1914) and Paul L. T. Héroult (French, 1863-1914), simultaneously and independently invented the same electrolytic process for obtaining aluminum metal from its oxide. A diagram of the kind of apparatus that is now used to obtain aluminum from its oxide is shown in Figure 10.2.

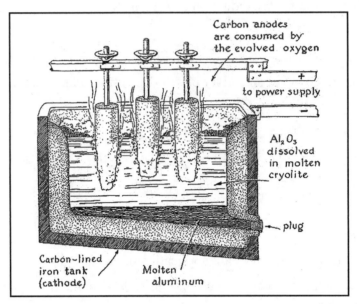

Figure 10.2. Diagram of an electrolytic cell used to produce aluminum metal.

In the electrolytic furnace shown in Figure 10.2, $Al_2O_3(s)$ is dissolved in a molten salt, Na_3AlF_6 (cryolite), at a temperature near 1000 °C and electrolyzed:

$$\text{cathode:} \quad Al^{3+}(solution) + 3e^- \rightarrow Al(l) \tag{10.14}$$

$$\text{anode:} \quad 2O^{2-}(solution) \rightarrow O_2(g) + 4e^- \tag{10.15}$$

$$\text{overall:} \quad 4Al^{3+}(solution) + 6O^{2-}(solution) \rightarrow 4Al(l) + 3O_2(g) \tag{10.16}$$

The species in the cryolite solution are more complex than equations (10.14) and (10.15) show, but the equations represent the electrochemical stoichiometry of the electrode reactions. At the high temperatures in the electrolytic furnace, the oxygen gas reacts with the carbon anodes to form $CO_2(g)$, so the anodes are consumed and have to be replaced periodically. Aluminum metal, melting point 660 °C, is more dense than the molten cryolite solution, so it sinks to the sloping bottom of the vessel and is drawn off daily.

10.12. Worked Example   *Electrolytic production of aluminum metal*

Suppose an electrolytic furnace like one of those shown here produces one metric ton, 1000 kg, of aluminum metal per day. What current, in amperes, must be used to produce this much metal?

Necessary information: The process inside each of these furnaces is what is shown in Figure 10.2. We will need the molar mass of Al, 26.98 g·mol⁻¹, and $F = 9.6485 \times 10^4$ C·mol⁻¹. We also need to remember that there are 60 s·min⁻¹, 60 min·hr⁻¹, and 24 hr·day⁻¹.

Strategy: Find the number of moles of Al produced per day and hence the number of moles of electrons required. Use the Faraday to convert moles of electrons to coulombs and then find the current that has to flow for one day to provide this many coulombs.

Implementation: The number of moles of Al produced per day is:

$$(\text{mol Al})\text{mol Al})\cdot\text{day}^{-1} \left(\frac{1000 \text{ kg Al}}{1 \text{ day}} \right) \left(\frac{1000 \text{ g}}{1 \text{ kg}} \right) \left(\frac{1 \text{ mol Al}}{26.98 \text{ g}} \right) = \frac{3.71 \times 10^4 \text{ mol Al}}{1 \text{ day}}$$

Each mole of aluminum produced requires 3 mol of electrons, equation (10.14); the number of coulombs of charge on these electrons is:

$$\text{coulombs}\cdot\text{day}^{-1} = \left(\frac{3.71 \times 10^4 \text{ mol Al}}{1 \text{ day}} \right) \left(\frac{3 \text{ mol } e^-}{1 \text{ mol Al}} \right) \left(\frac{9.6485 \times 10^4 \text{ C}}{1 \text{ mol } e^-} \right) = \frac{1.07 \times 10^{10} \text{ C}}{1 \text{ day}}$$

The current required to pass this many coulombs through the electrolytic cell in one day is:

$$\text{current} = \left(\frac{1.07 \times 10^{10} \text{ C}}{1 \text{ day}} \right) \left(\frac{1 \text{ day}}{24 \text{ hr}} \right) \left(\frac{1 \text{ hr}}{60 \text{ min}} \right) \left(\frac{1 \text{ min}}{60 \text{ s}} \right) = 1.24 \times 10^5 \text{ C}\cdot\text{s}^{-1} = 124{,}000 \text{ A}$$

Does the answer make sense? This is a very high current and may not seem reasonable. However, the furnaces shown in the figure do draw currents between 50,000 and 150,000 A. Contrast this with contemporary single-family houses whose electrical circuits carry a total of only 200 to 400 A. Aluminum electrolytic refining plants are usually located near an electricity generating plant, in order to avoid current losses through long distance transmission.

Copper metal is occasionally found in the Earth's crust in its metallic form, but the vast majority of copper is bound in its ores as oxides or sulfides. These are treated in various reduction-oxidation chemical processes to produce metal that is about 99% pure. One of the most important uses for copper metal is as an electrical conductor; almost all of the electrical wiring you see and use is copper. Copper that is only 99% pure conducts electricity much less well than

99.99% pure copper, so it must be further purified by electrolysis, as shown schematically in Figure 10.3.

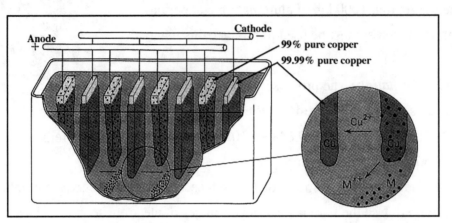

Figure 10.3. Diagram of an electrolytic cell used to purify copper.

The dots in the diagram represent impurities. During electrolysis, some of these such as Fe and Zn become oxidized to ions in solution. Others such as Ag, Au, Pt, and Pd are not oxidized and fall to the bottom as a sludge.

The reactions of copper in the electrolytic cell are:

$$\text{cathode:} \quad Cu^{2+}(aq) + 2e^- \rightarrow Cu(s) \tag{10.17}$$

$$\text{anode:} \quad Cu(s) \rightarrow Cu^{2+}(aq) + 2e^- \tag{10.18}$$

$$\text{overall:} \quad Cu(s) \text{ (from anode)} \rightarrow Cu(s) \text{ (to cathode)} \tag{10.19}$$

This overall reaction is denoted by the top arrow in the inset of Figure 10.3. The net effect is to transfer copper atoms from the impure anode (99% copper) to the cathode, which starts out as a thin sheet of pure copper and builds up to a thick sheet of 99.99% pure copper. The sludge that collects on the bottom of the cell is periodically collected and further refined to yield the valuable and expensive metals listed in the figure legend. In some cases, these recovered metals can be sold for enough to pay the entire cost of the electricity required to purify the copper.

10.13. **Check This**   *Electrolytic purification of copper metal*

The many cells shown here are being used to purify copper. In these cells the cathodes start out as sheets of pure copper about one meter square and less than 1 mm thick. When the cathodes are removed from the cell after 14 days of electrolysis, they are about 2 cm thick. If the original mass of a cathode is about 7 kg and the final mass is 170 kg, what current, in amperes, is required to purify the copper? Show your work clearly.

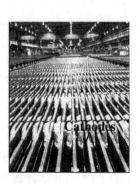

Reflection and projection

From the very beginning of the book, we have been using the electrical nature of matter to help explain many observable chemical properties. It is not at all surprising, therefore, to find that electricity interacts with matter to bring about chemical changes. The half reactions introduced in Chapter 6 help to make sense of the chemistry that occurs at the electrodes in an electrolytic cell when it is connected to a source of electrical current. At the cathode, electrons are supplied to the cell and a reducing half reaction occurs. At the anode, electrons leave the cell and an oxidizing half reaction occurs. The electric current is carried inside the cell by the movement of ions that maintain charge balance at both electrodes.

Half reactions also provide a way to keep track of the electrochemical stoichiometry of both the chemical species and the electrons that take part in the reactions. The key to this stoichiometry for electrons is the Faraday, 9.6485×10^4 C·mol^{-1}, the charge on a mole of electrons (or any other singly-charged particle). We can use current flow through a cell, 1 C·s^{-1} = 1 A, as a measure of the number of moles of electrons and relate moles of electrons to moles of chemical reaction products, as we did for electrodepositions.

The results of electrolytic reactions are all around you in the aluminum and copper and many electroplated objects you see and use daily. However, you rarely witness electrolytic reactions themselves. On the other hand, you use the reduction-oxidation reactions in galvanic cells all the time. These cells are the power sources for almost all portable electronic devices: calculators, laptop computers, cell phones, radios, CD and tape players, and hand-held computer games. In galvanic cells, we are interested in the electrical current created by chemical reactions. In the next section, we will begin our discussion of galvanic cells with some simple reduction-oxidation reactions that we can use to uncover further properties of electrochemical cells.

Section 10.2. Electric Current from Chemical Reactions

10.14. **Investigate This**   *How do you get electricity from a chemical reaction?*

(a) Do Investigate This 6.62 or review your notes and discussion from that investigation.

(b) Do this as a class investigation and work in small groups to discuss and analyze the results. Fill one well of a 24-well plate about two-thirds full of 0.10 M aqueous silver nitrate, $AgNO_3$, solution. Place a silver wire in the solution with one end sticking out of the well. Fill an adjacent well about two-thirds full of 0.10 M aqueous copper sulfate, $CuSO_4$, solution and place a copper wire in the solution with one end sticking out of the well. Connect one lead (pronounced "leed") from a digital multimeter to one of the wires and the other lead to the other wire. Set the multimeter to read current (amperes) and record the current reading from the meter.

(c) Roll a 1 × 5-cm rectangle of filter paper into a tight, 5-cm-long cylinder and fold it into a square "U" shape. Thoroughly wet the paper with a 10% aqueous solution of potassium nitrate, KNO_3. Place one leg of the "U" in the silver nitrate well and the other leg in the copper sulfate well and again record the current reading from the meter.

10.15. Consider This *How do you get electricity from a chemical reaction?*

(a) What were your current readings for parts (b) and (c) of Investigate This 10.14? Propose a molecular level explanation for why they were different.

(b) What were your results in Investigate This 6.62? What reaction did you see occurring in that investigation? How do you think that reaction is related to your observations in parts (b) and (c) of Investigate This 10.14?

In Chapter 6 and in the previous section, we wrote half reactions as either reductions or

| WEB Chap 10, Sect 10.2.1-2. Study interactive animations of reactions between two more metal-metal ion pairs. |

oxidations, depending upon the reaction we were trying to balance or the reaction that was going on at an electrode. From this point on, however, we will use the standard electrochemical convention: *half reactions are written as reductions*. For the silver-copper reaction from Investigate This 6.62, the half reactions are:

$$Ag^+(aq) + e^- \rightleftharpoons Ag(s) \tag{10.20}$$

$$Cu^{2+}(aq) + 2e^- \rightleftharpoons Cu(s) \tag{10.21}$$

10.16. Consider This *How are reduction half reactions combined?*

(a) Explain how reactions (10.20) and (10.21) can be combined, with appropriate stoichiometric factors, to give:

$$Cu(s) + 2Ag^+(aq) \rightarrow Cu^{2+}(aq) + 2Ag(s) \tag{10.22}$$

Reaction (10.22) is the overall reaction observed in Investigate This 10.14(a) [Investigate This 6.62].

(b) What happens to the electrons when the half reactions are combined as you did in part (a)? How is this analogous to what you had to do to get equation (10.5) in Section 10.1? Explain.

Separated half reactions

The net result of reaction (10.22) is to transfer two electrons from copper metal to two silver ions to produce metallic silver and one copper ion in solution. The overall reaction equation contains no electrons — free electrons do not exist in aqueous solutions. Is the idea of electron transfer just a convenient way to explain the products of the reaction? Here is where the *I Ching* quotation that opens this chapter is relevant. In Investigate This 10.14(b) and 10.14(c), we

> **WEB, Chap 10, Sect 10.2.3-6.** Study interactive animations of an experiment similar to that represented in Figure 10.4.

separated the components of the two half reactions, (10.20) and (10.21), in order to see if there is experimental evidence for electron transfer (a current flow) that unites the half reactions. The set-up for Investigate This 10.14(b) is shown schematically in Figure 10.4.

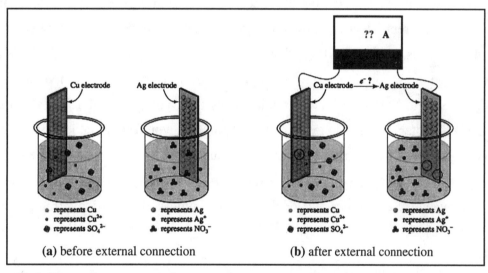

Figure 10.4. Does electron transfer occur between separate silver and copper wells?
In (b), red rings highlight the loss of one copper atom and gain of two silver atoms at the respective electrodes.

Figure 10.4 reminds us that one of the wells contains $Ag^+(aq)$ and $Ag(s)$, the reactant and product (excluding electrons) of reaction (10.20). The other well contains $Cu^{2+}(aq)$ and $Cu(s)$, the reactant and product of reaction (10.21). The external electrical connection in Figure 10.4(b) would allow electrons to move from one electrode to the other. A possible result is represented at the molecular level in Figure 10.4(b). A copper atom from the copper electrode loses two electrons to the rest of the metal and enters the solution as a $Cu^{2+}(aq)$ ion. The electrons travel through the external part of the circuit to the silver electrode, which gives it an extra two electrons. These electrons are transferred to two silver ions in the solution at the surface of the metal, thus converting them to silver atoms on the surface of the electrode.

10.17. Consider This   *What are the net charges on the solutions in Figure 10.4?*

(a) Explain how Figure 10.4 represents reaction equation (10.22).

(b) What is the net charge (number of positive charges minus number of negative charges) in the silver well in Figure 10.4(a)? in Figure 10.4(b)? What is the net charge in the copper well in Figure 10.4(a)? in Figure 10.4(b)? Explain how these charges are the result of the process just described.

Your result from Investigate This 10.14(b) shows that no current flows between the electrodes in the set-up shown in Figure 10.4(b). You know from your results in Investigate This 10.14(a) that reaction (10.22) occurs when copper metal is placed in contact with a solution containing silver ion. It seems as though the process we show in Figure 10.4 should be able to occur to produce a current (electron flow) through the meter. Why does no current flow?

You found in Consider This 10.17(b) that the process we imagined in Figure 10.4 leads to a net positive charge in the solution of $Cu^{2+}(aq)$ and $SO_4^{2-}(aq)$ ions and a net negative charge in the solution of $Ag^+(aq)$ and $NO_3^-(aq)$ ions. In Consider This 10.4, you analyzed a similar situation for the electrolysis of water. In order to avoid charge separation, you found that ions in the solution migrate to compensate for the possible build up of charge near the electrodes. In the set-up shown in Figure 10.4, there is no pathway for ions to migrate between the solutions. Electrons cannot flow between the electrodes outside the solutions unless the electrical *circuit* is completed within the solutions.

Salt bridge

We need a way to complete the circuit between the solutions, so we can test whether electron transfer occurs. When the reactants are in contact with one another, the ions are all in the same solution and can migrate freely; reaction (10.22) occurs and we can observe the products. What we need in our separated system is a pathway for the ions in the separate wells to interact with one another without mixing. This pathway is provided by a **salt bridge**, a connection (bridge) between two half-reaction systems that allows "salt" (equivalent net positive and negative ions) to move between the systems. In Activity 10.14(c), your salt bridge was a filter paper soaked in potassium nitrate solution, as represented in Figure 10.5.

Figure 10.5(a) is the same as Figure 10.4(a), except that there is a salt bridge between the

WEB Chap 10, Sect 10.2.7-9.
Study interactive animations to see how a salt bridge completes the cell electric circuit.

wells in Figure 10.5. Once again, we can imagine a transfer of two electrons from the copper side to the silver side of the set up. Now, when a $Cu^{2+}(aq)$ ion is formed, two nitrate ions, $NO_3^-(aq)$, can move out of the salt bridge to keep the $Cu^{2+}(aq)$ solution electrically neutral. And in the other

---

well, when two $Ag^+(aq)$ ions are reduced to silver metal, they can be replaced by $K^+(aq)$ ions from the salt bridge to keep the $Ag^+(aq)$ solution electrically neutral. These changes are represented in Figure 10.5(b). The net charges are zero in all three solutions; there is no charge separation.

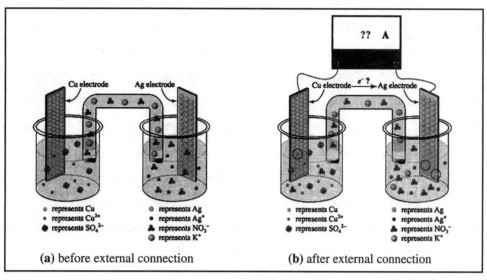

(a) before external connection   (b) after external connection

Figure 10.5. Does electron transfer occur with a salt bridge to complete the circuit?

10.18. Check This   *Ion migrations in a salt bridge*

   (a) Ions do not always move out of a salt bridge. For example, in Figure 10.5(b) a $Cu^{2+}(aq)$ ion might enter the salt bridge, instead of two $NO_3^-(aq)$ leaving it. Make a drawing showing how this alternative still makes the net charges in all three solutions zero. How would you detect the migration of $Cu^{2+}(aq)$ ions into the salt bridge?

   (b) Are there other ion migrations that could occur, together with electron transfers between the electrodes through the external circuit, that also maintain the electrical neutrality of all the solutions? If so, give examples. How could you detect such migrations?

   (c) Think once again about the electrophoresis apparatus in Figure 9.6, Chapter 9, Section 9.5. The samples on the gel are ionic; we detect proteins, the negatively charged ions in the sample. Explain how the gel in an electrophoresis apparatus acts much like a salt bridge. Does the salt bridge animation, WEB Chap 10, Sect 10.2.9, reinforce your explanation? Explain.

In Investigate This 10.14(c) you found that an electric current flows in the connection between the electrodes when the solutions are connected by a salt bridge. There is a net transfer of electrons between the electrodes through their external connection. An equivalent amount of electric charge is being carried by the movement of ions through the salt bridge and the solutions. The combination of electron flow between the electrodes outside the solution and the movement of ions in the solution completes the electric circuit.

Galvanic cells

The electrochemical cell you set up in Investigate This 10.14(c) is a galvanic cell. In a **galvanic cell** (also called a **voltaic cell**), a chemical redox reaction produces electrons that move through an electrical conductor connected between the two electrodes. As you have observed, the half reactions that make up an overall electrochemical cell reaction can be spatially separated, *if* ions can migrate between them through some form of salt bridge. The separate solutions and electrodes for each half reaction in a galvanic cell are called **half cells**.

Electrochemical cell notation

You can imagine how tedious it would be, if we had to draw something like Figure 10.5(a) or (b) every time we wanted to describe a new cell. There is a shorthand notation for electrochemical cells that we will use from now on. In this **line notation**, for electrochemical cells, the cell in Figure 10.5 is written on a single line like this:

$$Cu(s) \mid Cu^{2+}(aq, 0.1 \text{ M}) \parallel Ag^{+}(aq, 0.1 \text{ M}) \mid Ag(s) \tag{10.23}$$

A boundary between distinct chemical phases is denoted by a vertical line, |. For example, in cell (10.23), the boundary between the solid copper electrode and the $Cu^{2+}(aq)$ solution is shown. Since the half-cell solutions and the salt bridge solution are quite different, there is a boundary between the half-cell solution and the salt bridge solution at each end of the salt bridge. The bridge is shown as a double pair of vertical lines, ||. If, in a particular case, it matters what is in the salt bridge, then its contents are specified between the two lines. In our example, the contents are not significant, so the bridge appears simply as a pair of lines. The concentration and/or state of each species is indicated in parentheses after the species.

You could also imagine writing the copper-silver cell like this:

$$Ag(s) \mid Ag^{+}(aq, 0.1 \text{ M}) \parallel Cu^{2+}(aq, 0.1 \text{ M}) \mid Cu(s) \tag{10.24}$$

Although this cell representation doesn't look wrong, chemists have chosen, *by convention, to write the anode on the left in electrochemical line notation*, as in cell (10.23). If we assume that everyone uses this convention, we would assume that whoever wrote cell (10.24) meant to indicate that oxidation, the anodic reaction, is occurring in the half cell that is written on the left and reduction on the right. The overall cell reaction would then be:

$$2Ag(s) + Cu^{2+}(aq) \rightleftharpoons 2Ag^{+}(aq) + Cu(s) \tag{10.25}$$

Reaction (10.25) is the reverse of reaction (10.22). But we have seen above that it is reaction (10.22) that occurs when Cu(s) is placed in a solution containing $Ag^{+}(aq)$. Reaction (10.25) is misleading because it does not show the favored direction of the reaction. Therefore, we say that cell (10.24) is incorrect because it implies incorrect chemistry.

10.19. Worked Example   *Line notation for the electrolytic aluminum refining cell*

Use line notation to represent the electrolytic aluminum refining cell depicted in Figure 10.2 and accompanying text, including half reactions (10.14) and (10.15).

Necessary information: We learn from Figure 10.2 that both electrodes are carbon, that the cathode is in contact with a pool of molten aluminum, and that oxygen gas is produced at the anode. The solution in the cell is Al_2O_3 dissolved in molten cryolite. For simplicity, we assume that the ions of interest are Al^{3+} and O^{2-}.

Strategy: We write the anode on the left, solution in the middle, and cathode on the right.

Implementation: The line notation for the cell is:

$$\text{(anode)} \; C \mid O_2{\it (g)} \mid O^{2-}{\it (solution)}, \text{molten cryolite}, Al^{3+}{\it (solution)} \mid Al{\it (l)} \mid C \; \text{(cathode)} \quad (10.26)$$

Does the answer make sense? The species shown in half reactions (10.14) and (10.15) are all represented in our notation, as well as the physical set up shown in Figure 10.2. Note that all species are present in the same solution so there is no salt bridge in this cell.

10.20. Check This   *Interpreting line notation for a copper-zinc cell*

The line notation for a copper-zinc cell, with the metal ions each at 0.10 M concentration, is:

$$Zn{\it (s)} \mid Zn^{2+}{\it (aq, 0.10 M)} \parallel Cu^{2+}{\it (aq, 0.10 M)} \mid Cu{\it (s)} \qquad (10.27)$$

(a) Write the anodic and cathodic half reactions for this cell. Write the overall cell reaction. Explain the reasoning for your answers.

(b) If a piece of copper metal is placed in a solution of zinc ion, what do you predict will be observed? If a piece of zinc metal is placed in a solution of copper ion, what do you predict will be observed? Explain your reasoning.

We have learned how to set up a galvanic cell, an electrochemical cell in which a redox reaction produces a flow of electrons in an external circuit, and we have a succinct notation for describing these cells. The batteries you use to power flashlights, portable CD players, and cell phones operate on these same principles. Each consists of two electrodes – one at which an

> One meaning of "battery" is a group of the same things. In electricity, it used to mean a series of galvanic cells connected together to provide more power, but has now also come to mean an individual electrochemical cell as well as a set of connected cells.

oxidation occurs and another at which a reduction occurs. In order for electrons to flow between

---

the electrodes in the external part of the circuit (making the filament in a flashlight bulb glow and give off light, for example) there has to be a corresponding migration of ions inside the battery to complete the circuit.

To understand more about how these cells work, including how different combinations of half cells behave and what variables affect their performance, we need a quantitative measure that we can use to characterize electrochemical cells. The measure we use is the cell potential, which we often call "voltage." You are probably familiar with a variety of cells (batteries) that have different voltages. As we go on, we will look at what differentiates one cell from another and learn more about their chemistry.

Section 10.3. Cell Potentials

10.21. **Investigate This**   *What are cell voltages for different half-cell combinations?*

Do this as a class investigation and work in small groups to discuss and analyze the results. As shown in the diagram, fill one of the central wells of a 24-well plate, about two-thirds full of 10% KNO_3 solution. Fill a well adjacent to the KNO_3 well about two-thirds full of 0.10 M aqueous silver nitrate, $AgNO_3$, solution and insert a silver wire as an electrode. Fill two other adjacent wells about two-thirds full of 0.10 M aqueous zinc sulfate, $ZnSO_4$, and 0.10 M aqueous copper sulfate, $CuSO_4$, respectively. As electrodes, place a strip of zinc metal in the zinc ion solution and a copper wire in the copper ion solution. Use filter paper soaked in 10% KNO_3 solution to make salt bridge connections between the KNO_3 well and each of the other wells. This set up interconnects all three half cells through the central KNO_3 well.

(a) Connect the input leads from a digital voltmeter to the Cu and Zn electrodes. Note the meter reading, including its units and sign, and which lead is connected to the Cu electrode and which to the Zn. Record the cell voltage to the nearest millivolt. Reverse the connections and record the same information as previously.

(b) Repeat the entire procedure in part (a) for the other two pairs of half cells, Cu with Ag, and Zn with Ag.

10.22. **Consider This**   *Are cell voltages for different half-cell combinations related?*

(a) For each of the three half-cell combinations you investigated in Investigate This 10.21, you recorded two voltage readings from the digital voltmeter. How are the two readings for a combination related to one another? What was different about the set-up for the two readings? Are the answers to the preceding two questions related? Explain why or why not.

(b) Consider the positive voltages you measured for the three half-cell combinations. Is there any pattern in your results? If so, how might you explain it? Does the same pattern hold for the negative readings? Explain why or why not.

Cell potential, E

An **electrical potential difference** between two parts of an electrical system (between two electrodes, for example) is a measure of the potential energy available to move electrons between the two parts. Electrical potential difference is measured in **volts, V,** and is often called **voltage**. An electrochemical cell uses the free energy of a chemical redox reaction to produce an electrical potential difference between two electrodes. We will say more about the relationship between electrical potential difference and free energy in Section 10.5. The electrical potential difference in volts is symbolized by E, which stands for **electromotive force** (**emf**), the force that drives electrons from one electrode toward the other. E is also called the **cell potential**. An italic uppercase E is the accepted symbol for both cell potential (emf) and for internal energy. You can usually tell which is meant by the context where the symbol appears. In this chapter, E is used only for emf.

The sign of E

Measuring the cell potential, emf, of an electrochemical cell seems easy and straightforward. In Investigate This 10.21, you connected a digital voltmeter to the electrodes and read the meter for each half-cell combination. You also reversed the connections and found that the numeric reading for the cell was the same, but its sign was reversed. The meter reading (voltage) has a *sign* as well as a numeric value. The sign you read

> Usually, a positive reading on a meter will not have a sign shown (just as we usually write positive numbers without a plus sign), but a negative reading will have a negative sign shown.

depends upon how you connect the leads from the voltmeter to the electrodes. But reversing the connections of the measuring device, the voltmeter, can't affect the direction of the reaction occurring in the cell. This ambiguity about the sign of the cell potential could create communications

> **WEB Chap 10, Sect 10.3.1-2.**
> Use this animation to check your understanding of the conventions for cell potentials.

problems. We need a way assure that the signs we report for cell potentials will be interpreted the same way by everyone.

We choose *conventions* for connecting a voltmeter to the electrodes, so that the reading, E, will be positive. Voltmeters have inputs that are coded in some way to indicate which input lead should be connected to the anode and which to the cathode of the cell to be measured. Usually the input leads are color-coded, black to be connected to the anode and red to be connected to the cathode. On the meter, the inputs are often labeled "–" (or "common") for the input from the anode (from which electrons enter the external circuit) and "+" for the input from the cathode. When the input leads are connected correctly, the reading on the voltmeter will be positive, as illustrated for the zinc-copper cell in Figure 10.6. If you get a negative reading, you have connected the leads to the wrong electrodes; switching the connections produces the same reading, but with a positive sign (or no sign).

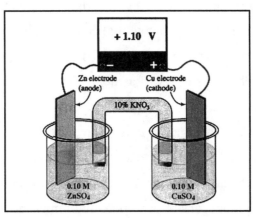

Figure 10.6. A zinc-copper cell correctly connected to measure its cell potential, E.

10.23. Consider This *What are the cell potentials and reactions in Investigate This 10.21?*

(a) Are your results (meter connections and sign and magnitude of E) for the zinc-copper cell in Investigate This 10.21 consistent with the illustration in Figure 10.6? Explain why or why not?

(b) Explain the relationship between the cell represented by the line notation in Check This 10.20 and the cell illustrated in Figure 10.6. What is the cell reaction for the cell in Figure 10.6? for the zinc-copper cell in Investigate This 10.21? Explain.

(c) Use your measurements from Investigate This 10.21 to find the cell potentials and write the cell reactions for the copper-silver and zinc-silver cells? Explain your reasoning.

Combining redox reactions and cell potentials

Any two half cells can be connected through a salt bridge to create an electrochemical cell. We can measure its cell potential and, by determining which electrode is the anode and which the cathode, can write the cell reaction that is producing this electron driving force, as you have

---

done for three combinations of half cells in Consider This 10.23. In previous chapters, you have seen several examples where we combined two chemical reaction equations to get a third reaction equation. The usual reason for doing this was to combine the known values of thermodynamic variables for the two reactions to get the values for the combination reaction. Let's see if we can do similar combinations of redox reactions to get cell potentials for unmeasured (or unmeasurable) reactions. We will develop and apply the procedure to the cells in Investigate This 10.21.

You constructed three cells:

$$Zn(s) \mid Zn^{2+}(aq,\ 0.10\ M) \parallel Cu^{2+}(aq,\ 0.10\ M) \mid Cu(s) \tag{10.27}$$

$$Cu(s) \mid Cu^{2+}(aq,\ 0.10\ M) \parallel Ag^{+}(aq,\ 0.10\ M) \mid Ag(s) \tag{10.28}$$

$$Zn(s) \mid Zn^{2+}(aq,\ 0.10\ M) \parallel Ag^{+}(aq,\ 0.10\ M) \mid Ag(s) \tag{10.29}$$

These cells are written correctly (anode —oxidation— on the left, as you found in Investigate This 10.21), so the cell reactions are, respectively:

$$Zn(s) + Cu^{2+}(aq) \rightleftharpoons Zn^{2+}(aq) + Cu(s) \tag{10.30}$$

$$Cu(s) + 2Ag^{+}(aq) \rightarrow Cu^{2+}(aq) + 2Ag(s) \tag{10.22}$$

$$Zn(s) + 2Ag^{+}(aq) \rightleftharpoons Zn^{2+}(aq) + 2Ag(s) \tag{10.31}$$

We will write the cell potentials for these three cell reactions as $E_{Zn|Cu}$, $E_{Cu|Ag}$, and $E_{Zn|Ag}$, respectively. The subscripts remind you which half cells have been combined in each case and which is the anode (written on the left).

We can combine any two of these reaction equations (with appropriate signs) to give the third. Let's combine reaction equations (10.30) and (10.22) to get (10.31). We have to subtract reaction equation (10.22) from (10.30):

$$Zn(s) + Cu^{2+}(aq) \rightleftharpoons Zn^{2+}(aq) + Cu(s) \tag{10.30}$$
$$\underline{-\{Cu(s) + 2Ag^{+}(aq) \rightarrow Cu^{2+}(aq) + 2Ag(s)\}} \tag{$-$10.22}$$
$$Zn(s) + 2Ag^{+}(aq) \rightleftharpoons Zn^{2+}(aq) + 2Ag(s) \tag{10.31}$$

If cell reactions can be combined to yield a new cell reaction, we might suppose that their cell potentials can be combined in the same way to get the cell potential for the new reaction. In the present case, the combination would be:

$$E_{Zn|Cu} - E_{Cu|Ag} = E_{Zn|Ag} \tag{10.32}$$

10.24. **Check This**   *Combining cell potentials and interpreting cell reactions*

(a) Are your cell potentials from Consider This 10.23 consistent with equation (10.32)? Explain why or why not.

(b) Use cell reaction equation (10.30) to explain what will be observed if a strip of zinc metal is placed in a solution of copper(II) ion. What will be observed if a strip of copper metal is placed in a solution of zinc(II) ion?

(c) Use cell reaction equation (10.31) to explain what will be observed if a strip of zinc metal is placed in a solution of silver(I) ion. What will be observed if a strip of silver metal is placed in a solution of zinc(II) ion?

(d) How are your predicted observations in part (c) related to your actual observations in Investigate This 6.62? Do reaction equations (10.22) and (10.31) help explain this relationship? Why or why not?

Equation (10.32) is a specific case of the general result: *the cell potential for a redox reaction that is the sum (or difference) of two other redox reactions is the sum (or difference) of the cell potentials for the two reactions.* Cell potentials are intensive quantities, like temperature and pressure. Each half cell can transfer electrons with a certain force or electric potential and the cell potential is the resultant of these two forces. The cell potential does not depend on the size of the half cells or on the amount of charge transferred (as long as the concentrations in the half cells don't change significantly). Combining one of these half cells with a different half-cell partner does not change its electron transfer potential. Electron transfer potentials for individual half cells —half-cell potentials— would be useful, if we could obtain them. We would then be able to combine half-cell potentials to calculate a cell potential without constructing an actual cell.

Reflection and projection

Electrochemical cells (galvanic cells) move the electrons that are transferred in a redox reaction through an external, electrically-conducting pathway. The interfaces between the external pathway and the oxidation and reduction that occur in the redox reaction are the cell electrodes, the anode and cathode. Oxidation occurs at the anode and the reduction at the cathode. In order for electrons to flow from the anode to the cathode outside the cell, there must be migrations of ions within the cell itself to complete the electric circuit and prevent charge separation.

We characterize an electrochemical cell by its cell potential, E, measured in volts. We measure E with a voltmeter, being careful to connect the meter to the cell electrodes correctly.

When two electrochemical cell reaction equations are combined (added or subtracted) to yield a third cell reaction equation, the cell potentials for the two reactions can be combined in the same way to yield the cell potential for the third reaction. Cell potentials are intensive quantities, so no stoichiometric factors enter into their combination to give the cell potential for the combination reaction. We can calculate the cell potential for any reaction that can be written as a combination of other reactions whose cell potentials are known. The calculated cell potential can then be used to make predictions about the combination reaction without having to carry it out or make a cell. The only problem with this approach is the need to find cell reactions with measured cell potentials that can be combined to yield the reaction of interest. It would be useful to have some standard way of cataloging redox reactions to make this search easier. The catalog we use is a list of standard half-cell potentials.

Section 10.4. Half-Cell Potentials: Reduction Potentials

No electrical potential can be measured for any single half reaction; only the cell potential for a combination of two half cells can be measured. For half cell potentials, the best we can get is a set of *relative* values, but these will allow us to make the combinations we discussed at the end of the previous section. We get these relative half-cell potentials by first choosing one half cell as a reference. Then we combine all other half cells with the reference and measure all their cell potentials in turn. Each cell potential is a combination of the half-cell potential for the reference and the half-cell potential for the one we are measuring. To generate a catalog of relative half-cell potentials, we *arbitrarily* assign a value for the half-cell potential of the reference half reaction, and then relate all others to this choice. Because we have chosen, by convention, to write all half reactions as reductions, half-cell potentials are called **reduction potentials**.

The standard hydrogen electrode

Chemists have chosen to use the **standard hydrogen electrode (SHE)** as the reference half cell and half reaction and to *assign* the SHE a standard reduction potential, $E°(H^+,H_2)$, of exactly 0 (zero) volts at 298 K. The half cell and half cell reaction are:

$$\text{Pt}(s) \mid \text{H}_2(g, \text{ 1 bar}) \mid \text{H}^+(aq, \text{ 1 M}) \qquad E°(\text{H}^+,\text{H}_2) \equiv 0 \text{ V} \qquad (10.33)$$

$$2\text{H}^+(aq) + 2e^- \rightleftharpoons \text{H}_2(g) \qquad\qquad\qquad\qquad\qquad (10.34)$$

This potential is labeled a **standard reduction potential**, denoted by the superscript °, because all species in the reaction are at their standard state concentrations (see Table 9.1, Chapter 9, Section 9.2).

The SHE, Figure 10.7, is made by bubbling hydrogen gas at one bar pressure (= 100 kPa ≈ 1 atm), over a specially prepared platinum electrode immersed in a solution with a hydronium ion concentration of unity, $[H^+(aq)] = 1$ M. The SHE is not easy to set up and maintain. Other reference electrodes, whose standard reduction potentials have been determined, are often used to measure new half-cell potentials. No matter how a half cell is measured in practice, its standard reduction potential is still given relative to the SHE.

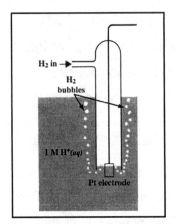

Figure 10.7. Representation of a standard hydrogen electrode.

Measuring a reduction potential

To measure the standard reduction potential for the silver-silver ion half cell, we would set up this cell (which is illustrated in Figure 10.8):

$$Pt(s) \mid H_2(g,\ 1\ bar) \mid H^+(aq,\ 1\ M) \parallel Ag^+(aq,\ 1\ M) \mid Ag(s) \qquad (10.35)$$

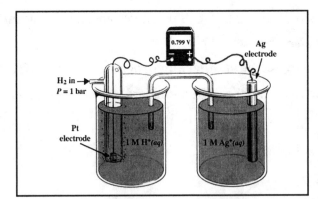

Figure 10.8. Measuring the standard reduction potential for the silver-silver ion half cell.

*By convention, **the SHE is taken as the anode for these determinations.*** Since the reaction at the anode is assumed to be an oxidation, we have to subtract the anode half reaction (10.34), which is written as a reduction, from the cathode half reaction (10.20) to get the reaction in the cell we have written and constructed. With appropriate stoichiometric factors to cancel the numbers of electrons we get:

$$2\{Ag^+(aq) + e^- \rightleftharpoons Ag(s)\} \qquad\qquad 2 \times (10.20)$$
$$- \{2H^+(aq) + 2e^- \rightleftharpoons H_2(g)\} \qquad\qquad - (10.34)$$
$$\overline{2Ag^+(aq) + H_2(g) \rightleftharpoons 2Ag(s) + 2H^+(aq)} \qquad\qquad (10.36)$$

The cell potential for the SHE-silver cell, $E_{H|Ag}$, is the difference between the reduction potentials for reactions (10.20) and (10.34):

$$E^°_{H|Ag} = E^°(Ag^+, Ag) - E^°(H^+, H_2) \qquad\qquad (10.37)$$

The reduction potentials for both half cells and the overall cell potential are shown as standard potentials, because all species in cell (10.35) are at their standard state concentration. Since $E^°(H^+, H_2)$ is defined as zero, we get:

$$E^°_{H|Ag} = E^°(Ag^+, Ag) \qquad\qquad (10.38)$$

Equation (10.38) exemplifies this rule: ***the standard reduction potential for any half cell is equal to the measured potential of a cell made by coupling the half cell of interest (with all species at standard state concentration) to the SHE to create a cell in which the SHE is taken to be the anode.*** The measured cell potential, $E^°_{H|Ag}$, of cell (10.35) is 0.799 V, as shown in Figure 10.8, so we have:

$$Ag^+(aq) + e^- \rightleftharpoons Ag(s) \qquad\qquad E^°(Ag^+, Ag) = 0.799\ V \qquad\qquad (10.39)$$

10.25. Check This   *Standard reduction potentials*

Assume that you have constructed this cell and find that the cell potential is 0.337 V:

$$Pt(s) \mid H_2(g,\ 1\ bar) \mid H^+(aq,\ 1\ M) \parallel Cu^{2+}(aq,\ 1\ M) \mid Cu(s) \qquad\qquad (10.40)$$

(a) Write the half reaction that occurs at the cathode. What is the standard reduction potential for this half reaction? Explain your reasoning.

(b) Use the information in this problem to explain what will be observed if a piece of copper metal is placed in a solution containing hydronium ions, say hydrochloric acid.

10.26. Investigate This   *Do metals react with acids?*

Do this as a class investigation and work in small groups to discuss and analyze the results. Fill each of five wells in one row of a 24-well plastic plate about two-thirds full of 1 M hydrochloric acid, HCl. Repeat with 1 M sulfuric acid, H_2SO_4, in five wells of another row. You have ten metal samples (as 2-cm lengths of wire or thin strips), two each of Ag, Cu, Fe, Mg, and

Zn. Place a different metal in each HCl well. Observe what happens and note any indication of reaction between a metal and the acid. Repeat with the five metals in H_2SO_4.

10.27. Consider This   *What are the reactions of metals with acids?*

   (a) What evidence for reactions between the metals and the acids did you observe in Investigate This 10.26? Did all five metals react the same way? Did the metals react the same way in both acids? How do you interpret the similarities and differences among the results?

   (b) Write reaction equations for those cases where you observe reaction. Explain your choice of reactants and products in each case.

Signs of reduction potentials

   We made an *arbitrary* choice of the SHE as our standard reference electrode. Are we guaranteed that it will really be the anode in every cell we make? This would be the same as saying that there is no reducing reaction we can couple with the SHE that will reduce $H^+(aq)$. That is, no substance would react with hydronium ion, $H^+(aq)$, to produce hydrogen gas, $H_2(g)$. But you know this is wrong. In Investigate This 10.26, you found that Fe, Mg, and Zn react vigorously with solutions containing hydronium ion, $H^+(aq)$ (acidic solutions) to produce a gas (hydrogen).

   Consider the cell we can construct with a SHE and a standard Zn^{2+}–Zn half cell:

$$Pt(s) \,|\, H_2(g, \text{ 1 bar}) \,|\, H^+(aq, \text{ 1 M}) \,\|\, Zn^{2+}(aq, \text{ 1 M}) \,|\, Zn(s) \qquad (10.41)$$

The cell reaction implied by writing the cell like this is:

$$Zn^{2+}(aq) + H_2(g) \rightleftharpoons Zn(s) + 2H^+(aq) \qquad (10.42)$$

The reaction you *observed* in Investigate This 10.26 is the reverse: zinc metal reacts with hydronium ion (acid solution) to produce hydrogen gas. If we measure the cell potential for cell (10.41) by connecting the anode input of our voltmeter to the SHE, the reading we get is -0.763 V. The negative reading tells us that we have connected the meter backwards (that the SHE is not truly acting as the anode). However, if we *apply the rule* above, we get:

$$Zn^{2+}(aq) + 2e^- \rightleftharpoons Zn(s) \qquad\qquad E°(Zn^{2+}, Zn) = -0.763 \text{ V} \qquad (10.43)$$

   We interpret this negative standard reduction potential as telling us that the driving force (the potential) for Zn^{2+} reduction, *relative to the SHE*, is unfavorable or weak. Conversely, the driving force for the reverse reaction, Zn oxidation, *relative to the SHE*, is quite strong, because it has a positive potential. This is the reaction you observed when you added zinc metal to acid.

10.28. **Check This**   *What are the signs of reduction potentials for other metals?*

Are the standard reduction potentials for Cu, Fe, and Mg positive or negative? What is the experimental evidence for your answers? State your reasoning clearly.

Table of standard reduction potentials

With the preceding interpretation of a negative cell potential, we can apply the rule for standard reduction potentials of half reactions in all cases and get both positive and negative values for standard reduction potentials. Appendix YY is table of standard reduction potentials and Table 10.1 gives a few of these that are relevant to this section. The sign depends upon the driving force for reduction relative to the SHE. Positive signs mean strong driving forces for reduction. The reactant in the half reaction is easily reduced. Negative signs mean weak driving forces for reduction, but strong driving forces for oxidation. The product in the half reaction is easily oxidized. The larger the absolute value of the standard reduction potential, the larger these driving forces. The values in the tables are arranged in order from the most positive to the most negative standard reduction potential.

Table 10.1.   Standard reduction potentials for a few half reactions.

Half-cell reaction	$E°$, V
$PbO_2 + SO_4^{2-} + 4H^+ + 2e^- \rightleftharpoons PbSO_4 + 2H_2O$	1.658
$Ag^+ + e^- \rightleftharpoons Ag$	0.799
$O_2 + 2H_2O + 4e^- \rightleftharpoons 4OH^-$	0.403
$Cu^{2+} + 2e^- \rightleftharpoons Cu$	0.337
$2H^+ + 2e^- \rightleftharpoons H_2 + 2H_2O$	0.000
$PbSO_4 + 2e^- \rightleftharpoons Pb + SO_4^{2-}$	–0.356
$Fe^{2+} + 2e^- \rightleftharpoons Fe$	–0.41
$Zn^{2+} + 2e^- \rightleftharpoons Zn$	–0.763
$Al(OH)_3 + 3e^- \rightleftharpoons Al + 3OH^-$	–2.33
$Mg^{2+} + 2e^- \rightleftharpoons Mg$	–2.38

The direction of redox reactions

Figure 10.9 shows how to use relative values of standard reduction potentials to predict redox reactions. Imagine a cell made by coupling two half cells (half reactions) from Appendix YY or

Table 10.1. The half cell with the more positive standard reduction potential (higher in the list) will be the cathode of the cell. The half reaction in this half cell will proceed as a reduction. The half cell with the less positive standard reduction potential (lower in the list) will be the anode of the cell. The half reaction in this half cell will proceed in reverse, as an oxidation. The standard cell potential is the difference between the more positive and less positive reduction potential. (In Section 10.6, we will take up the case that the concentrations in the cell are not standard concentrations. It is possible that a prediction based on standard reduction potentials could be wrong, but it is a good starting point.) The greater the difference in the standard reduction potentials for the half reactions, the greater the driving force for the overall reaction.

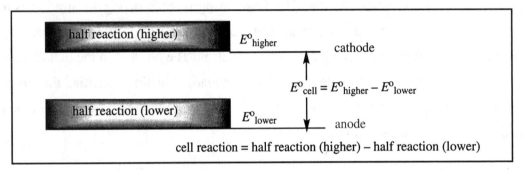

Figure 10.9. Predicting cell potential and cell reaction from standard reduction potentials.

The species on the left of a reduction half reaction is an **oxidizing agent**. The strongest oxidizing agents are the *reactants* in the reactions at the top of the listing in Appendix YY. The product of each half reaction is a **reducing agent**. The strongest reducing agents are the *products*

WEB Chap 10, Sect 10.7.1.
Check your understanding of combining reduction potential with this interactive exercise.

in the half reactions at the bottom of the listing. Thus, the prediction in the previous paragraph can also be stated this way: *an oxidizing agent (reactant) in a half reaction higher on the list has the potential to oxidize any reducing agent (product) lower on the list*. These ideas are noted in Appendix YY and applied to a practical electrochemical cell in Worked Example 10.29.

10.29. **Worked Example**   *The aluminum-air cell*

An aluminum-air cell consists of an aluminum metal electrode and a second electrically conducting electrode that is permeable to gases. Oxygen from the air passes through the permeable electrode and reacts at the surface exposed to the electrolyte solution. The solution in the cell is an aqueous solution of NaCl or NaOH. The half-cell reactions for an aluminum-air cell are:

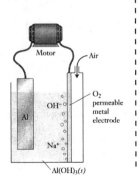

$$O_2(g) + 2H_2O(l) + 4e^- \rightleftharpoons 4OH^-(aq) \qquad\qquad (10.44)$$

$$Al(OH)_3(s) + 3e^- \rightleftharpoons Al(s) + 3OH^-(aq) \qquad\qquad (10.45)$$

Write the cell reaction and the line notation for the aluminum-air cell. What is the cell potential, if all components are at unit activity? As the cell is discharged, what changes occur?

Necessary information: From Table 10.1 (or Appendix YY), we need the standard reduction potentials for half reactions (10.44) and (10.45): 0.403 V and –2.33 V, respectively.

Strategy: Combine the standard reduction potentials by subtracting the less positive half reaction from the more positive (with appropriate stoichiometry to cancel the electrons). The combination gives the cell reaction and its standard cell potential. Once we know which electrode is the anode, we can write the line notation for the cell. The cell reaction gives the information we need to figure out the changes that occur as the reaction proceeds.

Implementation: The appropriate half-reaction combination is:

$$3\{O_2(g) + 2H_2O(l) + 4e^- \rightleftharpoons 4OH^-(aq)\} \qquad\qquad 3 \times (10.44)$$

$$-4\{Al(OH)_3(s) + 3e^- \rightleftharpoons Al(s) + 3OH^-(aq)\} \qquad\qquad -4 \times (10.45)$$

$$\overline{3O_2(g) + 4Al(s) + 6H_2O(l) \rightleftharpoons 4Al(OH)_3(s)} \qquad\qquad (10.46)$$

Combining the standard reduction potentials for the half cells gives the standard cell potential:

$$E^\circ_{O_2|Al} = E^\circ(O_2, OH^-) - E^\circ(Al(OH)_3, Al) = 0.403\ V - (-2.33\ V) = 2.73\ V \qquad (10.47)$$

Oxidation of the aluminum metal occurs, so the aluminum is the anode and the cell is:

$$Al(s) \,|\, Al(OH)_3(s) \,|\, \text{aqueous ionic solution} \,|\, O_2(g) \,|\, \text{inert metal} \qquad (10.48)$$

The ions in the solution are not part of the cell reaction equation. An ionic solution is necessary to facilitate ionic movement. Hydroxide ions migrate from the cathode, where they are produced, toward the anode, where they react to form $Al(OH)_3(s)$.

The cell reaction uses up the aluminum metal electrode and water from the solution to form $Al(OH)_3(s)$. Oxygen from the air is also used, but the air provides an inexhaustible source of oxygen. "Recharging" an aluminum-air cell involves replacing the spent aluminum electrode, adding more water, and removing the solid aluminum hydroxide that has formed.

Does the answer make sense? This picture shows an aluminum-air cell. Reduction potentials predict the cell reaction and standard cell potential for the combination of the two half reactions in this electrochemical cell. The cell reaction helps us understand the chemistry that occurs when the cell is used. These cells are used in some emergency lighting systems.

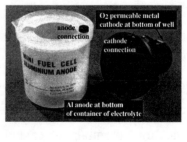

10.30. Check This   *The lead-acid battery*

(a) The half cell reactions for the lead-acid battery used in automobiles are:

$$PbO_2(s) + SO_4^{2-}(aq) + 4H^+(aq) + 2e^- \rightleftharpoons PbSO_4(s) + 2H_2O(l) \qquad (10.49)$$

$$PbSO_4(s) + 2e^- \rightleftharpoons Pb(s) + SO_4^{2-}(aq) \qquad (10.50)$$

Write the cell reaction. What is the cell potential, if all components are at unit activity? Use the standard reduction potentials in Table 10.1 (or Appendix YY). Modern automobiles almost all use 12-V batteries. How do you get 12 V from an automobile battery?

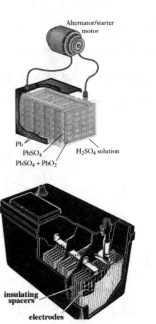

(b) The electrodes in the battery are grids of lead suspended in a sulfuric acid solution. When the battery is fully charged, one set of electrode grids is filled with $PbSO_4(s)$ and the other with a mixture of $PbSO_4(s)$ and $PbO_2(s)$. Write the line notation for the cell. As the cell is discharged, what changes occur at the electrodes?

(c) If the cell reaction is run in reverse, what changes occur at the electrodes? The reaction is run "in reverse" by applying an electrical potential that is larger than the cell potential and in the opposite direction. That is, the cell is treated like an electrolytic cell, as in Section 10.1. Remember that the cathode of an electrolytic cell is the electrode at which reduction occurs. Which electrode is this in this case? The car's alternator is the source of this potential and recharges the battery while the engine is running.

Reflection and projection

We have learned how to catalog redox reactions by defining standard reduction potentials, $E°$, for all half cells relative to the standard hydrogen electrode. A listing of these reduction potentials in order from most positive to most negative, Appendix YY, provides a tool for predicting the direction of redox reactions. In a redox reaction, the half reaction that is higher in the table will proceed as written, as a reduction. The half reaction lower in the table will go in reverse as an oxidation.

The standard reduction potentials are for all components of the half reaction at their standard state concentrations. So far, we have no way to figure out what will happen to cell or half-cell potentials when the concentrations are not standard concentrations. Also, up to now, our rules for combining cell reactions and cell potentials and for combining half reactions and standard reduction potentials to get standard cell potentials have been based on experimental results for observed chemical reactions and corresponding cell potentials. To go further in our analysis of electrochemical cells and redox reactions, we need to consider the thermodynamics of electrochemical cells. We will begin by considering the work available from electrochemical cells and the associated change in free energy for the cell reaction.

Section 10.5. Work From Electrochemical Cells: Free Energy

10.31. **Investigate This**   *How can we get work from a galvanic cell?*

   (a) Do this as a class investigation and work in small groups to discuss and analyze the results. Fill one well of a six-well plate about half full of aqueous 1.0 M zinc nitrate, $ZnSO_4$, solution. Bend a 1×8-cm piece of zinc sheet into an "L" so that the longer leg can rest flat on the bottom of the well with the shorter leg up and out of the well. Place the sheet in the well with the $ZnSO_4$ solution. Repeat this procedure in a second adjacent well, using a solution of aqueous 1.0 M copper nitrate, $CuSO_4$, solution and a 1×8-cm piece of copper sheet. Finish making a galvanic cell by connecting the half cells with a filter-paper salt bridge that has been soaked in 10% aqueous potassium nitrate, KNO_3 solution. Turn a digital multimeter to its voltage scale and properly connect the leads from a digital multimeter to the copper and zinc electrodes. Note the electrode to which each lead is attached. Record the reading, including its sign and units, to the nearest millivolt.

   (b) Disconnect one lead from the multimeter and then connect a small electric motor with an attached fan blade between this lead and the electrode, as shown schematically in Figure 10.5. Turn the multimeter to the current (amperes) scale and record the current reading in the circuit when the fan is going.

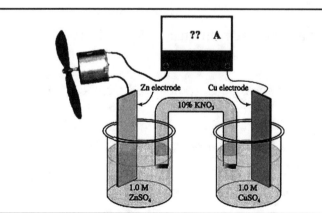

Figure 10.10. A zinc-copper galvanic cell connected to a motor and current meter.
The meter and motor are connected so that all current from the cell passes through both of them.

10.32. Consider This   *What reaction in the zinc-copper galvanic cell produces work?*

(a) Can the zinc-copper galvanic cell you made in Investigate This 10.31(a) — illustrated in Figure 10.10 — do useful work in the surroundings? What observations lead to your conclusion?

(b) In your zinc-copper galvanic cell, which electrode, copper or zinc, is the cathode? Which is the anode? Base your reasoning on your observations in Investigate This 10.31(a).

(c) When written as reductions, the half reactions in the copper and zinc half cells of your galvanic cell are:

$$Cu^{2+}(aq) + 2e^- \rightleftharpoons Cu(s) \tag{10.21}$$
$$Zn^{2+}(aq) + 2e^- \rightleftharpoons Zn(s) \tag{10.43}$$

What combination of equations (10.21) and (10.43) gives the overall reaction in your zinc-copper galvanic cell? What is the reasoning for your answer?

Electrical work

As you demonstrated in Investigate This 10.31(b), galvanic cells can do work on their surroundings when electrons flow between the electrodes. You harness this work when you use batteries to power portable electrical devices. You know that different devices require different numbers and different kinds of batteries. This is because different amounts of work are required to run each one. How much work is available from an electrochemical cell and what factors affect the available work?

An electrochemical cell uses the free energy of a chemical redox reaction to produce an electrical potential difference between two electrodes. We are interested in the work that the cell can produce in the external part of the circuit. This **electrical work**, w_{elec} (in joules, J), is defined as the product of the amount of charge (in

> Electrical measurements are usually more accurate than mechanical ones like acceleration, so joules are often measured in electrical units:
> 1 joule = 1 volt·coulomb

coulombs, C) that passes through the circuit times the electrical potential difference, E, (in volts, V) between the electrodes:

$$w_{elec} = (\text{amount of charge, C}) \cdot (\text{potential difference, V}) \tag{10.51}$$

10.33. Worked Example   *Electrical work*

A timer used to time a chemistry examination runs on a 1.35 V battery. The current through the timer is 0.385 mA (milliamp). How much electrical work does the cell produce to run the timer for a 50-minute examination? How many moles of electrons flow through the timer circuit?

Necessary information: We need equation (10.51) and need to know that there are 60 seconds in a minute, that 1000 mA = 1 A (= 1 C·s^{-1}), and that $F = 9.6485 \times 10^4$ C·mol^{-1}.

Strategy: Use the current and the time it flowed to determine the amount of charge that flowed and then substitute into equation (10.51) to get the work. Use F to convert the amount of charge to moles of electrons.

Implementation:

$$\text{current} = (0.385 \text{ mA})\left(\frac{1 \text{ A}}{1000 \text{ mA}}\right) = 3.85 \times 10^{-4} \text{ A} = 3.85 \times 10^{-4} \text{ C·s}^{-1}$$

$$\text{amount of charge} = (3.85 \times 10^{-4} \text{ C·s}^{-1})\cdot(50 \text{ min})\cdot(60 \text{ s·min}^{-1}) = 1.16 \text{ C}$$

$$w_{\text{elec}} = (1.16 \text{ C})\cdot(1.35 \text{ V}) = 1.57 \text{ J}$$

$$\text{mol } e^- = (1.16 \text{ C})\left(\frac{1 \text{ mol}}{96,485 \text{ C}}\right) = 1.20 \times 10^{-5} \text{ mol}$$

Does the answer make sense? Without some frame of reference, it's hard to tell whether one isolated result like this makes sense. As you calculate the electrical work under other conditions, keep this result in mind for comparison. In this case, the electricity is being used to run a tiny solid state device which requires little power, so a small amount of work (energy) makes sense.

10.34. Check This   *Electrical work*

Use your data for the zinc-copper cell voltage and the current drawn by the motor in Investigate This 10.31 to find out how much work the cell would have to produce to run the motor for one minute. How many moles of electrons would flow through the motor?

We can calculate the work *available* from an electrochemical cell without doing current and time measurements. Let n be the number of moles of electrons that flow in the external circuit per mole of the stoichiometric reaction in the cell. An equivalent definition of n is the number of electrons in each of the half reactions that are combined to give the overall cell reaction. For the zinc-copper cell we have been considering, there are two electrons in each half reaction, equations (10.21) and (10.43), so $n = 2$. Note that n does not have units. The amount of charge that flows per mole of reaction is nF coulombs. If E is the cell potential, the electrical work available per mole of cell reaction is:

$$w_{\text{elec}} \text{ [J·(mol of reaction)}^{-1}] = nFE \tag{10.52}$$

10.35. Consider This   *Why are there so many sizes of batteries?*

The display of batteries in this picture is bewildering in its variety. These cylindrical 1.5-volt batteries come in sizes from AAAA to D. Is there a difference in the amount of work available from these different batteries? How does equation (10.52) help answer this question?

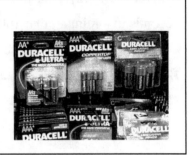

For an electrochemical cell with a certain E, we can obtain almost any amount of work by building a big enough cell and producing enough moles of reaction to get as many multiples of

> Some devices, especially those with motors, require a higher voltage than a single cell can provide, so you really do use a "battery" of cells (usually two to six) in series in most of your electronics. Even though smaller cells run down faster (fewer moles of reactants), most electronics use small cells to keep the overall mass and size down.

nFE as we want. In the real world, we would prefer not to use huge cells; consider the problems you would have with a "portable" CD player. The solution is to use half cells that can be connected to produce a substantial potential difference so that we can obtain a substantial amount of work without many moles of reaction. This is a practical reason why values of E and factors that affect them are so important. They will also prove to be important in understanding more about electron transfer reactions, including those in the electron transport pathway in living organisms.

10.36. Worked Example   *Amount of work available from an electrochemical cell*

The battery (cell) referred to in Worked Example 10.33 is a mercury cell, for which the overall cell reaction can be represented as:

$$Zn(s) + HgO(s) \rightarrow ZnO(s) + Hg(l) \tag{10.53}$$

If HgO(s) is the limiting reagent and the battery contains 1.0 g of HgO(s), how long can it run the timer in Worked Example 10.33?

Necessary information: We need equation (10.52) and the result from Worked Example 10.33: work required to run the timer for 50 minutes is 1.67 J. We also need the molar mass of HgO (216.6 g·mol^{-1}) and $F = 9.6485 \times 10^4$ C·mol^{-1}. We need to see that $n = 2$, two moles of electrons are transferred for each mole of reaction (10.53), since it takes two moles of electrons to oxidize one mole of Zn(s) (oxidation number 0) to one mole of zinc(II) (oxidation number 2+) in ZnO(s). The same reasoning applies to the number of moles of electrons required to reduce mercury(II) in HgO(s) to Hg(l).

Strategy: Use equation (10.52) to find the work available from one mole of reaction (10.53). The total work available from the cell is obtained from the total moles of reaction (= number of moles of HgO in the cell) and compared to the work required to run the timer for 50 minutes to find the total time the timer could be run.

Implementation: Since $n = 2$, the work available per mole of reaction is:

$$w_{elec} \text{ (per mol reaction)} = nFE = 2(9.6485 \times 10^4 \text{ C·mol}^{-1})·(1.35 \text{ V}) = 2.61 \times 10^5 \text{ J·mol}^{-1}$$

The number of moles of reaction available from the cell is:

$$\text{mol of reaction} = \text{mol of HgO} = (1.0 \text{ g HgO})\left(\frac{1 \text{ mol HgO}}{216.6 \text{ g}}\right) = 4.6 \times 10^{-3} \text{ mol}$$

The work available from the cell is:

$$w_{elec} = (2.61 \times 10^5 \text{ J·mol}^{-1})·(4.6 \times 10^{-3} \text{ mol}) = 1.2 \times 10^3 \text{ J}$$

The length of time that this amount of work will run the timer is:

$$\text{time} = (1.2 \times 10^3 \text{ J})\left(\frac{50 \text{ min}}{1.67 \text{ J}}\right) = 3.6 \times 10^4 \text{ min} = 25 \text{ days}$$

Does the answer make sense? The cell lasts long enough to time 720 50-minute exams. We expect the batteries (cells) in electronic devices to provide a considerable amount of use before replacement. It makes sense that the cell in this timer should last through this many uses.

10.37. Check This   *Amount of work available from an electrochemical cell*

In Consider This 10.32(c), you wrote the zinc-copper cell reaction:

$$\text{Zn}(s) + \text{Cu}^{2+}(aq) \rightarrow \text{Zn}^{2+}(aq) + \text{Cu}(s) \tag{10.30}$$

Assume that each well of your zinc-copper cell in Investigate This 10.31 contains 5.0 mL of solution. If you run the fan until one half of the $\text{Cu}^{2+}(aq)$ ion is used up, for approximately how long will the fan run? Show clearly how you get your answer.

Free energy and cell potential

The sign of E, the cell potential, tells us the direction of an electron transfer reaction. The magnitude of E is a measure of the electrical work, equation (10.52), available to us *from* the reaction. In Chapter 8, Section 8.8, we found that the **free energy** change for a reaction also tells us the direction and work available (the "free" energy) from the reaction. This parallelism suggests that there is a relationship between E and $\Delta G_{reaction}$ for redox reactions. The maximum work available *from* a system is $-\Delta G_{reaction}$ for the change. For electron transfer reactions we can write:

$$w_{\text{elec}} \ (\text{J·(mol of reaction)}^{-1}) = nFE = -\Delta G_{\text{reaction}} \tag{10.54}$$

$\Delta G_{\text{reaction}}$ in equation (10.54) is the free energy change per mole of stoichiometric reaction. If $\Delta G_{\text{reaction}}$ is negative, the reaction is favored in the direction written. Equation (10.54) shows that a negative $\Delta G_{\text{reaction}}$ means a favorable (positive) E for the reaction. Table 10.2 summarizes the relationships between $\Delta G_{\text{reaction}}$ and E. Under standard conditions, equation (10.54) shows that:

$$nFE° = -\Delta G°_{\text{reaction}} \tag{10.55}$$

Table 10.2.   Relationships among $\Delta G_{\text{reaction}}$, E, and favored direction of reaction.

$\Delta G_{\text{reaction}}$	E	reaction as written is
< 0	> 0	favored
> 0	< 0	not favored
$= 0$	$= 0$	at equilibrium

Combining redox reactions and cell potentials

The relationship between the cell potential and the free energy change in equation (10.54) can help us understand the basis for combining cell potentials that we introduced in Section 10.3. Let's take the following combination of cell reactions as an example:

$$Zn(s) + 2Ag^+(aq) \rightleftharpoons Zn^{2+}(aq) + 2Ag(s) \tag{10.31}$$

$$-2\{Fe(CN)_6^{4-}(aq) + Ag^+(aq) \rightleftharpoons Fe(CN)_6^{3-}(aq) + Ag(s)\} \qquad -2 \times (10.56)$$

$$\overline{Zn(s) + Fe(CN)_6^{3-}(aq) \rightleftharpoons Zn^{2+}(aq) + Fe(CN)_6^{4-}(aq)} \tag{10.57}$$

The free energy changes *per mole of reaction* combine in the same way as the reactions:

$$\Delta G_{\text{Zn|Fe}} = \Delta G_{\text{Zn|Ag}} - 2\Delta G_{\text{Fe|Ag}} \tag{10.58}$$

We can use the relationship between free energy and cell potential, equation (10.54) to rewrite equation (10.58) in terms of cell potentials:

$$-n_{\text{Zn|Fe}}FE_{\text{Zn|Fe}} = -n_{\text{Zn|Ag}}FE_{\text{Zn|Ag}} - 2(-n_{\text{Fe|Ag}}FE_{\text{Fe|Ag}}) \tag{10.59}$$

From the reaction equations, $n_{\text{Zn|Fe}} = n_{\text{Zn|Ag}} = 2$; that is, two moles of electrons are transferred per mole of cell reactions (10.31) and (10.57). However, $n_{\text{Fe|Ag}} = 1$, since only one mole of electrons is transferred in reaction (10.56). Substituting these n values in equation (10.59) gives:

$$-2FE_{\text{Zn|Fe}} = -2FE_{\text{Zn|Ag}} + 2FE_{\text{Fe|Ag}}$$

And dividing through by $-2F$, we get:

$$E_{\text{Zn|Fe}} = E_{\text{Zn|Ag}} - E_{\text{Fe|Ag}} \tag{10.60}$$

Thus, to calculate the free energy changes when we combine reactions, we need to take account of the number of times each reaction enters into the combination. However, as we found experimentally and then generalized in Section 10.3, the number of times each reaction enters into the combination is not relevant for the combination of cell potentials.

Reflection and projection

The flow of electrons from an electrochemical cell through an external circuit can be harnessed to do useful work. The amount of work is the product of the cell potential, E, and the amount of charge that flows. For one mole of cell reaction, the charge transferred is nF coulombs, where n is the number of moles of electrons that are transferred per mole of cell reaction. Since the free energy change for a system is also related to the work it can do on the surroundings, we can relate free energy change to cell potential, $nFE = -\Delta G_{reaction}$, which provides a fundamental thermodynamic basis for understanding electrochemical cells.

None of the cell potentials and free energy changes for cell reactions that we have written in this section have been standard cell potentials or standard free energies. The relationships in Table 10.2, for example, are for the reactions under whatever conditions exist in the cell or in a mixture of the reactants and products. Now we need to find out how to go from standard cell potentials and the related standard free energy changes to cell potentials and free energy changes for reactions under non-standard conditions.

Section 10.6. Concentration Dependence of Cell Potentials: The Nernst Equation

10.38. Investigate This   *Does the silver-copper cell potential depend on [Ag⁺]?*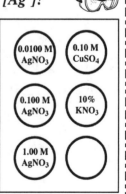

Do this as a class investigation and work in small groups to discuss and analyze the results. As shown in the diagram, fill one of the central wells of a 24-well plate about two-thirds full of 10% KNO_3 solution. Fill three other wells adjacent to the KNO_3 well about two-thirds full of 1.00 M, 0.100 M, and 0.0100 M aqueous silver nitrate, $AgNO_3$, solution, respectively. Place a clean silver wire in the 0.0100 M silver solution as an electrode. Fill a fourth adjacent well about two-thirds full of 0.10 M aqueous copper sulfate, $CuSO_4$, solution, and place a clean copper wire in the solution as an electrode. Use filter paper soaked in 10% KNO_3 solution to make salt bridge connections between the KNO_3 well and each of the other wells.

(a) Connect the input leads from a digital voltmeter to the electrodes so that the meter reading is positive. Note which lead is connected to the silver electrode and which to the copper. Record the cell potential to the nearest millivolt.

(b) Remove the silver electrode from the 0.0100 M silver solution, wipe it dry with a paper towel, place it in the 0.100 M silver solution, and measure the new cell potential.

(c) Remove the silver electrode from the 0.100 M silver solution, wipe it dry with a paper towel, place it in the 1.00 M silver solution, and measure the new cell potential.

10.39. Consider This   *How does the silver-copper cell potential depend on [Ag⁺]?*

(a) What were the cell potentials for the silver-copper cells in Investigate This 10.38(a), (b), and (c)? Which electrode was the anode in each cell?

(b) Reasoning from Le Chatelier's principle, would you predict changes in the silver-copper cell potential, if the concentration of the silver ion is changed? If so, what would be the direction of the change with increasing silver ion concentration? Clearly state your reasoning.

(c) Are your results from Investigate This 10.38 consistent with your prediction in part (b)? Explain why or why not.

Recall equation (9.44) from Chapter 9, Section 9.7:

$$\Delta G_{reaction} = \Delta G^{\circ}_{reaction} + RT\ln Q \tag{10.61}$$

The **reaction quotient**, Q, accounts for the actual concentrations in the system of interest and

If necessary, review the reaction quotient and its relationship to free energy in Chapter 9.

enables us to calculate $\Delta G_{reaction}$ from $\Delta G^{\circ}_{reaction}$ and the known concentrations. We can use the relationship in equations (10.54) and (10.55) to substitute cell potentials for free energy changes in equation (10.61):

$$-nFE = -nFE^{\circ} + RT\ln Q \tag{10.62}$$

Equation (10.62) is usually rearranged to:

$$E = E^{\circ} - \frac{RT}{nF}\ln Q \tag{10.63}$$

Equation (10.63) is called the Nernst equation, in honor of Walter Nernst (German chemical physicist, 1864-1941) who did extensive work on electrochemical systems and thermodynamics. The **Nernst equation** provides us what we seek for a cell reaction, a relationship among the standard cell potential, the concentrations in the cell, and the cell potential under non-standard concentration conditions.

Although equation (10.63) is useful as it stands, you will almost always see it (and use it) in an equivalent mathematical form:

$$E = E° - \frac{2.303RT}{nF}\log Q \tag{10.64}$$

We get equation (10.64) from equation (10.63) by substituting the mathematical equality: $\ln Q = 2.303 \log Q$. The advantage of equation (10.64) is that concentrations are almost always expressed in scientific notation as powers of ten and these are easier to manipulate if we are using base-ten logarithms (log). Equation (10.64) is particularly useful when dealing with hydronium ion concentrations, as we will see.

10.40. Check This   *Nernst equation at 25 °C (298 K)*

Show that, at 298 K ("room" temperature), the Nernst equation (10.64) becomes:

$$E = E° - \frac{(0.05916 \text{ V})}{n}\log Q \approx E° - \frac{(0.059 \text{ V})}{n}\log Q \tag{10.65}$$

Hint: Recall that $R = 8.314 \text{ J·K}^{-1}\text{·mol}^{-1}$.

10.41. Worked Example   *Nernst equation applied to a copper-cadmium cell*

Cell potential measurements at 25 °C on a series of copper-cadmium cells with varying cadmium ion concentration are given in this tabulation:

c, M	0.100	0.0500	0.0100	0.0050
E, V	0.737	0.745	0.767	0.775

Use a graphical method to determine whether these data are consistent with the Nernst equation and, if so, calculate $E°$ for the cell:

$$\text{Cd}(s) \,|\, \text{Cd}^{2+}(aq, c \text{ M}) \,\|\, \text{Cu}^{2+}(aq, 0.10 \text{ M}) \,|\, \text{Cu}(s) \tag{10.66}$$

Necessary information: We will need the Nernst equation (10.65) at 25 °C (298 K) and the appropriate the reaction quotient, Q.

Strategy: Write the cell reaction for cell (10.66) and use it to write Q. Use this Q to write the Nernst equation in a form that relates E linearly to a function of $(\text{Cd}^{2+}(aq))$. Plot the data to see if they fall on a straight line and, if so, use equation of the line to find $E°$ for the cell reaction.

Implementation: The cell reaction for cell (10.66) is:

$$\text{Cd}(s) + \text{Cu}^{2+}(aq) \rightleftharpoons \text{Cd}^{2+}(aq) + \text{Cu}(s) \tag{10.67}$$

The reaction quotient is:

$$Q = \frac{\left(Cd^{2+}(aq)\right)\left(Cu(s)\right)}{\left(Cd(s)\right)\left(Cu^{2+}(aq)\right)} = \frac{\left(Cd^{2+}(aq)\right)}{\left(Cu^{2+}(aq)\right)}$$

Recall that the concentration ratio for a pure solid is unity, so $(Cd(s)) = 1$ and $(Cu(s)) = 1$. The Nernst equation (for $n = 2$) is:

$$E = E^\circ - \frac{(0.059\ V)}{n}\log Q = E^\circ - \frac{(0.059\ V)}{2}\log\left\{\frac{\left(Cd^{2+}(aq)\right)}{\left(Cu^{2+}(aq)\right)}\right\} \qquad (10.68)$$

A $\log(x/y)$ term can be rewritten as a sum, $\log(x) + \log(1/y)$. Also, $\log(1/y) = -\log(y)$, so we can rewrite this Nernst equation in the form:

$$E = E^\circ + (0.030\ V)\cdot\log(Cu^{2+}(aq)) - (0.030\ V)\cdot\log(Cd^{2+}(aq))$$

This is the equation of straight line of E plotted as a function of $\log(Cd^{2+}(aq))$, with a slope = $-(0.030\ V)$ and an intercept = $E^\circ + (0.030\ V)\cdot\log(Cu^{2+}(aq))$. Figure 10.11 is a plot of the data.

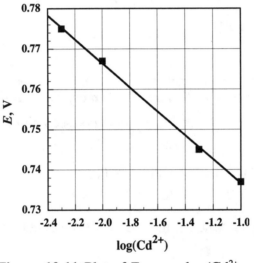

log(Cd^{2+})

Figure 10.11. Plot of E versus $\log(Cd^{2+}(aq))$.

The equation of the line in Figure 10.11 (generated by the computer plotting program) is:

$$E = -(0.0297\ V)\cdot\log(Cd^{2+}(aq)) + 0.707\ V$$

The slope of the line, $-(0.0297\ V) = -(0.030\ V)$, is the value predicted; the results are consistent with the Nernst equation. We can use the intercept and the known $(Cu^{2+}(aq))$ to calculate E°:

$$0.707\ V = E^\circ + (0.0296\ V)\cdot\log(Cu^{2+}(aq)) = E^\circ - 0.0296\ V$$

$$E^\circ = 0.737\ V$$

Does the result make sense? Combine the standard reduction potentials for the copper and cadmium half reactions in Appendix YY to see that our value of E° is correct. Another qualitative check on the data is to apply Le Chatelier's principle to the variation of the measured cell potentials. In the cell reaction, $Cd^{2+}(aq)$ is a reaction product, so we would predict that decreasing its concentration would favor the cell reaction as written, that is, would increase the cell potential. This is what we observe.

10.42. Check This   *Nernst equation applied to silver-copper cells*

   (a) Use a graphical method to determine whether your data from Investigate This 10.38 are consistent with the Nernst equation and, if so, calculate $E°$ for the silver-copper cell. Compare your experimental value for $E°$ with the one calculated from the data in Table 10.2. *Hint:* Recall that $\log(x)^a = a \cdot \log(x)$.

   (b) Is the *direction* of the variation in the cell potentials with silver ion concentration in Investigate This 10.38 consistent with Le Chatelier's principle? Explain why or why not?

10.43. Check This   *Three problems from the Web Companion*

   WEB   Chap 10, Sect 10.8.1-4. Work through the first three pages of this section of the *Web Companion*. Write out and explain the calculations required to solve the problems on page 4.

10.44. Investigate This   *Does the silver-quinhydrone cell potential depend on pH?*

   (a) Do this as a class investigation and work in small groups to discuss and analyze the results. As shown in the diagram, fill one of the central well of a 24-well plate, about two-thirds full of 10% KNO_3 solution. Fill a well adjacent to the KNO_3 well about two-thirds full of 0.100 M aqueous silver nitrate, $AgNO_3$, solution and insert a silver wire as an electrode. Fill two other adjacent wells about two-thirds full of pH 7.00 and pH 3.00 acid-base buffer solutions, respectively. To each buffer well add a tiny amount of 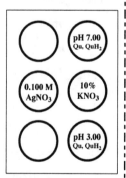 solid quinhydrone. Place a platinum wire in the pH 7.00 well as an electrode and to stir the solution. Use filter paper soaked in 10% KNO_3 solution to make salt bridge connections between the KNO_3 well and each of the other wells. Connect the input leads from a digital voltmeter to the electrodes so that the meter reading is positive. Note which lead is connected to the silver electrode and which to the platinum. Record the cell potential to the nearest millivolt.

   (b) Remove the platinum wire from the pH 7.00 quinhydrone well, wipe it off, and place it in the pH 3.00 quinhydrone well. Stir the solution in the well and again connect the digital voltmeter. Note which lead is connected to which electrode and record the cell potential.

10.45. **Consider This**   *How does the silver-quinhydrone cell potential depend on pH?*

(a) What were the cell potentials for the silver-quinhydrone cells in Investigate This 10.44? What would you expect the cell potential to be, if the quinhydrone half cell was prepared with a pH 5.00 solution? Explain the reasoning for your answer.

(b) Which electrode is the anode in Investigate This 10.44(a)? in 10.44(b)? Which electrode would be the anode if the quinhydrone half cell was prepared with a pH 5.00 solution? Explain the reasoning for your answer.

(c) **Quinhydrone** is a 1:1 combination of **quinone**, Qu, and **hydroquinone**, QuH_2, molecules. It is sparingly soluble in water and yields solutions that contain equal concentrations of quinone and hydroquinone:

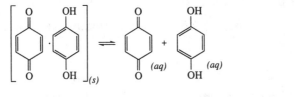

$$\text{Qu} \qquad \text{QuH}_2$$

(10.69)

Quinone can be reduced to hydroquinone in aqueous solution:

$$Qu(aq) + 2H^+(aq) + 2e^- \rightleftharpoons QuH_2(aq) \qquad (10.70)$$

(Note that this half reaction and the structures of quinone and hydroquinone are closely related to reaction (10.1) and the structures of ubiquinone and ubiquinol in living cells.) Write the line notation and the cell reaction for the cell you made in Investigate This 10.44(a). Be sure your answers are consistent with the identity of the anode and cathode for the cell.

(d) Does Le Chatelier's principle explain the variation of the cell potential with pH that you found in Investigate This 10.44? Explain why or why not.

The quinhydrone half cell is different than the others we have been constructing. The platinum electrode does not appear in half-cell reaction (10.70). In the quinhydrone half cell, there are electron acceptors, the Qu molecules, and electron donors, the QuH_2 molecules. For reaction (10.70), we can imagine that a QuH_2 molecule next to the surface of the platinum electrode donates two electrons to the metal and two protons to water to form hydronium ions, H_3O^+, and a Qu molecule; QuH_2 has been oxidized. The result is that the platinum electrode has two extra electrons, which flow to the silver electrode (where silver ions are reduced). The platinum electrode is the anode, as you found in Consider This 10.45(b). The role of the platinum electrode is to sense the potential created by the quinhydrone solution and provide a pathway for electrons to leave (or enter) the half cell.

The Nernst equation and pH

The chemistry of the cell reaction in Investigate This 10.44 is useful in photographic film developing. Photographic film contains tiny crystals of silver salts. Developing an exposed film involves reducing some of the silver ion, Ag^+, in these crystals to silver metal, Ag, which is half reaction (10.20). Some photographic developers contain hydroquinones as reducing agents, so we know that QuH_2 can reduce Ag^+. To get the developer reaction, we double half reaction (10.20) and subtract half reaction (10.70):

$$2[Ag^+(aq) + e^- \rightleftharpoons Ag(s)] \qquad\qquad 2 \times (10.20)$$

$$\underline{-[Qu(aq) + 2H^+(aq) + 2e^- \rightleftharpoons QuH_2(aq)] \qquad\qquad -(10.70)}$$

$$2Ag^+(aq) + QuH_2(aq) \rightarrow Ag(s) + Qu(aq) + 2H^+(aq) \qquad\qquad (10.71)$$

Reaction (10.71) is the silver-quinhydrone cell reaction you wrote in Consider This 10.45(c). Let's see if the Nernst equation applied to this reaction will help us understand our data from Investigate This 10.44. The reaction quotient for reaction equation (10.71) is:

$$Q = \left(\frac{(Ag(s))(Qu(aq))\left(H^+(aq)\right)^2}{\left(QuH_2(aq)\right)\left(Ag^+(aq)\right)^2} \right) \qquad\qquad (10.72)$$

We know that $(Ag(s)) = 1$. The stoichiometry of equation (10.64) shows that equal amounts of quinone and hydroquinone dissolve, so $(Qu(aq)) = (QuH_2(aq))$. Therefore, in equation (10.72), the *ratio* $(Qu(aq))/(QuH_2(aq)) = 1$. Substituting these values into equation (10.72) gives:

$$Q = \left(\frac{\left(H^+(aq)\right)^2}{\left(Ag^+(aq)\right)^2} \right) = \left(\frac{\left(H^+(aq)\right)}{\left(Ag^+(aq)\right)} \right)^2 \qquad\qquad (10.73)$$

Here, we have rewritten the quotient of squared terms as the square of the quotient of the terms. Substitution of this expression for Q into equation (10.65), with $n = 2$, gives:

$$E = E^\circ - \frac{(0.059\ V)}{n}\log\left(\frac{\left(H^+(aq)\right)}{\left(Ag^+(aq)\right)}\right)^2 = E^\circ - \frac{2\cdot(0.059\ V)}{2}\log\left(\frac{\left(H^+(aq)\right)}{\left(Ag^+(aq)\right)}\right)$$

$$E = E^\circ - (0.059\ V)\cdot\log\left(\frac{\left(H^+(aq)\right)}{\left(Ag^+(aq)\right)}\right)$$

$$E = E^\circ + (0.059\ V)\cdot\log(Ag^+(aq)) - (0.059\ V)\cdot\log(H^+(aq)) \qquad\qquad (10.74)$$

10.46. Check This   *Deriving equation (10.74)*

Show how equation (10.74) is obtained from the preceding equation.

We can substitute pH = –log(H$^+$(aq)) in the last term in equation (10.74) to get:

$$E = E° + (0.059 \text{ V}) \cdot \log(Ag^+(aq)) + (0.059 \text{ V}) \cdot pH = constant + (0.059 \text{ V}) \cdot pH \quad (10.75)$$

Equation (10.75) shows that the cell potential for a silver-quinhydrone cell is a linear function of pH and should increase by 0.059 V for every increase of one unit in the pH of the quinhydrone solution. Before the invention of the glass pH electrode that is used today, quinhydrone half cells were often used to obtain the pH of solutions.

10.47. Check This   *pH and cell potentials for silver-quinhydrone cells*

(a) Are your values for the cell potentials of the silver-quinhydrone cells in Investigate This 10.44 consistent with equation (10.75)? Explain your response.

(b) Is equation (10.75) consistent with your explanation based on Le Chatelier's principle in Consider This 10.45(d)? Explain why or why not.

(c) Use your data from Investigate This 10.44 to determine $E°$ for the silver-quinhydrone cell. Explain your method.

$E°$ and the equilibrium constant, K

Since $-nFE° = \Delta G°_{rxn} = -RT\ln K$, a numeric value for $E°$ enables us to calculate the equilibrium constant for a cell reaction:

$$\ln K = \frac{nFE°}{RT} \quad or \quad \log K = \frac{nFE°}{2.303RT} \quad (10.76)$$

You can also derive equation (10.76) from equation (10.64) by recalling that, at equilibrium, $\Delta G_{reaction} = 0$, so $E = 0 \ (= \Delta G_{reaction}/nF)$. The reaction quotient for a reaction at equilibrium is $Q_{eq} = K$, so equation (10.64) becomes equation (10.76). Dead batteries have cell potentials equal to zero. Dead cells have reached equilibrium; they can furnish no more work.

10.48. Worked Example *Equilibrium constant for reaction of silver ion with hydroquinone*

The standard cell potential for the silver-quinhydrone cell is 0.100 V. What is the equilibrium constant for the reaction of silver ion with hydroquinone at 25 °C.?

$$2Ag^+(aq) + QuH_2(aq) \rightleftharpoons Ag(s) + Qu(aq) + 2H^+(aq)$$

Necessary information: We need equation (10.76) and we need to recognize that this reaction is the cell reaction for a silver-quinhydrone cell, equation (10.71).

Strategy: Since the reaction of interest is the cell reaction in a silver-quinhydrone cell, substitute the standard cell potential into equation (10.76) and solve for K.

Implementation:

$$\ln K = \frac{nFE^\circ}{RT} = \frac{2 \cdot \left(96,485\ \text{C} \cdot \text{mol}^{-1}\right)(0.100\ \text{V})}{\left(8.314\ \text{J} \cdot \text{mol}^{-1} \cdot \text{K}^{-1}\right) \cdot (298\ \text{K})} = 7.79 \quad \therefore K = 2.4 \times 10^3$$

Does the answer make sense? We said that hydroquinone is used as a film developer to reduce silver ion. The relatively large value of the equilibrium constant for reaction (10.71) means that the products predominate at equilibrium, which is just what we expect for a reaction that is used to reduce silver ion in a practical application.

10.49. Check This   *Conditions for developing photographic film*

If you want to be sure that a good deal of the silver ion in a film is reduced by a hydroquinone developer, would you want the developer to be at a relatively low or relatively high pH? Explain. *Hint:* Write the equilibrium constant expression (or reaction quotient) for the developer reaction and see how pH affects the ratios of the other species in the system.

Section 10.7. Reduction Potentials and the Nernst Equation

Can we apply the Nernst equation to half-cell reactions and reduction potentials as well as cell reactions and cell potentials? Remember that standard reduction potentials, are really the potentials of complete cells in which the half cell of interest is coupled with the SHE. To see what we can learn about how half-cell potentials depend upon concentrations, let's apply the Nernst equation to a cell that couples a non-standard half cell with the SHE. As an example, consider this cell:

$$\text{Pt}(s) \mid \text{H}_2(g,\ 1\ \text{bar}) \mid \text{H}^+(aq,\ 1\ \text{M}) \parallel \text{Ag}^+(aq,\ c\ \text{M}) \mid \text{Ag}(s) \tag{10.77}$$

The cell reaction and Nernst equation (at 298 K) for reaction (10.77) are:

$$2\text{Ag}^+(aq) + \text{H}_2(g) \rightleftharpoons 2\text{Ag}(s) + 2\text{H}^+(aq) \tag{10.36}$$

$$E_{\text{H}|\text{Ag}} = E^\circ_{\text{H}|\text{Ag}} - \left(\frac{0.059\ \text{V}}{2}\right)\log Q_{\text{H}|\text{Ag}} \tag{10.78}$$

Rewrite the cell potentials in terms of the reduction potentials for this cell and write out $Q_{\text{H}|\text{Ag}}$:

$$E(\text{Ag}^+, \text{Ag}) - E^\circ(\text{H}^+, \text{H}_2) = E^\circ(\text{Ag}^+, \text{Ag}) - E^\circ(\text{H}^+, \text{H}_2) - \left(\frac{0.059\ \text{V}}{2}\right)\log\left(\frac{\left(\text{H}^+(aq)\right)^2}{\left(\text{H}_2(g)\right) \cdot \left(\text{Ag}^+(aq)\right)^2}\right)$$

$E^{\circ}(H^+, H_2)$ is written on the left side of this equation, because the cell we are considering has the SHE as the anode and its reduction potential is $E^{\circ}(H^+, H_2)$. Substituting $E^{\circ}(H^+, H_2) = 0$, $(H_2(g)) = 1$, and $(H^+(aq)) = 1$ into this equation gives:

$$E(Ag^+, Ag) = E^{\circ}(Ag^+, Ag) - \left(\frac{0.059 \text{ V}}{2}\right)\log\left(\frac{1}{\left(Ag^+(aq)\right)^2}\right) \tag{10.79}$$

10.50. Check This   *Simplifying equation (10.70)*

Show that equation (10.79) can be written as:

$$E(Ag^+, Ag) = E^{\circ}(Ag^+, Ag) - \left(\frac{0.059 \text{ V}}{1}\right)\log\left(\frac{1}{\left(Ag^+(aq)\right)}\right) \tag{10.80}$$

Nernst equation for a half reaction

Equation (10.80) is the way you would write the Nernst equation for the silver ion reduction half reaction, equation (10.20), if you left out the electrons. We can rationalize leaving out the electrons; they don't actually exist free in the solution, but only in the solid electrodes, which do not affect the numeric value of Q. Remembering that electrons are left out, you can use the Nernst equation to write the concentration dependence of a reduction potential. The value of n to use is the number of electrons in the half reaction equation.

10.51. Worked Example   *Oxygen gas reduction potential as a function of pH*

The half reaction and standard reduction potential for the reduction of oxygen gas are:

$$O_2(g) + 4H^+(aq) + 4e^- \rightleftharpoons 2H_2O(l) \qquad E^{\circ}(O_2, H_2O) = 1.229 \text{ V} \tag{10.81}$$

Use Le Chatelier's principle to predict whether the reduction potential for oxygen increases or decreases as the hydronium ion concentration decreases. Use the Nernst equation to get the reduction potential for oxygen at 298 K at pH 7 and at pH 14.

Necessary information: We need equation (10.65), the Nernst equation at 298 K.

Strategy: Apply Le Chatelier's principle to half reaction (10.81), recalling that the reduction potential is a measure of the driving force for the reaction. Write Q for the half reaction and use it in the Nernst equation with appropriate values for $(H^+(aq))$ to get $E(O_2, H_2O)$ at pH 7 and 14.

Implementation: $H^+(aq)$ is a reactant in the reduction half reaction, so reducing $[H^+(aq)]$ from 1 M (standard conditions) to 10^{-7} M (pH 7) will cause the reverse reaction to be favored and the reduction potential should decrease. A further reduction to 10^{-14} M (pH 14) will further reduce the potential.

The Nernst equation for half reaction (10.81), with $(H_2O(l)) = 1$, is:

$$E(O_2, H_2O) = E°(O_2, H_2O) - \left(\frac{0.059\ V}{4}\right)\log\left(\frac{1}{(O_2(g))(H^+(aq))^4}\right)$$

Under standard conditions, the oxygen gas pressure is 1 bar, so $(O_2(g)) = 1$ and we can rewrite the equation as:

$$E(O_2, H_2O) = E°(O_2, H_2O) - \left(\frac{0.059\ V}{4}\right)\log\left(\frac{1}{(H^+(aq))^4}\right)$$

$$E(O_2, H_2O) = E°(O_2, H_2O) - (0.059\ V)\log\left(\frac{1}{(H^+(aq))}\right)$$

$$E(O_2, H_2O) = E°(O_2, H_2O) - (0.059\ V)[-\log(H^+(aq))] = E°(O_2, H_2O) - (0.059\ V)pH$$

Therefore: $E(O_2, H_2O)$(at pH 7) $= 1.229\ V - (0.059\ V)\cdot(7) = 0.816\ V$

$E(O_2, H_2O)$(at pH 14) $= 1.229\ V - (0.059\ V)\cdot(14) = 0.403\ V$

Does the answer make sense? Le Chatelier's principle predicts a decrease in reduction potential with increasing pH (lower hydronium ion concentration) and this is what we calculate using the Nernst equation. In Appendix YY, the oxygen reduction potential is given for standard conditions, pH 0, and also for pH 7 where $E = 0.816$ V. The result at pH 14 represents reaction (10.81) with $(OH^-(aq)) = 1$ M. Recall from Section 6.10, Chapter 6, that, to convert a redox reaction equation in acidic solution to one in basic solution, we add enough hydroxide ions to convert all the hydronium ions to water and then cancel redundant waters. Applying this conversion to half reaction equation (10.81), we add four $OH^-(aq)$ to each side:

$$O_2(g) + 4H^+(aq) + 4OH^-(aq) + 4e^- \rightleftharpoons 2H_2O(l) + 4OH^-(aq)$$
$$O_2(g) + 4H_2O(l) + 4e^- \rightleftharpoons 2H_2O(l) + 4OH^-(aq)$$
$$O_2(g) + 2H_2O(l) + 4e^- \rightleftharpoons 4OH^-(aq) \qquad (10.82)$$

The standard reduction potential for half reaction equation (10.82) is denoted $E°(O_2, OH^-)$ and we know from the solution conditions that $E°(O_2, OH^-) = E(O_2, H_2O)$(at pH 14) $= 0.403$ V. Check this value in Appendix YY.

10.52. Check This   *Hydronium ion reduction potential as a function of pH*

The half reaction and standard reduction potential for the reduction of hydronium ion are:

$$2H^+(aq) + 2e^- \rightleftharpoons H_2(g) \qquad\qquad E°(H^+, H_2) = 0.000\ V \qquad (10.83)$$

Use Le Chatelier's principle to predict whether the reduction potential for the hydronium ion increases or decreases as the hydronium ion concentration decreases. Use the Nernst equation to get the reduction potential for the hydronium ion at 298 K at pH 7 and at pH 14. How do your results compare with the values in Appendix YY? Explain the comparisons you make.

10.53. Investigate This   *Can pH change the direction of a cell reaction?*

Do this as a class investigation and work in small groups to discuss and analyze the results. Use a set up similar to the one in Investigate This 10.44. Use three wells grouped around one containing 10% aqueous KNO_3 solution to prepare three half cells. Prepare the wells with (1) a solution that is 0.10 M in aqueous potassium ferricyanide, $K_3Fe(CN)_6$, and 0.10 M in aqueous potassium ferrocyanide, $K_4Fe(CN)_6$, with a platinum wire electrode, (2) a quinhydrone solution in 1 M HCl (pH ≈ 0) with a platinum wire electrode, and (3) a quinhydrone solution in a pH 7 buffer solution with a platinum wire electrode. Use filter paper salt bridges soaked in 10% aqueous KNO_3 solution to connect each half cell to the KNO_3 well.

Connect one lead of a digital voltmeter to the electrode in the $Fe(CN)_6^{3-}$-$Fe(CN)_6^{4-}$ solution and the other lead to the electrode in the pH ≈ 0 quinhydrone half cell. Record which electrode is the anode and the cell potential to the nearest millivolt. Switch the lead from the pH ≈ 0 quinhydrone half cell to the pH 7 quinhydrone half cell and repeat the recording.

10.54. Consider This

(a) Write the correct cell notation and cell reaction for each of the cells you tested in Investigate This 10.53.

(b) Does Le Chatelier's principle explain any differences between the reactions in part (a)? Explain why or why not.

Reduction potentials under biological conditions

In Consider This 10.45, you saw that the structures and redox reactions are similar for the quinone-hydroquinone pair and the ubiquinone-ubiquinol pair (UQ and UQH_2) from reaction (10.1). In Investigate This 10.53, you coupled the quinone-hydroquinone pair (the quinhydrone half cells) with a half cell containing iron complexes in their 3+ and 2+ states. In living cells, the ubiquinone-ubiquinol pair in mitochondria is also coupled to half reactions of iron complexes in their 3+ and 2+ states.

The cell potential for a cell that involves the quinone-hydroquinone pair depends upon the pH of the quinhydrone solution. The same is true for the ubiquinone-ubiquinol pair in biological systems. In Investigate This 10.53, the *direction* of the cell reaction was reversed by the change in pH from about 0 to 7. Standard reduction potentials are given for all species at unit concentration, which is pH 0 for hydronium ion, but most biological fluids have a pH near 7. Therefore, as you saw in Investigate This 10.53, using standard reduction potentials as a guide to the direction of reaction at pH 7 may be misleading. If we want to make predictions about redox reactions in biological systems, we have to correct the cell potentials to account for pH effects.

Worked Example (10. 51) and Check This (10.52) show that there is no problem doing this correction, but it means we lose some of the easy predictive power illustrated in Figure 10.9 for a table of standard reduction potentials. When we compare the half reactions that make up a redox reaction of biological interest, we would like to be able to look at a table and know that the half reaction higher in the table is *likely* to proceed as a reduction, in the reaction. A solution to this problem is a table of standard reduction potentials, $E^{\circ\prime}$, for half cells in which all components are at unit concentration, *except* $[H^+(aq)] = 10^{-7}$ M (pH 7). For biological systems, comparisons of $E^{\circ\prime}$ values give better predictions than E° values. The table of standard reduction potentials in Appendix YY gives values of both E° and $E^{\circ\prime}$ for many half reactions that involve the hydronium ion as a reactant or product. For some biological reactants, only $E^{\circ\prime}$ values are given. The reactions at low pH are irrelevant, if the biological molecules, such as proteins, do not function in high concentrations of hydronium ion.

10.55. Check This   *Predictions you can make about biological redox reactions*

(a) In aerobic organisms, a key metabolic reaction is the redox reaction between oxidized and reduced nicotinamide adenine dinucleotide (NAD^+, NADH) and oxidized and reduced ubiquinone (UQ, UQH_2). Use the values in Appendix YY to determine the standard cell potential at pH 7, $E^{\circ\prime}_{UQ|NAD}$, and $\Delta G^{\circ\prime}_{rxn}$ for this reaction:

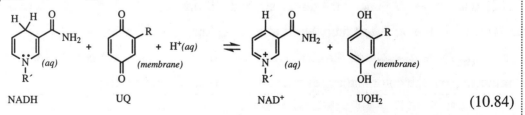

NADH   UQ   NAD$^+$ UQH$_2$ (10.84)

(b) Is reaction (10.84) favored in the direction written at pH 7? Explain your response. What is the equilibrium constant for reaction (10.84) at pH 7?

(c) Show that all electron pairs in the reactants are accounted for in the products.

Reflection and projection

The potential of an electrochemical cell depends on the concentrations of the ions and molecules in the cell solution(s). The direction of the dependence can be predicted by applying Le Chatelier's principle to the cell reaction and relating more positive cell potentials to a greater driving force toward products. Quantitatively, we can use the relationship of the free energy change for a cell reaction and the cell potential to obtain the Nernst equation, which relates the cell potential to the standard cell potential and the logarithm of the reaction quotient, Q, for the cell reaction. Standard cell potentials, like the standard free energy changes for cell reactions, can be used to calculate equilibrium constants for redox reactions. We can also apply the Nernst equation to half reactions by disregarding the electrons, which are never present free in the solution.

A special case that is particularly relevant to biological systems is the variation of reduction potentials with pH. Since biological redox reactions generally occur in solutions near pH 7, standard reduction potentials at pH 7, $E°'$, are used to make predictions for these reactions. Other changes in solution conditions, besides direct changes in the concentrations of species that appear in the redox reaction, can also affect cell potentials and reduction potentials. Metal-ion complexation, for example, is significant in biological systems. In the next section, we will discuss systems that contain several species that can be oxidized and reduced and see how the redox reactions can be coupled to one another.

Section 10.8. Coupled Redox Reactions

10.56. Investigate This   *Does copper ion react with sugars?*

Do this as a class investigation and work in small groups to discuss and analyze the results. Benedict's reagent is an aqueous solution of copper sulfate, $CuSO_4$, sodium carbonate, Na_2CO_3, and sodium citrate, $Na_3(C_6H_5O_7)$. Add 5 mL of Benedict's reagent to each of two small, clean, labeled test tubes. To one of the test tubes, add 2 drops of ethanol. Add one or two *small* crystals of glucose to the other test tube. Swirl to mix and dissolve the alcohol and sugars. Note the color and appearance of each mixture. Place the test tubes in a boiling water bath for about five minutes. Note the color and appearance of the mixtures after heating.

10.57. Consider This   *What is the reaction of copper ion with glucose?*

(a) What changes did you observe taking place in the solutions Benedict's reagent with ethanol and glucose in Investigate This 10.56? Did both compounds give the same results? If not, what were the differences?

(b) The colors of Cu^{2+} ions complexed in solution are at the blue end of the spectrum. Citrate ion is a good complexing agent for Cu^{2+} ions. The oxide of the Cu^+ ion, Cu_2O, is an insoluble red solid. How might you account for the changes that occurred in Investigate This 10.56? Explain as fully as possible.

The source of the electrons for our metabolic redox pathways is the food we eat. All of our food traces its origin to the glucose photosynthesized by green plants from carbon dioxide and water. To study the metabolic pathway that gathers electrons from our food is difficult without specialized equipment and instrumentation. As an alternative, we will focus on a simpler reaction of glucose that is not part of the metabolic pathway, but exemplifies its role as a reducing agent.

Glucose is oxidized to gluconic acid

In Investigate This 10.56, glucose reduced copper(II) to copper(I). The sugar is oxidized in the reaction (but not all the way to carbon dioxide and water). A carbon-containing molecule is usually most easily oxidized at the carbon that has the highest oxidation number. For glucose, this is carbon-1, the aldehyde carbon in the open-chain form of the molecule:

$$\text{HO} \diagdown \overset{\text{OH}}{\diagup} \diagup \overset{\text{OH}}{\diagdown}_{\text{OH}} \diagup \overset{\text{OH}}{\diagdown}_{\text{OH}} \diagup \overset{\text{H}}{\underset{O}{C}} \quad \text{carbon-1}$$

This carbon has an oxidation number of +2. The other carbons with –OH groups attached, the alcoholic carbons, have oxidation numbers of +1. Alcohols do not reduce the copper(II) in Benedict's reagent.

10.58. Check This   *Alcohols in Benedict's reagent*

What evidence do you have to support or refute the statement that alcohols do not reduce copper(II) in Benedict's reagent?

The oxidation of an aldehyde produces a carboxylic acid. The product of glucose oxidation at the aldehyde group is gluconic acid, which, at pH 7 and above, is present in its conjugate base form, gluconate ion. The half reaction for reduction of gluconate ion to glucose is:

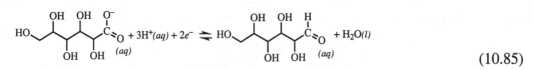

$$(10.85)$$

In abbreviated form, with R = $C_5H_{11}O_5$, reaction (10.85) is:

$$RCOO^-(aq) + 3H^+(aq) + 2e^- \rightleftharpoons RCHO(aq) + H_2O(l) \qquad E^{\circ\prime}(RCOO^-, RCHO) = -0.44 \text{ V} \quad (10.86)$$

> We have been using –C(O)O⁻ as the notation for the carboxylate ion, but in this section and in Appendix YY, for ease of writing, we use –COO⁻, another common notation.

Since the standard reduction potential at pH 7, is negative, –0.44 V, the oxidation of glucose, the reverse of reaction (10.86), is favored.

Benedict's reagent is basic (because it contains sodium carbonate), so we need to write the half reaction equation (10.86) in terms of hydroxide ion instead of hydronium ion. Using the method outlined in Chapter 6, Section 6.10 and recalled in Worked Example 10.51, we get for the reaction in basic solution:

$$RCOO^-(aq) + 2H_2O(l) + 2e^- \rightleftharpoons RCHO(aq) + 3OH^-(aq) \quad E^{\circ\prime}(RCOO^-, RCHO) = -0.44 \text{ V} \quad (10.87)$$

$E^{\circ\prime}(RCOO^-, RCHO)$ in basic solution is the reduction potential when $[OH^-(aq)] = 10^{-7}$ M. This is pH 7, so the reduction potential at pH 7 is the same for half reactions (10.86) and (10.87). At higher hydroxide ion concentrations, half reaction (10.87) is driven to the left, which decreases the reduction potential and makes the oxidation of glucose even more favorable. Now we need to show that copper(II) can oxidize glucose under the conditions in the Benedict's reagent.

The Cu²⁺-Cu⁺ redox system

In Investigate This 10.61, you observed a change from a clear, blue solution to a cloudy reddish-brown liquid when Benedict's reagent reacted with glucose. As you probably concluded in Consider This 10.57, the reaction is reduction of Cu^{2+} to Cu^+ (as its oxide precipitate). In the Benedict's reagent, the copper(II) is present as a citrate complex, $Cu(cit)^-(aq)$. The half reaction for reduction of the complex is:

$$Cu(cit)^-(aq) + e^- \rightleftharpoons Cu^+(aq) + cit^{3-}(aq) \qquad E^{\circ}(Cu(cit)^-, Cu^+) = -0.04 \text{ V} \qquad (10.88)$$

The standard free energy change for reaction (10.88) is $\Delta G^{\circ}(Cu(cit)^-, Cu^+) = nFE^{\circ}(Cu(cit)^-, Cu^+) =$ +4 kJ·mol⁻¹. The standard cell potential and free energy suggest that the reduction of copper(II) in this solution is not favored. However, we have not yet accounted for formation of copper(I) oxide, $Cu_2O(s)$, the reddish solid. Precipitation of $Cu_2O(s)$ lowers the concentration of $Cu^+(aq)$, thus driving reaction (10.88) to the right and making the reduction potential more positive.

10.59. Worked Example   *Standard reduction potential for copper(II) in Benedict's reagent*

The reaction of $Cu^+(aq)$ ion with hydroxide to precipitate $Cu_2O(s)$ can be represented as:

$$2Cu^+(aq) + 2OH^-(aq) \rightleftharpoons Cu_2O(s) + H_2O(l) \qquad K \approx 4 \times 10^{12} \qquad \Delta G^\circ_{reaction} = -72 \text{ kJ·mol}^{-1} \quad (10.89)$$

Determine the overall half reaction for copper(II) reduction in Benedict's reagent and the standard free energy change and standard reduction potential for the reduction.

Necessary information: We need equation (10.55) and equation (10.88) and its standard free energy change, +4 kJ·mol^{-1}.

Strategy: Combine reaction (10.89) with equation (10.88), using appropriate stoichiometric factors, to cancel $Cu^+(aq)$, and get the overall reaction for reduction of copper(II) to form solid copper(I) oxide. Combine the free energies of reactions (10.88) and (10.89) to find the free energy change for the overall reaction and thence its standard reduction potential.

Implementation: The appropriate combination of reactions (10.88) and (10.89) is:

$$2\{Cu(cit)^-(aq) + e^- \rightleftharpoons Cu^+(aq) + cit^{3-}(aq)\} \qquad 2\Delta G^\circ(Cu(cit)^-, Cu^+) = +8 \text{ kJ·mol}^{-1}$$

$$2Cu^+(aq) + 2OH^-(aq) \rightleftharpoons Cu_2O(s) + H_2O(l) \qquad \Delta G^\circ_{reaction} = -72 \text{ kJ·mol}^{-1}$$

$$2Cu(cit)^-(aq) + 2OH^-(aq) + 2e^- \rightleftharpoons Cu_2O(s) + 2cit^{3-}(aq) + H_2O(l) \qquad (10.90)$$

$$\Delta G^\circ(Cu(cit)^-, Cu_2O) = -64 \text{ kJ·mol}^{-1}$$

$$E^\circ(Cu(cit)^-, Cu_2O) = -\frac{\Delta G^\circ_{(Cu(cit)^-, Cu_2O)}}{nF} = -\frac{(-64 \times 10^3 \text{ J·mol}^{-1})}{2 \cdot 96,485 \text{ C·mol}^{-1}} = 0.33 \text{ V}$$

Does the answer make sense? Although we have no information to use to test whether the numerical results are correct, the reduction potential shows that the standard reduction potential for copper(II) citrate complex to copper(I) oxide is favorable and this is the reaction we observe.

10.60. Check This   *Reduction potential for copper(II) in Benedict's reagent*

(a) Use Le Chatelier's principle to predict the direction of change of the reduction potential, if the hydroxide ion concentration is reduced in the solution in which reaction (10.90) occurs.

(b) Write the reaction quotient for half reaction (10.90).

(c) The pH of Benedict's reagent is about 12. What is $E(Cu(cit)^-, Cu_2O)$ in this solution, if all other species are present at their standard state concentrations? Explain your approach. Is your result consistent with your prediction in part (a)? Explain why or why not.

Redox reaction in the Benedict's test

We can combine half reactions (10.87) and (10.90) to write the reduction of copper(II) by glucose in Benedict's reagent:

$$2Cu(cit)^-(aq) + RCHO(aq) + 5OH^-(aq) \rightleftharpoons Cu_2O(s) + RCOO^-(aq) + 2cit^{3-}(aq) + 3H_2O(l) \quad (10.91)$$

The cell potential for reaction (10.86) is:

$$E = E(Cu(cit)^-, Cu_2O) - E(RCOO^-, RCHO) \qquad (10.92)$$

In Check This 10.60, you found that $E(Cu(cit)^-, Cu_2O)$ at pH 12 is about 0.21 V. We do not have $E(RCOO^-, RCHO)$ at pH 12, but we know $E^{o\prime}(RCOO^-, RCHO) = -0.44$ V at pH 7 and we reasoned above that $E(RCOO^-, RCHO)$ at pH 12 is even more negative. The *minimum* value for the cell potential is:

$$E = 0.21 \text{ V} - (-0.44 \text{ V}) = 0.65 \text{ V}$$

The reduction of copper(II) by glucose in Benedict's reagent has a large driving force and explains the positive test for glucose as a reducing agent in Investigate This 10.56.

10.61. Investigate This   *Can you characterize the Blue-Bottle reaction?*

     Work in small groups on this investigation and discuss your results and hypotheses with the entire class. Your large, capped vial contains an aqueous solution that is 0.14 M glucose, 0.50 M potassium hydroxide and 10^{-5} M methylene blue. CAUTION: Solutions of hydroxide are caustic and can harm skin and clothing. Take care not to spill any of the solution, but work over a paper towel to catch any drops that may escape. Shake the vial two or three times. What, if any, changes do you observe? Watch to see whether anything further happens. When the solution has returned to its original appearance, shake the vial again and see if your observations are reproducible. Try variations. Uncap the vial and recap it before shaking. Shake the vial more than two or three times. Keep a record of the conditions and observations for all your trials.

10.62. Consider This   *Can you interpret the Blue-Bottle reaction?*

    (a) As a class, make a table of the various conditions used and observations recorded on the Blue-Bottle reaction in Investigate This 10.61. Try to reach a consensus about the factors that affect the reaction and the direction and magnitude of the effects.

    (b) Methylene blue is a dye used as a bacteriological stain and as an indicator in some chemical analyses. The dye is blue in its oxidized form (MB^+) and colorless in its reduced form (MBH):

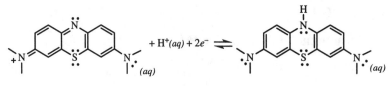

$$MB^+(aq) + H^+(aq) + 2e^- \rightleftharpoons MBH(aq) \qquad E^{o\prime}(MB^+, MBH) = 0.011 \text{ V} \qquad (10.93)$$

Using these properties of methylene blue and the redox properties of glucose in basic solution, can you develop a consistent explanation that accounts for the effects listed in part (a)?

The Blue-Bottle redox reaction(s)

Half reaction (10.93) accounts for the disappearance of the blue color that you observe in the Blue-Bottle reaction, Investigate This (10.61). There must be a reducing agent in the solution that reduces the blue, oxidized form of methylene blue, MB^+. You can make the blue color return by shaking the solution. There must be an oxidizing agent in the shaken solution that can oxidize the colorless, reduced form of methylene blue, MBH, back to the blue form. You probably have decided that the reducing agent is glucose (as in the Benedict's reagent reaction) and the oxidizing agent is molecular oxygen. If these choices are correct, they should be able to help us answer several questions about the observations.

10.63. Consider This   *Can you explain the timing of the Blue-Bottle reaction?*

(a) Why does the blue color persist for several seconds in the Blue-Bottle reaction and then suddenly fade away?

(b) Why does shaking bring the blue color back?

(c) How is the system able to repeat the cycle over and over?

(d) As you experimented with the Blue-Bottle reaction, you probably discovered that the more shakes you give the bottle, the longer the blue color persists, but it still seems to fade just as suddenly at the end. What is the explanation for these observations?

10.64. Check This   *Methylene blue, MB^+, reduction in basic solution*

The Blue-Bottle reaction solution is basic. Rewrite half reaction (10.93) to show the half reaction in terms of hydroxide ion instead of hydronium ion. What is $E°'(MB^+, MBH)$ for this half reaction? Explain how you get your answer.

The oxidized form of methylene blue, MB^+, is so highly colored that only a tiny amount is present in the Blue-Bottle reaction system to give the blue color. The disappearance of the blue color signals the reduction of this small amount of methylene blue by glucose, RCHO, which is present in substantial concentration:

$$MB^+(aq) + RCHO(aq) + 2OH^-(aq) \rightleftharpoons MBH(aq) + RCOO^-(aq) + H_2O(l) \qquad (10.94)$$

Using standard reduction potentials from equation (10.87) and Check This 10.64), we get the standard cell potential for reaction (10.94) at pH 7:

$$E^{\circ\prime}_{MB|RCHO} = E^{\circ\prime}(MB^+, MBH) - E^{\circ\prime}(RCOO^-, RCHO) = (0.011\ V) - (-0.44\ V) = 0.45\ V \quad (10.95)$$

10.65. Check This   *MB⁺–glucose redox reaction in basic solution*

How does $E^{\circ\prime}_{MB|RCHO}$ vary as a function of hydroxide ion concentration? Is reaction (10.94) more or less favored in more basic solutions? Give the reasoning for your answers.

Reaction (10.94) explains the disappearance of the blue color in the Blue-Bottle reaction, but the *way* it disappears is odd. In the blue solution, we start with the reactants in equation (10.94), so the reaction should continuously reduce methylene blue and you might expect to see the blue color fade continuously. In fact, the blue color persists for a time and then fades suddenly. Let us assume that reaction (10.94) is going on whenever MB⁺ and RCHO are present, that is, when the solution is blue. If this assumption is true, then the only way that the solution can remain blue is for the reduced form of methylene blue, MBH, to be oxidized back to the blue form as fast as it is formed. When the oxidizing agent has been used up, the solution quickly fades to colorless. Both the reduction of MB⁺ and oxidation of MBH seem to be fast. To explain the observations, the oxidation must be faster.

The oxidizing agent that explains the observations is oxygen gas dissolved in the solution:

$$O_2(g) + 2H_2O(l) + 4e^- \rightleftharpoons 4OH^-(aq) \qquad E^{\circ\prime}(O_2, OH^-) = 0.816\ V \qquad (10.96)$$

Shaking mixes the solution with the air in the container and some nitrogen and oxygen dissolve. Only a little bit of oxygen dissolves, but , since there is only a tiny bit of methylene blue present, a little oxygen is all we need to return all the MBH to the oxidized form. As the MB⁺ is reduced by glucose, the oxygen reoxidizes the MBH until the little bit of oxygen is used up. Then the dye color suddenly fades away. Another shake again adds oxygen and the process is repeated. The evidence for oxygen as the oxidizing agent seems to fit. We just need to be sure that oxygen can oxidize reduced methylene blue in basic solution.

10.66. Worked Example   *Oxidation of methylene blue by oxygen*

Write the reaction for oxidation of reduced methylene blue by oxygen in basic solution and determine $E^{\circ\prime}_{O2|MB}$. Does the driving force for the reaction increase or decrease as the solution becomes more basic? Estimate $E^{\circ}_{O2|MB}$ at 298 K when all species, including hydroxide ion, are at standard concentration.

Necessary information: We need the results from Check This 10.64 (reaction (10.93) in basic solution), the information from equation (10.96), and the Nernst equation at 298 K.

Strategy: Combine half reactions (10.93) and (10.96), both in basic solution, by subtracting the less positive half reaction (with appropriate stoichiometry to cancel the electrons). The combination gives the cell reaction and its standard cell potential. Applying Le Chatelier's principle to the cell reaction tells us the direction the cell potential changes as hydroxide ion concentration increases. Use the Nernst equation to find $E^°_{O2|MB}$, the potential under standard conditions with $[OH^-(aq)] = 1$ M (pH 14).

Implementation: The appropriate combination of half reactions and potentials is:

$$O_2(g) + 2H_2O(l) + 4e^- \rightleftharpoons 4OH^-(aq) \qquad\qquad E^{°\prime}(O_2, OH^-) = 0.816 \text{ V} \qquad (10.96)$$

$$-2\{MB^+(aq) + H_2O(l) + 2e^- \rightleftharpoons MBH(aq) + OH^-(aq)\} \quad -E^{°\prime}(MB^+, MBH) = -0.011 \text{ V} \quad -2\times(10.93)$$

$$\overline{O_2(g) + 2MBH(aq) \rightleftharpoons 2OH^-(aq) + 2MB^+(aq) \qquad\qquad E^{°\prime}_{O2|MB} = 0.805 \text{ V} \qquad (10.97)}$$

Hydroxide ion is a product of reaction (10.97), so, by Le Chatelier's principle, increasing $[OH^-(aq)]$ should decrease the cell potential. There will be a lower driving force for the reaction as written.

The Nernst equation for reaction (10.97) is:

$$E^{°\prime}_{O2|MB} = E^°_{O2|MB} - \left(\frac{0.059 \text{ V}}{4}\right)\log\left(\frac{(OH^-)^2(MB^+)^2}{(O_2(g))(MBH)^2}\right)$$

We have found that $E^{°\prime}_{O2|MB} = 0.805$ V, when $(OH^-(aq)) = 10^{-7}$ and all other species are at unit concentration. Substituting these values into this equation and solving for $E^°_{O2|MB}$ gives:

$$E^°_{O2|MB} = (0.805 \text{ V}) + \left(\frac{0.059 \text{ V}}{2}\right)\log(10^{-7}) = (0.805 \text{ V}) + (-0.21 \text{ V}) = 0.60 \text{ V}$$

Does the answer make sense? You know from your investigations that it is probably oxygen that oxidizes methylene blue in the Blue-Bottle reaction, so the substantial positive cell potential for this reaction makes sense. Le Chatelier's principle predicts that increasing the hydroxide ion concentration should decrease this potential and the Nernst equation confirms the prediction. The lower cell potential is still highly favorable for this oxidation.

10.67. Check This   *Oxidation of methylene blue by oxygen in the Blue-Bottle reaction*

Use the Nernst equation to estimate $E_{O2|MB}$ under the conditions of the Blue-Bottle reaction in Investigate This 10.61. Assume that the ratio of oxidized to reduced methylene blue is about 10 during the blue phase of the reaction and that the pH of the solution is 14. Assume the pressure of oxygen in the air is 20 kPa. (The standard pressure for a gas is 1 bar = 100 kPa.) Is the oxidation of MBH favored? Is your conclusion changed, if the ratio of oxidized to reduced methylene blue is about 1000? Explain.

Methylene blue couples glucose oxidation to oxygen reduction

The combination of reactions (10.94) and (10.97) alternately reduce and oxidize methylene blue. The net reaction is the sum of these reactions (with appropriate adjustment for stoichiometry) and the cell potential is the sum of the individual cell potentials:

$$O_2(g) + 2RCHO(aq) + 2OH^-(aq) \rightleftharpoons 2H_2O(l) + 2RCOO^-(aq) \quad E^{\circ\prime}_{O_2|RCHO} = 1.27 \text{ V} \quad (10.98)$$

Methylene blue does not appear in the net reaction expression (10.98). Without it, however, the reaction does not occur. Methylene blue **couples** the oxidation of glucose to the reduction of oxygen. **Coupled reactions**, which we introduced in Section 7.9, Chapter 7, are so common, especially in biochemical systems, that it is useful to have a better way to represent their pathways than with net reactions. Net reactions can be misleading, because they do not show the coupling that makes the net reaction possible. Figure 7.17 in Section 7.9, Chapter 7, is a mechanical analogy representing coupled reactions, but we need a more chemical representation.

Figure 10.12 represents the methylene blue coupling we have just analyzed. The species that is oxidized and the species that is reduced in each individual reaction are shown at the tails of two curved arrows. The arrows touch one another on their way to the respective oxidized and reduced products. The two tangent curves represent the reaction between these two reactants. Other species required for the reaction or produced by it are shown, respectively, at the tail and head of a straight arrow that is also tangent to the two curved arrows. The representation makes explicit the *cycling of the coupling species between its reduced and oxidized forms*. Each form is both a product (at the head of a curved arrow) and a reactant (at the tail of a curved arrow), so the arrows form a closed loop. Any species that cycles will not appear in the net reaction expression, as we found in equation (10.98).

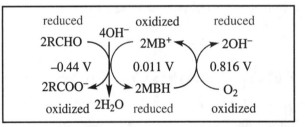

Figure 10.12. Methylene blue coupling of glucose oxidation to oxygen reduction.
The standard reduction potentials at pH 7 are given with each half reaction in the series. Oxidations are shown as blue curved arrows and reductions as red curved arrows.

10.68. Consider This   *What sort of redox coupling occurs in biological systems?*

In most organisms, the first stage in the oxidation of glucose is a series of reactions called **glycolysis** (*glyco* = sugar + *lysis* = breaking). The oxidizing agent is NAD^+ (see Check This 10.55). In the overall series of reactions, each molecule of glucose produces two molecules of pyruvate ion, $C_3H_3O_3^-$ (the carboxylate ion of pyruvic acid):

$$C_6H_{12}O_6 + 2NAD^+ \rightleftharpoons 2C_3H_3O_3^- + 2NADH + 4H^+ \qquad (10.99)$$

Some of the energy of the reaction is used to produce two molecules of ATP from ADP and phosphate ion, P_i: $2ADP + 2P_i \rightleftharpoons 2ATP$

Reaction (10.99) uses up NAD^+, which has to be regenerated in order to keep the reaction going to produce more ATP. Some organisms, such as yeast, regenerate the NAD^+ by alcoholic fermentation:

$$C_3H_3O_3^- + NADH + 2H^+ \rightleftharpoons CO_2 + CH_3CH_2OH + NAD^+ \qquad (10.100)$$

(a) Write the net reaction for conversion of glucose to carbon dioxide and ethanol by glycolysis followed by alcoholic fermentation. What happens to NAD^+ and NADH in the net reaction? How do you explain this result?

(b) Draw a coupling diagram, modeled on Figure 10.12, that represents the combination of reactions (10.99) and (10.100). What is the coupling species? Explain your response. Include the ATP production as a straight arrow, like the one in Figure 10.12.

In order to couple two redox half reactions, the reduction potential for the coupling half reaction has to be intermediate between the reduction potentials of the reactions that are coupled. We have illustrated this in Figure 10.12 for the Blue-Bottle reactions. Note that as you proceed left-to-right from one half reaction to the next, the potentials become successively more positive. Glucose, RCHO*(aq)*, can be oxidized by MB^+*(aq)* and, in turn, MBH*(aq)* can be oxidized by $O_2(g)$.

Coupled reactions in a biofuel cell

In an electrochemical **fuel cell** a fuel is oxidized at the anode and oxygen is reduced at the cathode. The reactions in a fuel cell are the same as combustions, except that a great deal of the free energy change of the reaction is converted to electrical energy instead of heat and light. Fuel cells require catalysts as a part of the electrodes to catalyze the desired oxidations and reductions. Hydrogen is the fuel in the fuel cells that were used to produce electrical power for the Apollo missions to the moon and are now used in the space shuttle. An advantage of hydrogen as the fuel is that the only product of the cell reaction is water (which is collected and used in the

An exploding fuel cell ruined the Apollo 13 mission and almost cost the astronaut's lives.

spacecraft). A disadvantage of present-day hydrogen fuel cells is that they have to be operated at quite high temperatures, in order to make the electrode catalysts work efficiently.

Since fuel cells are an efficient means to convert chemical energy to electrical energy, scientists are doing research to develop catalysts to oxidize more easily handled fuels such as alcohols or gasoline and to work at lower temperature. Biological catalysts, enzymes, offer one possible solution to these problems, since they work at the temperature of living organisms and usually act on readily available fuels, including glucose. The miniature glucose fuel cell shown in the chapter opening illustration and in Figure 10.13 is one such device.

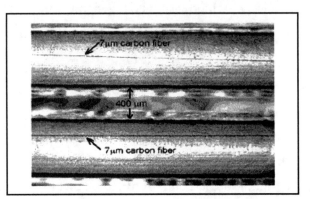

Figure 10.13. Photomicrograph of a miniature glucose-air fuel cell.
The connections of the carbon-fiber electrodes to the external circuit are not shown in this photograph.

In this glucose fuel cell, the electrodes are very fine carbon fibers that appear as thin lines dwarfed by the 1-mm wide wells in which they are suspended. Each electrode is coated with a polymeric material containing complexed osmium metal ions and the enzyme, glucose oxidase or laccase, required for the oxidation and reduction reactions, respectively. Many factors affect the electrical output of this cell, but you can use what you have learned to analyze the oxidation that produces the electrical energy.

10.69. Check This   *Reduction potentials in a glucose-air fuel cell*
The half reactions that power the glucose-air fuel cell shown in Figure 10.13 are:

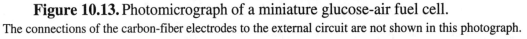

$$O_2(g) + 4H^+(aq) + 4e^- \rightleftharpoons 2H_2O(l) \qquad\qquad E^\circ(O_2, H_2O) = 1.229 \text{ V} \qquad (10.81)$$

$$RCOO^-(aq) + 3H^+(aq) + 2e^- \rightleftharpoons RCHO(aq) + H_2O(l) \quad E^{\circ\prime}(RCOO^-, RCHO) = -0.44 \text{ V} \quad (10.86)$$

(a) The solution in the cell is buffered at pH 5. Calculate the reduction potentials at 298 K for these reactions at this pH, assuming that all other species are at unit activity.

(b) What is the overall cell reaction and the cell potential under the conditions in part (a). Use Le Chatelier's principle to predict how the cell potential would change if the activity of oxygen were decreased.

Figure 10.14 shows a series of coupled redox reactions that illustrates how the biofuel cell in Figure 10.13 operates. The enzymes, glucose oxidase (GOx) and laccase (LAC), contain complexed metal ions that change oxidation number as they gain or lose electrons in the redox reactions they catalyze. In turn they lose or regain these electrons in their interactions with the osmium ions that are part of the polymer material in which the enzymes are embedded. The osmium ions on the cathode and anode have different reduction potentials because they are differently complexed with ligands from the polymer.

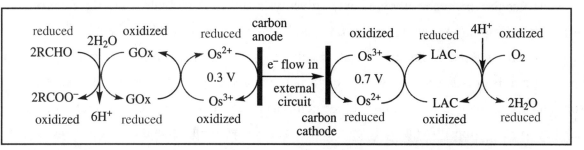

Figure 10.14. The coupled series of reactions in the biofuel cell in Figure 10.13.

GOx is the enzyme glucose oxidase and LAC is the enzyme laccase. The cell potential is about 0.4 V. Only the stoichiometries of the glucose oxidation and oxygen reduction are shown.

10.70. Check This   *Intermediate potentials in the biofuel cell*

(a) Use your results from Check This 10.69 to add to Figure 10.14 the reduction potentials for the gluconate-glucose and oxygen-water reactions.

(b) What would be the ideal reduction potentials for the glucose oxidase and laccase? Ideal potentials would provide about the same driving force for both the reduction and oxidation these enzymes undergo. Explain your choices.

(c) Approximately how efficient is this 0.4 V cell? That is, how much of the free energy released by the glucose-oxygen reaction can do work in the external circuit? Explain your reasoning.

Glucose oxidation in aerobic organisms

In aerobic organisms, including you, the pathway for glucose oxidation is glycolysis, net reaction (10.99), followed by a series of reactions called the tricarboxylic acid (TCA) cycle. The fate of the glucose carbons in this overall pathway is:

> The TCA cycle is also called the citric acid cycle, since citric acid (see Consider This 10.57) is the tricarboxylic acid in the cycle.

$$C_6H_{12}O_6 \xrightarrow{\text{glycolysis}} 2C_3H_3O_3^- \xrightarrow{\text{TCA cycle}} 6CO_2 \qquad (10.101)$$

The notations over the reaction arrows remind us that these are not individual reactions, but a series of reactions that has been given a collective name. Carbons are conserved in the pathway represented by expression (10.101), but oxygen and hydrogen are not. There is more oxygen in the product than in the reactant, so oxygen must come into the reaction from another source. Almost universally, when a source of oxygen or hydrogen atoms is required in living systems, the source is water. The stoichiometry of glucose oxidation could be written:

$$C_6H_{12}O_6 + 6H_2O \xrightarrow{\text{glycolysis + TCA cycle}} 6CO_2 + 24H^+ + 24e^- \qquad (10.102)$$

No molecular oxygen is involved in the glycolysis and TCA pathways, but the electrons ultimately end up on oxygen. **Electron transport** is the pathway by which electrons from glucose are transferred to oxygen. Not all the electrons take the same route to oxygen. However, since our aim is only to show how mitochondria use coupled redox reactions to extract energy from glucose oxidation, we will neglect the complications and reduce the electron transport pathway to a few steps. If you study metabolism further, you will amend and add to these steps, but the outcome is the same.

The electrons from glucose reduce NAD^+:

$$C_6H_{12}O_6 + 12NAD^+ + 6H_2O \xrightarrow{\text{glycolysis + TCA cycle}} 6CO_2 + 12NADH + 12H^+ \quad (10.103)$$

The net reaction represented by expression (10.103) requires more than a dozen steps with a different enzyme to catalyze each step. The electrons given up by glucose end up on NADH molecules in the matrix of mitochondria. The NADH reduces ubiquinone that is associated with the cytochrome bc_1 protein complex in the inner mitochondrial membrane (chapter-opening illustration):

$$UQ + NADH + H^+ \rightleftharpoons UQH_2 + NAD^+ \qquad (10.104)$$

Reaction (10.104) is reaction (10.84) rewritten with abbreviations for the biomolecules.

10.71. Check This   *Role of NAD⁺-NADH in the glucose oxidation pathway*

Assume that all the NADH produced by net reaction (10.103) undergoes reaction (10.104). What is the net outcome of reactions (10.103) and (10.104)? Write the net reaction that describes this outcome. What happens to NAD^+ and NADH in the net reaction? Explain the role that NAD^+ and NADH play in this part of the glucose oxidation pathway.

Within the cytochrome bc_1 complex, there is a redox reaction between UQH_2 and Fe^{3+} [recall reaction (10.1)] in a heme complex associated with cytochrome c_1:

$$2Fe^{3+}(\text{cyt } c_1) + UQH_2 \rightleftharpoons 2Fe^{2+}(\text{cyt } c_1) + UQ + 2H^+(\text{intermembrane space}) \qquad (10.105)$$

The protein complex is oriented in the membrane in such a way that the hydronium ions produced by reaction (10.105) end up in the intermembrane space of the mitochondrion. The net result of reactions (10.103) through (10.105) is to transfer hydronium ions across the membrane, as represented in Figure 10.15. The hydronium ion concentration in the matrix is lower than in the intermembrane space. This difference in hydronium ion concentrations sets up an **electrochemical gradient**, a difference in both electrical charge and pH across the inner mitochondrial membrane. The free energy stored in the electrochemical gradient is what other proteins in the mitochondrial membrane use to drive the synthesis of ATP from ADP and P_i. Recall once again the words from the *I Ching* at the beginning of this chapter, "…one must separate things in order to unite them."

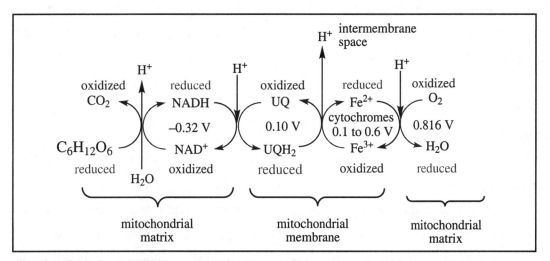

Figure 10.15. Coupled reactions in the aerobic glucose oxidation pathway.
The stoichiometry of the reactions is not represented on this diagram. All the reactions occur either in the mitochondrial matrix or inner mitochondrial membrane. The net effect of the oxidation of glucose is to move hydronium ions from the matrix to the intermembrane space of the mitochondrion.

Equation (10.105) shows that transfer of hydronium ions across the membrane will stop if all the Fe^{3+} in cytochrome c_1 is reduced. The rest of the electron transport pathway is a sequence of redox reactions involving other cytochromes containing both complexed iron and copper ions. At each step, the reduced product of the preceding step is reoxidized. The end of the pathway is a redox reaction between molecular oxygen and Fe^{2+} in cytochrome a_3:

$$O_2 + 4Fe^{2+}(cyt\ a_3) + 4H^+ \rightarrow 2H_2O + 4Fe^{3+}(cyt\ a_3) \qquad (10.106)$$

All the steps involving the cytochromes have been lumped together in Figure 10.16. Reaction (10.106), on the matrix side of the membrane, further reduces the hydronium ion concentration in the matrix and helps to maintain the electrochemical gradient across the membrane.

10.72. **Check This**   *Net reaction for oxygen reduction*

Beginning with one NADH molecule from reaction (10.103), the net reaction represented by Figure 10.15 is:

$$\text{NADH} + 3\text{H}^+(\text{matrix}) + \tfrac{1}{2}\text{O}_2 \rightarrow \text{NAD}^+ + 2\text{H}^+(\text{intermembrane space}) + \text{H}_2\text{O} \qquad (10.107)$$

(a) Write the sequence of reactions that gives net reaction (10.107).

(b) What is the standard cell potential for reaction (10.107) at pH 7? Explain how you get your answer.

Coupled reactions, including coupling that does not involve redox reactions, are common in living systems. The electron transport pathway is present in all aerobic organisms. A similar pathway is part of the photosynthetic process in green plants. The net result of the electron transport pathway in mitochondria is to create an electrochemical gradient that stores energy from the glucose oxidation. If you take a biochemistry course, you will learn how the electrochemical gradient is harnessed to produce the ATP you need to function and to continue to learn even more.

Reflection and projection

In this last section, we introduced the redox chemistry of glucose in living systems and simpler test tube reactions. This discussion provided further examples of how the Nernst equation helps us understand the dependence of cell potentials and reactions on the composition of the solution, and emphasized the role of coupled redox reactions. When we looked at the role of coupled reactions in living organisms, we left out or simplified many of the details of the pathway for storing the free energy of glucose oxidation as an electrochemical gradient. Our intent has been to show the importance of redox and electrochemical processes for living organisms and to apply the concepts developed for electrochemical cells to an analysis of redox reactions in living cells.

Analyses of equilibrium and electrochemical systems are probably the two most important applications of thermodynamics in chemistry. Questions about the direction and extent of reactions that we raised early in the book are qualitatively and quantitatively answered by these analyses. However, at least one large question remains: Why are some reactions so slow that they seem not to occur, even though their calculated free energy changes are favorable? The oxidation of glucose at room temperature is an example. Enormous amounts of glucose are synthesized by green plants every day; it becomes a part of their structure as cellulose or is stored as starch. The free energy for oxidation of glucose is large and negative and there is plenty of oxygen available in the atmosphere to react with it. Yet it can last for hundreds (even thousands) of years unless it is heated very hot or is eaten by an organism and broken down by the pathways we have just discussed. Thermodynamics cannot tell us how fast a favored process will occur. To understand the speeds of a chemical reactions, we need to know more about what affects the speeds and that is the topic of the next chapter.

Section 10.9. Outcomes Review

In this chapter, we introduced two kinds of electrochemical cells: electrolytic and galvanic. In electrolytic cells we use an electric current to produce chemical reactions by reducing and oxidizing species in solution. In galvanic cells oxidation and reduction half reactions are spatially separated and the free energy of a chemical reaction drives electrons through an external circuit, where they can be used to do work. The quantitative redox relationships that follow from electrochemical cell measurements also characterize redox reactions that occur between reactants mixed in the same solution. Redox reactions in living cells obey these same relationships, which can be used to help understand and explain observed stoichiometric relationships in organisms.

Check your understanding of the ideas in the chapter by reviewing these expected outcomes of your study. You should be able to:

•   use evidence from experimental observations to write probable half reactions for the reductions and oxidations taking place in an electrolytic cell [Section 10.1].

•   describe and draw molecular level diagrams of the processes going on and the flow of charge in an electrochemical cell [Sections 10.1 and 10.2].

•   use the Faraday and relationships among time, electric current, cell potential, and cell reaction stoichiometry to calculate the amounts of products from an electrolysis or the amount of work available from the reactants in a galvanic cell [Sections 10.1 and 10.5].

•   show how to connect two metal-metal ion half cells to make a galvanic cell and explain the role of each component of the cell [Section 10.2].

•   use the known cell reaction to identify the anode and cathode of an electrochemical cell [Sections 10.1, 10.2, 10.3, 10.4, and 10.5].

•   translate a physical cell set up to the conventional line notation for cells and *vice versa* [Section 10.2].

•   use the known sign for a galvanic cell potential to identify the direction of the cell reaction and the anode and cathode of the cell [Sections 10.3 and 10.4].

•   determine an unknown cell potential for a cell reaction by combining cell reactions with known cell potentials to give the desired cell reaction and its cell potential [Sections 10.4, 10.5, 10.6, 10.7, and 10.8].

- use a table of standard reduction potentials to predict the direction of any redox reaction (for which data are given) and determine its standard cell potential [Sections 10.4, 10.5, 10.6, 10.7, and 10.8].

- describe how an electrode senses the redox half reaction in a half cell in which the reduced and oxidized species are both present as dissolved ions and/or molecules [Section 10.6].

- determine free energy change for a cell reaction from cell potential and *vice versa* [Sections 10.5, 10.6, 10.7, and 10.8].

- apply Le Chatelier's principle to predict the direction of change of cell potentials as concentrations in the half cells are changed [Sections 10.6, 10.7, 10.8, and 10.10].

- use the Nernst equation, which relates the cell potential, the standard cell potential, and the reaction quotient for the cell reaction, to determine any one of these quantities, if the other two are known [Sections 10.6, 10.7, and 10.8].

- use the standard cell potential to determine the equilibrium constant for a cell reaction and *vice versa* [Sections 10.6 and 10.8].

- use the Nernst equation to convert standard reduction potentials to reduction potentials under non-standard conditions and *vice versa* [Sections 10.6, 10.7, and 10.8].

- use a table of standard reduction potentials at pH 7 to predict the direction and standard cell potential at pH 7 for redox reactions in biological systems [Sections 10.7 and 10.8].

- combine free energy changes for redox reactions, complexation equilibria, and solubility equilibria to get the free energy changes and reduction potentials for the net redox reactions that occur in systems with such interrelated reactions [Sections 10.8 and 10.10].

- convert a stepwise series of reactions to a diagram that shows the coupling of reactions and *vice versa* [Section 10.8].

- use standard reduction potentials to determine the probable sequence of a series of coupled redox reactions and *vice versa* [Section 10.8].

-

Section 10.10. EXTENSION — Cell Potentials and Non-Redox Equilibria

10.73. Investigate This   *Does addition of ethylenediamine affect E(Cu²⁺, Cu)?*

(a) Do this as a class investigation and work in small groups to discuss and analyze the results. Use a set up similar to the one in Investigate This 10.44. Use three wells grouped around one containing 10% aqueous KNO_3 solution to prepare three half cells. Fill two of the wells about two-thirds full of aqueous 0.10 M copper sulfate, $CuSO_4$, solution. Fill the third well about two-thirds full of aqueous 0.010 M $CuSO_4$ solution. Place a clean copper wire in each well as an electrode and use filter paper salt bridges soaked in 10% KNO_3 to connect each half cell to the KNO_3 well. Use a digital voltmeter to read the cell potential to the nearest millivolt and determine which electrode is the anode for each of the three cells you can make by combining pairs of half cells.

(b) To one of the wells containing 0.10 M $CuSO_4$ solution, add one drop of ethylenediamine (1,2-diaminoethane, $H_2NCH_2CH_2NH_2$) and stir the mixture with the electrode to make sure it is uniform. Record your observations. Use the digital voltmeter to read the cell potentials and determine which electrode is the anode for the two cells you can make by connecting the half cell with added ethylenediamine to each of the other half cells.

10.74. Consider This   *Why does addition of ethylenediamine affect E(Cu²⁺, Cu)?*

(a) Are the cell potentials you measured in Investigate This 10.73(a) consistent with one another? How do you interpret any differences you observe?

(b) What were the cell potentials you measured in Investigate This 10.73(b)? What was the direction of the change in cell potential from the values you got in part (a)? What do you think could be happening in the solution to cause a change in this direction?

Concentration cells

In Section 10.8 we combined redox equilibria (or standard cell potentials) and equilibria for non-redox reactions to get information about new reactions. In this section we will use cell potentials to determine equilibrium constants for non-redox reactions going on in the cell. The cells you prepared in Investigate This 10.73 are concentration cells. In a **concentration cell**, the half reactions in the half cells are identical and the cell potential depends upon the *difference* in concentrations in the half cells. The half-cell reaction and standard reduction potential for all of the half cells you made are:

$$Cu^{2+}(aq) + 2e^- \rightleftharpoons Cu(s) \qquad\qquad E^\circ = 0.337 \text{ V} \qquad\qquad (10.108)$$

A cell made with two identical copper half cells will have no driving force to transfer electrons in either direction. The reduction potential for both half cells is the same and the measured cell potential, $E_{Cu|Cu}$, is zero.

10.75. **Consider This**   *Do the copper concentration cells behave as expected?*

(a) Does your measured cell potential for identical copper half cells in Investigate This 10.73(a) support the prediction that the cell potential is zero? (Tiny differences between half cells can produce small cell potentials, but these are usually less than one or two millivolts.) What is $E°_{Cu|Cu}$ for identical copper half cells? Explain your answer.

(b) Do you get the same result for cells with non-identical copper half cells? Why or why not?

$Cu^{2+}(aq)$ is on the reactant side of half reaction (10.108), so Le Chatelier's principle tells us that the reduction potential will decrease, if $[Cu^{2+}(aq)]$ is decreased:

$$E(\text{higher } [Cu^{2+}]) > E(\text{lower } [Cu^{2+}])$$

The lower the concentration, the greater this difference. Since the half cell with the lower concentration has the lower reduction potential, its electrode will be the anode of the concentration cell, as you found in Investigate This 10.73(a). One of the cells you made was:

$$Cu(s) \,|\, Cu^{2+}(0.010 \text{ M}) \,\|\, Cu^{2+}(0.10 \text{ M}) \,|\, Cu(s) \tag{10.109}$$

The cell reaction is:

$$Cu^{2+}(0.10 \text{ M}) + Cu(s, anode) \rightleftharpoons Cu(s, cathode) + Cu^{2+}(0.010 \text{ M}) \tag{10.110}$$

Reaction (10.110) can be interpreted as a decrease in concentration of a more concentrated solution and an increase in concentration of a less concentrated solution. This change is analogous to diffusion of ions from higher toward lower concentration. We know that entropy and free energy favor this change and now we see that the cell potential for a concentration cell is another measure of this directionality. The Nernst equation, at 298 K, for reaction (10.110) is:

$$E_{Cu|Cu} = E°_{Cu|Cu} - \left(\frac{0.059 \text{ V}}{2}\right)\log\left(\frac{\text{lower } [Cu^{2+}]}{\text{higher } [Cu^{2+}]}\right) = 0 - \left(\frac{0.059 \text{ V}}{2}\right)\log\left(\frac{0.010 \text{ M}}{0.10 \text{ M}}\right) \tag{10.111}$$

$$E_{Cu|Cu} = -\left(\frac{0.059 \text{ V}}{2}\right)(-1) = 0.030 \text{ V}$$

10.76. **Check This**   *Cell potentials in Investigate This 10.73*

Do your results from Investigate This 10.73(a) support the prediction that the cell potential for the cells with unequal copper ion concentrations is about 0.030 V? Why or why not?

Cell potentials and complexed ions

An important point to recognize about half reaction (10.108) is that $Cu^{2+}(aq)$ is the reactant that is reduced. If other forms of copper in the 2+ oxidation state are present in the solution, they do not take part *directly* in the reaction. For example, $Cu^{2+}(aq)$ can form complexes with ethylenediamine, $H_2CH_2CH_2NH_2$ (abbreviated en):

$$Cu^{2+}(aq) + en(aq) \rightleftharpoons Cu(en)^{2+}(aq) \qquad (10.112)$$

$$Cu(en)^{2+}(aq) + en(aq) \rightleftharpoons Cu(en)_2^{2+}(aq) \qquad (10.113)$$

$$\overline{Cu^{2+}(aq) + 2en(aq) \rightleftharpoons Cu(en)_2^{2+}(aq)} \qquad (10.114)$$

You might have analyzed data from continuous variation experiments on this system in Chapter 6, Problem 6.29. The stepwise formation, reactions (10.112) and (10.113), is evident in such experiments. In Investigate This 10.73(b), addition of ethylenediamine to one of the copper half cells produced the magenta color of the $Cu(en)_2^{2+}(aq)$ complex. Complex formation "uses up" $Cu^{2+}(aq)$, so its concentration is lowered in the half cell to which ethylenediamine is added, as shown schematically in Figure 10.16.

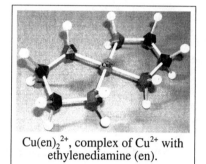

$Cu(en)_2^{2+}$, complex of Cu^{2+} with ethylenediamine (en).

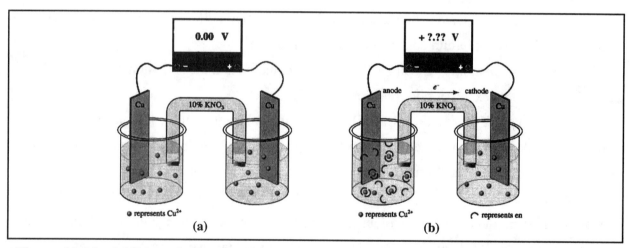

Figure 10.16. Addition of ethylenediamine to one half cell of a copper concentration cell.

10.77. Consider This   *How does addition of ethylenediamine affect [Cu²⁺(aq)]?*

(a) Use your results from Investigate This 10.73(a) to predict the *direction* of the effect on the cell potential, if adding ethylenediamine to one half of a copper concentration cell decreases the $[Cu^{2+}(aq)]$ in that half cell. State your reasoning clearly.

(b) Are your results from Investigate This 10.73(b) consistent with your prediction in part (a)? Explain why or why not.

(c) Write the line notation for the cell obtained by combining a half cell to which ethylenediamine is added with the identical half cell without ethylenediamine. Explain the reasoning for your choice of anode. Does Figure 10.16(b) correctly represent the cell you wrote? Explain why or why not.

We can use cell potentials to learn more about step-wise and overall equilibria, like those represented in reactions (10.112) through (10.114). Worked Example 10.78 shows how to use a cell potential measurement in a silver|silver concentration cell to calculate the overall equilibrium constant for the formation of the silver diammine complex, $Ag(NH_3)_2^+(aq)$:

$$Ag^+(aq) + 2NH_3(aq) \rightleftharpoons Ag(NH_3)_2^+(aq) \tag{10.115}$$

10.78. Worked Example   *K for formation of $Ag(NH_3)_2^+(aq)$*

Consider a cell set up like the one in Investigate This 10.73(b), except with silver half cells each containing 2.0 mL of aqueous 0.10 M solutions of silver nitrate, $AgNO_3$, and silver metal electrodes. When one drop (0.05 mL) of 15 M aqueous ammonia, NH_3, solution is added to one of the half cells, the measured cell potential is 0.325 V. The half cell containing ammonia is the anode. What is the equilibrium constant for reaction (10.115)?

Necessary information: We will need the Nernst equation and the stoichiometry and equilibrium constant expression for reaction (10.115).

Strategy: To get K for reaction (10.115), we need the concentrations of $[Ag^+(aq)]$, $[NH_3(aq)]$, and $[Ag(NH_3)_2^+(aq)]$ in the half cell to which ammonia is added. Use the cell potential and the Nernst equation to find $[Ag^+]$ (omitting the state designation). Then use the reaction stoichiometry and the amounts of reactants that are mixed in the half cell to find $[NH_3]$ and $[Ag(NH_3)_2^+]$, assuming that addition of ammonia makes a negligible change in the volume of the solution. Substitute all the concentrations into the equilibrium constant expression to get K.

Implementation: For this concentration cell, *n* is one, because silver ion reduction requires only one electron. Let (Ag^+) be the silver ion concentration ratio in the half cell with added ammonia. The Nernst equation for the concentration cell is:

$$E_{Ag|Ag} = 0.325 \text{ V} = -(0.059 \text{ V})\log\left(\frac{\text{lower }(Ag^+)}{\text{higher }(Ag^+)}\right) = -(0.059 \text{ V})\log\left(\frac{(Ag^+)}{0.10}\right)$$

Solve for the ratio of concentrations and then for (Ag^+) {= numerical value of $[Ag^+]$}:

$$\log\left(\frac{(Ag^+)}{0.10}\right) = \frac{0.325\ V}{-0.059\ V} = -5.51; \qquad \left(\frac{(Ag^+)}{0.10}\right) = 3.1 \times 10^{-6}$$

$$(Ag^+) = 3.1 \times 10^{-7}$$

Only a tiny amount of silver ion is left uncomplexed in solution. Essentially all the silver ion is present as $Ag(NH_3)_2^+$. We had 0.10 M silver ion to start with and all the silver must still be present, so we have:

$$(Ag(NH_3)_2^+) = 0.10$$

For the ammonia, we calculate the number of moles added and subtract the number of moles complexed with silver ion to find the number of moles left as $NH_3(aq)$ in solution.

$$\text{mol } NH_3 \text{ added} = (0.05 \times 10^{-3}\ L)(15\ M) = 7.5 \times 10^{-4}\ \text{mol } NH_3 \text{ added}$$

$$\text{mol } NH_3 \text{ complexed} = (2.0 \times 10^{-3}\ L)(0.10\ M)\left(\frac{2\ \text{mol } NH_3}{1\ \text{mol complex}}\right)$$

$$= 4.0 \times 10^{-4}\ \text{mol } NH_3 \text{ complexed}$$

$$\text{mol } NH_3 \text{ unreacted} = (\text{mol } NH_3 \text{ added}) - (\text{mol } NH_3 \text{ in the complex})$$

$$= 3.5 \times 10^{-4}\ \text{mol } NH_3 \text{ unreacted}$$

$$[NH_3] = \left(\frac{3.5 \times 10^{-4}\ \text{mol}}{2.0 \times 10^{-3}\ L}\right) = 0.18\ M; \qquad (NH_3) = 0.18$$

The equilibrium constant expression and equilibrium constant for reaction (10.115) are:

$$K = \left(\frac{(Ag(NH_3)_2^+)}{(Ag^+)(NH_3)^2}\right)_{eq} = \frac{(0.10)}{(3.1 \times 10^{-7})(0.18)^2} = 1.0 \times 10^7$$

Does the answer make sense? We do not know whether the numerical answer is correct until we look in a table of stability constants for metal ion complexes to see that it is. However, we know that the equilibrium constant will be large (much greater than unity), because essentially all of the silver ion gets complexed, that is, reaction (10.115) greatly favors the products at equilibrium. Cell potentials can be used to determine equilibrium constants for reactions that affect a cell concentration, but are not, themselves, redox reactions. This is a powerful method for determining equilibrium constants, especially those involving metal ions.

10.79. Check This   *K for formation of Cu(en)$_2^{2+}$(aq)*

Use your data from Investigate This 10.73 to determine the equilibrium constant for reaction (10.114). The density of liquid ethylenediamine, $C_2H_8N_2$, is 0.9 $g \cdot mL^{-1}$ and one drop is about 0.05 mL. How does the equilibrium constant for this complexation reaction compare to that for

---

the formation of the silver diammine complex? Which is the stronger complex? Give the reasoning for your answer.

Index of Terms

Chapter 10 Problems

Section 10.1. Electrolysis

10.1.   Passing an electric current through an aqueous solution of sodium chloride, NaCl, containing universal pH indicator produces gases and changes in the color of the indicator at both electrodes. The observations are different for low and high concentrations of salt.

   (i)   In a 0.1 M NaCl solution, electrolysis produces a basic solution at the cathode and an acidic solution at the anode. The anode gas does not burn; both gases are odorless.

   (ii)   In a 6 M NaCl solution, electrolysis produces a basic solution at the cathode and the solution at the anode loses color (is bleached). The gas produced at the anode causes a choking sensation when inhaled.

(a)   Write the half reactions occurring at the cathode and anode in the dilute solution. Show how your reactions explain all the observations.

(b)   Repeat part (a) for the concentrated solution.

10.2.   **(a)**   When iodine, I_2, dissolves in a clear, colorless aqueous solution of potassium iodide, KI, a clear, orange solution of triiodide, I_3^- (which can be thought of as $I_2 \cdot I^-$), is formed. When an aqueous solution of KI is electrolyzed, bubbles of gas are formed at the cathode and the solution around the anode becomes orange. What are the cathodic and anodic half reactions and net cell reaction for this electrolysis? Explain your reasoning.

(b)   Sodium thiosulfate, $Na_2S_2O_3$, added to a triiodide solution reacts immediately with I_3^- to produce iodide ion, I^-, and decolorizes the solution. If a little $Na_2S_2O_3$ is added to a KI solution before electrolysis, bubbles begin to form immediately at the cathode when electrolysis is begun, but the solution around the anode remains clear and colorless for a period of time and then begins to turn orange. Are these observations consistent with the reactions you wrote in part (a)? Explain why or why not.

10.3.   Elemental sodium is too reactive to be found naturally in an uncombined form. The elemental metal can be produced by the electrolysis of molten sodium chloride.

(a)   How long would it take to produce 23 g of metallic sodium if a 12 A current is passed through a molten sodium chloride electrolysis cell? Explain.

(b)   Chlorine gas is also produced in this electrolysis. Identify the half reactions occurring at the cathode and anode. Write the net cell reaction. How much chlorine gas is produced during the electrolysis in part (a). Explain your reasoning.

10.4. The Dow process is used to produce pure magnesium metal by electrolysis of molten magnesium chloride. Chlorine gas is also produced in the process.

(a) Write the half reactions that occur at the cathode and anode.

(b) Which of the half reactions in part (a) is the oxidation and which the reduction? Is your identification consistent with the definition of the cathode and anode of an electrochemical cell? Explain.

(c) What is the net cell reaction in the Dow process?

(d) How many kilograms of magnesium metal could be produced if a 95,000 A current is passed through a molten magnesium chloride cell for 8.0 hours?

(e) A temperature of almost 1000 K is required to melt magnesium chloride. Wouldn't it be easier to electrolyze the solid salt at a lower temperature? Why isn't this done?

10.5. Gold is often electroplated onto other, less expensive, metals to produce gold-plated jewelry. The electrolyte solution usually contains gold(III) as its cyanide complex, $Au(CN)_4^-(aq)$. The reaction at the surface to be plated is:

$$Au(CN)_4^-(aq) + 3e^- \rightarrow Au(s) + 4CN^-(aq)$$

(a) In the electroplating cell, the object to be gold plated is one electrode and a sheet of gold is the other. Which is the cathode and which the anode in this cell? What are the cathodic and anodic half reactions? Explain your responses.

(b) How many grams of gold can be electroplated on a bracelet by a 2.5 A current passing through the electroplating cell for 7.5 minutes?

(c) Commercial electroplating cells are not 100% efficient; some of the electrical energy is dissipated as heat and in side reactions. If the bracelet in part (b) gained 0.65 g during the plating, how efficient was the process? Explain.

10.6. Two electrolysis cells were set up in series so that all the current that passed through the first also passed through the second. The first cell contained an aqueous solution of silver nitrate and silver electrodes. The second cell contained an aqueous solution of a cobalt complex ion and platinum electrodes. After current had passed for some time, the cathodes in each cell were removed, washed, dried, and weighed. The silver cathode had gained 0.2789 g and the platinum cathode had gained 0.0502 g.

(a) Based on the results from the silver cell, how many moles of electrons had passed through the cells? Explain.

(b) What is the oxidation number of the cobalt in the cobalt complex? Explain the reasoning for your answer.

10.7. How much time will it take to electroplate 0.0353 g of chromium from an aqueous potassium dichromate, $K_2Cr_2O_7$, solution, with a current of 0.125 A passing through the cell? Explain the reasoning for your answer.

10.8. When Thomas Edison first sold electricity commercially near the end of the nineteenth century, he needed a method to measure the amount of electricity each customer used, so he invented a way, based on electrolysis, to measure electrical use. A fraction of the current that entered the customer's house or business passed through an electrolysis cell containing a solution of zinc sulfate and zinc electrodes. Each month, a "meter" reader would visit, remove the cathode, wash and dry it, weigh it, and then calculate the coulombs used.

(a) If 8% of the current passed through such a meter and the cathode increased in mass by 57 g in a 30-day period, how many coulombs had the customer used? On the average, what was the current flow (amperes) during this month? Clearly explain how you get your answers.

(b) The electricity that Edison sold was direct current (DC), that is, one of the prongs of a plug was always positive and the other negative. Today our electricity is supplied as alternating current (AC) with the polarity (positive and negative signs) changing sixty times per second. Would Edison's meter work today? Why or why not?

Section 10.2. Electric Current from Chemical Reactions

10.9. Write the cathodic and anodic half reactions, the net cell reaction, and the correct line notation for a cell based on each of these unbalanced reaction equations. Explain your reasoning.

(a) $Mn(s) + Ti^{2+}(aq) \rightleftharpoons Mn^{2+}(aq) + Ti(s)$

(b) $U(s) + V^{2+}(aq) \rightleftharpoons U^{3+}(aq) + V(s)$

(c) $Zn(s) + Ni^{2+}(aq) \rightleftharpoons Zn^{2+}(aq) + Ni(s)$

(d) $Mg(s) + Cr^{3+}(aq) \rightleftharpoons Mg^{2+}(aq) + Cr(s)$

10.10. Write the cathodic and anodic half reactions and the net cell reaction for the galvanic cells represented by these correct line notations. Explain your reasoning.

(a) $Cu(s) \,|\, Cu^{2+}(aq) \,\|\, Cu^+(aq) \,|\, Cu(s)$

(b) $Co(s) \,|\, Co^{2+}(aq) \,\|\, Ag^+(aq) \,|\, Ag(s)$

(c) $Al(s) \,|\, Al^{3+}(aq) \,\|\, Fe^{2+}(aq) \,|\, Fe(s)$

(d) $Pb(s) \,|\, Pb^{2+}(aq) \,\|\, Cu^{2+}(aq) \,|\, Cu(s)$

10.11. Consider a cell made by placing a manganese rod, Mn(s), in a solution of manganese(II) sulfate, $MnSO_4(aq)$, that is connected by a salt bridge to a solution of chromium(II) sulfate, $CrSO_4(aq)$, containing a chromium rod, Cr(s). The manganese electrode is negative. Write the cathodic and anodic half reactions, the net cell reaction, and the correct line notation for the cell. Which direction do anions migrate in the cell? Which direction do electrons travel through a wire connecting the electrodes? Explain.

10.12. When a clean, shiny strip of magnesium metal, Mg, is immersed in a clear, blue aqueous solution of copper sulfate, $CuSO_4$, the metal in contact with the solution quickly turns dark and the color of the solution begins to fade.

(a) Write a net ionic reaction for this process and show how it explains the observations.

(b) Sketch the set up of an electrochemical cell that would take advantage of this reaction to provide a flow of electrons in an external circuit. Write the cathodic and anodic half reactions, the net cell reaction, and the correct line notation for this cell.

10.13. When a strip of iron metal, Fe, is immersed in an aqueous solution of chromium(III) chloride, $CrCl_3$, no apparent changes occur. When a strip of chromium metal, Cr, is immersed in an aqueous solution of iron(II) nitrate, $Fe(NO_3)_2$, a dark deposit forms on the surface of the metal in contact with the solution.

(a) Sketch the set up of an electrochemical cell that would take advantage of the information above to provide a flow of electrons in an external circuit. Explain the rationale for your set up.

(b) Write the cathodic and anodic half reactions, the net cell reaction, and the correct line notation for your cell. Explain why it is correct.

10.14. If you draw power too rapidly from a flashlight battery by connecting it to a device that draws a lot of current, the battery quickly "runs down." If you remove the battery and let it rest for a while, it often seems to recover and be usable once again. The rapid drain on the battery causes it to become "polarized" and resting allows it to depolarize.

(a) What do you think happens in the battery that makes it polarized? What happens when it depolarizes? Relate your explanation to Figures 10.4 and 10.5. (The dictionary defines polarization as the separation of positive and negative charge.)

(b) The "solution" inside a flashlight battery is a thick paste of ionic and molecular solids in water. Ions do not move as rapidly in this medium as they do in an aqueous solution. Is this condition consistent with your explanation in part (a)? Explain.

10.15. You can make a simple galvanic cell by sandwiching a thick piece of filter paper soaked with salt solution between a sheet of copper and a sheet of zinc. An ammeter connected to the metal electrodes shows that an electrical current is produced by this cell. (You may have seen "lemon batteries" or "potato batteries" that use the same design, two different metals stuck into the same ionic solution – the liquid in and around the cells in the lemon or potato.) As current flows the zinc electrode loses mass and hydrogen gas is produced at the copper electrode.

(a) What are the half reactions going on at each electrode in this cell? Explain your reasoning.

(b) What is the net cell reaction in this cell?

(c) Write the correct line notation for this cell. Explain your choice of cathode and anode.

10.16. Consider an electrochemical cell based on this spontaneous reaction:

$$Zn(s) + Cl_2(g) \rightarrow Zn^{2+}(aq) + 2Cl^-(aq)$$

The chlorine gas bubbles into the cell solution and reacts at the surface of a graphite (carbon) electrode. Both electrodes dip into the same solution.

(a) What are the half reactions going on in this cell?

(b) Which electrode is the anode in this cell? Explain the reasoning for your choice.

(c) Make a sketch of this cell and write the line notation that describes it.

(d) Which direction do electrons move in a wire connecting the two electrodes?

(e) How many moles of $Cl_2(g)$ must react to produce an electric current of 0.12 amp for exactly 7 days? Explain the reasoning for your answer.

Section 10.3. Cell Potentials

10.17. The measured voltage of an electrochemical cell made by connecting an $Fe^{2+}|Fe$ half cell with a $Pb^{2+}|Pb$ half cell is positive when the anode input from the voltmeter is connected to the lead metal electrode. What is the cell reaction? Explain how you get your answer. What experiment(s) could you do to test whether your cell reaction is correct?

10.18. Consider the cell potentials under standard conditions, $E°$, for these reactions:

$$Ni(s) + 2Ag^+(aq) \rightleftharpoons Ni^{2+}(aq) + 2Ag(s) \qquad E° = 1.05 \text{ V}$$
$$Cu(s) + 2Ag^+(aq) \rightleftharpoons Cu^{2+}(aq) + 2Ag(s) \qquad E° = 0.46 \text{ V}$$

Show how to obtain the following cell reaction and its standard cell potential.

$$Ni(s) + Cu^{2+}(aq) \rightleftharpoons Ni^{2+}(aq) + Cu(s)$$

10.19. When a clean strip of lead is placed in a 0.2 M aqueous solution of silver nitrate, the surface of the metal in contact with the solution soon turns dark and then small silvery needles begin to grow on the surface. This behavior is similar to what you observe when a strip of copper metal is immersed in an aqueous solution of silver nitrate.

(a) Using these data, can you predict what will happen if a strip of copper is placed in a solution of lead nitrate? What will happen if a strip of lead is placed in a solution of copper nitrate? Explain your responses.

(b) These cells were made and the cell potentials measured:

$Pb_{(s)} | Pb^{2+}_{(aq, 0.2 M)} \| Ag^+_{(aq, 0.2 M)} | Ag_{(s)}$ $E = 0.90$ V

$Cu_{(s)} | Cu^{2+}_{(aq, 0.2 M)} \| Ag^+_{(aq, 0.2 M)} | Ag_{(s)}$ $E = 0.44$ V

What is the cell potential for a cell made by combining the $Pb^{2+}|Pb$ and $Cu^{2+}|Cu$ half cells? Write the cell in correct line notation. What is the net cell reaction in this cell? Explain how you get your answers.

(c) Given the information in part (b), what are your answers to part (a)? Are your answers different than before? Explain why or why not.

10.20. Consider a cell made by placing a nickel wire, $Ni_{(s)}$, in a 0.12 M aqueous solution of nickel(II) sulfate, $NiSO_{4(aq)}$, that is connected by a salt bridge to a 0.12 M aqueous solution of lead(II) nitrate, $Pb(NO_3)_{2(aq)}$, containing a strip of lead, $Pb_{(s)}$. When the black (anode) lead from a digital voltmeter is connected to the lead electrode, the meter reading is –0.10 V.

(a) Write the cathodic and anodic half reactions, the net cell reaction, and the correct line notation for this cell. Clearly explain your reasoning.

(b) If a strip of lead were placed in a 0.12 M aqueous solution of nickel(II) nitrate, what do you predict would be observed? Explain.

Section 10.4. Half-Cell Potentials: Reduction Potentials

10.21. What is the standard cell potential for the cell you sketched in Problem 10.12? Explain.

10.22. Consider an electrochemical cell made by connecting a $Ag^+|Ag$ half cell to a $Fe^{2+}|Fe$ half cell *via* a salt bridge.

(a) Which electrode will be the cathode and which the anode? Write the correct line notation for this cell. Explain how you arrive at your answers, including any assumptions.

(b) Write the net cell reaction for this cell and show how to calculate its standard cell potential.

10.23. List these metals, Al, Ca, Mg, Na, and Zn, in order of increasing strength as reducing agents. Explain how you decide the order.

10.24. List these molecular species, Br_2, Cl_2, F_2, I_2, and O_2, in order of increasing strength as oxidizing agents. Explain how you decide the order.

10.25. Explain whether each of these statements describes a spontaneous or nonspontaneous process. For those that are not spontaneous, rewrite the statement to describe the spontaneous process. Write a net cell reaction and calculate the standard cell potential for the spontaneous process in each case.

(a) Chromium metal, Cr*(s)*, reduces lead(II) ion, Pb^{2+}*(aq)* to lead metal, Pb*(s)*.

(b) In basic solution mercury metal, Hg*(l)*, reduces cadmium hydroxide, $Cd(OH)_2$*(s)*, to cadmium metal, Cd*(s)*, and forms mercury(II) oxide, HgO*(s)*.

(c) Nickel(II) ion, Ni^{2+}*(aq)*, is reduced to nickel metal, Ni*(s)*, by hydrogen gas, H_2*(g)* in acidic solution.

(d) Sodium metal, Na*(s)*, reduces water, H_2O*(l)*, at pH 7 to form hydrogen gas H_2*(g)*.

10.26. Halogens are somewhat soluble in solutions of their corresponding halide, bromine, Br_2, in aqueous Br^-, for example. These halogen-halide solutions of chlorine, bromine, and iodine are, respectively, very pale yellow-green, orange, and red. Solutions of the halide ions alone are clear and colorless. Mixing halogen-halide solutions with different halide solutions gave these results:

       (i) $Cl_2/Cl^- + Br^- \rightarrow$ orange solution

       (ii) $Cl_2/Cl^- + I^- \rightarrow$ red solution

       (iii) $Br_2/Br^- + Cl^- \rightarrow$ orange solution

       (iv) $Br_2/Br^- + I^- \rightarrow$ red solution

(a) Write the redox reaction, if any, that occurs in each case. Use the table of reduction potentials to explain why the reactions you write should occur.

(b) Predict the result you would observe, if you mixed an I_2/I^- solution with a chloride ion solution. Explain the basis of your prediction.

(c) Predict the result you would observe, if you bubbled chlorine gas, Cl_2*(g)*, into an aqueous solution of sodium bromide, NaBr*(aq)*. Explain the basis of your prediction.

10.27. One commercial source of bromine, Br_2*(aq)*, is brine that contains sodium bromide, NaBr*(aq)*. Might it be possible to oxidize the bromide to bromine by bubbling oxygen gas, O_2*(g)*, through acidified brine? Explain why or why not.

10.28. Before 1886, when Charles Hall and Paul Heroult discovered how to produce aluminum by electrolysis, the metal was extremely expensive. It was produced chemically by a redox reaction between molten aluminum chloride, $AlCl_3$, and sodium metal (which was also quite expensive but could be obtained by electrolysis – see Problem 10.3):

$$AlCl_3(l) + 3Na(s) \rightarrow Al(s) + 3NaCl(s)$$

What is the cell potential that corresponds to this redox reaction? What assumption(s) do you have to make to get your answer?

10.29. Consider a cell made by placing a sheet of zinc metal, $Zn(s)$, in an aqueous solution of zinc sulfate, $ZnSO_4(aq)$, that is connected by a salt bridge to an aqueous solution of sodium chloride, $NaCl(aq)$, containing a coil of platinum wire, $Pt(s)$, over which chlorine gas, $Cl_2(g)$, is bubbled.

(a) Which electrode is the cathode and which the anode in this cell? Explain the reasoning for your choice.

(b) Write the cathodic and anodic half reactions and the net cell reaction for this cell. Explain.

(c) Write the correct line notation for this cell and show how to calculate its standard cell potential.

10.30. Samples containing iron can be analyzed by converting all the iron to iron(II) ion, $Fe^{2+}(aq)$, in acidic aqueous solution and then titrating the iron(II) with a solution of potassium permanganate, $KMnO_4(aq)$. The permanganate ion, $MnO_4^-(aq)$, oxidizes the iron(II) to iron(III) ion, $Fe^{3+}(aq)$; the manganese(VII) in permanganate is reduced to the manganese(II) ion, $Mn^{2+}(aq)$. Permanganate ion is an intense violet color and the products of the reaction are essentially colorless, so the equivalence point of the titration is signaled when the sample solution first turns very light pink.

(a) Write the balanced redox reaction for the titration reaction.

(b) What is the standard cell potential for this reaction? Explain.

(c) How many moles of permanganate ion are required to react with one mole of iron(II) ion? If 23.46 mL of 0.0200 M $KMnO_4(aq)$ is required to titrate a sample containing iron(II) ion, how many moles of iron are in the sample. Show your work.

10.31. One way to synthesize small quantities of chlorine gas for use in the laboratory is to drop hydrochloric acid solution, $HCl(aq)$, onto crystals of potassium permanganate, $KMnO_4(s)$.

(a) Write the balanced redox reaction for this synthesis reaction.

(b) What is the standard cell potential for this reaction? Explain.

10.32. Use the data in Appendix YY to predict whether a reaction will or will not occur between each of these pairs of reactants. Explain the basis of each prediction. Write balanced redox reaction equations and give standard cell potentials for those that will occur.

   (a) $Fe^{3+}(aq)$ and $Br^-(aq)$

   (b) $Fe^{3+}(aq)$ and $I^-(aq)$

   (b) $Fe^{2+}(aq)$ and $Br_2(aq)$

   (d) $Fe^{2+}(aq)$ and $Ag^+(aq)$

10.33. Use the data in Appendix YY to predict whether these reactions will or will not occur. Explain the basis of each prediction. Write balanced redox reaction equations and give standard cell potentials for those that will occur.

   (a) $Cu^{2+}(aq) + H_2O_2(aq) \rightarrow Cu(s) + O_2(g)$

   (b) $Cu(s) + NO_3^-(aq) + H^+(aq) \rightarrow Cu^{2+}(aq) + NO(g)$

   (c) $MnO_2(s) + I^-(aq) \rightarrow Mn^{2+}(aq) + I_2(aq)$

   (d) $Cr_2O_7^{2-}(aq) + CH_3CH_2OH(aq) + H^+(aq) \rightarrow Cr^{3+}(aq) + CH_3CHO(aq)$

Section 10.5. Work From Electrochemical Cells: Free Energy

10.34. Consider a cell in which the cell reaction is:

$$Zn(s) + Pb^{2+}(aq) \rightarrow Zn^{2+}(aq) + Pb(s)$$

Under the conditions in the cell, the measured cell potential is 0.660 V.

(a) What is the free energy change, $\Delta G_{reaction}$, for the cell reaction under these conditions? Is this reaction spontaneous under these conditions? Explain.

(b) If 0.125 mol of $Zn(s)$ react in the cell, what is the maximum value of the work produced by the cell? Explain the reasoning for your answer.

(c) What mass of Pb(s) is formed when this amount of work is produced? Explain.

10.35. In a copper-zinc cell (see Consider This 10.32), zinc metal is oxidized to zinc ion, so the zinc electrode loses mass when electrons are drawn from the cell. In a large copper-zinc cell, a new zinc electrode had a mass of 486.5 g and after long use was found to have a mass of 215.8 g.

(a) How many moles of electrons had the cell produced during this time? Explain.

(b) About how much electrical work had the cell produced during this time? Explain the reasoning and the assumptions you make to get your answer. *Hint:* Use information from Figure 10.6 or Appendix YY.

10.36. Consider a cell made by placing a silver wire, Ag(s), coated with solid silver chloride, AgCl(s), into a 1 M aqueous hydrochloric acid solution, H^+(aq) and Cl^-(aq), that also contains a platinum electrode, Pt(s), over which hydrogen gas, H_2(g), at 1 bar pressure is bubbled. The cell potential, measured with the anode lead from the digital voltmeter connected to the platinum electrode, is 0.22 V.

(a) Write the line notation for this cell. Explain your reasoning.

(b) Write the cathodic and anodic half reactions (as reductions) and the net cell reaction for this cell. Explain your choices of reaction.

(c) Use the data above to determine the standard potential for this cell and the standard free energy change for the reaction, $\Delta G°_{reaction}$. Explain your reasoning.

(d) What are the standard reduction potentials for the half reactions? Explain.

10.37. The standard free energy change, $\Delta G°_{reaction}$, is –418.8 kJ for this redox reaction:
$$O_2(g) + 2H_2S(aq) \rightleftharpoons 2H_2O(l) + 2S(s)$$

(a) What are the half reactions (as reductions) that make up this overall reaction?

(b) What is the standard cell potential for the reaction? Explain how you get your result.

10.38. Oxalic acid, HOOCCOOH(s), is often used to determine the concentration of permanganate ion, MnO_4^-(aq), solutions, using this redox reaction:
$$5HOOCCOOH(s) + 2MnO_4^-(aq) + 6H^+(aq) \rightarrow 10CO_2(g) + 2Mn^{2+}(aq) + 8H_2O(l)$$
The standard free energies of formation, $\Delta G°_f$, for MnO_4^-(aq) and Mn^{2+}(aq), are –425 kJ·mol^{-1}, and –223 kJ·mol^{-1}, respectively, and values for the other species are in Appendix XX.

(a) Show how to calculate the standard free energy change for permanganate-oxalic acid reaction. Is the reaction spontaneous under standard conditions? Explain.

(b) Calculate the standard cell potential for this reaction. Explain your method.

(c) Show how to determine the standard reduction potential for this half reaction:
$$2CO_2(g) + 2H^+(aq) + 2e^- \rightleftharpoons HOOCCOOH(s)$$

10.39. Write the cathodic and anodic half cell reactions (as reductions) and the net cell reaction for each of these cells. Also determine the standard cell potential and standard free energy change for each cell reaction. Explain the reasoning for your answers.

(a) $Cr(s) \mid Cr^{2+}(aq) \parallel Cr^{3+}(aq) \mid Cr(s)$

(b) $Pt(s) \mid Fe^{3+}(aq), Fe^{2+}(aq) \parallel Cr_2O_7^{2-}(aq), Cr^{3+}(aq) \mid Pt(s)$

(c) $Zn(s) \mid Zn^{2+}(aq) \parallel Fe^{2+}(aq) \mid Fe(s)$

(d) $Hg(l) \mid Hg_2Cl_2(s) \mid Cl^-(aq) \parallel Hg_2^{2+}(aq) \mid Hg(l)$

Section 10.6. Concentration Dependence of Cell Potentials: The Nernst Equation

10.40. Consider this experimental set up: A thin sheet of copper metal is immersed in a beaker containing a 1 M aqueous solution of copper sulfate. A thin sheet of iron is immersed in another beaker containing a 1 M solution of iron(II) nitrate. The two beakers are connected by a salt bridge and then the metal sheets are connected by an electrically conducting wire.

(a) Which metal sheet will lose mass and which will gain mass? Explain the reasoning for your answer.

(b) Which solution will become more concentrated in its metal ion? Explain.

(c) At 298 K, what is the initial electrical potential between the two pieces of metal? Will this potential increase, decrease, or remain constant as the reactions proceed? Explain your reasoning.

10.41. The measured cell potential is positive for this cell:

$$Ni(s) \mid Ni^{2+}(aq, 0.1 \text{ M}) \parallel Fe^{3+}(aq, 0.1 \text{ M}) , Fe^{2+}(aq, 0.1 \text{ M}) \mid Pt(s)$$

(a) What is the net cell reaction? Explain your reasoning.

(b) Will the cell potential increase, decrease, or stay the same, if the concentration of Fe^{2+} is increased? Explain your reasoning.

(c) Will the cell potential increase, decrease, or stay the same, if the concentration of Ni^{2+} is increased? Explain your reasoning.

(d) If a strip of nickel metal is immersed in an equimolar aqueous solution of Fe^{3+} and Fe^{2+} ions, what, if any, reaction would you expect to occur. If a reaction occurs, what ions would you expect to find in the solution? What would be the relative amounts of the ions? Explain.

10.42. The measured cell potential, E, for this cell is 0.225 V at 298 K:

$$Cd(s) \mid Cd^{2+}(aq, \text{?? M}) \parallel Ni^{2+}(aq, 1.0 \text{ M}) \mid Ni(s)$$

(a) Write the cathodic and anodic half reactions (as reductions) and the net cell reaction for this cell. Explain your choices.

(b) What is $E°$ for this cell? Explain.

(c) What is the concentration of cadmium ion, $[Cd^{2+}(aq)]$, in this cell. Explain the method you use to get your answer.

(d) What is the equilibrium constant for the net cell reaction? Explain your reasoning.

10.43. Check the appropriate box in the table to indicate the effect of each change on the potential of this cell:

$$Pt(s) \mid H_2(g) \mid H^+(aq), SO_4^{2-}(aq) \mid PbSO_4(s) \mid Pb(s).$$

In each case, explain the reasoning for your choice using the net cell reaction, Le Chatelier's principle, and/or the Nernst equation.

change in the cell	increase	decrease	no effect
increase in pH of the solution			
dissolving Na_2SO_4 in the solution			
increase in size of the Pb electrode			
decrease in H_2 gas pressure			
addition of water to the solution			
increase in the amount of $PbSO_4$			
dissolving a bit of NaOH in the solution			

10.44. Check the appropriate box in the table to indicate the effect of each change on the potential of this cell:

$$Cu(s) \mid Cu^{2+}(aq) \parallel Cr_2O_7^{2-}(aq), Cr^{3+}(aq), H^+(aq) \mid Pt(s)$$

In each case, explain the reasoning for your choice using the net cell reaction, Le Chatelier's principle, and/or the Nernst equation.

change in the cell	increase	decrease	no effect
decrease in pH in the Cr solution			
decrease in size of the Cu electrode			
addition of water to the Cu solution			
addition of water to the Cr solution			
dissolving $Cr(NO_3)_3$ in the Cr solution			
dissolving $K_2Cr_2O_7$ in the Cr solution			

10.45. The pH of a solution was measured by using a sample of the solution as the electrolyte in a quinhydrone half cell coupled to a silver half cell:

$$Pt(s) \mid H_2Qu(aq, c), Qu(aq, c), H_3O^+(pH) \parallel Ag^+(aq, 0.10 \text{ M}) \mid Ag(s)$$

The measured cell potential, E, is 0.256 V at 298 K.

(a) What is the standard cell potential, $E°$, for this cell? Explain.

(b) Show how to determine the pH of the solution.

10.46. Consider a cell in which the net reaction is: $Zn(s) + Hg_2^{2+}(aq) \rightleftharpoons Zn^{2+}(aq) + Hg(l)$.

(a) Write the correct line notation for the cell.

(b) When the ionic concentrations are $[Hg_2^{2+}(aq)] = 0.010$ M and $[Zn^{2+}(aq)] = 0.50$ M, the cell potential is 1.51 V, measured at 298 K. Show how to use these date to calculate the standard cell potential, E°, and the equilibrium constant, K, for the cell reaction.

(c) If 1 g of mercury is poured into 100. mL of an aqueous 1.0 M solution of zinc sulfate, what will be the concentrations of ions in the solution at equilibrium? Explain your reasoning.

10.47. A cell potential, E, of 0.25 V was measured for this cell at 298 K:

$$Sn(s) \mid Sn^{2+}(aq,\ 0.10\ M) \parallel Sn^{4+}(aq,\ 0.010\ M)\ ,\ Sn^{2+}(aq,\ 1.0\ M) \mid Pt(s)$$

(a) Write the net cell reaction and corresponding reaction quotient, Q, for this cell. Explain your reasoning.

(b) Using only the data in this problem, show how to calculate the standard cell potential, E°, and standard free energy change, $\Delta G^{\circ}_{reaction}$, for the cell reaction.

(c) What are the equilibrium constant expression and equilibrium constant, K, for the reaction? Explain.

(d) If 1 g of tin metal is added to 50. mL of an aqueous 0.005 M solution of tin(IV) ion, what will be the concentrations of tin species in the solution at equilibrium? Explain your reasoning.

10.48. The saturated calomel electrode (S.C.E.) is a half cell that is often used as a reference for measuring the reduction potentials of other half cells. (Calomel is an old name for mercury(I) chloride, Hg_2Cl_2.) The half-cell reaction and reduction potential for the S.C.E. are:

$$Hg_2Cl_2(s) + 2e^- \rightleftharpoons 2Hg(l) + Cl^-(\textit{saturated aqueous KCl solution}) \qquad E = 0.241\ V$$

When an S.C.E. is connected to a $Co^{2+}|Co$ half cell in which the concentration of Co^{2+} is 0.050 M, the measured cell potential at 298 K is 0.561 V and the cobalt half cell is the anode.

(a) What is E for this $Co^{2+}|Co$ half cell? Show how you get your answer.

(b) Write the net cell reaction and determine the free energy change, $\Delta G_{reaction}$, for the reaction in this cell. Explain your reasoning.

(c) What are E° and $\Delta G^{\circ}_{reaction}$ for a $Co^{2+}|Co$ half cell? Explain how you get your result.

10.49. Mercury batteries (cells) are often used in electronics, watches, and medical devices like pacemakers. The half reactions (as reductions) in these cells can be represented as:

$$ZnO(s) + H_2O(l) + 2e^- \rightleftharpoons Zn(s) + 2OH^-(aq)$$

$$HgO(s) + H_2O(l) + 2e^- \rightleftharpoons Hg(l) + 2HO^-(aq)$$

A digital voltmeter connected to a mercury cell, measures a voltage of 1.35 V, when the anode input is connected to the electrode corresponding to the zinc metal.

(a) What is the cell reaction? Explain how you get your answer.

(b) Le Chatelier's principle suggests that a cell's potential should decrease as the concentrations of products builds up and reactants get depleted. Why? However, the voltage of a mercury cell is quite constant until almost the end of its useful life. Does the cell reaction you wrote in part (a) help explain this observation? Why or why not?

(c) Write the reaction quotient, Q, for the cell reaction in part (a). Does the reaction quotient help explain the cell's constant potential? Why or why not?

10.50. Consider a cell in which the net reaction is: $Al(s) + Fe^{3+}(aq) \rightleftharpoons Al^{3+}(aq) + Fe(s)$. A cell potential, E, of 1.59 V was measured at 298 K for this cell when $[Fe^{3+}(aq)] = 0.0050$ M and $[Al^{3+}(aq)] = 0.250$ M.

(a) Write the net cell reaction and corresponding reaction quotient, Q, for this cell. Explain your reasoning.

(b) Using only the data in this problem, show how to calculate the standard cell potential, $E°$, and standard free energy change, $\Delta G°_{reaction}$, for the cell reaction.

(c) What are the equilibrium constant expression and equilibrium constant, K, for the reaction? Explain.

(d) The thermite reaction, $2Al(s) + Fe_2O_3(s) \rightarrow Al_2O_3(s) + 2Fe(l)$, is a spectacular reaction that produces molten iron as a product. Use your result in part (b) to estimate the free energy change for the thermite reaction. Explain your reasoning. The thermal energy required to melt iron (starting from room temperature) is about 81 kJ·mol^{-1}. Does it make sense that molten iron is produced? Explain.

10.51. Consider this electrochemical cell: $Mg(s) \,|\, Mg^{2+}(aq, 0.60 M) \,\|\, Cu^{2+}(aq, 0.60 M) \,|\, Cu(s)$

(a) What is the cell potential for this cell at 298 K? Explain your reasoning.

(b) Suppose you connect the electrodes with a conducting wire and allow the cell reaction to reach equilibrium. What is the cell potential at equilibrium? What assumption(s) must you make about the cell contents? Explain.

(c) What is the $[Mg^{2+}(aq)]/[Cu^{2+}(aq)]$ ratio at equilibrium? Give your reasoning.

10.52. Consider this design for a zinc-chlorine cell:

$$Zn(s) \mid Zn^{2+}(aq, \text{ sat'd } ZnCl_2), Cl^-(aq, \text{ sat'd } ZnCl_2) \mid Cl_2(g, P) \mid Pt(s)$$

(a) What is the standard cell potential at 298 K for the net reaction in this cell? Explain how you get your answer.

(b) Write the reaction quotient expression, Q, for this cell reaction.

(c) Assume that solid zinc chloride, $ZnCl_2(s)$, has been dissolved to make the electrolyte solution in this cell and that some solid remains undissolved in the cell. Approximately 3 mol of $ZnCl_2(s)$ dissolve per liter of saturated solution. If chlorine gas is bubbled over the platinum electrode at 1.0 bar pressure, what is the cell potential for this cell? Explain your reasoning. (At these high concentrations, it is unlikely that calculations based on ideal solution behavior will give the correct cell potential, but it will probably be in the right direction.)

(d) How many moles of $Zn(s)$ will have to react to produce 75 kJ of work from this cell? How many moles of $Cl_2(g)$ will react? Explain.

(e) What are the concentrations of ions in the electrolyte solution after 75 kJ of work are produced by this cell. Explain. *Hint:* Remember that the electrolyte solution is saturated with $ZnCl_2(s)$.

10.53. A silver-zinc cell can be represented as:

$$Zn(s) \mid ZnO(s) \mid KOH(aq, 40\%) \mid Ag_2O(s) \mid Ag(s)$$

(a) What is the cathodic half reaction? Explain why you write your reaction as you do.

(b) What is the anodic half reaction? Explain.

(c) What is the net cell reaction?

(d) What is the purpose of the electrolyte, the KOH solution, in this cell? Is your response consistent with the net cell reaction in part (c)? Explain.

Section 10.7. Reduction Potentials and the Nernst Equation

10.54. Use Le Chatelier's principle to predict how the reduction potential for methylene blue reduction, $MB^+(aq) + H^+(aq) + 2e^- \rightleftharpoons MBH(aq)$, reaction (10.93), should vary as the pH of the half reaction is varied. This table gives reduction potential data for methylene blue at different pH's. Do these data confirm your prediction? Explain why you answer as you do.

pH	0	1	2	3	4	5	6	7	8	9	10
E, V	0.53	0.47	0.38	0.29	0.21	013	0.07	0.01	–0.03	–0.07	–0.11

10.55. Household bleach is an approximately equimolar solution of hypochlorite ion, OCl^-, and chloride ion, Cl^-, at about pH 8. (Sodium ion is the cation.) Addition of acid to bleach will produce hypochlorous acid, HClO, which is a strong oxidizing agent:

$$2HClO(aq) + 2H^+(aq) + 2e^- \rightleftharpoons Cl_2(g) + 2H_2O(l) \quad E^\circ = 1.630 \text{ V}$$

(a) Labels on bleach warn you never to add acid, because "hazardous gas can be released." What is present for hypochlorous acid to oxidize? What is the hazardous gas produced? Write the net redox reaction that occurs and determine the standard free energy change for the reaction. Explain your reasoning.

(b) What is the equilibrium constant expression and equilibrium constant for the net redox reaction you wrote in part (a)?

(c) What is the standard reduction potential for the above reaction at pH 14 and 298 K? Show how you get your answer.

(d) Why do you not have to worry about the net redox reaction in part (a) when the bleach solution is basic? Show clearly why the reaction is not favored under the conditions in a bottle of bleach. The concentration of hypochlorite ion in bleach is a little less than one molar.

10.56. (a) What is the cell potential for this cell at 298 K?

$$Pb(s) \mid Pb^{2+}(aq, 1.0 \text{ M}) \parallel H^+(aq, 1.0 \text{ M}) \mid H_2(g, 1 \text{ bar}) \mid Pt(s)$$

(b) Suppose that enough sodium sulfate, $Na_2SO_4(s)$, is dissolved in the lead ion solution to make the final sulfate ion concentration 0.50 M. Lead sulfate, $PbSO_4(s)$, is not a very soluble salt, $K_{sp} = 6.3 \times 10^{-7}$. Much of the lead ion will be precipitated and the concentration left in the solution will be determined by the final concentration of sulfate ion and the solubility product. What is the concentration of lead ion in this solution? Explain how you get your answer.

(c) What is the reduction potential for the lead half cell under the conditions in part (b)? Explain.

(d) What is the cell potential for this cell under the conditions in part (b)? Explain.

10.57. The measured cell potential, E, for this cell (the Clark cell) is 1.435 V at 298 K:

$$Zn(aq, 0.100 \text{ M}) \mid Zn^{2+}(aq, 0.100 \text{ M}), SO_4^{2-}(aq, 0.100 \text{ M}) \mid Hg_2SO_4(s) \mid Hg(l)$$

Use this information, together with data from Appendix YY, to calculate the standard reduction potential for this half reaction: $Hg_2SO_4(s) + 2e^- \rightleftharpoons 2Hg(l) + SO_4^{2-}(aq)$. Clearly explain your method.

10.58. Reduced nicotinamide adenine dinucleotide, NADH, absorbs light in the near ultraviolet region of the spectrum, around 360 nm. The oxidized form, NAD^+, does not absorb light at these wavelengths. A standard procedure for determining ethanol concentrations in the blood is to react the sample with NAD^+ and use a spectrophotometer to measure the amount of NADH that is formed:

$$NAD^+(aq) + CH_3CH_2OH(aq) \rightleftharpoons NADH(aq) + CH_3CHO(aq) + H^+(aq) \qquad E^{\circ\prime} = -0.12 \text{ V}$$

(a) An enzyme, alcohol dehydrogenase, in the reaction mixture can catalyze this reaction in both directions. What is the standard free energy change for this reaction at pH 7? Which direction of reaction is favored under standard conditions at pH 7? Explain.

(b) The reaction mixture also contains a reagent that reacts with aldehydes, but not alcohols. What is the purpose of this reagent? How does its presence affect the free energy for the reaction? Use the reaction quotient to explain your answer.

(c) How would changes in the pH of the reaction mixture affect the free energy? Is there any reason that pH change could not be used to make the reaction favorable for the analysis? (Recall the denaturing affect of pH changes on many proteins.)

10.59. A possible method for producing the work needed to power electronic devices implanted in our bodies is to implant zinc and platinum electrodes with the device. In oxygenated and ionic body fluids the zinc metal would be oxidized at the anode and oxygen would be reduced at the cathode.

(a) Write the cathodic and anodic half reactions and the net cell equation for such a cell.

(b) What is the standard cell potential for the cell reaction in part (a) under biological conditions?

(c) If a current of 35 μA (1 μA = 10^{-6} A) is drawn continuously from such a cell, how often would a 4.5 g Zn electrode have to be replaced? Does this seem like a reasonable length of time for an implant to last before replacement or maintenance? Explain.

10.60. (a) Show how to use standard reduction potentials and standard free energies of formation to find the standard free energy changes for each of these three reactions and their sum.

    (i) $NO_3^-(aq) + 4H^+(aq) + 3e^- \rightleftharpoons NO(g) + 2H_2O(l)$

    (ii) $NO(g) + {}^1\!/_2O_2(g) \rightleftharpoons NO_2(g)$

    (iii) $H_2O(l) \rightleftharpoons {}^1\!/_2O_2(g) + 2H^+(aq) + 2e^-$

    ———————————————————————

    (iv) $NO_3^-(aq) + 2H^+(aq) + e^- \rightleftharpoons NO_2(g) + H_2O(l)$

(b) What is the standard reduction potential for half reaction (iv)? Explain.

10.61. Another way to synthesize small quantities of chlorine gas for use in the laboratory (see Problem 10.31) is to drop hydrochloric acid solution, $HCl(aq)$, onto manganese dioxide, $MnO_2(s)$.

(a) Write the balanced redox reaction for this synthesis reaction.

(b) What is the standard cell potential for this reaction? Is the reaction spontaneous under standard conditions? Explain.

(c) If 6 M hydrochloric acid is used, what are the reduction potentials for the two half reactions that make up the redox reaction you wrote in part (a)? What is the cell potential for the overall reaction? Is the reaction spontaneous under these conditions? Explain.

Section 10.8. Coupled Redox Reactions

10.62. The vanilla used as a flavoring in foods contains vanillin, an aldehyde:

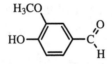

(a) Most aldehydes give a positive Benedict's test. Write the equation for the reaction of Benedict's reagent with vanillin.

(b) A drop of "pure vanilla extract" from the supermarket shelf, gives a positive Benedict's test. The ingredients label reads "vanilla bean extractives in water, alcohol, corn syrup." Can you assume that the bottle contains vanillin? Why or why not?

10.63. Many techniques for chemical analysis can be considered to be coupled reactions. For example, bleach (hypochlorite solution) can be analyzed using two reactions in sequence:

    (i) $HClO(aq) + 3I^-(aq) + H^+(aq) \rightleftharpoons Cl^-(aq) + I_3^-(aq) + H_2O(l)$

    (ii) $I_3^-(aq) + 2S_2O_3^{2-}(aq) \rightleftharpoons 3I^-(aq) + S_4O_6^{2-}(aq)$

(Bleach cannot be analyzed directly by reaction with thiosulfate, $S_2O_3^{2-}$, because the reaction leads to a variety of products that depend upon the reaction conditions.)

(a) Are both of these reactions favored in the direction written? Show how you obtain your answers.

(b) What is the coupling half reaction in this sequence? What property(ies) does this half reaction have that makes it a good coupling reaction for these redox reactions? Explain.

(c) What is the net reaction for this sequence?

(d) A 5.00-mL sample of bleach was reacted with iodide. 32.56 mL of 0.200 M thiosulfate solution were required to react stoichiometrically with the product. What is the concentration of hypochlorite in the bleach? Show how you get your answer.

10.64. In Investigate This 6.86, Chapter 6, Section 6.11, you reacted an ammoniacal, basic solution of silver ion with an aldehyde and observed the formation of silver metal, as a mirror, on the inside surface of the test tube. The ammoniacal, basic silver ion solution is called Tollens' reagent and the appearance of the silver mirror is a positive Tollens' test for aldehydes. If glucose, RCHO, had been used instead of the aldehyde, you would have observed the same reaction:

(i) $2Ag^+(aq) + RCHO(aq) + 3OH^-(aq) \rightleftharpoons 2Ag(s) + RCOO^-(aq) + 2H_2O(l)$

Tollens' reagent has a pH of about 12. In basic solution, silver ion precipitates as silver oxide. The reaction can be represented as:

(ii) $2Ag^+(aq) + 2OH^-(aq) \rightleftharpoons Ag_2O(s) + H_2O(l)$ $K \approx 10^{16}$

To make Tollens' reagent, ammonia is added to complex $Ag^+(aq)$ [as $Ag(NH_3)_2^+(aq)$] and keep its concentration *just low enough* that it doesn't precipitate as the oxide.

(a) What is highest concentration of silver ion, $[Ag^+(aq)]$, that can be present in the Tollens' reagent solution without precipitating by reaction (ii)? Show your reasoning.

(b) What is the reduction potential for silver ion, $E(Ag^+, Ag)$, if the silver ion is at the concentration you calculated in part (a)?

(c) What is the minimum cell potential for reaction (i), if RCHO is glucose? What is the free energy change per mole of reaction? Show your reasoning clearly.

10.65. The standard reduction potentials, $E°'$, for the cytochromes in the electron transport pathway in mitochondria are given alphabetically in the following list. All the half reactions involve iron(III), complexed by heme and the protein, being reduced to iron(II). What is the order of the cytochromes, from the one that oxidizes ubiquinol to the one that is oxidized by oxygen? Give the reasoning for your order.

cytochrome	$E°'$, V
a	0.29
a_3	0.55
b	0.077
c	0.254
c_1	0.22

10.66. The reduction of 3-phosphoglycerate, 3PG (a carboxylate, RCOO⁻), to glyceraldehyde-3-phosphate, G3P (an aldehyde, RCHO), by NADPH, is required in photosynthesis:

(i)
$$\begin{array}{ccc}
\underset{HC-OH}{\overset{O}{\underset{|}{\overset{\diagdown}{C}}\diagup O^-}} \quad + NADPH + 2H_3O^+ & \rightleftharpoons & \underset{HC-OH}{\overset{O}{\underset{|}{\overset{\diagdown}{C}}\diagup H}} \quad + NADP^+ + 3H_2O \\
H_2C-OPO_3{}^{2-} & & H_2C-OPO_3{}^{2-}
\end{array}$$

3PG = RCOO⁻   G3P = RCHO

NADPH is almost identical to NADH, except for an extra phosphate group (noted by the "P") in NADPH. (Organisms generally use NADH as a reducing agent in energy-producing pathways and NADPH as a reducing agent in synthetic pathways.)

(a) What is the standard free energy change for reaction (i) at pH 7 and 298 K? Is the reaction favored in the direction written? Explain why or why not.

(b) An alternative pathway to reaction (i) is a sequence of two reactions that first produce 1,3-diphosphoglycerate, 1,3-DPG (RCOO-PO₃²⁻) and then reduce it to G3P:

(ii) $RCOO^- + ATP^{4-} \rightleftharpoons RCOO\text{-}PO_3{}^{2-} + ADP^{3-}$

(iii) $RCOO\text{-}PO_3{}^{2-} + NADPH + H^+ \rightleftharpoons RCHO + NADP^+ + HOPO_3{}^{2-}$

What is the standard free energy change for reaction (iii) at pH 7 and 298 K?

(c) What is the net reaction for the pathway in part (b)? What is the standard free energy change for the net reaction? The standard free energy change for reaction (ii) is about 19 kJ·mol⁻¹. Is the net reaction favored in the direction written? Explain why or why not.

(d) The conditions in a photosynthesizing plant cell are far from standard. Assume that they are approximately these: [ATP]/[ADP] = 10; [NADPH]/[NADP⁺] = 10; and [HOPO₃²⁻] = 10^{-2} M. Calculate the [G3P]/[3PG] ratio at equilibrium under these conditions for reaction (i) and the net reaction you wrote in part (c). Which pathway produces the greater amount of G3P at equilibrium? Explain why.

(e) Why are the reactions in part (b) called "coupled?"

Section 10.10.   EXTENSION — Cell Potentials and Non-Redox Equilibria

10.67. Consider a half cell consisting of a silver wire placed in a solution saturated with silver thiocyanate, AgSCN(s), to which KSCN has been added to make [SCN⁻(aq)] = 0.10 M. When this half cell is connected via a salt bridge to a standard hydrogen electrode, the cell potential at 298 K is 0.45 V with the SHE as the anode.

(a) What is the concentration of silver ion in the silver half cell? Explain your reasoning.

(b) What is the solubility product, K_{sp}, for AgSCN(s). Explain.

10.68. Consider a half cell that contains a metal electrode, an insoluble salt of the metal cation, and a solution of the anion of the salt, for example, $Ag(s)$, $AgCl(s)$, and $Cl^-(aq)$. The half reaction and half cell can be written as:

$$AgCl(s) + e^- \rightleftharpoons Ag(s) + Cl^-(aq) \quad \text{and} \quad Ag(s) \,|\, AgCl(s \,|\, Cl^-(aq)$$

(a) Show that this half reaction is the sum of the silver ion-silver half reaction and the solubility reaction for silver chloride.

(b) Calculate the standard free energy changes for the two reactions you added in part (a). From Table 9.4, the solubility product, K_{sp}, for silver chloride is 1.8×10^{-10}. What is the standard free energy change for the half reaction above? Explain.

(c) What is the standard reduction potential for the half reaction above? How does your result compare with the value given in Appendix YY?

10.69. The measured cell potential, E, for this cell is 0.608 V at 298 K:

$$Ag(s) \,|\, AgBr(s) \,|\, Br^-(aq,\ 0.050\ M) \,\|\, Ag^+(aq,\ 0.200\ M) \,|\, Ag(s)$$

(a) What is the solubility product for $AgBr(s)$? Explain your reasoning. *Hint:* See Problem 10.68.

(b) Show how to calculate the standard reduction potential for this half reaction:

$$AgBr(s) + e^- \rightleftharpoons Ag(s) + Br^-(aq)$$

10.70. A silver-silver concentration cell was set up with 10.0 mL of 0.050 M silver ion, Ag^+, in each half cell. The potential of this cell was 0.000 V at 298 K. After 0.500 g of sodium thiosulfate pentahydrate, $Na_2S_2O_3 \cdot 5H_2O$, was dissolved in one of the half cells and thoroughly mixed, the cell potential was 0.618 V and the half cell to which the thiosulfate was added was the anode. Silver ion forms a complex with thiosulfate:

$$Ag^+(aq) + 2S_2O_3^{2-}(aq) \rightleftharpoons Ag(S_2O_3)_2^{3-}(aq)$$

(a) Write the line notation for this concentration cell. Explain.

(b) What is the equilibrium constant for complex formation? Show how you get your answer.

(c) Show how to calculate the standard reduction potential for this half reaction:

$$Ag(S_2O_3)_2^{3-}(aq) + e^- \rightleftharpoons Ag(s) + 2S_2O_3^{2-}(aq)$$

10.71. Plot the cell potential for this cell at 298 K as a function of pH in the range 0 to 14.

$$Pt(s) \,|\, H_2(g,\ 1\ bar) \,|\, H^+(aq,\ pH) \,\|\, Cl^-(aq,\ 1.00\ M) \,|AgCl(s) \,|\, Ag(s)$$

(a) Does your plot look the way you expected? Explain why or why not.

(b) If the cell potential is 0.435 V, what is the pH in the hydrogen half cell? Explain.

10.72. (a) Write the Nernst equation for the reduction of methylene blue, reaction (10.93).

(b) Plot the data in Problem 10.54 to find out whether they obey the Nernst equation you wrote for methylene blue reduction. Under what conditions, acidic or basic, are the data most consistent with the Nernst equation? Explain why you answer as you do.

(c) The structure of the reduced form of methylene blue shows that the molecule has two amine groups that are proton acceptors, that is, the reduced form is a base. The oxidized form is a very much weaker base. Why? Do these properties help explain your plot? Show why or why not.

10.73. You can think of solutions in the mitochondrial matrix and intermembrane space as the solutions in the two half cells of a concentration cell that depends on pH (such as two $H^+|H_2$ half cells) connected by a salt bridge. The electrical potential difference between the two solutions is the same as it would be between two actual half cells.

(a) If the pH in the matrix is one unit higher than the pH in the intermembrane space, what is the potential difference (cell potential) between them? Explain your method.

(b) What is the free energy change equivalent to this potential difference?

(c) What should be the spontaneous direction of transfer of hydronium ions in this system? Are the cell potential in part (a) and the free energy change in part (b) consistent with this direction? Explain why or why not.

General Problems

10.74. The Daniell or gravity electrochemical cell shown in the diagram was widely used to provide electricity for doorbells and railroad telegraph offices in the 19th century. A less dense aqueous solution of zinc sulfate, $ZnSO_4$, floats on a more dense aqueous solution of copper sulfate, $CuSO_4$, and serves to keep the half-cell solutions separated. Ions can still diffuse across the interface to prevent charge separation. The zinc electrode is the anode in this cell; the negative sign shows that electrons leave the cell at this electrode.

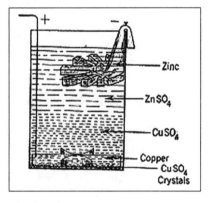

(a) What is the cell reaction? If the measured potential difference between the electrodes is 1.15 V, how much work can the cell do for one mole of cell reaction?

(b) What changes in cation concentrations occur as the cell is used? What ion(s) probably migrate across the interface to maintain charge neutrality in the solutions? What is the purpose of the crystals of $CuSO_4$ at the bottom of the container?

(c) The voltage of the cell slowly decreases as it is used. What do you think might cause the decrease in electrical potential? Consider Le Chatelier's principle and the changes that occur as the cell is used.

10.75. The standard cell potential, $E°$, for the Daniell cell, Problem 10.74, is 1.10 V.

(a) As it is usually set up, a fresh cell has a cell potential of about 1.15 V. What is the concentration ratio, $\dfrac{\left[Zn^{2+}(upper\ layer)\right]}{\left[Cu^{2+}(lower\ layer)\right]}$, for this cell? Show how you get your answer.

(b) The solid copper sulfate pentahydrate, $CuSO_4 \cdot 5H_2O$, at the bottom of the container (usually called a "battery jar") keeps the lower solution saturated with copper sulfate. The density of this solution is about 1.27 $g \cdot mL^{-1}$. What is the concentration (molarity) of Cu^{2+} in this solution? What is the concentration of Zn^{2+} ion in the upper solution of a fresh cell? Explain how you get your answers.

(c) Assume that the volumes of the solutions in the upper and lower layers of the cell are both about 2.5 liters. If the data from Problem 10.35 apply to this cell, what is the concentration of Zn^{2+} ion in the upper solution of the used cell? Show your reasoning.

(d) What is the cell potential for the used cell? Explain. Can you think of any reason(s) why the cell might no longer be useful?

10.76. It seems like a lot of trouble to construct a salt bridge to connect two half reactions. Why not simplify the system by setting up one of the half reactions in a tube with a porous glass disc at the end (readily available from scientific suppliers), the other in a beaker, and immersing the porous end of the tube in the solution in the beaker, as shown in this diagram? This will allow migration of ions through the porous disc with no need for a salt bridge.

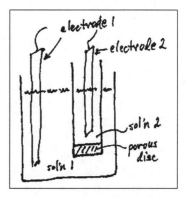

(a) What advantages can you think of for this arrangement? what disadvantages? What happens when species migrate between the solutions?

(b) Can you think of reaction(s) for which this set up would work well?

(c) Can you think of reaction(s) for which this set up would not work well?

10.77. Explain how the Al-air cell, Worked Example 10.29, can be thought of as a fuel cell. An effort is being made to develop this cell into the power source for cell phones and other electronic devices. The cell would never need charging, but you would have to carry a supply of fuel, in case the cell ran out. Would if be safe to carry around a supply of fuel for the cell. Explain why or why not.

Chapter 11. Reaction Pathways

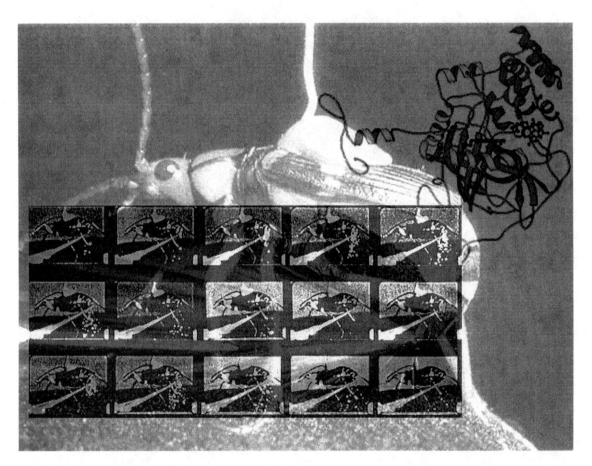

Superimposed on an image of a bombardier beetle is part of a stroboscopic movie showing the beetle emitting its defensive spray (the tiny white droplets in some frames). The time interval between these pictures is 0.25 ms ($^1/_{4000}$th of a second). The chemistry of the beetle's defense involves hydrogen peroxide reactions and decomposition. The molecular structure is a protein, the enzyme catalase, which catalyzes decomposition of hydrogen peroxide to water and oxygen. The active enzyme has four of these molecules stacked together.

Chapter 11. Reaction Pathways

> "… the ultimate objective of a kinetic study is to arrive at a reaction mechanism."
>
> *Keith J. Laidler (1916-   ),* Reaction Kinetics

The series of stroboscopic pictures on the facing page represent a time study of the way bombardier beetles defend themselves from predators. Within a millisecond of being provoked, the beetle sprays a hot (100 °C), foul-smelling mist at its attacker. The discharge of spray is accompanied by an audible "pop" which also startles the attacker. The challenge for chemists (and biologists) was to discover what chemical reaction (or reactions) produces the energy required to heat the beetle's spray to the boiling point of water and how this reaction can occur so quickly. The energy question is one that can be answered by the thermodynamics discussed in Chapter 7. The question of speed requires an understanding of reaction pathways, or mechanisms, as they are called in the opening quotation. These are the topic of this chapter.

The reactants the beetle uses are common in aerobic organisms, hydrogen peroxide, H_2O_2, and hydroquinones. In most parts of a cell, hydrogen peroxide can do damage to other molecules, so an enzyme called catalase (the structure shown on the facing page) catalyzes its decomposition to water and oxygen before it can do any harm. Catalase is found in almost all organisms. The decomposition of hydrogen peroxide is quite exothermic (as you found in Chapter 7 Section 7.7) and, if it takes place fast enough, can heat an aqueous reaction solution to its boiling point. We shall investigate the pathway required to produce this speed.

Studying the change of systems as a function of time is a part of all sciences. Physicists study acceleration, biologists study the growth and decline of populations, geologists study the motion of continents, and chemists study chemical reactions, the changes of one set of substances into another. Each discipline measures *rates*, the quantitative time dependence of the changes they are interested in studying. These studies usually lead to *rate laws* which describe the changes and their rates as functions of the variables that affect the rates, such as the concentrations of the reactants in the bombardier beetle. Chemists who study rates use rate laws and other information from the study of reactions to understand the pathways by which reactions occur, so that they can control reactions and design useful new reactions.

> *Kinetic* and related terms (from Greek *kineein* = to move) also appears in many other contexts: kinetic energy is the energy associated with motion; kinematics is the branch of physics that deals with motion; kinesthesiology is the study of muscular movement in animals; and telekinesis is the yet-to-be-proved movement of objects by use of thought alone.

In chemistry, the study of reaction rates and reaction pathways is often called **kinetics**, as you find in Laidler's quotation that opens the chapter. When chemists speak about the kinetics of a reaction, they mean the combination of rate(s) and pathway(s) that lead to the observed chemical changes. Kinetics, like rate, reminds us of change as a function time.

Section 11.1. Pathways of Change

11.1. Investigate This   *What happens when phenolphthalein and hydroxide mix?*

(a) Work in small groups to investigate this question and analyze the results. You have a capped vial containing about 5 mL of either a 1.5 M or 0.75 M aqueous solution of sodium hydroxide, NaOH, and a thin-stem plastic pipet containing a 0.1% (3×10^{-3} M) solution of phenolphthalein acid-base indicator. CAUTION: Solutions of hydroxide are caustic and can harm skin and clothing. Take care not to spill any of the solution, but work over a paper towel to catch any drops that may escape. Observe and record any changes that occur when you uncap the vial, add *one drop* of the phenolphthalein solution, recap the vial, and swirl to mix the base and indicator. Observe the mixture for a minute or two until you think no more change will occur.

(b) Repeat the addition of *one drop* of phenolphthalein and your observations to see if they are reproducible.

11.2. Consider This   *What reactions occur when phenolphthalein and hydroxide mix?*

(a) What did you observe when phenolphthalein and hydroxide solutions were mixed in Investigate This 11.1? Were the changes reproducible? Explain how they were or were not.

(b) Were any of the observed changes what you expected? Why did you expect them? Were any of the observed changes surprising? Why were you surprised?

(c) How did the observations in the 1.5 M and 0.75 M NaOH solutions compare? How do you interpret the similarities and differences?

You have observed reactions that occur rapidly and others that take more time. For example, in our analysis of the Blue-Bottle reaction, Chapter 10, Section 10.8, the observations could be explained by competition between two reactions: oxidation of reduced methylene blue by oxygen and reduction of oxidized methylene blue by glucose. The oxidation is very rapid; reduced methylene blue is reoxidized so fast that we can detect no change in its concentration (blue color) until the oxygen in the solution is essentially all gone. Then the slower reduction reaction can compete favorably and we see the blue solution fade to colorless. Evidently the concentrations of various reactants can make a difference in the reactions we observe and is one of the factors we study to understand reaction pathways.

Effect of reactant concentration on rate

In Investigate This 11.1, you observed that phenolphthalein turns red the instant it is mixed with a solution containing hydroxide ion:

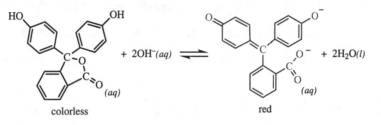

$$+ 2OH^-(aq) \rightleftharpoons \qquad + 2H_2O(l)$$

colorless   red   (11.1)

Reaction (11.1) is an acid-base proton transfer reaction. The instantaneous color change from colorless to red is your experimental evidence that the reaction is very fast. Experiments requiring specialized instruments have shown that proton transfer reactions are about as fast as reactions in solution can be. These reactions occur almost every time the proton donor (acid) and proton acceptor (base) meet one another in the solution. The **rate** of a chemical reaction is a measure of how rapidly the products are formed or the reactants disappear, so knowledge of the rate of a reaction can help us understand the molecular-level pathway of the reaction, as this proton-transfer example shows. In Section 11.2, we will examine how rates are measured and quantified.

The proton transfer reaction is so rapid that you cannot detect any difference between the reaction in different base concentrations, but you observed a slower change in the mixture of phenolphthalein and hydroxide ion as the initial red color of the solution faded to colorless in a minute or two. Addition of more phenolphthalein restored the red color, which again faded to colorless. Apparently the red form of phenolphthalein undergoes some reaction that produces a colorless product or products. You also found that the color disappeared more quickly when the hydroxide concentration was higher. This result suggests that hydroxide ion is involved in a reaction with phenolphthalein that produces a colorless product. Since the hydroxide ion and the red form of phenolphthalein are both negatively charged, they repel one another and only infrequently come close enough to react. Thus, it makes sense that a reaction between them would be slow, but faster when there are more hydroxide ions, so that the number of possible interactions between the reactants is increased . A reaction that is consistent with these observations and interpretations is:

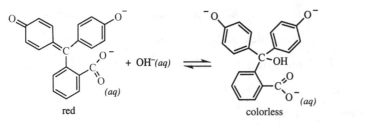

red   colorless   (11.2)

***11.3. Consider This** Are our pathways consistent with the phenolphthalein structure?*

(a) One Lewis structure (without nonbonding electrons) for the red form of phenolphthalein in basic solution is shown in equations (11.1) and (11.2). Draw another Lewis structure that has the same energy, but with one of the negative charges on a different oxygen atom. Can you draw more than one such structure?

(b) Recall that when electrons in a molecule have a larger volume to move in, the energy of the molecule is lower. The electron energy levels will also be closer together and the molecule will be able to absorb electromagnetic radiation (light) at lower energies (longer wavelengths). How do these ideas and your structures from part (a) help to explain the color change for phenolphthalein when it is dissolved in basic solution?

(c) Compare the structures of the colorless form of phenolphthalein in equation (11.1) and the colorless product in equation (11.2). Does the comparison help you explain why the product of reaction (11.2) is colorless? Why or why not?

(d) Another Lewis structure for the red form of phenolphthalein is shown here. Does this structure help you visualize the reaction between the red form of phenolphthalein and hydroxide ion as the reaction of an electrophile with a nucleophile? Explain. If necessary, review Chapter 6, Sections 6.7 and 6.8.

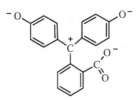

(e) Is the Lewis structure in part (d) of lower, higher, or the same energy as those you drew in part (a)? Explain your answer.

(f) All the possible Lewis structures you can draw for a molecule contribute to the overall properties of the molecule. Higher energy structures contribute less to the properties than lower

---

energy structures. Do these ideas and your answer to part (e) add to our explanation for why the reaction of the red form of phenolphthalein with hydroxide might be slow? Explain.

***11.4.* Investigate This**   *Does temperature affect chemical reactions?*

Do this as a class investigation and work in small groups to discuss and analyze the results. In two clear, plastic cups or beakers prepare two water baths: a room temperature bath and a bath at about 35 °C. Place a capped plastic vial containing about 5 mL of a 1.0 M aqueous solution of sodium hydroxide, NaOH, in each of the baths and let them sit for a few minutes to reach the temperature of the bath. CAUTION: Solutions of hydroxide are caustic and can harm skin and clothing. Take care not to spill any of the solution, but work over a paper towel to catch any drops that may escape. Uncap each vial, add *one drop* of a 0.1% solution of phenolphthalein, recap, invert to mix, and replace in the temperature bath. Record how long it takes for the red phenolphthalein color to disappear in each sample. Repeat the addition of phenolphthalein solution if you wish to test the reproducibility of your results.

11.5. Consider This   *How does temperature affect chemical reactions?*

(a) Did temperature affect the rate of the decolorizing reaction between phenolphthalein and hydroxide ion in Investigate This 11.4? If so, did the rate of reaction increase or decrease as temperature increased? Explain the reasoning for your answer.

(b) Did you expect to see any effect of temperature on the rate of the acid-base reaction when phenolphthalein was first added to the basic solution? Explain why or why not.

Effect of temperature on rate

Most chemical reactions go faster at higher temperatures. This is what you found for the decolorizing reaction of phenolphthalein with hydroxide in Investigate This 11.4. Since increasing the temperature of a system increases its energy, we can probably conclude that the energy of a reacting system plays a role in the rate of reaction. Knowing what this role is will help us relate the temperature dependence of reactions to the possible pathways they take from reactants to products. What other factors might affect reaction rates?

***11.6.* Investigate This**   *How do added species affect the decomposition of H_2O_2?*

Do this as a class investigation and work in small groups to discuss and analyze the results. Add about 10 mL of 3% ($\approx$ 1 M) aqueous hydrogen peroxide solution to each of nine small test tubes containing, respectively, about 0.5 g of: (1) nothing, (2) raw potato, (3) raw apple, (4) uncooked beef or liver, (5) dried baker's yeast, (6) solid potassium iodide (KI), (7) solid potassium chloride (KCl), (8) solid ferric chloride or nitrate ($FeCl_3 \cdot 6H_2O$ or $Fe(NO_3)_6 \cdot 9H_2O$), and (9) solid cupric sulfate ($CuSO_4 \cdot 6H_2O$). Record what you observe happening in each test tube. Gently touch each test tube to see if there is any evidence for energy transfer to or from the mixture.

11.7.   **Consider This**   *What is the effect of added species on the decomposition of H_2O_2*

    (a) Did you observe evidence for chemical reactions occurring in any of the mixtures in Investigate This 11.6? In which mixtures? Were there any patterns in your observations?

    (b) What reaction(s) do you think occurs in these mixtures? Is(Are) the reaction(s) the same in all cases? Do you have any information from previous investigations to help answer these questions? Explain the reasoning for your answers.

Effect of catalysts on rate

In Investigate This 11.6, you observed that 3% hydrogen peroxide alone, test tube (1), shows no evidence of reaction to form new substances. In several of the other test tubes you observed the evolution of gas (bubbles) and perhaps changes in temperature or changes in color not attributable to the colored solids. Let's restrict our attention to the formation of gas, which occurred in several of the systems.

In Chapter 7, Section 7.7, we investigated the decomposition of hydrogen peroxide in the presence of yeast. There, we found experimentally that the gaseous product is oxygen:

$$2H_2O_2(aq) \rightarrow O_2(g) + 2H_2O(l) \qquad\qquad\qquad (11.3)$$

The gas evolved by the samples in Investigate This 11.6, is oxygen. Thus, the reaction is the same, the decomposition of $H_2O_2(aq)$, in all cases. However, the reaction rate depends on the presence of other substances in the reaction mixture. With nothing added to the solution, the decomposition is too slow to observe on a time scale of a few minutes. When certain other substances are added to the solution, you observed rapid hydrogen peroxide decomposition.

What causes the difference in rate with the added ingredients? Evidently something in the other ingredient in these samples increased the reaction rate and made it observable. If the only *net* change that occurs in these samples is the more rapid hydrogen peroxide decomposition, we say that the other ingredients **catalyze** reaction (11.3). Evidently, a variety of salts and biological substances, can act as **catalysts** for this reaction. We will consider catalysis in more detail in

---

Sections 11.4, 11.xx, and 11.yy. Knowing how a reaction rate is affected by catalysts helps us understand the possible pathway(s) for the reaction.

11.8. **Consider This**   *What catalyzes hydrogen peroxide decomposition?*

(a) In Investigate This 11.6, potassium iodide, KI, was one of the added substances that catalyzed $H_2O_2(aq)$ decomposition. What experimental evidence do you have that shows it was the added iodide ion, $I^-(aq)$, not the added potassium ion, $K^+(aq)$, that affected the reaction? Explain your reasoning.

(b) One use of drugstore hydrogen peroxide solution is as an antiseptic (*anti* = against + *sepsis* = rotting, putrefaction). When peroxide is applied to cuts and other open wounds, it fizzes vigorously. Is this what you might expect (predict) to happen? Why or why not?

By carrying out a few simple investigations using only visual observations and comparisons, you have been able to learn a good deal about the rates of chemical reactions. Some reactions, such as acid-base proton transfers, are so fast that they seem to go instantaneously under all conditions. Other reactions, such as the electrophilic center of the red form of phenolphthalein reacting with a nucleophile, hydroxide ion, are slow enough to permit easy observation of the effects of changing reaction conditions. Still other reactions, such as the decomposition of aqueous hydrogen peroxide, go so slowly that no reaction is detectable, unless other species, catalysts, are added to the reaction mixture.

You have found that the concentration of species in a solution, the temperature of the reaction mixture, and the addition of catalysts can affect reaction rates. In order to understand how these effects are used to learn more about reaction pathways, we need to get numerical values for rates, so they can be compared quantitatively. However, the ultimate purpose of these experiments is not to generate numbers, but, as the quotation at the beginning of the chapter reminds us, to learn more about how and why reactions go the way they do.

Section 11.2. Measuring and Expressing Rates of Chemical Change

To quantify **rate**, we define it as the change in some variable related to the reaction as a function of time, that is, the ***extent of change per time***. We'll use the $H_2O_2(aq)$ decomposition, reaction (11.3), as our first example for quantifying the change that occurs in a chemical reaction as a function of time. Since a gas is produced in the reaction, we might choose to measure the volume of gas produced (at constant pressure) or the pressure of the gas produced (at constant

> **WEB Chap 11, Sect 11.2.1-6.**
> Work through this interactive analysis of another example reaction.

volume) at various times after the reaction starts. It's easier to automate the collection of pressure data, as shown in Figure 11.1, so we will use that as our first example.

Figure 11.1. Apparatus for measuring the pressure change in the $H_2O_{2(aq)}$ decomposition.

The reaction is carried out in a flask connected to a pressure sensor. An electrical signal from the sensor goes to a data interface where the electronics interpret the signal as a pressure in kPa. The data interface output goes to a computer where the data are stored and graphed as they are being produced. In one such experiment, we added 0.025 g of dried yeast to 15 mL of 0.175 M $H_2O_2(aq)$ in the reaction flask just before connecting it to the pressure sensor. Atmospheric pressure in the system before reaction was 102.7 kPa. Data points were taken every second, so they form a continuous curve on a graph at this scale, as shown in Figure 11.2.

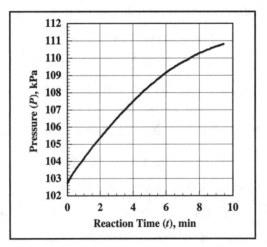

Figure 11.2. *P vs. t* curve for $O_2(g)$ production from $H_2O_2(aq)$ decomposition.

11.9. Consider This *What are the pressure changes in* $H_2O_2(aq)$ *decomposition?*

(a) What is the pressure change (in kPa) in the 2-minute interval from 0 to 2 minutes in the reaction system described by Figure 11.2? What is the change in the 2-minute interval from 4 to 6 minutes? Explain how you get your answers.

(b) Are the pressure changes in part (a) the same or different? Do you see any problem for determining the decomposition reaction rate? Explain why or why not.

(c) WEB Chap 11, Sect 11.2.2. Explain how the graphic on this page of the *Web Companion* is related to the questions in parts (a) and (b) here. Also explain the reasoning for your answer to the fill-in-the-blank box.

Rate of reaction

The curve plotted in Figure 11.2 shows the increase in pressure in the reaction flask as a function of time as $O_2(g)$ is produced by $H_2O_2(aq)$ decomposition. The extent of the change in pressure, ΔP, in a given time interval, Δt, is the rate of oxygen production during that interval:

$$\text{rate of } O_2(g) \text{ production} = \frac{\Delta P}{\Delta t} = \text{slope of the } P \text{ vs. } t \text{ curve} \qquad (11.4)$$

Equation (11.4) reminds us that the rate of oxygen production is equal to the **slope** of the *P* vs. *t* curve. However, your results in Consider This 11.9 show that ΔP is different for the same time interval at different times during the reaction. It is obvious that the curve in Figure 11.2 is not linear, but bends downward as the reaction time increases; in other words, the slope of the curve decreases as the reaction proceeds. This is shown quantitatively in Figure 11.3.

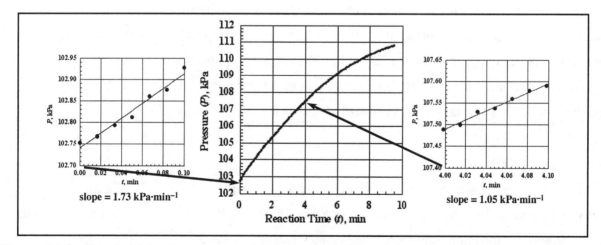

Figure 11.3. Slopes of the *P* vs. *t* curve for $O_2(g)$ production during $H_2O_2(aq)$ decomposition.
The smaller plots show the points for 0.10 minute (6 second) intervals starting at the beginning of the reaction and at 4 minutes. The slopes are for the best straight line through the experimental points.

> ### 11.10. Check This   *Slopes of reaction rate curves*
>
> WEB Chap 11, Sect 11.2.3. If we used calculus to analyze the data in Figure 11.2, we would express the rate as dP/dt, the slope of the *tangent* to the curve at any point. Explain how the animation on this page of the *Web Companion* illustrates this sort of analysis. How are the results related to Figure 11.3? For our purposes, small finite differences, equation (11.4), that approximate the tangent slope are completely adequate.

Initial reaction rates

If the rate changes as a reaction proceeds, what rate should we use to characterize the reaction? There is no universal answer to this question. Often we choose to use the rate determined as near to the beginning of the reaction as possible. This is called the **initial reaction rate**. An advantage of this choice is that almost none of the reactants will have been used up, so it is easy to characterize the concentrations or amounts of reactant present. A more subtle advantage is that almost no products will have been formed. As products build up in a reaction, they often affect the reaction. The most obvious problem for a reversible reaction is that the products can react to re-form the reactants. The major disadvantage of initial reaction rates is linked to the advantages. If only a tiny amount of the reactants have reacted or products been formed, measuring the small changes can be a challenge. If they can be measured, initial reaction rates are the easiest to interpret.

> ### 11.11. Consider This   *What effect does the reverse reaction have on reaction rate?*
>
> **(a)** If you are following a reversible reaction by measuring the formation of a product that can react to re-form the reactants, how will your measurements be affected? Will the rate of product formation appear slower or faster than if the reaction were not reversible? Explain.
>
> **(b)** Do you think that reaction (11.3) might go in reverse? Why or why not? If it did, would that be a possible explanation for its observed decrease in rate with time?

Reaction rate units

The initial reaction rate for $O_2(g)$ production shown in Figure 11.3 is 1.73 kPa·min^{-1}. Any convenient time units can be used to express a rate, but seconds are most often used in chemistry and biology. For the experiment we have been analyzing, we have:

$$\text{rate of } O_2(g) \text{ production} = (1.73 \text{ kPa·min}^{-1})\left(\frac{1 \text{ min}}{60 \text{ s}}\right) = 2.88 \times 10^{-2} \text{ kPa·s}^{-1} \quad (11.5)$$

Any convenient units can be used to express what is changing as a function of time in a reaction, but, when possible, we usually try to use moles, the fundamental unit of amount in chemistry.

11.12. Worked Example   *Changing pressure units to moles*

The volume of gas in the reaction flask and tubing in Figure 11.1 is 2.7×10^2 mL and the results in Figure 11.2 are for a reaction at 295 K. What is the rate of $O_2(g)$ production in mol·s^{-1}?

Necessary information: We describe gases with the ideal gas equation, $PV = nRT$, where the gas constant, R, is 8.314 kPa·L·mol^{-1}·K^{-1}. The rate of $O_2(g)$ production is 2.88×10^{-2} kPa·s^{-1}.

Strategy: To convert kPa to moles, find n/P, that is, the ratio of number of moles of gas corresponding to the pressure of the gas in the given volume. Rearrange the ideal gas equation to get this ratio, substitute known values to find a numeric value for the ratio, and then use that value to convert rate in kPa·s^{-1} to mol·s^{-1}.

Implementation: For our apparatus we have:

$$\frac{n}{P} = \frac{V}{RT} = \frac{2.7 \times 10^{-1} \text{ L}}{(8.314 \text{ kPa} \cdot \text{L} \cdot \text{mol}^{-1} \cdot \text{K}^{-1})(295 \text{ K})} = 1.10 \times 10^{-4} \text{ mol·kPa}^{-1}$$

Converting rate in kPa·s^{-1} to mol·s^{-1} gives:

$$\text{rate of } O_2(g) \text{ production} = (2.88 \times 10^{-2} \text{ kPa·s}^{-1})\left(\frac{1.10 \times 10^{-4} \text{ mol}}{1 \text{ kPa}}\right) = 3.2 \times 10^{-6} \text{ mol·s}^{-1}$$

Does the answer make sense? A mole of gas at room temperature and atmospheric pressure occupies about 25 L. The volume of gas we are measuring is one-hundredth this much and the pressure of the $O_2(g)$ produced is only a few kPa, compared to atmospheric pressure of 101 kPa. The combination of small volume and low pressure means that a total of much less than one mole of gas is produced by the reaction, so the low rate of production makes sense.

11.13. Check This   *Changing pressure units to moles and assessing stoichiometry*

(a) After about 35 minutes the pressure in the reaction represented in Figure 11.2 stops increasing (no more $O_2(g)$ is produced); the pressure in the flask remains constant at 114.5 kPa. What is the pressure of the $O_2(g)$ produced by the reaction? Explain the reasoning for your answer.

(b) Use your result from part (a) to find the number of moles of $O_2(g)$ produced by the reaction represented in Figure 11.2. Show clearly how you obtain your answer.

(c) If the decomposition reaction (11.3) goes to completion, how many moles of $O_2(g)$ would have been produced by the $H_2O_2(aq)$ in the reaction flask? Within the uncertainty of the data, what can you conclude about the reaction going to completion? Explain your reasoning.

When measuring a reaction rate, we almost always choose to follow the change that is most convenient to measure, the pressure of gas in our present example. But how do we relate what we measure to other changes in the reacting system? For example, how is the rate of production of $O_2(g)$ related to the rate of decomposition (loss) of $H_2O_2(aq)$ in the experiment we have been analyzing? To answer this question, the fundamental relationship we need is the stoichiometry of reaction (11.3), which tells us that one mole of $O_2(g)$ is produced for every two moles of $H_2O_2(aq)$ that decompose. Before we can relate the rate expressed in (mol O_2 produced)·s^{-1} to the rate expressed in (mol H_2O_2 decomposed)·s^{-1}, we need to know that, by convention: ***all rates are expressed as positive quantities***. Since $H_2O_2(aq)$ is being used up in this reaction, Δ(mol H_2O_2) is a negative quantity. Therefore, in the rate expression, Δ(mol H_2O_2) is given a negative sign so that the *rate* will be positive:

$$\frac{-\Delta\left(\text{mol } H_2O_2\right)}{\Delta t} = -\frac{\Delta\left(\text{mol } H_2O_2\right)}{\Delta t} = 2\left(\frac{\Delta\left(\text{mol } O_2\right)}{\Delta t}\right) \qquad (11.6)$$

11.14. Consider This   *How are the stoichiometries of rate expressions related*

   (a) Why is the factor of two required in equation (11.6) to equate the rate expressions for $H_2O_2(aq)$ decomposition? Explain your reasoning.

   (b) WEB Chap 11, Sect 11.2.4. Explain how the concentration *vs.* time plots are related to the correct rate expressions. If the change in concentration of the iodide ion, I^-, could be measured in this reaction, how would the rate be expressed in terms of $\Delta[I^-]$? Explain.

Equation (11.6) makes it look like the "rate" for the reaction is different, depending on what product or reactant species is being considered. We need to define a single rate that characterizes a reaction, no matter what reactant or product change is measured. Consider a general reaction:

$$aA + bB \rightarrow cC \qquad (11.7)$$

The rate of reaction (11.7), expressed in moles, is defined in this way:

$$\text{rate of reaction (11.7)} \equiv -\left(\frac{1}{a}\right)\frac{\Delta\left(\text{mol } A\right)}{\Delta t} = -\left(\frac{1}{b}\right)\frac{\Delta\left(\text{mol } B\right)}{\Delta t} = \left(\frac{1}{c}\right)\frac{\Delta\left(\text{mol } C\right)}{\Delta t} \qquad (11.8)$$

For the rest of the chapter, when we refer to the **reaction rate**, we will mean rate defined like this, where we have accounted for the stoichiometry of the reaction. We will still sometimes refer to the rate of formation of a product or rate of disappearance of a reactant, but, to get the reaction rate, we have to account for the stoichiometry of that reactant or product in the reaction equation.

11.15. Check This   *Rate of $H_2O_2(aq)$ decomposition*

(a) What is the rate for reaction (11.3) expressed in terms of $\Delta(\text{mol } H_2O_2)$? What is the rate expressed in terms of $\Delta(\text{mol } O_2)$? Explain why you write the rate expressions as you do.

(b) For the reaction represented by Figure 11.2, what is the numeric value of the initial rate, expressed in moles? Explain your choice.

Rates expressed in molarity

The rates of reactions in solution are almost always expressed as changes in molarity, rather than changes in moles. This makes sense because we want a value for the rate that can be applied to any size system: a living cell, a test tube, or an industrial-scale bioreactor. Molarity is an intensive variable (independent of the size of the system); rates expressed as $M \cdot s^{-1}$ will be independent of the amount of reaction mixture we are considering. If the volume of reacting solution is v L, the rate of hydrogen peroxide decomposition, reaction (11.3), expressed in terms of hydrogen peroxide molarity, $[H_2O_2(aq)]$, is:

$$\text{rate} = -\left(\frac{1}{2}\right)\frac{\Delta[H_2O_2(aq)]}{\Delta t} = -\left(\frac{1}{2}\right)\left(\frac{\Delta(\text{mol } H_2O_2)}{\Delta t}\right)\left(\frac{1}{v\,L}\right) \tag{11.9}$$

11.16. Worked Example   *Rate of $H_2O_2(aq)$ decomposition in $M \cdot s^{-1}$*

What is the numeric value in $M \cdot s^{-1}$ for the initial rate of $H_2O_2(aq)$ decomposition for the reaction represented by Figure 11.2? If the rate of the reaction were the same as this initial rate for the entire reaction, how long would it take to decompose all the $H_2O_2(aq)$ in the sample?

Necessary information: The volume and concentration of the reacting solution are 0.015 L and 0.175 M, respectively. The initial reaction rate, defined in equation (11.8) and calculated in Check This 11.15, is 3.2×10^{-6} mol·s^{-1}.

Strategy: To get the rate in $M \cdot s^{-1}$, divide the rate in mol·s^{-1} by the volume of reacting solution, as shown in equation (11.9). Dividing the molarity of the solution by the rate in $M \cdot s^{-1}$, gives the time in seconds required for all the $H_2O_2(aq)$ in the sample to react, if the reaction were to go continuously at the initial rate.

Implementation:

$$\text{rate (in } M \cdot s^{-1}) = \frac{3.2 \times 10^{-6} \text{ mol} \cdot s^{-1}}{0.015 \text{ L}} = 2.1 \times 10^{-4} \, M \cdot s^{-1}$$

Time required for all the $H_2O_2(aq)$ in the sample to react at this rate:

$$\text{time for complete reaction} = \frac{0.175 \text{ M}}{2.1 \times 10^{-4} \, M \cdot s^{-1}} = 8.3 \times 10^2 \text{ s} \approx 14 \text{ min}$$

Does the answer make sense? We know that the reaction slows down as it proceeds so 14 minutes will be an *underestimate* of the time the reaction actually takes to go to completion. Recall from Check This 11.12 that complete reaction takes about 35 minutes, so our result here is consistent with the experimental observations.

11.17. Check This   *Rate of $H_2O_2(aq)$ decomposition in a bombardier beetle*

In one species of bombardier beetle, a 7.4 M hydrogen peroxide solution reacts in about a millisecond, 10^{-3} s. (See the chapter opening illustration for the experimental evidence.) What is the approximate rate, in $M \cdot s^{-1}$, of the decomposition reaction in the beetle? How does this rate compare to the initial rate we measured (by analyzing Figures 11.2 and 11.3) in the presence of yeast? How might you explain any difference?

Reflection and projection

We used just visual observations and comparisons on reacting systems to discover three factors that affect reaction rates: the concentrations of reacting species, the temperature of the reaction mixture, and the presence of catalysts, substances that increase the rate of reaction. The data necessary to determine the rate of a reaction are measurements of the change in some variable related to the reaction as a function of time. The rate of reaction is the slope of the curve on a variable *vs.* time plot of the data. We approximate the slope as the change in the variable over a small interval of time. The rate at the beginning of a reaction, the initial rate, is especially useful for analysis of the reaction, because only a tiny amount of the reactants have been used up and not enough of the products have been formed to interfere with or change the reaction of interest.

Rates are always expressed as positive quantities, so the change with time for a reactant that is used up is given a negative sign. Numeric values for the rate of a reaction can be given in any convenient units that express the amount of a product formed (or a reactant used up) per unit time of reaction. You can use the stoichiometry of the reaction to convert rates from one unit to another.

We know that changing the concentrations of reactants can change the rate of a reaction. In the next sections, we will see how to correlate these changes to discover more about the details of reaction pathways.

Section 11.3. Reaction Rate Laws

Dependence of $H_2O_2(aq)$ decomposition rate on $[H_2O_2(aq)]$

The experiment we analyzed in Section 11.2 used yeast to catalyze $H_2O_2(aq)$ decomposition. Because yeast is a living organism, it is a bit difficult to control exactly how much of what we add is active in catalyzing the decomposition. In order to make the results easier to compare, in this section we will analyze results from a series of $H_2O_2(aq)$ decomposition experiments catalyzed by iodide ion, $I^-(aq)$, added as potassium iodide, KI. Initial rates of reaction from plots like the one in Figure 11.3 give the data in Table 11.1. As you see, the initial reaction conditions, the concentrations of $H_2O_2(aq)$ and $I^-(aq)$, are different from one experiment to another. There are many ways to analyze data like these, but you should start out by looking for trends and patterns and then consider how you might quantify any patterns you find.

> **WEB Chap 11, Sect 11.3.1-6.**
> Work through this continuation of the interactive analysis of another example reaction.

Table 11.1.   Initial concentrations and rates of $H_2O_2(aq)$ decomposition catalyzed by $I^-(aq)$.

Experiment	$[H_2O_2(aq)]_0$, M	$[I^-(aq)]_0$, M	$-\left(\dfrac{1}{2}\right)\dfrac{\Delta[H_2O_2(aq)]}{\Delta t}$, $(M \cdot s^{-1})$
1	0.0850	0.100	0.7×10^{-4}
2	0.170	0.100	1.4×10^{-4}
3	0.250	0.100	2.1×10^{-4}
4	0.350	0.100	2.9×10^{-4}
5	0.350	0.200	6.0×10^{-4}
6	0.350	0.300	9.0×10^{-4}
7	0.350	0.400	11.7×10^{-4}

One trend that is easy to see is that the initial rates increase as you read down the table. Also, the initial concentration of $H_2O_2(aq)$, $[H_2O_2(aq)]_0$, increases down the table for experiments 1 through 4. For these four experiments, $[I^-(aq)]_0$ is the same, so it appears that another pattern is increasing rate with increasing $[H_2O_2(aq)]$ when $[I^-(aq)]$ is constant. Looking more quantitatively at the data, note that the rate about doubles when the $[H_2O_2(aq)]$ doubles; compare experiments 1 and 2 or 2 and 4, for example. If the rate and $[H_2O_2(aq)]$ are directly proportional, then a plot of the rate *vs.* $[H_2O_2(aq)]_0$ should be a straight line. Figure 11.4 is a plot of these data from experiments 1 through 4 in Table 11.1.

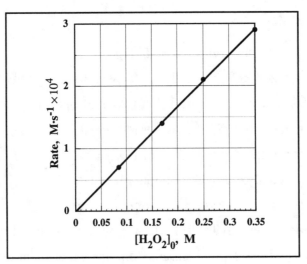

Figure 11.4. Initial rate of $H_2O_2(aq)$ decomposition as a function of $[H_2O_2(aq)]_0$.

The plot in Figure 11.4 looks linear and the computer-generated equation for the best straight line through the data points is:

$$-\left(\frac{1}{2}\right)\frac{\Delta[H_2O_2(aq)]}{\Delta t} = (8.3 \times 10^{-4}\ s^{-1})\ [H_2O_2(aq)] - 0.007 \times 10^{-4}\ M \cdot s^{-1} \qquad (11.10)$$

Within the uncertainty of the data (the rates are only reproducible to about $\pm\ 0.1 \times 10^{-4}\ M \cdot s^{-1}$), the intercept, $-\ 0.007 \times 10^{-4}\ M \cdot s^{-1}$, is zero; we can write the rate as a function of $[H_2O_2(aq)]$ as:

$$-\left(\frac{1}{2}\right)\frac{\Delta[H_2O_2(aq)]}{\Delta t} = (8.3 \times 10^{-4}\ s^{-1})\ [H_2O_2(aq)] \qquad (11.11)$$

This result makes sense; when there's no H_2O_2, its rate of decomposition is zero.

Equation (11.11), the dependence of the rate of $H_2O_2(aq)$ decomposition on concentration of hydrogen peroxide, $[H_2O_2(aq)]$, in the solution, can explain the curvature of the plot in Figure 11.2, if the decomposition catalyzed by yeast also depends on $[H_2O_2(aq)]$. As the decomposition proceeds, $[H_2O_2(aq)]$ decreases and equation (11.11) predicts that the reaction should slow down, as we found in Section 11.2. The concentration dependence predicted by equation (11.11) can be observed in a single experiment that is run far enough to use up a significant amount of peroxide.

***11.18.* Consider This**   *How does the rate of $H_2O_2(aq)$ decomposition depend on $[I^-(aq)]$?*

(a) Consider the data in Table 11.1 that show the rate of $H_2O_2(aq)$ decomposition at different initial concentrations of iodide ion, $[I^-(aq)]_0$. What patterns and relationships do you find?

(b) Does a relationship analogous to equation (11.11) relate the rate of reaction to $[I^-(aq)]$? If so, find the quantitative dependence of rate on $[I^-(aq)]$.

(c) What do your results in parts (a) and/or (b) tell you about the reaction rate when $[I^-(aq)]$ goes to zero? Does your answer make sense? Explain.

11.19. Check This   *Rates of the* **Web Companion** *reaction*

WEB   Chap 11, Sect 11.3.1-2. Work through these two pages and then answer these questions. If necessary, review Chap 11, Sect 11.2.6, for the technique used to study this reaction.

(a) Clearly explain the reasoning for your choice of the relationship between rate and Δt on the first page.

(b) What is the ratio of rates of reaction when $[(NH_4)_2S_2O_8]$ is doubled? Explain and cite specific examples.

(c) What is the ratio of rates of reaction when [KI] is doubled? Explain and cite specific examples.

Rate laws

Our analysis leading to equation (11.11) and your results in Consider This 11.18 show that the initial rate of decomposition of $H_2O_2(aq)$ is directly proportional to both $[H_2O_2(aq)]$ and $[I^-(aq)]$, that is:

$$-\left(\frac{1}{2}\right)\frac{\Delta[H_2O_2(aq)]}{\Delta t} \propto [H_2O_2(aq)] \qquad \text{(when } [I^-(aq)] \text{ is constant)} \qquad (11.12)$$

$$-\left(\frac{1}{2}\right)\frac{\Delta[H_2O_2(aq)]}{\Delta t} \propto [I^-(aq)] \qquad \text{(when } [H_2O_2(aq)] \text{ is constant)} \qquad (11.13)$$

These results suggest that we might combine the proportionalities into a single rate equation that includes both:

$$-\left(\frac{1}{2}\right)\frac{\Delta[H_2O_2(aq)]}{\Delta t} = k\,[H_2O_2(aq)][I^-(aq)] \qquad (11.14)$$

An equation like (11.14) that expresses *the dependence of the rate of a reaction on the concentrations that affect the rate* is called a **rate law**. The proportionality constant, k, in the rate law is called the **rate constant** for the reaction. The relationships expressed in equations (11.11), (11.12), and (11.13), as well as your analysis in Consider This 11.18, are all contained in the rate law, equation (11.14).

11.20. Worked Example   *Rate constant for $H_2O_2(aq)$ decomposition catalyzed by $I^-(aq)$*

Use the numeric value in equation (11.11), which is the result from Figure 11.4, to find the numeric value of the rate constant, k, in equation (11.14).

Necessary information: We need the value 8.3×10^{-4} s^{-1} from equation (11.11) and we need $[I^-(aq)]_0 = 0.100$ M for all the reactions represented by the data in Figure 11.4.

Strategy: Comparison of equation (11.14) with equation (11.11) shows that the numeric factor in equation (11.11) can be equated to $k\,[I^-(aq)]_0$ for reactions with $[I^-(aq)]_0$ held constant at 0.100 M. Solve the equation for k.

Implementation: Write and solve the equation:

$$8.3 \times 10^{-4}\ s^{-1} = k\,[I^-(aq)]_0 = k\,(0.100\ M); \quad \therefore\ k = \frac{8.3 \times 10^{-4}\ s^{-1}}{0.100\ M} = 8.3 \times 10^{-3}\ M^{-1}\cdot s^{-1}$$

Does the answer make sense? Substitute this value of k and data for the experiments in Table 11.1 into equation (11.14) to find out if you successfully predict the observed rates. Also consider the units of k, $M^{-1}\cdot s^{-1}$. When multiplied by two concentration terms with units of molarity, M, the result is a value with units of $M\cdot s^{-1}$, which are the units of rate, as they must be.

11.21. Check This   *Rate constant for $H_2O_2(aq)$ decomposition catalyzed by $I^-(aq)$*

(a) Use the numeric value you found in Consider This 11.18(b) for the dependence of the rate of $H_2O_2(aq)$ decomposition on $[I^-(aq)]$ to find the numeric value of the rate constant, k, in equation (11.14).

(b) How does your value in part (a) compare with the value for k we got in Worked Example 11.20? Is the comparison what you expected? Explain why or why not.

All rate laws are experimental. Note that the rate law, equation (11.14) could not have been predicted or derived from the stoichiometry of reaction (10.3). The stoichiometric equation involves two molecules of hydrogen peroxide as reactants and iodide ion does not appear in the equation at all. The rate law, however, includes the concentrations of both hydrogen peroxide and iodide ion, and there is no hint that two hydrogen peroxides are involved. This is a specific example of a general rule: *there is no necessary relationship between the stoichiometry of a reaction and the rate law for the reaction.* The stoichiometric equation for a reaction is simply a bookkeeping device to keep track of the overall change that occurs. The rate law, as we will see in Section 11.4, is determined by the pathway the reactants take as they change to products.

11.22. Check This   *Rate law for the* **Web Companion** *reaction*

WEB   Chap 11, Sect 11.3.3-6. Work through the determination of the rate law for this reaction. Is this rate law another example of the general rule about the relationship between rate laws and stoichiometry? Explain why or why not; be specific.

Reaction order

A further bit of nomenclature for rate laws has to do with the exponent or power to which each of the concentration terms is raised. For example, although we usually do not show exponents of unity, we could write equation (11.14) as:

$$-\left(\frac{1}{2}\right)\frac{\Delta[H_2O_2(aq)]}{\Delta t} = k\,[H_2O_2(aq)]^1[I^-(aq)]^1 \tag{11.15}$$

The power to which the concentration of some reactant concentration in a rate law is raised is called the **order** of the reaction with respect to that reactant. In expression (11.15) [as well as in (11.14)], the reaction is first order with respect to $H_2O_2(aq)$ and also first order with respect to $I^-(aq)$. The **overall reaction order** is the sum of the orders with respect to the individual species. The decomposition of $H_2O_2(aq)$ in the presence of $I^-(aq)$ is second order overall. *The order of a reaction with respect to any of its components must be determined experimentally.*

In our example, we take the linear dependence of the rate on $[H_2O_2(aq)]$, Figure 11.4 and equations (11.10) and (11.11), as evidence that the rate depends on the first power of $[H_2O_2(aq)]$. Orders with respect to individual species may be zeroth order (no dependence on the concentration), first order (by far the most common and the one we will focus on in this chapter), or second order (dependence on the square of the concentration, which is relatively rare). Non-integral orders are also found, but we will not introduce any such examples.

11.23. Worked Example   *Rate law for reaction of OCl⁻(aq) with I⁻(aq) in basic solution*

Hypochlorite ion, $OCl^-(aq)$, and iodide ion, $I^-(aq)$, react in basic solution:

$$OCl^-(aq) + I^-(aq) \rightarrow OI^-(aq) + Cl^-(aq) \tag{11.16}$$

The graph represents the formation of $OI^-(aq)$ as a function of time during the first second of reactions carried out in aqueous 1.0 M NaOH solution with the specified initial concentrations of $OCl^-(aq)$ and $I^-(aq)$. What rate law for reaction (11.16) can be deduced from these data?

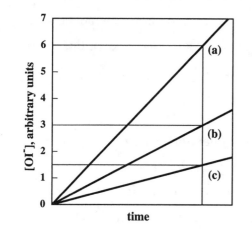

(a) $[OCl^-]_0 = 2.0 \times 10^{-3}$ M; $[I^-]_0 = 2.0 \times 10^{-3}$ M

(b) $[OCl^-]_0 = 2.0 \times 10^{-3}$ M; $[I^-]_0 = 1.0 \times 10^{-3}$ M

(c) $[OCl^-]_0 = 1.0 \times 10^{-3}$ M; $[I^-]_0 = 1.0 \times 10^{-3}$ M

Necessary information: The graphical results and the concentrations are all we need.

Strategy: Take the relative slopes of the concentration *vs.* time curves (lines) as a measure of the relative initial rates of reaction. Compare the ratios of relative rates to the ratios of concentrations to determine the dependence of the rate on the concentrations.

Implementation: On the graph, we have drawn horizontal lines from the experimental curves to the [OI⁻] (vertical) axis at the same time of reaction for each curve. These horizontal lines intersect the axis at 1.5, 3.0, and 6.0, respectively, for reactions (c), (b), and (a). We can write the rate of reaction (b), for example, as:

$$\text{rate (b)} = \frac{\Delta[\text{OI}^-]}{\Delta t} = \frac{[\text{OI}^-]_t - [\text{OI}^-]_0}{\Delta t} = \frac{3.0 - 0}{\Delta t} = \frac{3.0}{\Delta t}$$

The rates of the other two reactions are obtained in the same way and, since the time is the same in each case, the rates of reaction are in the ratios:

$$\frac{\text{rate (b)}}{\text{rate (c)}} = \frac{3.0}{1.5} = 2.0; \qquad \frac{\text{rate (a)}}{\text{rate (b)}} = \frac{6.0}{3.0} = 2.0; \qquad \frac{\text{rate (a)}}{\text{rate (c)}} = \frac{6.0}{1.5} = 4.0$$

The difference between reactions (b) and (c) is that the initial [OCl⁻] is doubled in (b). Since the rate is also doubled by this change, the reaction rate is directly proportional to [OCl⁻].

The difference between reactions (a) and (b) is that the initial [I⁻] is doubled in (a). Since the rate is also doubled by this change, the reaction rate is directly proportional to [I⁻].

With this information we can write the rate law as:

$$\text{rate} = \frac{\Delta[\text{OI}^-]}{\Delta t} = k[\text{OCl}^-][\text{I}^-] \tag{11.17}$$

Does the answer make sense? If you compare reaction (a) to reaction (c), you see that both concentrations are doubled in (a) compared to (c) and the rate of (a) is four times the rate of (c). This is what the rate law predicts: if both concentrations are doubled the rate is quadrupled. For this reaction system, we need to be careful to specify that our result applies only to basic solutions, because, if the two reactants are mixed in neutral or acidic solutions, elemental iodine, I_2, is one of the products. Evidently OH⁻*(aq)* and/or H⁺*(aq)* ions affect the pathway of this redox reaction, as we saw for some other reactions in Chapter 10, Sections 10.7 and 10.8.

11.24. Check This   *Order of reaction for [OH⁻(aq)] with phenolphthalein*

(a) In Investigate This 11.1, you observed the red color of phenolphthalein in solutions of two different concentrations of base. The intensity of the color in these solutions is directly related to the concentration of the dianion whose structure is shown in equation (11.1). If you had measured the concentration of the dianion, $[phth^{2-}]$, as a function of time in your experiments, which one of these plots could best represent your results during the first few seconds of the reaction? Explain your reasoning and tell which line corresponds to your result in the more concentrated base.

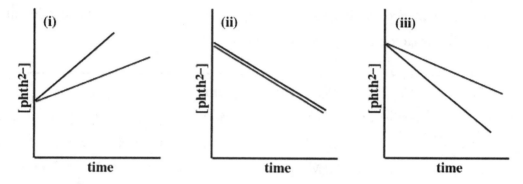

(b) How would you write the rate law for the dependence of the phenolphthalein reaction on $[OH^-(aq)]$? You will have to make an assumption about the order of the reaction with respect to $[OH^-(aq)]$. Clearly explain the reasoning for your choice.

Section 11.4. Reaction Pathways or Mechanisms

Once we have the rate law for a reaction, what does it tell us about the pathway or mechanism of the reaction? The **reaction pathway** or **reaction mechanism** is our attempt to describe in as much detail as possible the changes in arrangements of electrons and atomic cores that lead from reactants to products. Reaction mechanisms help us understand why reactions of similar compounds can lead to different kinds of products and how to make modifications to obtain desirable products. To design more efficient and environmentally sound processes for making the materials we depend on in our daily lives, we need to know the mechanisms of the reactions involved. Knowledge of biological reaction mechanisms is critical to the search for new drugs and therapies to treat disease.

The first step in developing a reaction mechanism is determining which molecules in the system have to come together to bring about the reaction. Your first thought might be that the stoichiometry of the reaction would provide this information. But, the reaction analyzed in Section 11.3, $I^-(aq)$-catalyzed decomposition of $H_2O_2(aq)$, already shows you that this isn't true. You know from that analysis that the reaction depends on the amount of iodide ion in the solution, $[I^-(aq)]$. It seems reasonable to suppose that $I^-(aq)$ must come together with the other reactant, $H_2O_2(aq)$, in order to have an effect on the reaction. But $I^-(aq)$ does not appear in the stoichiometric equation (11.3) for $H_2O_2(aq)$ decomposition.

The rate law for the $I^-(aq)$-catalyzed decomposition of $H_2O_2(aq)$, equation (11.14), *does* involve $[I^-(aq)]$ and $[H_2O_2(aq)]$. But we know that two molecules of $H_2O_2(aq)$ react for every molecule of $O_2(g)$ produced and the rate law suggests only a one-to-one meeting of $H_2O_2(aq)$ and $I^-(aq)$. Developing reaction pathways or mechanisms that can untangle apparent contradictions like these is often a challenge. We have to find a step-by-step description of the reaction that is consistent with *both* the stoichiometry and the rate law.

11.25. Investigate This   *Do reaction changes occur in steps?*

Do this as a class investigation and work in small groups to discuss and analyze the results. To each of six wells of a 24-well plate add 1.0 mL of an aqueous 0.10 M nickel sulfate, $NiSO_4$, solution. Then add the volumes of 0.50 M aqueous ethylenediamine, en, $H_2NCH_2CH_2NH_2$, and water shown in this table.

well #	1	2	3	4	5	6
en, mL	0.0	0.2	0.4	0.6	0.8	1.0
water, mL	1.4	1.2	1.0	0.8	0.6	0.4

Use an overhead projector to project the plate and record the colors of the solutions in each well.

11.26. Consider This   *Does reaction of $Ni^{2+}(aq)$ with ethylenediamine occur in steps?*

(a) How many moles of $Ni^{2+}(aq)$ ion are present in each well in Investigate This 11.25? How many moles of ethylenediamine (en) are added to each well? What is the ratio of number of moles of en to moles of $Ni^{2+}(aq)$ in each well?

(b) Based on your analysis in part (a) and observations in Investigate This 11.25, what is the stoichiometry of the reaction of $Ni^{2+}(aq)$ with en? Explain the reasoning for your answer.

(c) What, if any, evidence do you have that the reaction of $Ni^{2+}(aq)$ with en occurs in steps? Explain your reasoning.

Reactions occur in steps

Let's consider why most reactions must take place in a series of steps. Formation of the metal-ligand complex (see Chapter 6, Section 6.6) between $Ni^{2+}(aq)$ (or $Ni(H_2O)_6^{2+}$) and ethylenediamine (en) is an apparently simple reaction. When solutions of $Ni^{2+}(aq)$ and excess en(aq) are mixed, reaction occurs immediately (as fast as the reactants are mixed):

$$Ni(H_2O)_6^{2+} + 3en(aq) \rightarrow Ni(en)_3^{2+}(aq) + 6H_2O(l) \tag{11.18}$$

(This is a dynamic equilibrium system, but for the purposes of this discussion we are interested only in the formation of the complex.) The likelihood that the four reacting species, one $Ni(H_2O)_6^{2+}$ and three en(aq) will come together simultaneously is very small. If the reaction depended upon such meetings, it would be much slower than is observed. A series of meetings and reactions between two species at a time is much more likely. These individual steps in the overall reaction are called **elementary reactions**. Elementary reactions take place just as they are written; their rate laws can be written from their stoichiometry. For example, a likely first step (elementary reaction) in overall reaction (11.18) is:

$$Ni(H_2O)_6^{2+} + en(aq) \rightarrow Ni(H_2O)_4(en)^{2+}(aq) + 2H_2O(l) \tag{11.19}$$

The elementary reaction (11.19) represents an encounter (meeting) between two species to form a combined species. The probability of an encounter between two reactants in a mixture is directly proportional to the concentration of each species, so the rate of encounters depends on the product of their concentrations, that is, $[Ni(H_2O)_6^{2+}][en(aq)]$ in our example. However, as we will discuss in Section 11.6, not all encounters lead to reaction. Therefore, we have to include a proportionality constant, the rate constant, to account for the fraction of encounters that result in formation of product. In this case, the rate law for formation of the combined species is:

$$\frac{\Delta\left[Ni(H_2O)_4(en)^{2+}\right]}{\Delta t} = k[Ni(H_2O)_6^{2+}][en(aq)] \tag{11.20}$$

11.27. Consider This   *What are the elementary reactions for Ni(en)$_3^{2+}$(aq) formation?*

(a) Write a series of elementary reactions of two species at a time that will add up to yield net reaction (11.18). You can think of your series of reactions as the pathway or mechanism for the reaction.

(b) Is your experimental evidence from Investigate This 11.25 consistent with the mechanism you wrote in part (a)? Explain why or why not.

Bottlenecks: rate-limiting steps

If a reaction takes place in a stepwise fashion, its overall rate cannot be any faster than the slowest elementary reaction in the sequence. A mechanical analogy may make this idea easier to understand. Consider connecting four garden hoses together to make one long hose to reach to the back of a garden: three possible ways to connect the hoses are shown in Figure 11.5. The sections of hose have diameters of: 3/8 in, 1/2 in (two of these), and 3/4 in. The rate of water flow in the individual hoses is two gallons per minute in the 3/8-in hose, four gallons per minute through the 1/2-in hoses, and eight gallons per minute in the 3/4-in hose.

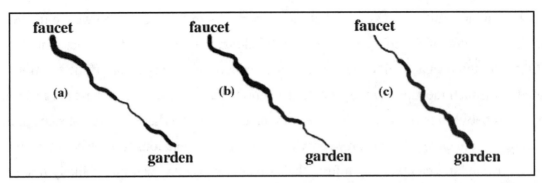

Figure 11.5. Possible connections of four hoses to make one long hose.

11.28. Consider This   *What is the water flow through each connection in Figure 11.5?*

Which connection in Figure 11.5 would give the largest flow of water to the garden? Is there another connection (not shown) that would provide an even larger flow of water? Give the reasoning for your responses.

The flow of water through the four connected hoses in Figure 11.5 is two gallons per minute, no matter where you put the smallest diameter, 3/8-in, hose. No more water can get through the entire hose than the amount that can get through this smallest diameter section. The 3/8-in section of hose is a **bottleneck** in the flow; it is the *rate-limiting* section of the hose. In a sequential series of elementary reactions, the overall reaction cannot be faster than the slowest step in the sequence. This slowest step is the **rate-limiting** or **rate-determining** step in the elementary reaction series. *The rate law describes the rate-determining step.* The ions and molecules that appear in the rate law are those that must have come together before and/or during the rate-determining step. The rate law is one of the most important pieces of experimental evidence we have to work with in developing a reaction pathway or mechanism.

Other ions and molecules that take part in the reaction *after* the rate-determining elementary reaction will not appear in the rate law because these subsequent steps are faster than the rate-

> **WEB Chap 11, Sect 11.4.1-6.**
> Analyze animated reaction pathways for one consistent with experimental observations.

determining step. However, to be an acceptable mechanism, the sum of all the elementary reactions in the sequence must give the overall reaction with the correct experimental stoichiometry.

This is how a reaction mechanism can be consistent with both the rate law and the reaction stoichiometry.

Applying the concept of rate-limiting steps

Let's see how these ideas about a rate-limiting step can be applied to the $I^-(aq)$-catalyzed decomposition of $H_2O_2(aq)$. Our rate law suggests that the rate-limiting step in the reaction is an

---

elementary reaction between $H_2O_2(aq)$ and $I^-(aq)$, since these are the species that appear in the rate law. A possible rate-determining reaction is:

$$H_2O_2(aq) + I^-(aq) \rightarrow \left[\begin{array}{c} \ddots \ddots \\ \textrm{:I:} \\ \textrm{H—O—O:} \\ \textrm{H} \end{array} \right]^- \rightarrow HOI(aq) + OH^-(aq) \qquad \textrm{slow}$$

$$(11.21)$$

11.29. Consider This   *What are the properties of the intermediate in reaction (11.21)?*

(a) What is the formal charge on each atom in the intermediate shown in reaction (11.21)? Show how you get your answers.

(b) In Chapter 6, Section 6.8, we said that "intermediates undergo electron and atomic core redistributions to reduce the number of formal charges; these redistributions lead to the observed products." Is this statement applicable to this intermediate? Show why or why not.

(c) Is reaction (11.21) a redox reaction? If so, which reactant is the reducing agent and which the oxidizing agent? Which atoms change oxidation number. Give your reasoning clearly.

The intermediate shown in reaction (11.21) probably has only transient existence and represents one stage in a continuous transition as the two reactants come together and the iodide ion displaces or substitutes for the hydroxide ion. This is a reversible reaction that favors the reactants (the reverse reaction is rapid), so only a small amount of the products are formed. However, the $HOI(aq)$ formed can react rapidly with $H_2O_2(aq)$:

$$HOI(aq) + H_2O_2(aq) \rightarrow O_2(g) + I^-(aq) + H^+(aq) + H_2O(l) \qquad \textrm{rapid} \qquad (11.22)$$

In neutral solution, pH 7, reaction (11.22) has an $E^{o\prime}$ of about 0.5 V, so the driving force, free energy change, for reaction is large. The hydronium ion, $H^+(aq)$, formed in reaction (11.22) reacts rapidly with the hydroxide ion formed in reaction (11.21):

$$H^+(aq) + OH^-(aq) \rightarrow H_2O(l) \qquad \textrm{rapid} \qquad (11.23)$$

11.30. Check This   *Summation of elementary reactions*

Show that the sum of elementary reactions (11.21), (11.22), and (11.23) is the overall reaction (11.3) for the decomposition of hydrogen peroxide to produce molecular oxygen and water. Why does $I^-(aq)$ not appear in the stoichiometric equation?

11.31. **Worked Example**   *Pathway for reaction of OCl⁻(aq) with I⁻(aq) in basic solution*

In Worked Example 11.23, we analyzed some rate data for this reaction:

$$OCl^-(aq) + I^-(aq) \rightarrow OI^-(aq) + Cl^-(aq) \tag{11.16}$$

Further experiments at different concentrations of hydroxide ion show that the reaction is slower at higher concentrations of base. The revised rate law is:

$$\frac{\Delta[OI^-]}{\Delta t} = k_{expt} \frac{[OCl^-][I^-]}{[OH^-]} \tag{11.24}$$

A suggested mechanism for stoichiometric reaction (11.16) is:

$$OCl^-(aq) + H_2O(l) \rightleftharpoons HOCl(aq) + OH^-(aq) \qquad \text{rapid equilibrium, } K_{eq} \tag{11.25}$$

$$HOCl(aq) + I^-(aq) \rightarrow HOI(aq) + Cl^-(aq) \qquad \text{slow, } k \tag{11.26}$$

$$HOI(aq) + OH^-(aq) \rightleftharpoons OI^-(aq) + H_2O(l) \qquad \text{rapid equilibrium} \tag{11.27}$$

The sum of the series of reactions (11.25), (11.26), and (11.27) is the stoichiometric reaction (11.16). Rapid equilibrium (11.27) greatly favors the products; $OI^-(aq)$ is a weak base. Show that this pathway is consistent with the rate law and find out how k_{expt}, the rate constant derived from experiments, is related to K_{eq} for reaction (11.25) and k for the rate-limiting step (11.26).

Necessary information: We need the elementary reactions and the rate law with which to compare our predictions.

Strategy: Write the rate law for the elementary rate-limiting step (11.26). Express all concentrations in this rate law in terms of known concentrations in the solution and compare the result to the experimental rate law.

Implementation: The rate law for elementary reaction (11.26) is:

$$\frac{\Delta[HOI^-]}{\Delta t} = \frac{\Delta[OI^-]}{\Delta t} = k[HOCl(aq)][I^-(aq)] \tag{11.28}$$

The rate of formation of $OI^-(aq)$ can be equated to the rate of formation of $HOI(aq)$ because reaction (11.27) is fast. $OI^-(aq)$ is produced as fast as $HOI(aq)$ is made.

$HOCl(aq)$ is formed in the equilibrium reaction (11.25), so we write the equilibrium constant expression (in terms of molarities) and solve it for $[HOCl(aq)]$:

$$K_{eq} = \frac{[HOCl(aq)][OH^-(aq)]}{[OCl^-(aq)][H_2O(l)]}; \qquad [HOCl(aq)] = K_{eq}\frac{[OCl^-(aq)][H_2O(l)]}{[OH^-(aq)]}$$

We have kept the concentration of water in these expressions, in order to keep track of all the reactants. Substitute this expression for $[HOCl(aq)]$ into equation (11.28):

$$\frac{\Delta[OI^-]}{\Delta t} = kK_{eq}\frac{[OCl^-(aq)][H_2O(l)]}{[OH^-(aq)]}[I^-(aq)] = kK_{eq}[H_2O(l)]\frac{[OCl^-(aq)][I^-(aq)]}{[OH^-(aq)]} \quad (11.29)$$

The rate law, equation (11.29), has exactly the same form as the experimental rate law equation (11.24) with $k_{expt} = kK_{eq}[H_2O(l)]$.

Does the answer make sense? The pathway fits both the stoichiometry of the reaction and the rate law. One way to interpret the rate law is to sum all the atoms and ions that appear in the numerator and subtract those in the denominator to give the atoms and ions of the reactants in the rate-limiting step. In this case we have $(ClH_2IO_2{}^{2-}) - (HO^-) = (ClHIO^-)$, which is correct for the sum of HOCl and I^-, the reactants in the rate-limiting step. Note that our data cannot test whether water is actually involved in the pathway, as shown, because its concentration does not vary in these solutions.

11.32. Check This *Pathway for $I^-(aq)$-catalyzed decomposition of $H_2O_2(aq)$*

Another possible pathway for the $I^-(aq)$-catalyzed decomposition of $H_2O_2(aq)$ is:

$$H_2O_2(aq) + H_2O(l) \rightleftharpoons H_3O_2{}^+(aq) + OH^-(aq) \qquad \text{rapid equilibrium, } K_{eq} \qquad (11.30)$$

$$H_3O_2{}^+(aq) + I^-(aq) \rightarrow HOI(aq) + H_2O(l) \qquad \text{slow, } k \qquad (11.31)$$

$$HOI(aq) + H_2O_2(aq) \rightarrow O_2(g) + I^-(aq) + H^+(aq) + H_2O(l) \qquad \text{rapid} \qquad (11.22)$$

$$H^+(aq) + OH^-(aq) \rightarrow H_2O(l) \qquad \text{rapid} \qquad (11.23)$$

(a) Show that the sum of elementary reactions (11.30), (11.31), (11.22), and (11.23) is the overall reaction (11.3) for the decomposition of hydrogen peroxide to produce molecular oxygen and water.

(b) What might the intermediate structure formed in reaction equation (11.31) look like? Use the intermediate structure shown in reaction equation (11.21) to guide your thinking.

(c) What does this pathway predict for the rate law for decomposition of $H_2O_2(aq)$? Explain how you get your answer.

(d) How does your predicted rate law from part (c) compare to the experimental rate law, equation (11.14)? If they are different, do you have enough information about the reaction to choose between them? Why or why not? If not, what experiments would you propose that might provide the information necessary to decide which pathway the reaction takes? *Hint:* If you do not vary the concentration of a species to determine its effect on the rate, you cannot exclude the possibility that it takes part in the reaction.

You have now seen how a slow, rate-limiting step that is consistent with the experimental rate law for a reaction can be combined with a series of rapid steps to yield the overall observed stoichiometry of the reaction. When all the steps of a reaction mechanism are summed, intermediate products from one step are reactants in subsequent steps and ultimately cancel out to leave you only the stoichiometric reactants and products. The protonated form of hydrogen peroxide, $H_3O_2^+(aq)$, is in this category in the reaction mechanism in Check This 11.32.

Some essential reactants enter the reaction at an early stage and are formed as products at a later stage, so they cancel out of the overall stoichiometry. The iodide ion, $I^-(aq)$, falls into this category in the $I^-(aq)$-catalyzed decomposition of hydrogen peroxide. Now we can define a **catalyst** as a reactant that is essential to the pathway or mechanism of a reaction and often appears in the rate law, but not in the overall stoichiometry of the reaction. Catalysts do not act in some mysterious way, but are part of the mechanism of the catalyzed reaction and behave like any other reactant.

11.33. **Check This**   *Correlating animated reaction pathways with symbolic reactions*

WEB   Chap 11, Sect 11.4.1. For each of the four proposed pathways (mechanisms) describe how the animation illustrates the reaction equations written to symbolize the pathway. Be as specific as possible with your descriptions.

Reaction mechanisms are tentative

An important point about our proposed reaction mechanisms is that they are tentative. *We usually can't prove that a reaction mechanism we write is the only one that satisfies all the data*, though we can *reject* proposed mechanisms that don't fit all the data for the reaction of interest. Even though we can't necessarily prove that a particular mechanism is correct or unique, we use it anyway to help predict the course of reactions that haven't been tried. We also use mechanisms as a basis for modifying reactions in directions that could be useful to attain some practical goal, such as producing more of a desired product and less of an undesired by-product. If we find that our predictions don't work, then we have to go back and see where our proposed pathway might need modification, based on the new experimental results, and try again.

Reflection and projection

Changes in the concentrations of species in a reaction mixture affect the rate of the reaction. We analyzed a few dependencies quantitatively to obtain rate laws, equations that relate the rate of a reaction to the concentrations of species in the reaction mixture. In the examples presented, the rates were directly dependent on the concentration of some of the reactants. These rate laws are first order in their dependence on these reactants. Other reaction orders you may encounter are zeroth order (no dependence on the concentration) and, very occasionally, second order (dependence on the square of the concentration). You have also seen that reaction orders may be negative (rate is inversely dependent on the concentration). Reaction orders and rate laws are determined experimentally. There is no necessary relationship between the stoichiometry of a reaction and the experimental rate law for the reaction.

Assuming that molecules, atoms, and/or ions must come together to react, we introduced the pathway or mechanism of a reaction as a sequence of such elementary reaction events. A reaction mechanism must be consistent with *both* the stoichiometry and the observed rate law for a reaction. Developing plausible reaction mechanisms is one of the hardest and most creative tasks in chemistry. We have tried to show that each of the elementary reactions we proposed is chemically reasonable, but you probably would have had difficulty developing the sequences without help. Figuring out what steps make chemical sense and how to put these together to construct a reaction pathway requires much experience with reactions and with mechanisms that others have constructed. At this stage of your study of chemistry, it's appropriate for you to be able to:

- derive a rate law from experimental rate data.
- show that a pathway fits an experimental rate law (or other experimental evidence).
- write a pathway that is parallel to one you already know.

In the next section, we will consider a more general way to analyze first order reactions and then go on to use the results to investigate what we can learn from the temperature dependence of rates that can help us further characterize reaction pathways.

Section 11.5. First Order Reactions

Because they are simple to analyze, we have so far focused exclusively on initial rates of reactions. But it isn't difficult to analyze irreversible first order reactions throughout the entire reaction. The results will help you better understand radioactive decay curves, such as Figure 3.14 in Chapter 3, Section 3.4, as well as chemical reactions.

Radioactive decay

In Chapter 3, Section 3.4, we introduced **radioisotopes**, elemental nuclei that spontaneously emit alpha or beta particles. Isotopes (both radioactive and nonradioactive) are used extensively in chemical and biological studies as well as in medical applications. Many of the radioisotopes used for these purposes decay to other isotopes by emitting a beta particle (electron); for example:

$$^{14}\mathrm{C}^{6+} \rightarrow {}^{14}\mathrm{N}^{7+} + e^{-} \tag{11.32}$$

$$^{3}\mathrm{H}^{1+} \rightarrow {}^{3}\mathrm{He}^{2+} + e^{-} \tag{11.33}$$

$$^{35}\mathrm{S}^{16+} \rightarrow {}^{35}\mathrm{Cl}^{17+} + e^{-} \tag{11.34}$$

All pathways for radioactive decay have in common that the rate law for decay is first order:

$$-\frac{\Delta N}{\Delta t} = k_1 N \tag{11.35}$$

In equation (11.35), N is the number of unstable nuclei present at a given time, ΔN is the number of nuclei that decay in a short time interval Δt, and k_1 is the first order rate constant for the reaction. During radioactive decay, N changes with time, but, if we choose a short enough time interval, not very many nuclei will decay and N will be nearly constant. We can rearrange equation (11.35) to:

$$-\frac{\Delta N}{N} = k_1 \Delta t \tag{11.36}$$

For a given time interval, Δt, the right-hand side of equation (11.36) is a constant. This means that the left-hand side also must be a constant. We interpret this relationship to mean that the *fraction* of nuclei that decay in a given time interval is the same, no matter how many nuclei are present. This is a general result *for all first order reactions: the fraction of reactant that reacts in a given time interval is the same for a particular reaction no matter what the concentration of the reactant.* For example, if $^{1}/_{1000}$th of some reactant reacts in the first minute of its reaction, then $^{1}/_{1000}$th of the remaining reactant reacts in the next minute and so on throughout the reaction.

Progress of a first order reaction with time

When we analyze related pairs of data points, such as concentration as a function of time in a reaction, we usually try to express the relationship in a way that gives a linear plot of the data. Here we will show an algebraic way to do this for a first order reaction, using radioactive decay as the example. To make the next few equations easier to write, we will make a substitution:

$$-\frac{\Delta N}{N} = f = k_1 \Delta t$$

The new variable, f, is the fraction of the reactant that changes (decays) in time Δt.

We have specified that Δt and, hence, ΔN must be tiny changes. In the limit of infinitesimal changes, dt and dN, we can use calculus to calculate what happens over a finite reaction time:

$$-\int_{N_0}^{N_t} \frac{\mathrm{d}N}{N} = \int_0^t k_1 \mathrm{d}t$$

$$-(\ln N_t - \ln N_0) = k_1(t - 0)$$

$$\ln N_t = \ln N_0 - k_1 t$$

or $\quad \ln\!\left(\frac{N_t}{N_0}\right) = -k_1 t$

If this mathematics is familiar to you, skip the algebraic derivation and start again at the text following equation 11.40

Let's call the number of unstable nuclei we start with in our sample (at time zero) N_0. At the end of the first time interval, Δt, the number of nuclei that have decayed is the number we started with, N_0, times the fraction of these that decayed, f:

number of nuclei that decay in 1st time interval = $N_0 \cdot f$

The number of nuclei left undecayed is the number we started with minus the number that decay:

N_1 = number of nuclei left after 1st time interval = $N_0 - N_0 \cdot f = N_0(1 - f)$

Continuing for another time interval, we get:

number of nuclei that decay in 2nd time interval = $N_1 \cdot f$

N_2 = number of nuclei left after 2nd time interval = $N_1 - N_1 \cdot f = N_1(1 - f)$

$$N_2 = [N_0(1 - f)](1 - f) = N_0(1 - f)^2 \tag{11.37}$$

11.34. Check This   *Algebra of exponential change*

(a) Explain why the number of nuclei that decay in the 2nd time interval is $N_1 \cdot f$.

(b) Show how equation (11.37) is derived from the preceding equation.

We can generalize this argument to the nth time interval to get:

$$N_n = N_0(1 - f)^n \tag{11.38}$$

Equation (11.38) shows how the number of nuclei varies with the number (n) of time intervals that have passed. We can convert equation (11.38) to a linear form by taking the logarithm of both sides:

$$\ln N_n = \ln N_0 + n \ln(1 - f)$$

To make this relationship even simpler, we take advantage of the fact that $\ln(1 - f) \approx -f$, if f is small. (Try it on your calculator for $f = 0.1$, 0.01, and 0.001 to see how small f must be.) Our whole analysis is based on choosing a time interval, Δt, that makes f ($= k_1 \cdot \Delta t$) small. The approximation gives:

$$\ln N_n = \ln N_0 - n \cdot f \tag{11.39}$$

If we follow the nuclear decay reaction for a time t, the number (n) of time intervals (Δt) that elapse is: $n = {}^t/_{\Delta t}$. The number of nuclei remaining undecayed at time t, N_t, is N_n. Substituting these equivalences into equation (11.39) gives the variation of $\ln N$ with reaction time, t:

$$\ln N_n = \ln N_t = \ln N_0 - n \cdot f = \ln N_0 - ({}^t/_{\Delta t})(k_1 \Delta t)$$

$$\ln N_t = \ln N_0 - k_1 t \tag{11.40}$$

Thus, for first order nuclear decay, the natural logarithm of the number of nuclei remaining, $\ln N_t$, is a linear function of time.

Measurement of radioactive decay

For a beta-emitting isotope, the number of beta particles emitted is proportional to N_t, the number of nuclei left undecayed at time, t. To determine the number of beta-emitting nuclei that are present in a sample, we count how many beta particles the sample is emitting with a **scintillation counter**. The sample is put in a liquid that gives off flashes of light when a beta particle hits it and the light flashes (scintillations) are counted by a photodetector. The number of

> **WEB Chap 3, Sect 3.4.5-9**
> Review radionuclide counting
> and half-life in this interactive
> animated experiment.

counts per minute, cpm, that are detected is a measure of the amount of a radioisotope in the sample. Figure 11.6 shows plots of the data from a scintillation counter as a function of time for a

sample containing ^{32}P, phosphorus-32, a radioisotope that is often used for studying nucleic acids because it can be incorporated into their sugar-*phosphate* backbones.

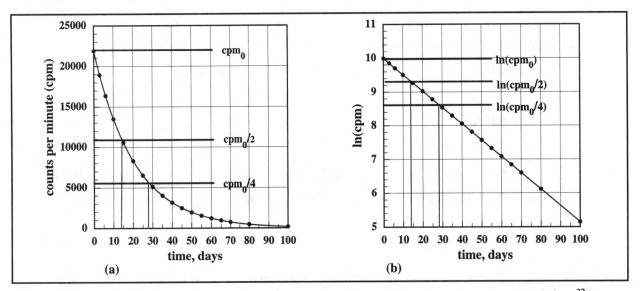

Figure 11.6. Plots of scintillation counter data as a function of time for a sample containing ^{32}P.
The count at time zero was 21897 cpm. All counts are corrected for background radiation so the counts represent the ^{32}P decay only. The slope of the line in (b) is -4.85×10^{-2} day^{-1}.

Compare Figure 11.6(a) with Figure 3.14. You will see the same exponential decay of the radioactivity in both curves. Figure 3.14 is a generalized curve of percent sample remaining as a function of numbers of half lives, which is applicable to any first order decay. Figure 11.6(a) is a specific radioactive decay with sample data, cpm, as a function of time. The horizontal red lines mark the number of counts per minute when the fraction of radioactive nuclei present is 1.00, 0.50, and 0.25, that is, at 0, 1, and 2 half lives, respectively. Figure 11.6(b) shows the natural logarithm of these same data, ln(cpm), plotted as a function of time. As equation (11.40) predicts, the data fall on a straight line. These data demonstrate how half lives and first order rate constants can be measured.

11.35. **Check This**   *Half life for beta emission from ^{32}P*

(a) Use the data in Figure 11.6(a) to find the half-life for beta emission from ^{32}P. Can you get more than one value for comparison and better precision? Show clearly how you get your answers.

(b) Is a stable isotope formed as a product of the beta decay of ^{32}P? Explain how you decide.

Half life and first-order rate constant

The slope of the line in Figure 11.6(b) is $-k_1$, where k_1 is the first-order rate constant for the beta emission reaction. We can use equation (11.40) to see how the half life for the reaction and the rate constant are related. The time required for a sample to decay from N_0 nuclei to $N_0/2$ nuclei is a half life, $t_{1/2}$. Substitution into equation (11.40) gives:

$$\ln(N_0/2) = \ln N_0 - k_1 t_{1/2}$$

This equation can be rearranged and solved:

$$k_1 t_{1/2} = \ln N_0 - \ln(N_0/2) = \ln\left(\frac{(N_0)}{(N_0/2)}\right) = \ln 2 = 0.693 \qquad (11.41)$$

Equation (11.41) shows that the product of the rate constant and half life for a first order reaction is a constant equal to ln2 or 0.693.

11.36. **Worked Example**   *Half life for beta emission from ^{32}P*

Use the slope of the line in Figure 11.6(b) to find the half life for beta emission from ^{32}P.

Necessary information: We need equation (11.41) and the slope of the line, -4.85×10^{-2} day^{-1}, in Figure 11.6(b).

Strategy: Substitute the known value for k_1 in equation (11.41) and solve for $t_{1/2}$.

Implementation:

$$k_1 t_{1/2} = (4.85 \times 10^{-2} \text{ day}^{-1})t_{1/2} = 0.693; \qquad \therefore \ t_{1/2} = \frac{0.693}{4.85 \times 10^{-2} \text{ day}^{-1}} = 14.3 \text{ day}$$

Does the answer make sense? In Check This 11.35, you read the Figure 11.6(a) graph to get the time required for a sample to decay to half its beginning value. You probably got a value of 14 days for the half life, so the value here makes sense. The slope of the straight line plot in Figure 11.6(b) — the same data as for Figure 11.6(a) — gives a more precise half life because all the decay data are used, not just two points.

11.37. Check This *First order decay constant and radiocarbon, ^{14}C, dating*

(a) The half life for beta decay of ^{14}C is 5730 years. What is the first order decay constant for this decay?

(b) A basket woven of twigs was found in an archaeological dig; a sample from the twigs contained 0.213 as much carbon-14 as present-day living plants. Use equation (11.40) and your result from part (a) to find the age of the basket. *Hint:* If necessary, review Chapter 3, Section 3.4, especially Worked Example 3.36 and Check This 3.37.

First-order chemical reactions

Many reactions that are first order in a reactant of interest can be analyzed in exactly the same way as radioisotope decays by rewriting equation (11.40) for that reactant in terms of the dimensionless ratio of its molar concentration to the standard concentration, (reactant):

$$\ln(\text{reactant})_t = \ln(\text{reactant})_0 - kt \tag{11.42}$$

A plot of the logarithm of the concentration of reactant left unreacted as a function of time will be a straight line with a slope equal to minus the numerical value of k, the first order rate constant. For example, we can test the data for the yeast-catalyzed decomposition of $H_2O_2(aq)$ from Section 11.2 to see whether the reaction is first order in $[H_2O_2(aq)]$ (or $(H_2O_2(aq))$, which is numerically equal), as we found in Section 11.3 for the $I^-(aq)$-catalyzed decomposition of $H_2O_2(aq)$. The yeast-catalyzed reaction is likely to depend on the amount of yeast present, but, since yeast is a catalyst, we can assume that its concentration does not change during the reaction.

11.38. Worked Example *Rate law for yeast-catalyzed decomposition of $H_2O_2(aq)$*

Data for $(H_2O_2(aq))$ as a function of time for yeast-catalyzed decomposition of $H_2O_2(aq)$ are:

time, min	0	1	2	3	4	5
$(H_2O_2(aq))$	0.175	0.154	0.136	0.119	0.104	0.091

These data are calculated from the data in Figure 11.2 by using the reaction stoichiometry to convert the amount of oxygen formed to the amount of $H_2O_2(aq)$ remaining unreacted. Use a graphical method to test whether the reaction is first order with respect to $(H_2O_2(aq))$ and, if so, determine the experimental rate constant, k_{expt}, for the reaction.

Necessary information: In addition to the data in the problem, we need equation (11.42).

Strategy: Assume the reaction is first order in $(H_2O_2(aq))$ and plot $\ln(H_2O_2(aq))$ *vs. t* to see if the data give a straight line. If so, use the slope to find k_{expt} for the reaction.

Implementation: The data for the plot and the plot are:

time, min	$\ln(H_2O_2(aq))$
0	-1.743
1	-1.871
2	-1.994
3	-2.125
4	-2.259
5	-2.399

The plot is a straight line with a slope -1.31×10^{-1} min^{-1}. Thus, the reaction is first order in $(H_2O_2(aq))$ (or $[H_2O_2(aq)]$) and $k_{expt} = 1.31 \times 10^{-1}$ min^{-1} = 2.18×10^{-3} sec^{-1}.

Does the answer make sense? The hydrogen peroxide decomposition catalyzed by iodide is first order in $[H_2O_2(aq)]$, so it is not surprising to find that the reaction catalyzed by yeast is also. Since the rate of the reaction probably depends on the amount or concentration of yeast in the solution, it's likely that the rate law has the form: rate = $k[\text{yeast}][H_2O_2(aq)]$. Thus, $k_{expt} = k[\text{yeast}]$, but we have no information about [yeast], so cannot estimate the rate constant, k, in the rate law. The best we can do is compare k_{expt} with the constant in equation (11.11), 8.3×10^{-4} s^{-1}, for the iodide catalyzed reaction. This constant also included the concentration of the catalyst. The values are comparable, so we can have some confidence that we are in the right ballpark.

Analysis of the catalyzed hydrogen peroxide decompositions is relatively easy because the only reactant concentration that is changing is $[H_2O_2(aq)]$. The other reactant concentration in the rate law equation, [catalyst], remains constant throughout the reaction. Sometimes, even when a reactant is actually consumed in a reaction, we can make its initial concentration large enough, compared to the other reactants, that its concentration changes only a negligible amount during the reaction. This technique, making one or more of the reactant concentrations so large that they change only a small amount during the reaction, is called **flooding**.

You have already studied a system flooded with one reactant, the reaction between the colored form of phenolphthalein and hydroxide ion. In the reactions you carried out in Investigations 11.1 and 11.4, you added one drop (about 0.05 mL) of millimolar concentration phenolphthalein solution to several milliliters of molar concentration hydroxide ion. While the concentration of the colored dianion of phenolphthalein decreased during the reaction, the concentration of hydroxide ion remained constant. In Check This 11.24 you used dependence on the constant initial concentrations of hydroxide ion to find the order with respect to $[OH^-(aq)]$. We can take advantage of flooding by hydroxide ion to determine whether the disappearance of the red phenolphthalein dianion is first order, as reaction equation (10.2) implies.

11.39. Check This   *Rate law for the reaction of phenolphthalein with OH⁻(aq) ion*

(a) In a 1.0 M solution of NaOH, the decay of the color of the phenolphthalein dianion was followed colorimetrically at a wavelength absorbed by the red dianion. The absorbance data here are directly proportional to the $[phth^{2-}(aq)]$ in the solution. Use a graphical method to find the order of the reaction with respect to $[phth^{2-}(aq)]$. Clearly explain your procedure.

time, min	0	0.1	0.2	0.3	0.4	0.5	0.6	0.7	0.8
absorbance	0.467	0.296	0.192	0.128	0.086	0.058	0.039	0.028	0.021

(b) Use your result from part (a) and your conclusion from Check This 11.24 to write a rate law for the reaction of phenolphthalein with hydroxide ion. Is your rate law consistent with reaction (11.2) being an elementary, rate-determining reaction? Explain why or why not.

(c) If it is possible, use your rate law in part (b) and the data in part (a), to determine the rate constant for this reaction. If it is not possible, explain why not and what information you lack.

Reflection and projection

In previous sections, we emphasized kinetic analyses based on initial rates of reaction. In this section, we have expanded the scope of kinetic analyses to include data from more than the first short time interval of a reaction. We mathematically manipulated (in the language of the calculus, "integrated") the differential rate law for first order reactions to find a graphical way to analyze rate data that would yield straight line plots with slopes related to the rate constants for the reactions. For a first order reaction, a plot of the logarithm of the first-order reactant species as a function of time is a straight line, if the concentrations of all other species in the reaction are constant. We also related the half life and rate constant for first-order reactions and applied this relationship to data on radioactive decays, such as radioisotope dating studies. Finally, we introduced the flooding technique that permits us to follow a reaction for a substantial time with the concentrations of most of the relevant reaction species held approximately constant.

The end result of all our approaches is a rate law which we hope will help us understand the pathway (mechanism) for the reaction of interest. We have tried to develop reaction pathways based on elementary reactions that make sense based on the chemistry that you have studied in the previous chapters. The thing that may not make much sense is the great variability in reaction rates. Reactants come together at about the same rate in any reaction (in solution). If, then, reactions occur when reactants come together, why don't all reactions go at about the same rate? The effect of temperature on reaction rate can help us answer this question.

Section 11.6. Temperature and Reaction Rates

11.40. **Investigate This**   *Are UV-sensitive beads affected by temperature?*

Do this as a class investigation and work in small groups to discuss and analyze the results. UV-sensitive beads are plastic beads that change color when irradiated with an ultraviolet light (including the sun) and then return to the original color (usually white) when the source of UV light is removed. In Styrofoam® plastic cups, prepare three temperature baths: ice-water, room temperature water, and warm (about 35 °C) water. Irradiate three identical UV-sensitive beads to change their color. Simultaneously, place one colored bead in each temperature bath. Record the temperature of the bath and how long it takes each bead to return to the original color. Repeat to test whether the results are reproducible.

11.41. **Consider This**   *How are UV-sensitive beads affected by temperature?*

In Investigate This 11.40, did the time required for the UV-sensitive beads to return to their original color vary with temperature? If so, at what temperature was the rate of return fastest? slowest? How might you explain your results?

The Arrhenius equation

As you found in Investigate This 11.4 and 11.40, reactions have higher rates at higher temperatures. If the products of a reaction don't change, we assume that the rate law has the same form at all temperatures studied. The only factor that can be responsible for the increasing rate with increasing temperature is an increasing rate constant. By the last quarter of the 19th century, the effect of temperature on the rates of many reactions had been studied. Enough data had been obtained to make it possible to show that the temperature dependence (temperature in kelvin) of the rate constants could be expressed in this form:

$$\ln k = (\ln A) - \frac{B}{T} \tag{11.43}$$

In equation (11.43), A and B are constants that are not functions of temperature and are different for each reaction. You will more often see equation (11.43) in its equivalent exponential form:

$$k = A\, e^{-\frac{B}{T}}$$ (11.44)

Equation (11.44) is called the **Arrhenius equation** in honor of Svante Arrhenius (Swedish

> To convert equation (11.43) to equation (11.44), take the antilogarithm (exponential) of both sides of equation (11.43):
> $$e^{\ln k} = k$$
> $$e^{(\ln A) - B/T} = e^{\ln A}\, e^{-B/T} = A\, e^{-B/T}$$

chemist, 1859-1927), who did much of the analysis that led to this formulation. The units of A are the same as the units of the rate constant, which depends upon the rate law for the reaction.

Since an exponent (or power) can have no units, the units of B and T must cancel; therefore the units of B are kelvin, K.

11.42. Consider This   *What does the Arrhenius equation tell us?*

(a) Show that equations (11.43) and (11.4) give increasing values of k as T increases.

(b) What value of B will give a rate constant at 303 K (30 °C) that is two times larger than the rate constant at 293 K (20 °C)?

11.43. Worked Example   *Temperature dependence of phenolphthalein-hydroxide reaction*

In an investigation like Investigate This 11.4, the times required for the same amount of phenolphthalein color to disappear in a colorimeter at 34.5, 20.4, and 1.3 °C were 47, 93, and 373 s, respectively, in 1.0 M solutions of NaOH. The times were reproducible within about 5-10%. What is the value of B in the Arrhenius equation for reaction 11.2?

Necessary information: The same amount of phenolphthalein dinegative ion reacted in each reaction and the concentration of base was the same in each case. Your results from Check This 11.24 and 11.39 show that the reaction is first order in both $[OH^-(aq)]$ and $[phth^{2-}(aq)]$:

$$-\frac{\Delta[phth^{2-}(aq)]}{\Delta t} = k[phth^{2-}(aq)][OH^-(aq)]$$ (11.45)

Strategy: Use a graphical method to obtain B as the slope of a $\ln k$ *vs.* $1/T$ plot by taking advantage of the fact that the *slopes* of $\ln k$ *vs.* $1/T$ and $\ln(ak)$ *vs.* $1/T$ plots are the same, if a is constant. In these reactions, $[OH^-(aq)]$ is the same in each reaction and is so large that it does not vary; the reactions are flooded with $OH^-(aq)$. Thus, equation (11.45) can be treated as a first order equation in the disappearance of $[phth^{2-}(aq)]$, and we can use equation (11.42) to get ak and hence $\ln(ak)$.

Implementation: For this reaction, equation (11.42) can be written as:

$$\ln[phth^{2-}(aq)]_t - \ln[phth^{2-}(aq)]_0 = \ln\!\left(\frac{[phth^{2-}(aq)]_t}{[phth^{2-}(aq)]_0}\right) = -k[OH^-(aq)]t$$

The reaction time we measure, t, is the time required for the same change in absorbance of each solution. Thus, the ratio of final to initial concentration is the same in all three reactions, so the logarithm of the ratio is also the same: $-k[OH^-(aq)]t = -$(constant). (The constant is negative because the ratio is less than one and the logarithm of a value less than one is negative.) Rearranging, we have:

$$\frac{k\left[OH^-(aq)\right]}{(\text{constant})} = ak = \frac{1}{t}$$

Substituting our experimental data and doing the calculations gives the values in this table and plotted on the accompanying graph. The bars on each point represent the uncertainty introduced by the uncertainty in the time measurements.

T, K	$1/T$, K^{-1}	t, s	$ak = 1/t$, s^{-1}	$\ln(ak)$
274.5	0.00364	373	0.0027	-5.92
293.4	0.00341	93	0.0108	-4.53
307.7	0.00325	47	0.0213	-3.85

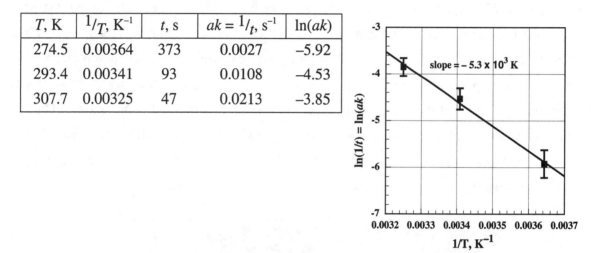

The slope of the line on the graph is $-B$, so $B = 5.3 \times 10^3$ K.

Does the answer make sense? Arrhenius found that plots of $\ln k$ vs. $1/T$ were straight lines and our data give a straight line plot, which makes sense. However, we do not know what B means nor do we have any other values with which to compare our result. Our next task is to find out what the parameters A and B in the Arrhenius equation mean.

11.44. Check This   *Temperature dependence of UV-sensitive bead decolorizing reaction*

(a) In Investigate This 11.40, you measured the times required for the colored form of UV-sensitive beads to return to their original color at three different temperatures. What is the value of B in the Arrhenius equation for this decolorizing reaction?

(b) How does the value of B for the UV-sensitive bead decolorizing reaction compare to value for the phenolphthalein reaction with hydroxide? Which reaction is more affected by a change in temperature? Explain the reasoning for your answer.

The Arrhenius equation provides a good description of the temperature dependence of reaction rates, but leaves us with three questions: What is the meaning of A? What is the meaning of B? Why does the equation have this particular form? These questions are interrelated, but let's take them up in order.

The frequency factor, A

Throughout our discussion of reaction pathways, we have said that reacting species must come together in order to react. For a gas-phase reaction, this means that the species must collide with one another. Such encounters (interactions) between the reactants are quite fleeting and any reaction has to occur within the short time of a single collision. The situation in solutions, which are of greater interest to us, is quite different.

Reactants in solution are surrounded by solvent molecules and are constantly being jostled about by their nearest neighbors. We can picture the reactant molecules as being surrounded by a cage of solvent molecules, as in Figure 11.7. As the jostling continues, the reactant molecule sooner or later moves to a different solvent cage and thus moves about randomly in the solution. The motion of two reactant molecules with respect to one another is a diffusion process. The number of encounters between reactants in solution depends upon how rapidly molecules diffuse in the solution. A is usually called the **frequency factor**, because it accounts for the frequency of encounters between molecules in the solution.

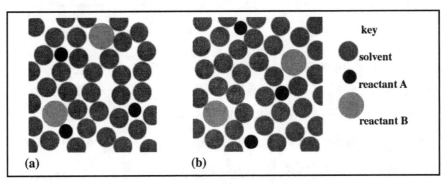

Figure 11.7. Representation of reactant diffusion in solution.
Panel (b) represents the solution a few picoseconds (10^{-12} s) later than panel (a).

The size and shape of molecules and ions affects their diffusion through a solvent. However, most species move at about the same speed, so they encounter one another with about the same frequency. For a reaction that takes place on every encounter between two reactants in aqueous

> The reaction between hydronium and hydroxide ions is the fastest reaction in aqueous solution, with a rate constant of 1.3×10^{11} M^{-1}s^{-1}. The mechanism of diffusion of these ions is probably different. See Figure 2.26, Chapter 2, Section 2.12 and the accompanying discussion.

solution, the frequency factor, A, is $10^9 - 10^{10}$ $M^{-1}s^{-1}$. These are said to be **diffusion controlled reactions**, because they occur as rapidly as the reactants can diffuse together.

Figure 11.7(a) shows a reactant pair of molecules together in a cage of solvent molecules and Figure 11.7(b) shows that the pair has left the cage without reacting. If reaction does not occur every time a reactant pair is trapped together in a solvent cage, the rate constant for the reaction will be less than the diffusion-controlled rate constant. Reaction might not occur because, in order to react, the reactants have to come together in just the right orientation. This effect will show up as a frequency factor, A, less than the diffusion-controlled limit. Although chemists try to model and calculate these effects theoretically, it is more common to determine the frequency factor experimentally, compare it to the diffusion-controlled limit, and then try to figure out what orientation effects could make the frequency factor less than the limiting value.

11.45. Consider This   *How do frequency factors for similar reactions compare?*

The frequency factors for reactions (11.46) and (11.47) studied in acetone solutions were 2.5×10^4 and 3.2×10^6 $M^{-1}s^{-1}$, respectively. How might orientation effects account for the low values compared to the diffusion-limited value. How might you account for the factor of about 100 difference between the two values? Build molecular models of the reactants, if that will help you formulate your explanations.

$$(C_2H_5)_3N\text{:} + C_2H_5I \rightarrow [(C_2H_5)_4N]^+I^- \text{ (a salt)} \tag{11.46}$$

(a salt)   (11.47)

For example, when a reaction requires an electrophilic center in one molecule to approach a nucleophilic center in another, the other parts of both molecules can interfere and inhibit their approach simply by being in the way. If two pairs of such reactants are compared and the interfering parts of the molecules are smaller or more compact in one pair, then we would expect the interferences to be smaller and the frequency factor to be larger for that pair. These are the sort of effects you find in Consider This 11.45.

Activation energy for reaction

Another reason a reaction might not occur when the reactants are caged together, is that the reactants lack the energy to react. For example, the pathway for reaction (11.47) requires the nucleophilic electron pair on the nitrogen to begin to bond to the electrophilic carbon bonded to the iodine atom. This interaction weakens the carbon-iodine bond and leads to a partial negative charge on the iodine atom and partial positive charge on the nitrogen, as shown for the activated complex in Figure 11.8, the activation energy diagram for this reaction.

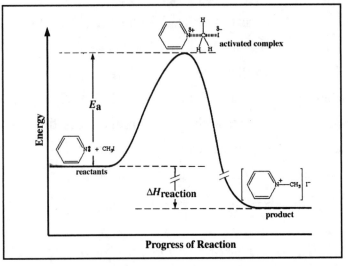

Figure 11.8. An activation energy diagram for reaction (11.47).
In order to emphasize the activation barrier, E_a and $\Delta H_{reaction}$ are not drawn to scale.

For reaction (11.47), $E_a = 58$ kJ·mol^{-1} and $\Delta H_{reaction} = -98$ kJ·mol^{-1}. Both values depend somewhat on the polarity of the solvent, because uncharged polar molecules react to give ions and there is a good deal of solvent reorganization during the reaction.

The **activated complex** is the highest energy species formed as reactants progress toward products in an elementary reaction. The nucleophilic amine and electrophilic iodide have to come together with enough energy to bring about the simultaneous making and breaking of bonds that forms the activated complex shown in Figure 11.8 for this reaction. The **activation energy**, E_a, for a reaction is the *net* energy required for the reactants to form activated complexes, which can then go on to form products (or go back to reactants).

The "progress of reaction" that we are following from left to right on an activation energy diagram is a fuzzy concept. Many different electronic and atomic core rearrangements occur during a reaction, so defining how far the changes have progressed is a complex problem. For our purposes, these details are unimportant. The energies of the reactants, activated complex, and products and a sense of continuous change as the reaction progresses are enough.

11.46. **Check This**   *Activation energy diagrams*

(a) Activation energy diagrams combine both kinetic and thermodynamic information for a reaction. Using the information in Figure 11.8, explain whether reaction (11.47) is exothermic or endothermic.

(b) Sketch the activation energy diagram for a reaction that has the opposite overall thermodynamics from that shown in Figure 11.8.

(c) If reaction (11.47) is reversible, how would the activation energy for the reverse reaction compare to the activation energy shown in Figure 11.8? Explain with a diagram.

(d) Imagine an endothermic reaction that can occur by two pathways, one catalyzed and the other not catalyzed, with two different activation energies. Sketch and label an activation energy diagram that shows the energy as a function of the progress of reaction for both pathways.

Pairs of reactants that come together with too little energy may progress part way toward products (which corresponds to being part way up the barrier in Figure 11.8), but will not be able to form the activated complex and go on to products. While caged by solvent molecules, the pair might gain enough energy from their collisions with solvent molecules to be able to form the activated complex and undergo reaction. If they don't gain enough energy, they ultimately leave the cage unreacted, as shown for the pair in Figure 11.7. Thus, the rate constant for a reaction depends on how many encounters of reactant pairs have an energy E_a or greater. To find the proportion of such encounters, we need to know about energy distributions in molecular systems.

Molecular energy distributions. In Section 11.1, when we first looked at the increase in reaction rate with increase in temperature, we said that it is likely that the temperature effect is related to the energy of the reacting system, which increases with temperature. Thus we can guess that B, in the temperature dependent part of the Arrhenius equation, has something to do with energy. Since a minimum energy, E_a, is necessary for reaction to occur, it is likely that B and E_a are related. To understand the relationship, we need to examine the distribution of energy among the molecules in a system at different temperatures shown in Figure 11.9.

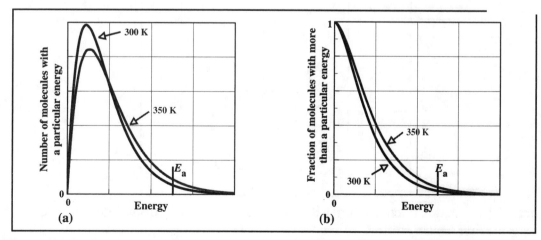

Figure 11.9. Energy distributions for a sample of molecules at two temperatures.
(a) is the number of molecules with a given energy. (b) is the fraction of molecules with *more* than a given energy.

For any value on the energy axis in Figure 11.9(a), the value on one of the curves is the number of molecules in the sample that have that energy. Though the two curves have somewhat different shapes, the *total* area under each curve, which is equal to the number of molecules in the sample, is the same for both temperatures. The curves show that the distribution of energies is broader at the higher temperature, that is, there are more molecules with higher energies. Therefore, the average molecular energy, which is related to the energy where the curve is a maximum, is a bit higher at the higher temperature. Figure 11.9 makes the relationship between temperature and energy of the molecules in a system quantitative.

The energy distribution shown by the plots in Figure 11.9 is called the **Maxwell-Boltzmann distribution**, because the theory describing the distribution was developed by James Clerk Maxwell (Scottish physicist, 1831-1879) and Ludwig Boltzmann (Austrian physicist, 1844-1906). Many kinds of experiments confirm the predictions represented in Figure 11.9. The plots in the figure are for the translational energies of molecules in a gas, but the conclusions we reach are applicable also to the energies of encounters in a solution.

An energy, E_a, is marked on the graphs in Figure 11.9. For the system represented, all encounters with energies higher than this can produce reaction. The number of such encounters is given by the area under each curve to the right of E_a. The *fraction* of such encounters is given by the ratio of this area to the total area under the curve and is shown in Figure 11.9(b) and shown enlarged at the high energy end of the plot in Figure 11.10. What you should observe from all the figures is that the fraction of all encounters with energy E_a or higher is larger for the higher temperature system.

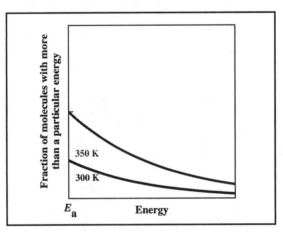

Figure 11.10. Fraction of encounters with energy greater than E_a from Figure 11.9(b).

11.47. Consider This *Are there always more encounters with E > E_a at higher temperature?*

The position of E_a in Figure 11.11 was chosen arbitrarily at some high energy value. Is the observation that "the fraction of all encounters with energy E_a or higher is larger for the higher temperature system" true for any value of E_a? Clearly explain the reasoning for your response.

Our generalization of this important idea is that *the fraction of molecules with energies greater than a given high value, E_a, increases rapidly as the temperature goes up.* The quantitative expression of this relationship, is the **Boltzmann equation**:

$$\text{fraction of molecular encounters with energy } E_a \text{ or greater} = e^{-\frac{E_a}{RT}} \qquad (11.48)$$

Equation (11.48) is expressed on a molar basis: E_a has units of $J \cdot mol^{-1}$ and R, the gas constant, is $8.314 \ J \cdot mol^{-1} \cdot K^{-1}$.

Accounting for the fraction of molecular encounters that have enough energy to produce reaction, equation (11.48), and the frequency of encounters, A, that have the appropriate orientation for reaction, gives for the rate constant:

$$k = A e^{-\frac{E_a}{RT}} \qquad (11.49)$$

Thus, we have identified the meaning of B and the relationship between E_a and B in the Arrhenius equation (11.44): $B = E_a/R$. We can use this relationship to convert the B values we have calculated to activation energies, E_a. We use equation (11.49), the form of the Arrhenius equation you will usually see, to calculate and compare rate constants under different conditions.

11.48. Worked Example *Activation energy for the phenolphthalein-hydroxide reaction*

What is the activation energy for the decolorization reaction between phenolphthalein dianion and hydroxide ion, reaction equation (11.2)?

Necessary information: We need the result from Worked Example 11.43, $B = 5.3 \times 10^3$ K, and the equation $B = E_a/R$.

Strategy: Solve this equation for E_a and substitute the experimental value of B.

Implementation:

$$E_a = B \cdot R = (5.3 \times 10^3 \text{ K})(8.314 \ J \cdot K^{-1} \cdot mol^{-1}) = 44 \times 10^3 \ J \cdot mol^{-1} = 44 \ kJ \cdot mol^{-1}$$

Does the answer make sense? Most reactions that occur in solution at or near 300 K (room temperature) have activation energies less than 100 $kJ \cdot mol^{-1}$, so the result for this reaction fits that pattern. At room temperature there is such a small fraction of encounters with energies of 100 $kJ \cdot mol^{-1}$ or greater that reactions with higher activation energies cannot get over the energy

---

barrier. Such reactions require higher temperatures and/or catalysts that provide a lower activation energy pathway for reaction.

11.49. Check This   *Activation energy for the UV-sensitive bead decolorizing reaction*

Use your results from Check This 11.40 to find the activation energy for the UV-sensitive bead decolorizing reaction.

11.50. Worked Example   *Rate constant from Arrhenius equation variables*

The activation energy for reaction (11.47) is 58 kJ·mol^{-1}. What is the rate constant, k, for the reaction at 0 °C (273 K) and 50 °C (323 K)?

Necessary information: We need the activation energy, the Arrhenius equation (11.49), and the frequency factor, 3.2×10^6 M^{-1}s^{-1}, from Consider This 11.45.

Strategy: Substitute the known values of T, A and E_a into equation (11.49) to get k at the desired temperatures.

Implementation:

$$k \text{ (at 273 K)} = (3.2 \times 10^6 \text{ M}^{-1}\text{s}^{-1})\, e^{-\frac{58 \times 10^3 \text{ J·mol}^{-1}}{(8.314 \text{ J·mol}^{-1}\cdot\text{K}^{-1})(273 \text{ K})}} = 2.6 \times 10^{-5} \text{ M}^{-1}\text{s}^{-1}$$

$$k \text{ (at 323 K)} = (3.2 \times 10^6 \text{ M}^{-1}\text{s}^{-1})\, e^{-\frac{58 \times 10^3 \text{ J·mol}^{-1}}{(8.314 \text{ J·mol}^{-1}\cdot\text{K}^{-1})(323 \text{ K})}} = 1.3 \times 10^{-3} \text{ M}^{-1}\text{s}^{-1}$$

Does the answer make sense? Rates increase with temperature because the rate constant increases with temperature. Here the higher temperature rate is about 50 times faster.

11.51. Check This   *Rate constant from Arrhenius equation variables*

(a) If the activation energy for reaction (11.46) is 50 kJ·mol^{-1}, what is the rate constant, k, for the reaction at 0 °C (273 K) and 50 °C (323 K)?

(b) Is reaction (11.46) or (11.47) faster at 273 K? at 323? Explain your reasoning.

11.52. Check This   *Variation in rates of room temperature reactions*

(a) A common rule of thumb is that room temperature reactions approximately double in rate for every 10 °C increase in temperature. (See Consider This 11.42(b).) Does reaction (11.46) or (11.7) or both obey this "rule?" Show how you get your answer.

(b) What does this doubling rule imply about the activation energy of room temperature reactions? Give the reasoning for your answer.

Section 11.7. Light: Another Way to Activate a Reaction

11.53. **Investigate This**   *What factors affect the color of UV-sensitive beads?*

Do this as a class investigation and work in small groups to discuss and analyze the results. In Styrofoam® plastic cups, prepare three temperature baths: ice-water, room temperature water, and warm (about 35 °C) water. Place an identical UV-sensitive bead in each cup. Place the cups under a source of ultraviolet light so that each bead gets the same amount of illumination. Note and record how dark the color of the bead becomes in each bath. (How can you check to make sure any differences you see are not due to differences between beads?)

11.54. **Consider This**   *How does temperature affect the color of UV-sensitive beads?*

(a) In Investigate This 11.53, did the darkness or depth the color developed by the UV-sensitive beads change continually over time or reach some constant color in each bath.

(b) Was the depth of the color developed by the UV-sensitive beads affected by their temperature? Describe the relationship between the depth of color and the temperature. What explanation can you offer for this relationship?

In Section 11.6, you analyzed the temperature dependence of the decolorizing reaction following illumination of UV-sensitive beads. In Investigate This 11.53, you focused your attention on the color-forming process to see how much color develops under the same amount of illumination at different temperatures. The color development is greater at lower temperatures, which would lead you to conclude that the color-forming reaction cannot be described by the Arrhenius equation for temperature dependence of reactions. The temperature dependence we discussed in Section 11.6 was based on the exponential Boltzmann factor for the distribution of thermal energies in a system. If the energy for a reaction is supplied in a different way, then this factor is not applicable.

The energy for the reaction that produces the color change in the UV-sensitive beads is provided by ultraviolet light. In order to transfer energy to a molecule, a photon of light has to be absorbed by the molecule. The UV-sensitive beads are white (or colorless when melted) when they have not been exposed to ultraviolet light. This means the molecules responsible for the color change do not absorb light in the visible region of the spectrum. They must absorb light in the ultraviolet, which gives them much more energy than is available thermally at room temperature.

11.55. Check This   *Energy of ultraviolet light photons*

The lower wavelength (higher energy) limit of the visible region of the spectrum is about 400 nm. What is the energy of a 400 nm photon? How does this energy compare to the activation energies we found for reactions in Section 11.6? *Hint*: The Planck relationship, $E = \mathrm{h} \cdot \mathrm{c}/\lambda$, from Chapter 4 gives the energy of a single photon. You will need the energy of a mole of photons to compare with molar activation energies.

Rate of photochemical reactions

Your result in Check This 11.55 shows that the energy of ultraviolet light photons is enough, over 300 kJ·mol⁻¹, to break chemical bonds. Molecules that absorb a photon of this light will be highly activated and able to undergo chemical reactions. Chemical reactions initiated by absorption of light are called **photochemical reactions**. Since gaining light energy does not depend on encounters among molecules, we would not expect the rates of photochemical reactions to be temperature dependent. What will the rates depend on? Take the reaction of the molecules in the UV-sensitive beads going from their colorless (*L*) to colored (*C*) form as an example:

$$L \xrightarrow{\ h\nu\ } C \tag{11.50}$$

$$\frac{\Delta[C]}{\Delta t} = aI\phi\,[L] \tag{11.51}$$

The $h\nu$ over the reaction arrow in equation (11.50) reminds us that the reaction is initiated by light energy. In the rate law, equation (11.51), I is the intensity (mol photons·L⁻¹·s⁻¹) of light entering the sample (the bead) and a is fraction of this light absorbed by a unit concentration (1 M) of the *L* molecules. The actual concentration of colorless reactant molecules in the bead is accounted for by their concentration, [*L*]. Thus, the product, $aI[L]$, is the concentration of activated molecules formed by absorption of light each second. In most photochemical reactions, only a fraction of the activated molecules, ϕ, go on to give product. The rest of the activated molecules lose their

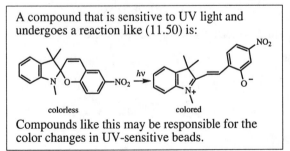

A compound that is sensitive to UV light and undergoes a reaction like (11.50) is:

colorless   colored

Compounds like this may be responsible for the color changes in UV-sensitive beads.

energy in various ways and return to the initial form, **L**. The quantity ϕ, the **quantum yield** for the reaction, is a measure of the number of product molecules produced per quantum of light energy absorbed. Quantum yields for elementary reactions like this range from zero (no photochemical reaction) to unity (all photons yield product).

Competing reactions and the steady state

The variables in the photochemical rate law, equation (11.51), do not depend on temperature. Thus, the rate of the photochemical reaction, formation of **C** from **L**, does not depend on temperature. Why, then, do you observe a temperature effect on the depth of the color formed by the photochemical reaction? We must account for the reverse thermal reaction:

$$C \rightarrow L \tag{11.52}$$

This is the reaction you observed in Investigate This 11.40 and for which you determined an activation energy in Check This 11.49. The rate law for elementary reaction (11.52) is:

$$-\frac{\Delta[C]}{\Delta t} = k[C] \tag{11.53}$$

In Investigate This 11.53, you observed that the beads in each bath quickly reached an unchanging depth of color (different at the different temperatures). This means that some constant concentration of the colored form, $[C]$, was present at each temperature. Since reactions (11.50) and (11.52) are going on simultaneously, an unchanging $[C]$ means that a **steady state** has been reached: the rates of formation and decay, equations (11.51) and (11.53), respectively, are equal:

$$\frac{\Delta[C]}{\Delta t} = -\frac{\Delta[C]}{\Delta t}; \qquad aI\phi\,[L] = k[C] \tag{11.54}$$

When the concentration of a species that forms and decays simultaneously remains constant in a system, a steady state has been reached with the formation and decay reactions going on at the same rate.

We can rearrange equation (11.54) to get the steady-state ratio of the colored to colorless forms of the reactant molecules:

$$\left(\frac{[C]}{[L]}\right)_{\text{steady state}} = \frac{aI\phi}{k} \tag{11.55}$$

The numerator on the right of equation (11.55) is a constant, independent of temperature, for a given system, such as the UV-sensitive beads, and a given light intensity. The rate constant, k, in the denominator is temperature dependent.

11.56. **Check This**   *The affect of temperature on the color of UV-sensitive beads*

(a) Does the temperature dependence of k for reaction (11.52), which you found in Check This 11.49, explain the temperature dependence of the steady-state $[C]/[L]$ ratio for the UV-sensitive beads you observed in Investigate This 11.53? Why or why not?

(b) What variable(s) in equation (11.55) could you test to see whether the equation correctly predicts the behavior of the steady-state $[C]/[L]$ ratio.

Light and life

Photochemical reactions are essential for life, as we know it. All animals on the surface of the Earth and most in the seas depend upon photosynthesis in microorganisms and plants for food and to maintain an atmosphere of oxygen. The first steps in photosynthesis are reduction-oxidation reactions initiated by the absorption of visible-light photons from sunlight. The outcome of these reactions can be represented as:

$$2H_2O(l) \xrightarrow{\text{hv, chloroplasts}} O_2(g) + 4H^+(aq) + 4e^- \tag{11.56}$$

The notation on the reaction arrow reminds us that the reaction requires light and the chloroplasts in the plant. The electrons are not "free" in the chloroplast, but are captured as reduced forms of molecules like nicotinamide adenine dinucleotide phosphate, NADPH (Check This 10.55, Chapter 10, Section 10.7 and Problem 10.66). These reduced molecules in turn are used to reduce carbon dioxide to glucose as part of a series of reactions that does not involve light.

11.57. **Check This**   *Visible-light photons and oxidation of water*

(a) The standard reduction potential for reduction of oxygen to water (the reverse of reaction (11.56) is 0.816 V at pH 7. What is the standard free energy change for reaction (11.56)? Explain.

(b) The free energy to drive reaction (11.56) is supplied by the energy from visible-light photons, wavelengths in the 400-700 nm range. Is one photon of visible light enough to drive the reaction under standard conditions? Show your reasoning clearly.

Note, once again, how important water is to life. In addition to all the other essential characteristics that we have pointed out, water is the reducing agent required to make the food we eat as well as the oxygen we need to oxidize this food to provide the energy for life.

Reflection and projection

Most reactions are temperature dependent and increase in rate as the temperature increases. The Arrhenius equation, $k = A\,e^{-\frac{E_a}{RT}}$, is the quantitative expression of the temperature dependence of reaction rate constants. The molecular basis of the Arrhenius equation depends upon encounters among molecules that lead to reaction. We assume that molecules have to come together in order to react and that their orientation with respect to one another is important in determining whether they react. This orientation requirement leads to Arrhenius frequency factors, A, smaller than the diffusion-controlled encounter frequency.

As reactants interact and change to products, many electron and atomic core rearrangements must occur. These usually require a net input of energy, the activation energy, E_a, in order to form an activated complex, which can proceed to the product(s). To react, pairs of reactant molecules that encounter one another must have energy equal to or greater than E_a. The distribution of energy among molecules at a given temperature is given by the Maxwell-Boltzmann distribution. This distribution leads to the Boltzmann equation which we use to find the fraction of encounters with energies greater than or equal to E_a. The fraction leads to the exponential term in the Arrhenius equation. The temperature dependence of a reaction rate is a consequence of the activation energy for the reaction.

Another way that reactants can obtain enough energy to form the activated complex is by absorbing light. The energy available from visible and ultraviolet photons is enough to break molecular bonds and can initiate reactions at room temperature that would require heating the system to hundreds or even thousands of degrees to initiate thermally. Photochemical reactions are relatively insensitive to temperature changes, since the energy provided by the photons is usually so much greater than is available thermally.

Reactions that require high temperatures to form the activated complex can often occur under milder conditions, if a catalyst provides a lower energy pathway to the products. Biological reactions must occur under mild conditions, so almost all biochemical reactions are catalyzed by proteins called enzymes which are the subject of Section 11.10. However, a catalyst cannot make a reaction occur that is not thermodynamically possible. In the next section we will discuss the connections between kinetics and thermodynamics.

Section 11.8. Thermodynamics and Kinetics

11.58. Investigate This   *What more can we learn about H$_2$O$_2$(aq) decomposition?*

Do this as a class investigation and work in small groups to discuss and analyze the results. Measure and record the temperature of 100 mL of 3% aqueous hydrogen peroxide, H$_2$O$_2$(aq), solution in a Styrofoam® cup. Add a packet of dried yeast to the solution and use the thermometer to mix the yeast into the solution. Record the temperature and any other observed changes in the system for 3-4 minutes.

11.59. Consider This   *What can you conclude about H$_2$O$_2$(aq) decomposition?*

(a) In Investigate This 11.58, you have again decomposed H$_2$O$_2$(aq) (as you first did in Investigate This 7.32, Chapter 7, Section 7.7). Is the reaction exothermic or endothermic? Assume that the specific heat of the solution is 4.18 kJ·g^{-1} and that 3% H$_2$O$_2$(aq) is 0.88 M. What is $\Delta H_{reaction}$ for the decomposition? Explain your method.

(b) What is the sign of $\Delta S_{reaction}$ for H$_2$O$_2$(aq) decomposition, reaction (11.3)? Explain.

(c) Is $\Delta G_{reaction}$ positive or negative for reaction (11.3)? Explain your reasoning.

(d) What can you say about the rate of reaction (11.3) in the absence and presence of yeast? Is this an expected result? What is the basis for your expectation?

Through the first seven chapters of this book, we puzzled over the underlying basis for directionality in chemical reactions. We found that energy (or enthalpy) did not provide the criterion, because both exothermic and endothermic reactions are observed to occur. Finally, in Chapter 8, we found that net entropy change (or free energy change), a measure of probability, is the property we can use to predict the direction of change in a system. In Chapter 9, Section 9.1, when we discussed how to identify systems at equilibrium, we considered the example of wood (largely glucose), whose oxidation is greatly favored, but which exists unchanged for centuries surrounded by a sea of oxygen. Under the proper conditions, application of a match flame, for example, wood can be oxidized to form carbon dioxide and water as combustion products. Aqueous solutions of hydrogen peroxide provide another example of a reaction, decomposition to oxygen and water, that is highly favored, as you showed in Consider This 11.59(c). This reaction is so slow, however, that it is almost unobservable over hours or even days. Under proper conditions, for example, the presence of a catalyst such as yeast or any of the others you found in Investigate This 11.6, the reaction proceeds rapidly.

These two examples, and many others like them, show that thermodynamically favored (spontaneous) reactions are not necessarily observed, unless other factors are also favorable. In this chapter, we have examined these other factors that affect the rates of chemical reactions. In order to proceed along the pathway from reactants to products, the reactants have to come together in the appropriate orientation and have sufficient energy to surmount one or more energy barriers along the pathway.

In the combustion of wood, a match flame provides a source of high temperature that gives oxygen enough energy to react with the molecules in the wood to produce the intermediate reactive species that go on to react with one another and more oxygen. These processes produce large amounts of thermal energy which heats more oxygen to produce more reaction and so on. Thus, once initiated, the oxidation is self-sustaining, which is why one match can create a great deal of combustion. To stop the combustion, you can remove one or the other of the reactants, fuel or oxygen. Or you can remove enough energy, by pouring water on the fire, for example, to halt the self-sustaining initiation processes. The reaction is still thermodynamically favored, but the reactants are not together and/or there is not enough energy to initiate it.

The decomposition of hydrogen peroxide is quite exothermic, as you found in Consider This 11.59(a) and as is shown in Figure 11.11(a). Also shown in the figure is the high activation energy for the uncatalyzed decomposition of hydrogen peroxide. Therefore, even though the decomposition is thermodynamically favored (by both enthalpy and entropy), the activation energy makes it very slow at room temperature.

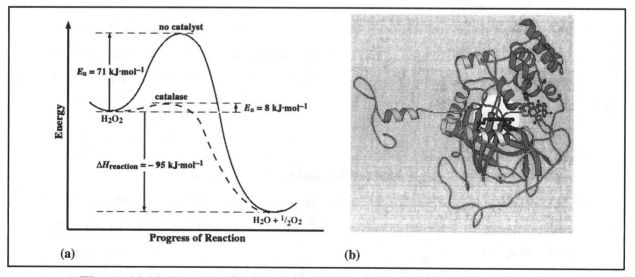

Figure 11.11. Factors affecting the rate of decomposition of $H_2O_2(aq)$.

(a) Activation energy diagram comparing E_a for the uncatalyzed and catalyzed reaction. (b) Heme group in the catalase structure highlighted. View is from the edge of the flat ring that complexes the iron ion which interacts with H_2O_2 and helps provide a lower activation energy pathway for the decomposition.

The enzyme catalase (contained in yeast and most other living tissues and organisms) provides an alternative reaction pathway for hydrogen peroxide decomposition. Along this pathway the peroxide molecule interacts within the protein structure with the iron in the heme group (see Chapter 6, Figure 6.10) highlighted in Figure 11.11(b) which lowers the activation energy for decomposition by almost a factor of ten.

11.60. Worked Example   *Uncatalyzed decomposition of $H_2O_2(aq)$*

The uncatalyzed decomposition of $H_2O_2(aq)$ shows first order kinetics with a rate of about 10^{-8} M·s^{-1} for a 1 M solution at room temperature, 298 K. What is the first order rate constant and the Arrhenius frequency factor, A, for this reaction?

Necessary information: We need the first order rate law, the Arrhenius equation (11.49), and the activation energy, 71 kJ·mol^{-1} from Figure 11.11(a).

Strategy: Use the first order rate law to find k_1, the first order rate constant, and then substitute this result and the activation energy into the Arrhenius equation and solve for A.

Implementation: Write and solve the first order rate law for k_1:

$$\text{rate} = 10^{-8} \text{ M·s}^{-1} = k_1[H_2O_2(aq)] = k_1(1 \text{ M}); \qquad \therefore k_1 = 10^{-8} \text{ s}^{-1}$$

The Arrhenius equation is:

$$k_1 = 10^{-8} \text{ s}^{-1} = A\, e^{-\left(\frac{71{,}000 \text{ J·mol}^{-1}}{(8.314 \text{ J·mol}^{-1}\cdot\text{K}^{-1})(298 \text{ K})}\right)} = A\cdot(3.6 \times 10^{-13})$$

$$A = \frac{10^{-8} \text{ s}^{-1}}{3.6 \times 10^{-13}} = 3 \times 10^{5} \text{ s}^{-1}$$

Does the answer make sense? The Arrhenius frequency factors for first order reactions are often about 10^{14} s^{-1}, so this value for $H_2O_2(aq)$ decomposition is quite low and in combination with the high activation energy makes the decomposition very slow, as observed. See Check This 11.61(b).

11.61. Check This   *Catalase-catalyzed decomposition of $H_2O_2(aq)$*

(a) At 1 M concentrations of $H_2O_2(aq)$ and catalase, the rate of $H_2O_2(aq)$ decomposition is about 10^7 M·s^{-1} at 298 K. What is the second order rate constant and the Arrhenius frequency factor for this reaction?

(b) The "uncatalyzed," first-order decomposition of $H_2O_2(aq)$ may actually be catalyzed by dust particles and/or trace amounts of ions (or even the walls of the container) that are almost impossible to remove from the solutions. If, in Worked Example 11.60, the apparent first order

---

rate constant is actually $k_1 = k_2$[trace impurity], what is the second order Arrhenius frequency factor for [trace impurity] = 10^{-4} M (approximately one part per million impurity)? Does the answer make sense for a second order reaction in solution? Explain why or why not.

For the examples we have chosen, combustion of wood (glucose) and decomposition of hydrogen peroxide, the free energy changes are so large and negative that the reaction at equilibrium (if it can be attained) greatly favors the products. For practical purposes, these reactions can be considered to go in the forward direction only. For other reactions, including many in living systems, the free energy changes are not so large and significant concentrations of both reactants and products are present at equilibrium (if it can be attained). Whether or not equilibrium is attained depends on the rates of the reactions. Note that a catalyst does not change the thermodynamics of a reaction. The enthalpy, entropy, and free energy changes for the overall reaction are the same in the catalyzed and uncatalyzed reaction, so equilibrium in a system is not affected by the presence of a catalyst, except that it is attained more quickly.

A **reversible reaction** takes the same pathway (in reverse) from products to reactants as from reactants to products. If this pathway can be characterized by activation energy diagrams like Figures 11.8 and 11.11(a), which apply to elementary reactions, then the activation energies for the forward, E_a(forward), and reverse, E_a(reverse) reactions can be related through $\Delta H_{\text{reaction}}$:

$$\Delta H_{\text{reaction}} = E_a(\text{forward}) - E_a(\text{reverse}) \tag{11.57}$$

This relationship, which is demonstrated in Figure 11.12, assumes (as we usually have) that energy and enthalpy changes are numerically equivalent for changes in these systems.

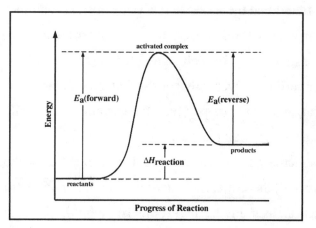

Figure 11.12. Relationship among energies of activation and enthalpy of reaction.

Equation (11.57) provides another connection between the thermodynamics and kinetics of a reaction. In some cases the temperature dependence of the forward and reverse reactions can be studied independently. The differences in activation energies in these cases are consistent with calorimetric measurements of the reaction enthalpies. Now we'll use our knowledge of rates and equilibrium to develop another connection between these two important concepts.

11.62. Consider This   *What more can activation energy diagrams tell us?*

(a) Is the reaction represented in Figure 11.12 exothermic or endothermic? Explain.

(b) Show that equation (11.57) is valid for this activation energy diagram.

(c) If the temperature is increased, what is the effect on the forward and reverse reactions? Which reaction is more affected? Explain the reasoning for your answers.

When the elementary reaction represented in Figure 11.12 attains equilibrium, the rates of the forward and reverse reactions must be equal, since the ratio of products to reactants, the equilibrium constant expression, is unchanging. For simplicity, let the forward and reverse rate laws be k_f[reactant] and k_r[product], respectively. Thus, at equilibrium:

$$k_f[\text{reactant}]_{eq} = k_r[\text{product}]_{eq} \tag{11.58}$$

Equation (11.58) can be rearranged to connect kinetic and thermodynamic constants:

$$\left(\frac{[\text{product}]}{[\text{reactant}]}\right)_{eq} = \frac{k_f}{k_r} = K_{eq} \tag{11.59}$$

The numeric value of the ratio of molar concentrations in equation (11.59) is equivalent to the numeric value of the thermodynamic K_{eq}.

Thus, for elementary reactions, the equilibrium constant can be expressed as the ratio of the forward to the reverse rate constant for the reaction.

In Section 11.6, we found that the increase in a rate constant with increase in temperature depends on the activation energy for the reaction — the larger the activation energy the greater the change in the rate constant for a given change in temperature. Let's apply this finding to the rate constants for the forward and reverse reactions of an elementary reaction system at equilibrium. Assume that, at a given temperature, the system is in equilibrium with the forward and reverse rates equal, as in equation (11.58).

If the temperature is increased, the rate constant for the reaction with the higher activation energy will change more and, therefore, the reaction with higher activation energy will get faster. For an endothermic reaction, as in Figure 11.12, this is the forward reaction, so, as temperature increases, the rate constant ratio will change (primes are the higher temperature values):

$$\frac{k_f{}'}{k_r{}'} > \frac{k_f}{k_r} \tag{11.60}$$

The new rate constants will mean different forward and reverse reaction rates, unless the product-to-reactant ratio changes.

> **11.63. Check This**   *Imbalance of forward and reverse reactions*
>
> **(a)** Show why the changed ratio in equation (11.60) leads to an imbalance of the forward and reverse rate when the temperature is increased for an exothermic system at equilibrium.
>
> **(b)** Show what the ratio would be and how the argument would go for an exothermic reaction, as in Figure 11.8.

In order to restore equilibrium with equal forward and reverse rates, some more reactant must go to product, so that the rates are again equal, $k_f'[\text{reactant}]_{eq}' = k_r'[\text{product}]_{eq}'$. Comparing equations (11.60) and (11.59), shows that we must have:

$$\left(\frac{[\text{product}]}{[\text{reactant}]}\right)_{eq}' > \left(\frac{[\text{product}]}{[\text{reactant}]}\right)_{eq} \qquad \text{and} \qquad K_{eq}' > K_{eq}$$

Thus, the kinetic analysis predicts that increasing the temperature of an endothermic reaction system at equilibrium will increase the numeric value of the equilibrium constant and increase the ratio of products to reactants.

Le Chatelier's principle predicts that an endothermic reaction at equilibrium will respond to a temperature increase (addition of energy) by going forward to use up some of the added energy and hence to turn more reactant into product. Thus, the kinetic prediction and Le Chatelier's prediction (which is consistent with equilibrium thermodynamics) agree and we find that kinetic and thermodynamic arguments are consistent with one another.

Section 11.9. Outcomes Review

In this final chapter, we tried to bring together much of what you already had learned in order to propose chemically reasonable pathways for chemical reactions. In addition, we introduced new concepts, based on measurements of reaction rates, to help in developing and understanding reaction pathways. We found that rates of reaction can be described in terms of rate laws that involve the concentrations of species in the reaction system and a proportionality constant, the rate constant. The species in the rate law are those that enter the reaction pathway before or during the rate-limiting step for the reaction.

The rate constants for reactions are temperature dependent and their increase with increasing temperature is described by the Arrhenius equation. The form of the Arrhenius equation is a consequence of the necessity for reactants to encounter one another in the right orientation and with enough energy to overcome an activation energy barrier and of the distribution of energy among encounters in the reacting system. Absorption of light by a reactant molecule is a way to initiate reactions that require more energy than is available thermally at room temperature.

Finally, we have to keep in mind that only reactions that are thermodynamically favorable, under the conditions specified, can occur spontaneously. Whether they will actually be observed to occur depends upon the kinetics of the reaction pathway(s) available.

Check your understanding of the ideas in this chapter by reviewing these expected outcomes of your study. You should be able to:

• predict the usual direction of the effects of changing concentration, changing temperature, or presence of catalysts on the rate of a reaction [Section 11.1].

• determine the initial rate of a reaction in units of $M \cdot s^{-1}$ from data for the concentration (or a property that is directly proportional to concentration, such as gas pressure or absorbance) of a reactant or product as a function of time [Section 11.2].

• use initial rate and concentration data to determine the order of a reaction with respect to the concentrations of species in the solution and write the rate law for the reaction [Section 11.3].

• explain in words and/or with drawings how the rate of a reaction depends only on the rate of the rate-limiting step in the reaction pathway [Section 11.4].

• derive the rate law for a reaction whose pathway (mechanism), including knowledge of the rate-limiting step, you are given [Sections 11.4 and 11.5].

• propose a pathway (mechanism) for a reaction that is similar or analogous to one whose pathway you know [Sections 11.4 and 11.5].

• use data for concentration (or a property that is directly proportional to concentration) of a reactant or product as a function of time for a first order reaction to determine the rate constant and half life for the reaction [Section 11.5].

• use rate constants and/or half lives for first order reactions to determine how long a sample has been reacting, given some known initial amount or the amount remaining after a known time [Section 11.5].

• explain what is meant by flooding a reaction and use rate and concentration data for a reaction that is flooded with respect to a species to find the order of reaction with respect to the concentration of that species and/or find out whether the disappearance of another species is first order [Section 11.5].

• use the temperature variation of the rate constant (or variables directly proportional to the rate constant, such as rates with the same concentrations of all species) to determine the activation energy for a reaction [Section 11.6].

- use rate constant and activation energy values to determine the Arrhenius frequency factor for a reaction [Section 11.6].

- construct and interpret an activation energy diagram, given information about the activation energy and enthalpy change for the reaction [Section 11.6].

- describe how encounters between reactants in solution differ from collisions between reactants in the gas phase [Section 11.6].

- describe the origin of the temperature dependence of the rate constant and why the dependence is so strong when the *average* energy of molecular encounters does not increase so rapidly [Section 11.6].

- explain how light can initiate reactions that would not otherwise occur at low (room) temperature [Section 11.7].

- describe how competing photochemical and thermal reactions can lead to a steady state concentration of a reactive species in a system [Section 11.7].

- explain why some reactions that are highly favored thermodynamically are not observed to occur [Section 11.8].

- describe factors that can be changed to provide favorable kinetics for spontaneous reactions that are not otherwise observable [Section 11.8].

- relate the temperature dependence of an equilibrium constant to the temperature dependences of the forward and reverse rate constants for the reaction [Section 11.8].

- describe the Michaelis-Menten pathway for enzyme-catalyzed reactions and use kinetic data to characterize these reactions, especially their behavior as a function of substrate and enzyme concentrations [Section 11.10].

- describe enzyme specificity in terms of functional and substrate specificity and relate specificity to the characteristics of the active sites of enzymes [Section 11.10].

Section 11.10.   EXTENSION — Enzymatic Catalysis

11.64. **Investigate This   How fast is the reaction of OH⁻(aq) with dissolved CO_2?**

 (a) Do this as a class investigation and work in small groups to discuss and analyze the results. Add 40 mL of ice-cold carbonated water and 2-3 drops of bromthymol blue indicator solution to a 50-mL erlenmeyer flask containing a magnetic stir bar. Record the color of the solution. Stir the mixture at moderate speed with the magnetic stirrer and quickly add 1 mL of 4 M aqueous sodium hydroxide, NaOH, solution to the stirred mixture. Record the time required for the solution to reach its final color.

(b) Repeat the procedure, but add 1 mL of a solution of the enzyme carbonic anhydrase just before the base is added. The amount of enzyme in 1 mL of the enzyme solution is about the same as in one drop (about 50 μL) of your blood.

11.65. Consider This   *What limits the rate of the reaction of OH⁻(aq) with dissolved CO_2?*

When carbon dioxide dissolves in water to give $CO_2(aq)$, some of it reacts with water to form carbonic acid:

$$CO_2(aq) + H_2O(l) \rightleftharpoons (HO)_2CO(aq) \tag{11.61}$$

Reaction (11.61) is reversible; at equilibrium, most of the CO_2 is present as $CO_2(aq)$. Hydroxide ion, $HO^-(aq)$, added to this solution reacts with the carbonic acid to give the hydrogen carbonate (bicarbonate) ion:

$$(HO)_2CO(aq) + OH^-(aq) \rightarrow H_2O(l) + HOCO_2^-(aq) \quad \text{(fast)} \tag{11.62}$$

(a) Water saturated with $CO_2(g)$ at one atmosphere pressure is approximately 0.1 M in total of $CO_2(aq)$ plus $(HO)_2CO(aq)$. How many moles of $CO_2(aq)$ plus $(HO)_2CO(aq)$ are present in the solutions you used in Investigate This 11.64?

(b) How many moles of $OH^-(aq)$ did you add to each solution in Investigate This 11.64? How do the numbers of moles of $OH^-(aq)$ added compare to the number of moles of $CO_2(aq)$ plus $(HO)_2CO(aq)$?

(c) Do your results suggest that reaction (11.61) in the forward direction is fast or slow? Explain the reasoning for your answer.

(d) What effect does addition of the enzyme have on the reaction? What reaction(s) is the enzyme affecting? Explain how you know.

Enzymes and rate laws

Enzymes, like other catalysts, have to interact with the reactants in order to affect the reaction rate. This is just like we've seen for $I^-(aq)$ in reaction (11.21). Therefore, we expect that the rate of an enzyme-catalyzed reaction like reaction (11.61), the hydration of dissolved carbon dioxide to produce carbonic acid, will depend upon the amount of enzyme present. Figure 11.13 shows plots of product formation as a function of time for different concentrations of enzyme in an enzyme-catalyzed reaction.

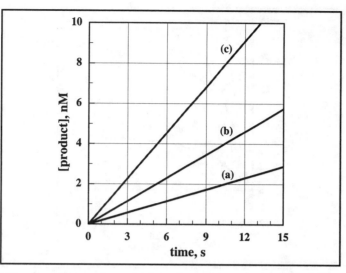

Figure 11.13. Product formation *vs.* time in an enzyme-catalyzed reaction.
The relative enzyme concentrations for experiments (a), (b), and (c) are 1, 2, and 4, respectively, with all other conditions the same.

The data in Figure 11.13 can be used to determine the initial rate of the reaction:

$$\text{initial rate} = V_0 = \left(\frac{\Delta(\text{product})}{\Delta t} \right)_{\text{initial}} \tag{11.63}$$

V_0 is the usual symbol for the initial rate of enzyme-catalyzed reactions. You will also see a lower case "v", instead of V_0, used for initial reaction rate (velocity).

11.66. Check This   *Partial rate law for a enzymatic reaction*

Use the data from Figure 11.13 to determine the order of the reaction with respect to the enzyme. Write the rate law that relates the initial rate of the reaction, V_0, to the enzyme concentration, [E]. Give your reasoning clearly.

Substrate concentration effects

In enzyme-catalyzed reactions, enzymes are almost always present at much lower concentrations than their **substrates**, the reactant compounds whose reactions they catalyze. For reasons that we will discuss below, enzymes are such efficient catalysts that it takes only tiny concentrations to catalyze the reactions necessary to maintain an organism. The consequences of low enzyme concentrations are kinetic results like those in Figure 11.14. Each of the two plots in the figure represents a series of initial rate determinations as the initial concentration of substrate, $[S]_0$ is varied. A different enzyme concentration is used in each series.

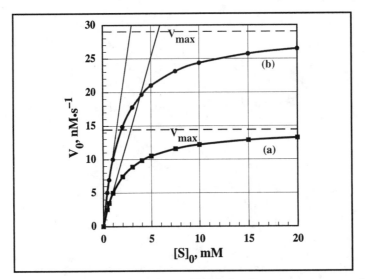

Figure 11.14. Initial rate, V_0, as a function of initial substrate concentration, $[S]_0$.
The enzyme concentration in the (b) series was double that in the (a) series of reactions.

The straight line from the origin through the first few points of each series in Figure 11.14 shows that, at low $[S]_0$, the rate increases linearly with $[S]_0$:

$$V_0 \propto [S]_0 \tag{11.64}$$

Thus, the reaction is first order with respect to substrate concentration at low $[S]_0$. At high $[S]_0$, the initial rates approach a constant value:

$$V_0 = V_{max} \tag{11.63}$$

The limiting rate, V_{max}, for each series of reactions is shown by the dashed horizontal lines in Figure 11.14. (At $[S]_0 = 50$ mM, the initial rates for each series are within 3% of their limiting values.) If a reaction reaches a limiting rate that does not increase with further increase in substrate concentration, we say that the reaction is **saturated**, that is, it is going as fast as it can with the concentration of enzyme present in the system. The data in Figure 11.14 show that the limiting rate, V_{max}, depends on the concentration of enzyme present in the reaction mixture.

11.67. Consider This *What are the initial and maximum rate laws for an enzyme reaction?*

   (a) When an enzymatic reaction is saturated, $V_0 = V_{max}$, what is the order of the reaction with respect to the substrate? Explain your reasoning. *Hint:* Review the discussion where reaction order was introduced in Section 11.3.

   (b) Use the data from Figure 11.14 to determine how the limiting rate, V_{max}, depends on the concentration of enzyme.

   (c) Use your results in part (a) and (b) to write a rate law for the rate of an enzymatic reaction at saturation, that is, the order with respect to enzyme and substrate concentrations.

(d) How does the initial rate of the reaction, V_0, depend on the concentration of enzyme at low substrate concentration? What is the rate law for the rate of an enzymatic reaction at low substrate concentration? That is, show the order with respect to enzyme and substrate concentrations. Explain your reasoning.

Catalysts other than enzymes can also show these saturation effects. We interpret saturation to mean that there are only a limited number of sites for catalysis available to the substrate. When the concentration of substrate is high enough, the sites are all occupied; the reaction of another substrate molecule can't be catalyzed until one of the sites becomes available. The situation is analogous to a supermarket with several checkout stations. When there are only a few shoppers (substrate) in the store, the rate at which shoppers leave with their purchases is proportional to the number of them in the store, since they will find an open checkout station (enzyme) whenever they are finished. When the number of shoppers is large, there will be lines at all checkout stations, and the rate at which shoppers leave will be constant and set by the number of checkout stations available. If more checkout stations are made available, the maximum rate at which shoppers leave will be greater, but will still reach saturation if a large enough number of shoppers is in the store.

Michaelis-Menten mechanism for enzyme catalysis

Between the two extremes at low and high substrate concentration, the initial-rate data in Figure 11.13 fall on a curve. A mechanism of enzyme action that describes the whole curve quantitatively was proposed in 1913 by two German chemists, Leonor Michaelis (1875-1949) and Maud Menten (1879-1960), who were studying the rate of hydrolysis of sucrose catalyzed by the yeast enzyme invertase:

> The names of enzymes almost all end with "-ase," as in cata*lase*, invert*ase*, and so on.

$$C_{12}H_{22}O_{11}(aq) + H_2O(l) \xrightarrow{\text{invertase}} C_6H_{12}O_6(aq) + C_6H_{12}O_6(aq) \qquad (11.66)$$

$$\text{sucrose} \qquad\qquad\qquad\qquad \text{glucose} \qquad \text{fructose}$$

Enzyme-catalyzed reactions are almost always discussed in terms of the Michaelis-Menten mechanism or some variation of it). Understanding the mechanism is important for understanding the control and direction of reactions in living cells.

Near the beginning of the 20th century several chemists postulated that an enzyme, E, and its substrate, S, formed an enzyme-substrate complex, E-S, prior to reaction. The **Michaelis-Menten mechanism** postulates that this first step is rapidly reversible and that the rate-limiting step in the reaction is the reaction of E-S to produce the reaction product, P. The simplest formulation of the Michaelis-Menten mechanism with all species in aqueous solution is:

$$E + S \rightleftharpoons E\text{-}S \qquad \text{(fast equilibrium with equilibrium constant} = K) \qquad (11.67)$$

$$E\text{-}S \rightarrow E + P \qquad \text{(slow, rate-determining step with rate constant} = k) \qquad (11.68)$$

The mechanism can be shown pictorially as:

$$(11.69)$$

Many enzyme reactions are reversible, that is, the enzyme can catalyze the reaction of P to give back S. The Michaelis-Menten mechanism applies in this simple form only to initial reaction rates when almost no P has been formed and we don't have to consider the reverse reaction.

The equilibrium constant expression, in terms of concentrations, for reaction (11.67) at the beginning of the reaction is:

$$K = \frac{[E\text{-}S]}{[E][S]_0} \qquad (11.70)$$

The sum of the concentrations of enzyme in both forms, [E] and [E-S], has to equal the total concentration added to the reaction, $[E]_{tot}$:

$$[E]_{tot} = [E] - [E\text{-}S] \qquad (11.71)$$

11.68. Check This   *Concentration of enzyme-substrate complex*

     Combine equations (11.70) and (11.71) and solve to get [E-S] in terms of K, $[S]_0$, and $[E]_{tot}$.

Since the rate of the reaction is determined by reaction (11.68), the rate-limiting step, we can write the initial rate of formation of product, P, as:

$$\frac{\Delta[P]}{\Delta t} = V_0 = k\,[E\text{-}S] \qquad (11.72)$$

We want rate laws for observed changes expressed in terms of measurable quantities and we usually can't measure [E-S] experimentally. However, we can substitute your result for [E-S] from Check This 11.68 into (11.72), to get:

$$V_0 = k \left(\frac{[E]_{tot}[S]_0}{\left(\frac{1}{K}\right) + [S]_0} \right) \qquad (11.73)$$

Limiting cases for the Michaelis-Menten mechanism

     Equation (11.73) looks a bit complicated, but let's look at the limiting cases, low $[S]_0$ and high $[S]_0$. When we run the reactions at lower and lower $[S]_0$, we reach a point where $[S]_0 \ll 1/K$. For this low $[S]_0$ case, $1/K + [S]_0 \approx 1/K$ and the initial rate equation (11.73) becomes:

$$V_0 \text{ (low } [S]_0) \approx k \left(\frac{[E]_{tot}[S]_0}{1/K} \right) = k \, K \, [E]_{tot}[S]_0 \qquad (11.74)$$

If the amount of enzyme is held constant, all the factors on the right-hand side of equation (11.74) are constant except $[S]_0$. The initial rate of reaction increases linearly with $[S]_0$, as shown by the straight lines through the low $[S]_0$ points in Figure 11.14. The slopes of those lines are directly proportional to the concentration of enzyme in the reaction, as equation (11.74) predicts and you found in Consider This 11.67(d).

If we run our reactions at very high $[S]_0$, then $[S]_0 \gg 1/K$. For this high $[S]_0$ case, $1/K + [S]_0 \approx [S]_0$ and equation (11.73) becomes:

$$V_0 \text{ (high } [S]_0) \approx k \left(\frac{[E]_{tot}[S]_0}{[S]_0} \right) = k \, [E]_{tot} \qquad (11.75)$$

For constant amounts of enzyme, both factors on the right-hand side of (11.75) are constant, so the initial rate is a constant (directly proportional to the concentration of enzyme used):

$$V_0 \text{ (high } [S]_0) = V_{max} = k \, [E]_{tot} \qquad (11.76)$$

11.69. Consider This   *How do you interpret enzyme reaction rate data?*

(a) What initial substrate concentration conditions, low $[S]_0$ or high $[S]_0$, do you think were used in the enzyme-catalyzed reactions whose results are shown in Figure 11.13? Explain the reasoning for your answer.

(b) Which rate law, equation (11.74) or (11.75), is consistent with the result you got in Consider This 11.66? State your reasoning clearly.

Determining and interpreting K

Now we have a molecular level interpretation of the experimental results represented by Figures 11.13 and 11.14. The Michaelis-Menten mechanism and rate law, expression (11.73), provide a way to categorize enzyme-substrate interactions by their values of $1/K$. In equation (11.73), substitute V_{max} for $k \, [E]_{tot}$ [from equation (11.76)]:

$$V_0 = \frac{V_{max}[S]_0}{\left(1/K \right) + [S]_0} \qquad (11.77)$$

Now consider the case when V_0 has reached half its maximum value, $V_0 = 1/2 V_{max}$:

$$1/2 V_{max} = \frac{V_{max}[S]_{01/2}}{\left(1/K \right) + [S]_{01/2}} \qquad (11.78)$$

$[S]_{01/2}$ reminds you that the conditions are for $V_0 = \frac{1}{2}V_{max}$. Divide through both sides of equation (11.78) by V_{max} and rearrange the result to give:

$$\frac{1}{K} = [S]_{01/2} \tag{11.79}$$

You can determine $\frac{1}{K}$ or K from an experimental value for $[S]_{01/2}$.

11.70. Consider This   *What is K for the enzyme data in Figure 11.13?*

   (a) What is K for the enzyme and substrate whose initial rate data are shown in Figure 11.14? Explain how you get your answer.

   (b) Are your units for K (expressed in concentrations) appropriate for equation (11.67)? Explain why or why not.

To see how to interpret values of K, return to reaction (11.67) and its equilibrium constant expression (11.70). Reaction (11.67) is the *association* of the enzyme and substrate to give the enzyme-substrate complex. The larger the equilibrium constant, K, for reaction (11.67), the larger the amount of association. Another way to think about this association is the larger the equilibrium constant, the more tightly the enzyme and substrate are bound in their complex and the smaller the $[S]_{01/2}$ required to half saturate the reaction, as shown in Figure 11.15. Tight binding is usually a result of good *fit* of the shape and polarity of the substrate with the complementary shape and polarity of the catalytic site of the enzyme, which is discussed below.

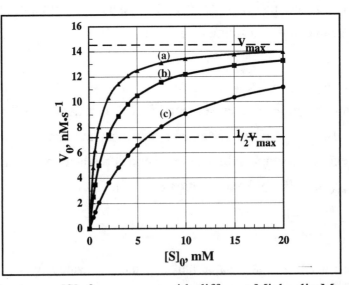

Figure 11.15. Initial rates *vs.* $[S]_0$ for enzymes with different Michaelis-Menten K values.
Enzymes (a), (b), and (c) were reacted with their substrates under conditions that give the same V_{max}, values.

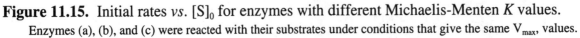

11.71. **Worked Example**   *Michaelis-Menten K values*

What are the Michaelis-Menten K values for the three enzymes, (a), (b), and (c), for which data are shown in Figure 11.15.

Necessary information: We need equation (11.79) and the data in the figure.

Strategy: Find $[S]_{01/2}$ (= $[S]_0$ at $^1/_2V_{max}$) for each enzyme and then take the inverse to get K.

Implementation: The $^1/_2V_{max}$ rate is shown by a dashed line in the figure. The corresponding values of $[S]_{01/2}$ are about 0.9 (a little less than 1), 2, and 6 mM for enzymes (a), (b), and (c), respectively. The K values are 1.1, 0.5, and 0.2 mM^{-1}, for enzymes (a), (b), and (c), respectively.

Does the answer make sense? Answer this question in Check This 11.72.

11.72. **Check This**   *Tightness of substrate binding*

Which of the three enzymes represented in Figure 11.15 binds its substrate most tightly? Explain your reasoning.

Enzyme specificity

Also related to the fit of enzyme and substrate is the specificity of enzyme catalysis. There are two ways that enzymatic catalysis is specific. **Functional specificity** means that an enzyme usually catalyzes only a specific reaction (or class of reactions). Catalase, for example, catalyzes the decomposition $H_2O_2(aq)$ to give $O_2(g)$ and H_2O, reaction (11.3) and invertase catalyzes the hydrolysis of sucrose to fructose and glucose, reaction (11.66), but neither enzyme catalyzes other kinds of reactions. Enzymatic catalysis also shows **substrate specificity**, that is, most enzymes can accommodate only one or a very few specific substrates. For example, catalase decomposes only $H_2O_2(aq)$ and urease catalyzes only urea hydrolysis:

$$NH_2CONH_2(aq) + 2H_2O \xrightarrow{\text{urease}} 2NH_4^+(aq) + CO_3^{2-}(aq) \qquad (11.80)$$

Other enzymes can be less specific and accommodate several similar substrates but the range is generally fairly limited.

Enzyme active sites

The **active site** of an enzyme is the region where the interactions between the enzyme and the substrate catalyze the reaction of the substrate(s). The structure of the active site is the basis for both the functional and substrate specificity. Many detailed structures for enzymes and their active sites are known, mostly by x-ray diffraction analysis. A representation of the structure of catalase is shown as part of the chapter opening illustration. The details of the structures are complex; the pictures we need are simpler schematic diagrams, such as Figure 11.16, that make clearer the kinds of interactions that are important for functional and substrate specificity.

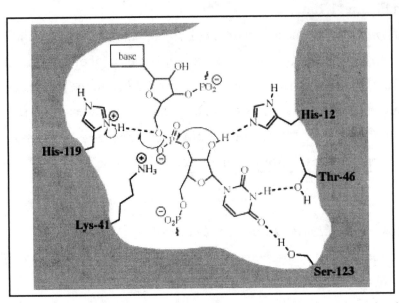

Figure 11.16. Representation of substrate binding and active site interactions in ribonuclease. The solid blue represents the interior surface of the active site of the enzyme. Five of the amino acid side groups that are essential for binding and catalysis in the active site are shown. The abbreviations stand for histidine, lysine, serine, and threonine and the numbers are the positions of the amino acids along the 124 amino-acid protein chain.

All active sites are in clefts or folds of the overall protein structure, as we have tried to represent in Figure 11.16, not on the surface of the protein. The substrate and other species involved in the reaction have to move into the cleft, in order for the catalyzed reaction to occur. Within the active site cleft are amino-acid side groups

> We sometimes say that an enzyme "recognizes" its substrate(s). Be cautious using terms like this that give cognitive properties to the enzyme. Recognition between enzyme and substrate requires a matching of favorable interactions and is governed by the random motions of the molecules with respect to one another.

that interact with functional groups on the reacting substrates. An example is shown at the lower right in Figure 11.16 for the interaction of one of the ribonucleic acid, RNA, bases with polar groups in the active site of ribonuclease, an enzyme that cuts up RNA. Other places on the surface of the cleft have non-polar side groups that interact with non-polar (hydrophobic) parts of the substrate structure. The combination of these interactions holds the substrate and other reactants in a particular orientation that is favorable for reaction.

To simplify our illustrations, we often represent active-site interactions two-dimensionally, as in Figure 11.16, but remember that the site is three-dimensional and its shape is very important in accommodating substrates. Only certain substrates will have a structure that can be bound by the arrangement of binding positions in the active site.

The arrangement of amino-acid side groups is also responsible for catalyzing the reaction, usually by disturbing the electronic distribution of the reactants in ways that make them more susceptible to reaction. For ribonuclease in Figure 11.16 folding of the protein chain brings amino acids from near both ends of the chain together as part of the active site. When pairs of electrons move as shown by the three curved arrows, a P–O bond is broken and the phosphodiester backbone of the ribonucleic acid molecule (in red) is broken. This is the function of the enzyme.

Within or very near the active site, many enzymes have non-protein groups that help bring about electronic reorganizations favorable to product formation. These non-protein groups include complexed metal ions (such as zinc ion in carbonic anhydrase), metal-containing groups (such as heme with a bound metal ion in catalase), or other relatively small molecules (such as vitamins) that are part of the catalytic pathway.

Catalase activity and the bombardier beetle

Catalase is one of the most active enzymes known. A mole of catalase active sites can decompose about 10^7 mole of $H_2O_2(aq)$ in one second. The ratio of moles of substrate reacted per second to moles of enzyme active sites is called the **turnover number** for the enzyme.

11.73. **Check This**   *Catalase activity in one millisecond*

How many moles of $H_2O_2(aq)$ can be decomposed by one mole of catalase active sites in one millisecond, 10^{-3} s? Explain.

There are many species of bombardier beetle, which vary in the amount of defensive spray they discharge. For the beetle shown in the chapter opening illustration, about 0.4 μL (1 μL = 10^{-6} L) of 7.4 M $H_2O_2(aq)$ solution, Check This 11.17, reacts for one burst of spray. The number of moles of $H_2O_2(aq)$ that react is: $(0.4 \times 10^{-6} \text{ L·burst}^{-1})\cdot(7.4 \text{ M}) = 3 \times 10^{-6}$ mol·burst^{-1}.

11.74. **Worked Example**   *Moles of catalase required by a bombardier beetle*

How many moles of catalase active sites are required to decompose 3×10^{-6} mol $H_2O_2(aq)$ in one millisecond (one burst from the beetle)?

Necessary information: We need your result from Check This 11.73; in one millisecond, one mole of enzyme active sites can decompose 10^4 mol of $H_2O_2(aq)$.

Strategy: Use the ratio of moles of active sites to moles of $H_2O_2(aq)$ decomposed in one millisecond (burst) to convert the moles of $H_2O_2(aq)$ decomposed per burst to moles of active sites required.

Implementation:

$$(3 \times 10^{-6} \text{ mol } H_2O_{2}(aq)\cdot\text{burst}^{-1})\left(\frac{1 \text{ mol active sites}}{10^4 \text{ mol } H_2O_2(aq)\cdot\text{burst}^{-1}}\right) = 3 \times 10^{-10} \text{ mol active sites}$$

Does the answer make sense? The turnover number for the enzyme is large and the amount of substrate, $H_2O_2(aq)$, that reacts is small, so it makes sense that only a small amount of enzyme will be required.

11.75. Check This   *Mass of catalase required by a bombardier beetle*

(a) The catalase structure shown in the chapter opening illustration has one active site in a protein with a molar mass of about 60,000 gram. Assuming catalase is the enzyme in the reaction chamber of the beetle, use the information from Worked Example 11.74 to determine the mass of the protein required by the beetle. Explain.

(b) Is it reasonable for a beetle with a mass of a few grams to contain the mass of protein you calculated in part (a)? Explain the reasoning for your answer.

As you have found throughout this book, chemistry and chemical phenomena are part of everything you do and part of everything around you, including the amazing workings of an insect like the bombardier beetle. We hope you will continue to use the chemistry you have learned to help you better understand the world and its workings.

Index of Terms

Chapter 11 Problems

Section 11.1. Pathways of Change

11.1. Suggest two examples of reactions not included in this chapter that are very slow, that require days or years or longer to complete. Suggest two examples of reactions not included in this chapter that are very fast, that are complete in minutes or seconds.

11.2. In Investigate This 11.1 you investigated the time it took the indicator color to fade when phenolphthalein is added to a solution of sodium hydroxide.

(a) Would you predict the color to fade more or less quickly for a 2.0 M solution compared to a 1.0 M solution of sodium hydroxide? Explain your reasoning.

(b) Would you predict the color to fade more or less quickly for a 0.10 M solution compared to a 1.0 M solution of sodium hydroxide? Explain your reasoning.

11.3. Examine the structures from equation 11.1:

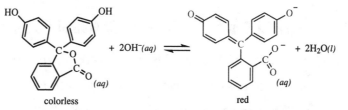

What differences do you seen in the structures before and after adding OH$^-$(aq)? What kind of reaction (see Chapter 6) is this? Explain your choice.

11.4. What do we mean when we talk about the rate of a chemical reaction?

11.5. The bombardier beetle shown in the chapter opening illustration is tethered to a stiff wire by a blob of wax on its back. In the stroboscopic study, the tethered beetle is touched with a pair of forceps (the white wedge in each picture) in order to induce it to discharge. The discharge appears as clusters of white dots in three or four successive pictures.

(a) What is the time between discharges? What is the rate of discharge (discharges·sec^{-1})? Explain your reasoning.

(b) The normal human ear is sensitive to vibrations (sounds) in the range from about 20 to 20000 cycles per second. Could the rate of discharge of the beetle be responsible for the sound that accompanies its discharge. Explain.

11.6. In Investigate This 11.1, why does it take longer for the phenolphthalein to go to colorless than it does for the phenolphthalein to turn red? Is this explained by the structures of the two forms of phenolphthalein?

11.7. Consider this reaction beginning with AgCl(s) and then adding the NH$_3(aq)$ solution:

$$AgCl(s) + 2NH_3(aq) \rightleftharpoons Ag(NH_3)_2^+(aq) + Cl^-(aq)$$

    (a) What happens to the concentration of ammonia over time?

    (b) What happens to the amount of solid silver chloride over time?

    (c) What happens to the amount of the silver diamine, Ag(NH$_3$)$_2^+$, ion over time?

11.8. Chemical reactions often occur faster at higher temperatures. What happens to molecules as their temperature increases? Does this help explain why reactions occur faster at a higher temperature? Explain why or why not.

11.9. Investigate This 11.6 showed that some substances could help catalyze, speed up, the decomposition of hydrogen peroxide, H$_2$O$_2$. Investigate This 11.1 showed that reactant concentration could affect the rate of reaction. How could you test to see if a substance, either a reactant or a potential catalyst, changed the rate of a reaction? Describe an experiment that you might conduct.

Section 11.2. Measuring and Expressing Rates of Chemical Change

11.10. Consider a solution reaction that bubbles as the solution goes from clear and colorless to clear blue. Suggest three possible ways you might measure the rate of the process going on in the solution. Explain how your choices are related to the rate.

11.11. Consider the reaction: N$_2(g)$ + 3H$_2(g)$ → 2NH$_3(g)$

    (a) Sketch a concentration *vs.* time graph for how you predict the concentrations of N$_2$, H$_2$, and NH$_3$ change over time, beginning with a mixture of N$_2$ and H$_2$.

    (b) Write an expression for the rate of ammonia, NH$_3$, production in terms of gas pressure, P_{NH_3}, change. Use equation 11.4 to guide you.

    (c) Write an expression for the rate of ammonia production in terms of concentration, [NH$_3(g)$], change.

    (d) If N$_2$ is reacting at a rate of 0.15 M·min^{-1}, what is the rate of ammonia production in these same units?

    (e) If N$_2$ is reacting at a rate of 0.15 M·min^{-1}, what is the rate at which H$_2$ disappears?

    (f) If N$_2$ is reacting at a rate of 0.15 M·min^{-1}, what is the rate of the reaction written above? Explain.

11.12. Graph these data collected for a reaction that can be symbolized as: $A \rightarrow 2B$

time, s	[A], mol·L^{-1}
0.0	1.000
10.0	0.833
20.0	0.714
30.0	0.625
40.0	0.555

(a) Calculate the rate of disappearance of **A** at 10.0, 20.0, and 30.0 seconds by approximating the slope of a tangent line at each of these times.

(b) What happens to the rate in part (a) over time? Is this what you might have expected for this reaction? Explain your reasoning.

(c) Calculate the average rate of disappearance of **A** during the time period from 0.0 to 10.0, 10.0 to 20.0, 20.0 to 30.0, and 30.0 to 40.0 seconds.

(d) How do the average rates in part (c) compare to the instantaneous rates you calculated in part (a)? Is there any pattern to the comparisons? Explain.

11.13. Under suitable conditions, the reaction, $CH_3OH + H^+ + Cl^- \rightarrow CH_3Cl + H_2O$, occurs in a solution of methanol and hydrochloric acid. The progress of the reaction can be followed by determining the concentration of hydronium ion, [H$^+$], as a function of reaction time.

time, min	[H$^+$], M
0	2.12
90	1.95
200	1.74
360	1.54
720	1.19

(a) Graph these data and calculate the average rate of disappearance of H$^+$ for each time interval.

(b) What do your graph and the average rates in part (a) tell you about the rate of the reaction as a function of time? Explain.

(c) On the same graph as in part (a), plot the concentration of chloromethane, [CH$_3$Cl$_{(aq)}$], as a function of time in this reaction. Explain your reasoning.

11.14. **(a)** For each of these reactions, write the rate of reaction expressed in terms of the changes in pressure for each reactant and product in the reaction.

       (i) $2NO_2(g) \rightarrow NO(g) + NO_3(g)$

       (ii) $2NOBr(g) \rightarrow 2NO(g) + Br_2(g)$

       (iii) $2N_2O_5(g) \rightarrow 4NO_2(g) + O_2(g)$

(b) For each reaction, what is the relationship of the rate of formation of each product to the rate of decomposition of the reactant? How, if at all, are these relationships connected to the reaction rate expressions you wrote in part (a)? Explain your reasoning.

11.15. 2-chloro-2-methylpropane, $(CH_3)_3CCl$, reacts in a solution of water and acetone to give 2-methyl-2-propanol, $(CH_3)_3COH$. When a small amount of hydroxide ion, OH^-, is present in the solution, we can write the overall initial reaction as:

$$(CH_3)_3CCl + OH^- \rightarrow (CH_3)_3COH + Cl^-$$

In a series of experiments, this reaction was followed by adding bromphenol blue acid-base indicator and timing how long it took for the indicator to change color (blue to yellow) showing that all the hydroxide ion had reacted. The results for different initial concentrations of hydroxide ion, $[OH^-]_0$, with all other conditions the same were:

expt #	$[OH^-]_0$, M	time to color change, s
1	0.0025	35
2	0.0030	43
3	0.0015	22

(a) What is the rate of disappearance of the hydroxide ion for each of the experiments? Explain your reasoning.

(b) What is the rate of disappearance of the 2-chloro-2-methylpropane, $(CH_3)_3CCl$, in each of the experiments? What do these results suggest about the dependence of the reaction rate on the concentration of hydroxide ion in the solution? Explain your reasoning.

11.16. **(a)** If the rate of formation of oxygen from ozone, $2O_3(g) \rightarrow 3O_2(g)$, is 1.8×10^{-3} M·s^{-1} at a certain temperature, what is the rate of decomposition of ozone under these conditions? Explain your reasoning.

(b) What is the reaction rate for this reaction? Explain.

11.17.   WEB   Chap 11, Sect 11.2.6. To measure the initial rate of the reaction of iodide ion, $I^-(aq)$, with persulfate ion, $S_2O_8^{2-}(aq)$, a small amount of thiosulfate ion, $S_2O_3^{2-}(aq)$, is added to the reaction mixture. See the "Summary of the Reactions" on this page of the *Web Companion* to see why the solution remains clear and colorless until all the thiosulfate has reacted and then suddenly changes color.

(a) This reaction with added thiosulfate is like the changes in the Blue-Bottle reaction, Investigate This 10.61, Chapter 10, Section 10.8, which remains blue for a time and then suddenly changes to colorless. The coupled reactions responsible for the observations on the Blue-Bottle reaction are shown in Figure 10.12. Make a sketch, patterned after Figure 10.12, that shows the coupling of reactions in the iodide-persulfate-thiosulfate system.

(b) What is the coupling species in the iodide-persulfate-thiosulfate system? What reactions are being coupled?

(c) What are the relative rates of the coupled reactions in part (b)? What is the experimental evidence and how do you reason from it to get your response?

(d) During the time the reaction mixture remains colorless, how does the concentration of iodide ion compare with its initial concentration? Explain.

(e) During the time the reaction mixture remains colorless, there is no *net* formation of iodine, $I_2(aq)$. However, the plot shown in the *Web Companion* is labeled with a $\Delta[I_2]$ during this time period, Δt. Why? To what change does this $\Delta[I_2]$ correspond? Explain.

(f) As you consider the points in part (e), do you think the graph might be somewhat misleading? Explain why or why not and, if it's misleading, suggest a way to fix it.

11.18. This reaction, $2CO(g) \rightarrow CO_2(g) + C(s)$, which occurs at high temperatures, can be studied by adding $CO(g)$ to a hot reaction vessel and following the decrease in total pressure, P_{total}, of the gases in the constant volume reactor

time, s	P_{total}, kPa
0	33.2
400	31.6
1000	29.8
1800	27.9

(a) Calculate the average rate of disappearance of $CO(g)$ in $kPa \cdot s^{-1}$ for each time interval. Explain your reasoning. *Hint:* P_{total} is proportional to the number of moles of gas.

(b) Calculate the average rate of appearance of $CO_2(g)$ for each time interval. Explain.

Section 11.3. Reaction Rate Laws

11.19. If a reaction is second order in the concentration of reactant **A** and first order in the concentration of reactant **B**, how will the rate of reaction change if the concentrations of **A** and **B** are both halved? Explain your reasoning.

11.20. For the reaction, $CH_3Br(aq) + OH^-(aq) \rightarrow CH_3OH(aq) + Br^-(aq)$, the initial rate of reaction was found to decrease by a factor of two when the concentration of hydroxide ion, $[OH^-(aq)]$, was halved. An increase in the concentration of bromomethane, $[CH_3Br(aq)]$, by a factor of 1.4 increased the rate by a factor of 1.4.

(a) What is the order of this reaction with respect to $[OH^-(aq)]$? Explain your reasoning.

(b) What is the order of this reaction with respect to $[CH_3Br(aq)]$? Explain.

(c) What is the overall order of this reaction? Write the rate law for this reaction. Explain your reasoning.

11.21. Consider this gas-phase reaction: $2NO(g) + Cl_2(g) \rightarrow 2NOCl(g)$. Studies show that the initial rate of disappearance of $NO(g)$ is eight times faster when the concentrations of both reactants are doubled and twice as fast when the concentration of chlorine alone is doubled.

(a) What is the order of this reaction with respect to $[Cl_2(g)]$? Explain your reasoning.

(b) What is the order of this reaction with respect to $[NO(g)]$? Explain.

(c) What is the overall order of this reaction? Write the rate law for this reaction. Explain your reasoning.

11.22. The rate of a certain reaction is proportional to the concentration of a reactant and the concentration of a catalyst. During experimental runs on this reaction, the concentration of the catalyst remains constant and the measured initial rates of reaction can be used to determine a rate constant, 8.9×10^{-5} s^{-1}, for the first order disappearance of the reactant.

(a) If the catalyst concentration is 5.0×10^{-3} M, what is the true rate constant for the overall second order reaction? Explain your reasoning.

(b) What would be the observed initial rate of reaction if the reactant concentration is 0.12 M and the catalyst concentration is 3.5×10^{-3} M? Explain.

11.23. The initial rate of a reaction that is first order in each of two reactants was found to be 7.65×10^{-4} M·s^{-1}, when the initial concentration of each reactant was 0.050 M. What is the numeric value of the rate constant for this reaction? Explain your solution.

11.24. The reaction, $2NO(g) + O_2(g) \rightarrow 2NO_2(g)$, was studied by following the initial rate of disappearance of nitric oxide, $-(\Delta NO(g)/\Delta t)_0$, for different initial reactant concentrations.

expt #	$[NO(g)]_0$, M	$[O_2(g)]_0$, M	$-(\Delta NO(g)/\Delta t)_0$, $M \cdot s^{-1}$
1	0.0125	0.0183	0.0202
2	0.0250	0.0183	0.0803
3	0.0125	0.0370	0.0409

(a) What is the order of the reaction with respect to $[NO(g)]$? Explain your reasoning.

(b) What is the order of the reaction with respect to $[O_2(g)]_0$? Explain your reasoning.

(c) What is the reaction rate (in $M \cdot s^{-1}$) for this reaction in each experiment? Explain.

(d) Write the rate law for the reaction and determine the numerical value of the reaction rate constant. Explain your approach and the units of the rate constant.

(e) If the reaction were carried out with $[NO(g)]_0 = [O_2(g)]_0 = 0.0200$ M, what should be the observed rate of disappearance of nitric oxide? Explain.

11.25. In Problem 11.15 some experimental details and data were given for a study of this net reaction: $(CH_3)_3CCl + OH^- \rightarrow (CH_3)_3COH + Cl^-$. Here are more data from this study for experiments in which the initial concentration of hydroxide ion, $[OH^-]_0$, was the same, 0.0025 M, and the initial concentration of 2-chloro-2-methylpropane, $[(CH_3)_3CCl]_0$, was varied.

expt #	$[(CH_3)_3CCl]_0$, M	time to color change, s
1	0.015	35
4	0.030	18

(a) What is the rate of disappearance of the hydroxide ion for each of the experiments? Explain your reasoning.

(b) What is the rate of disappearance of the 2-chloro-2-methylpropane, $(CH_3)_3CCl$, in each of the experiments? What do these results suggest about the dependence of the reaction rate on the concentration of 2-chloro-2-methylpropane, $[(CH_3)_3CCl]$, in the solution? Explain your reasoning.

(c) Based on your results from part (b) and from Problem 11.15, write a rate law for this reaction. Explain your reasoning.

(d) What is the numerical value of the rate constant in the rate law you wrote in part (c)? Explain.

11.26. WEB Chap 11, Sect 11.3.3-6. In Check This 11.22, you worked through these pages of the *Web Companion* to find the rate law for this reaction:

$$S_2O_8^{2-}(aq) + 2I^-(aq) \rightarrow I_2(aq) + 2SO_4^{2-}(aq)$$

(a) Can you use the data given on these pages of the *Web Companion* to determine the numerical value for the reaction rate constant in the rate law? If so, what is the value? If not, explain clearly why not.

(b) If you knew that each of the samples in these experiments was initially 0.0015 M in thiosulfate ion, $S_2O_3^{2-}(aq)$, would your answer to part (a) be different? If so, explain specifically how it would be different. If not, explain why not.

11.27. In aqueous solution, permanganate ion, $MnO_4^-(aq)$, reacts with chromium(III) ion, $Cr^{3+}(aq)$: $MnO_4^-(aq) + Cr^{3+}(aq) \rightarrow Mn^{4+}(aq) + CrO_4^{2-}(aq)$. The reaction rate can be studied by measuring the time required for the concentration of $CrO_4^{2-}(aq)$ to increase from zero to 0.020 M. The results for different initial concentrations of the two reactants are:

expt #	relative $[MnO_4^-(aq)]_0$	relative $[Cr^{3+}(aq)]_0$	time to $[CrO_4^{2-}(aq)] = 0.020$ M
1	1	1	23 min
2	2	1	11 min
3	1	0.5	45 min

(a) What type of reaction (see Chapter 6) is this? Explain how you know.

(b) How are the times in the table related to the rates of the reaction? Explain.

(c) What is the reaction order with respect to the concentration of each of the reactants? How do your know?

(d) Write the rate law for the reaction.

(e) In experiment #3, if the relative $[MnO_4^-(aq)]_0$ had also been changed to 0.5, how long would it have taken for $[CrO_4^{2-}(aq)]$ to reach 0.020 M. Explain your reasoning.

11.28. Ammonia decomposes on a hot tungsten filament: $2NH_3(g) \rightarrow N_2(g) + 3H_2(g)$. The reaction can be followed by measuring the increase in pressure, ΔP, in the system.

time, s	0	100	200	400	600	800	1000
ΔP, kPa	0	1.46	2.94	5.85	8.81	11.68	14.61

(a) Plot these data and try to deduce the order of reaction with respect to the pressure (or concentration) of ammonia. State the reasoning for your answer.

(b) The initial pressure of ammonia in the system was 26.00 kPa. What is the initial rate of reaction in kPa·s^{-1}? What is the rate constant for the reaction? Explain your reasoning.

11.29. In a series of experiments, the decomposition of hydrogen peroxide, $H_2O_2(aq)$, to water and oxygen gas, $2H_2O_2(aq) \rightarrow H_2O(l) + O_2(g)$, was studied using a titration method to determine the concentration of reactant left after 30 minutes of reaction, $[H_2O_2(aq)]_{30}$. The data for various starting concentrations, $[H_2O_2(aq)]_0$ are given in this table.

expt #	$H_2O_2(aq)]_0$, M	$[H_2O_2(aq)]_{30}$, M
1	0.874	0.812
2	0.356	0.331
3	0.589	0.549

(a) What is the average rate of disappearance of hydrogen peroxide in each of these experiments? Explain.

(b) What is the order of the reaction with respect to $[H_2O_2(aq)]$? Explain your reasoning.

(c) Write the rate law for this reaction. Show how to calculate the rate of reaction for each experiment and the rate constant for the reaction.

(d) If another experiment was done under the same conditions, but starting with 0.234 M hydrogen peroxide, what do you predict would be the concentration of unreacted peroxide after 30 minutes of reaction? Explain.

11.30. In aqueous acidic solution, hydrogen peroxide oxidizes iodide ions to elemental iodine and is, itself, reduced to water. In separate experiments, with all other conditions the same, doubling the initial hydrogen peroxide concentration or doubling the iodide ion concentration caused the initial rate of formation of iodine to double. The rate of formation of iodine was four times higher at pH 1.4 compared to pH 2.0, with all other conditions the same.

(a) Write the balanced equation for this redox reaction.

(b) What are the orders of reaction with respect to $[H_2O_2(aq)]$, $[I^-(aq)]$, and $[H^+(aq)]$? Write the rate law for this reaction. Explain how the rate data lead to these conclusions.

(c) If, under particular conditions, the rate of formation of iodine is 2.5×10^{-3} M·s^{-1}, what is the rate of disappearance of hydrogen peroxide? of iodide? Explain.

(d) What affect would increasing the pH have on the rate constant for the reaction? Explain your answer.

(e) If an initial reaction solution is diluted by a factor of two by addition of water, what is the affect on the rate constant for the reaction? How is the rate of the reaction affected? Explain the reasoning for your answers.

11.31. The decomposition reaction for azomethane is: $CH_3NNCH_3(g) \rightarrow CH_3CH_3(g) + N_2(g)$.

(a) If the initial pressure of azomethane in a reaction vessel is 10.0 kPa, what will be the total pressure in the vessel after one-quarter of the azomethane has decomposed? *Hint:* How will the total number of moles in the flask change if x moles of azomethane decompose? How is the total pressure in the vessel related to the total number of moles?

(b) Azomethane was introduced into a reaction vessel at 300 °C. The initial pressure of azomethane was 20.0 kPa and after 2.00 minutes the pressure in the vessel had increased to 21.0 kPa. What fraction of the azomethane had decomposed in this time. What was the initial rate of decomposition (in $kPa·s^{-1}$)? Explain.

(c) In another experiment at 300 °C, the initial pressure of azomethane was 5.00 kPa and the total pressure in the vessel reached 5.25 kPa after 2.00 minutes. What was the initial rate of decomposition of azomethane in this case? How does rate compare to what you found in part (b)? Show how to use your results to determine the order of the reaction with respect to azomethane. Write the rate law for the reaction and calculate the numeric value of the rate constant. Explain your approach.

Section 11.4. Reaction Pathways or Mechanisms

11.32. In Worked Example 11.31, we stated that "the sum of the series of reactions (11.25), (11.26), and (11.27) is the stoichiometric reaction (11.16)." Show that this statement is true.

11.33. Possible pathways for the reaction of nitrogen dioxide with carbon monoxide are:

    (i) $NO_2(g) + CO(g) \rightarrow NO(g) + CO_2(g)$ $k_{(i)}$

    (ii) $NO_2(g) + NO_2(g) \rightleftharpoons NO(g) + NO_3(g)$ fast equilibrium, $K_{(ii)}$

           $NO_3(g) + CO(g) \rightarrow NO_2(g) + CO_2(g)$ slow, $k_{(ii)}$

    (iii) $NO_2(g) + NO_2(g) \rightleftharpoons NO(g) + NO_3(g)$ slow equilibrium, $K_{(iii)}$

           $NO_3(g) + CO(g) \rightarrow NO_2(g) + CO_2(g)$ fast

(a) What is the net reaction for each pathway?

(b) Determine the rate law for the reaction predicted by each of the pathways. Explain your reasoning.

(c) What rate experiment(s) would you suggest performing to distinguish among these pathways? What would be the difference(s) in the predicted result(s) from your experiment(s) for each pathway?

11.34. Consider the gas phase reaction of nitric oxide with chlorine to form NOCl:

$$2NO_{(g)} + Cl_{2(g)} \rightleftharpoons 2NOCl_{(g)}$$

The experimentally-determined rate law for this reaction is: rate $=[NO_{(g)}]^2[Cl_{2(g)}]$. Although this rate law is consistent with a one-step, termolecular (three molecule) reaction, a simultaneous collision among three molecules is not very probable. An alternative pathway has been proposed for which the first step is:

$$NO_{(g)} + Cl_{2(g)} \rightleftharpoons NOCl_{2(g)} \qquad \text{fast, equilibrium, } K$$

(a) Propose a second, rate-limiting step that is consistent with the overall stoichiometry of the reaction.

(b) What overall rate law would be predicted for this two-step pathway? Explain.

(c) How might researchers distinguish experimentally between this alternative pathway and the one-step termolecular pathway?

11.35. In Problems 11.15 and 11.25 some experimental details and data were given for a study of this net reaction: $(CH_3)_3CCl + OH^- \rightarrow (CH_3)_3COH + Cl^-$. A proposed mechanism for this net reaction in a solvent mixture of acetone and water is:

$$(CH_3)_3CCl \rightleftharpoons (CH_3)_3C^+ + Cl^- \qquad \text{slow equilibrium, } K$$
$$(CH_3)_3C^+ + H_2O \rightarrow (CH_3)_3COH_2^+ \qquad \text{fast}$$
$$(CH_3)_3COH_2^+ + OH^- \rightarrow (CH_3)_3COH + H_2O \qquad \text{fast}$$

(a) Show that this mechanism gives the observed net reaction in the presence of hydroxide ion in solution.

(b) What is the reaction rate law predicted by this mechanism? Show your reasoning.

(c) Is the rate law you wrote in part (b) consistent with the one you wrote in Problem 11.25(c). Explain why or why not.

11.36. A possible mechanism for the reaction studied in Problem 11.13 is (the reactants and products are solvated by the mixed solvent, but the solvation is not indicated here):

$$CH_3OH + H^+ \rightleftharpoons CH_3OH_2^+ \qquad \text{fast equilibrium, } K$$
$$CH_3OH_2^+ + Cl^- \rightarrow CH_3Cl + H_2O \qquad \text{slow, } k$$

(a) What is the reaction rate law predicted by this mechanism? Explain your reasoning.

(b) When more data for this reaction, like those in Problem 11.13, were analyzed, the analysis showed that the reaction rate appeared to depend on the square of the hydronium ion concentration, $[H^+]^2$. When the hydrochloric acid, HCl, concentration was doubled, the rate of reaction quadrupled. Is this result consistent with the rate law you wrote in part (a)? If so, show how. If not, show why not.

11.37. Consider these three pathways proposed for the gas phase decomposition of dinitrogen pentoxide, $N_2O_5(g)$:

(i) $N_2O_5(g) \rightleftharpoons NO_2(g) + NO_3(g)$ fast equilibrium, $K_{(i)}$

       $NO_2(g) + NO_3(g) \rightarrow NO(g) + NO_2(g) + O_2(g)$ slow, $k_{(i)}$

       $NO(g) + NO_3(g) \rightarrow 2NO_2(g)$ fast

(ii) $N_2O_5(g) \rightleftharpoons NO_2(g) + NO_3(g)$ fast equilibrium, $K_{(ii)}$

       $NO_3(g) + N_2O_5(g) \rightarrow N_2O_4(g) + NO_2(g) + O_2(g)$ slow, $k_{(ii)}$

       $N_2O_4(g) \rightarrow 2NO_2(g)$ fast

(iii) $N_2O_5(g) \rightarrow NO_2(g) + NO_3(g)$ slow, $k_{(iii)}$

       $NO_3(g) + N_2O_5(g) \rightarrow 3NO_2(g) + O_2(g)$ fast

(a) What is the net reaction represented by each of these pathways? Explain your reasoning. *Hint:* Sometimes an elementary reaction has to occur more than once in order to satisfy the stoichiometry of the other elementary reaction(s).

(b) What is the rate law for $N_2O_5(g)$ decomposition predicted by each of these pathways? Explain.

(c) In a series of experiments, the initial rate of disappearance of $N_2O_5(g)$ was found to be a linear function of the initial concentration of $N_2O_5(g)$. What is the experimental rate law suggested by these results? Explain your reasoning.

(d) Does your rate law from part (c) allow you to rule out any of the proposed pathways? If so, which one(s) and why? If not, why not?

(e) What other experiments can you suggest that might help to distinguish among the proposed pathways? What distinguishing results would you expect?

11.38. In a continuing study of the system discussed in Problems 11.15, 11.25, and 11.35, a series of experiments was carried out in which salt, sodium chloride, was dissolved in reaction mixtures which were otherwise identical. The initial concentration of hydroxide ion, $[OH^-]_0$, was 0.0025 M in all these experiments.

expt #	[NaCl], M	time to color change, s
1	0	35
5	0.85	35
6	1.71	43
7	2.56	95
8	3.42	268

(a) What is the rate of disappearance of the hydroxide ion for each of the experiments? Explain your reasoning.

(b) What is the rate of disappearance of the 2-chloro-2-methylpropane, $(CH_3)_3CCl$, in each of the experiments? What is the affect of dissolved sodium chloride on the rate of this reaction? Explain your reasoning.

(c) What species in the solution do you think is having the affect you found in part (b)? Show how this affect provides evidence in support of the reaction mechanism in Problem 11.31. Be as specific as possible in your explanation.

11.39. In Sections 11.2. 11.3, and 11.4 we discussed the decomposition of hydrogen peroxide catalyzed by iodide ion in pH 7 solutions. In acidic solutions, as you found in Problem 11.30, the reaction is quite different: $H_2O_2(aq) + 2I^-(aq) + 2H^+(aq) \rightarrow I_2(aq) + 2H_2O(aq)$. The rate law for the reaction is:

$$\text{rate} = k[H_2O_2(aq)][I^-(aq)][H^+(aq)]$$

(a) Hydrogen peroxide is a weak Brønsted-Lowry base (like water). In acidic solution, what might be a reasonable reaction for $H_2O_2(aq)$ to undergo? Explain your reasoning. *Hint:* See Check This 11.32.

(b) Write the rate limiting step for reaction in acidic solution, assuming it is analogous to the rate limiting step in pH 7 solution. *Hint:* See Check This 11.32.

(c) We said that $E^{o'} = 0.5$ V provides a strong driving force for reaction (11.22) in pH 7 solution:

$$HOI(aq) + H_2O_2(aq) \rightarrow O_2(g) + I^-(aq) + H^+(aq) + H_2O(l)$$

What affect will a decrease in pH of the solution have on the cell potential for this reaction? What is the cell potential for this reaction at pH 0? *Hint:* Use Le Chatelier's principle and the Nernst equation.

(d) Use the data in Appendix YY to find the cell potential for this reaction at pH 0:

$$HOI(aq) + I^-(aq) + H^+(aq) \rightarrow I_2(aq) + H_2O(l)$$

(e) Although thermodynamically favorable reactions are not necessarily fast, if the reactions in parts (c) and (d) both can proceed, which is more favored at pH 0? Could this comparison explain the difference between the reactions of $H_2O_2(aq)$ with $I^-(aq)$ at pH 0 and pH 7? Show why or why not.

(f) Use the results from the preceding parts to suggest a reaction pathway for the reaction of $H_2O_2(aq)$ with $I^-(aq)$ at pH 0 that fits the rate law and the stoichiometry.

11.40. **(a)** You can interpret reactions (11.21) and (11.22) as a pair of coupled reactions. What pair of reactants serves to couple the two reactions? Explain.

(b) Make a sketch, patterned after Figures 10.12, 1014, and 10.15, that shows the coupling of reactions in this $I^-(aq)$-catalyzed decomposition of $H_2O_2(aq)$.

(c) In Chapter 10, Section 10.8, we said that coupling species do not appear in the net reaction for the coupled series of reactions. Is that true in this case as well? Explain.

11.41. Two possible pathways for the gas phase reaction of iodine and hydrogen are:

    (i) $H_2(g) + I_2(g) \rightleftharpoons 2HI(g)$ $k_{(i)}$

    (ii) $I_2(g) \rightleftharpoons I(g) + I(g)$ fast equilibrium, $K_{(ii)}$

           $H_2(g) + I(g) + I(g) \rightarrow 2HI(g)$ slow, $k_{(ii)}$

(a) Show that both pathways predict the same rate law for the reaction.

(b) Pathway (ii) involves a collision among three particles. Why is this reaction slow? Suggest alternative two-body collisions that would give the same result.

11.42. The net reaction of atmospheric ozone with nitric oxide is:

$$O_3(g) + NO(g) \rightarrow NO_2(g) + O_2(g)$$

The experimental rate law for this reaction is:

$$\frac{\Delta[O_3(g)]}{\Delta t} = k[O_3(g)][NO(g)]$$

Among the mechanisms proposed for this reaction are:

    (i) $O_3(g) + NO(g) \rightarrow O(g) + NO_3(g)$ slow, $k_{(i)}$

           $O(g) + O_3(g) \rightarrow 2O_2(g)$ fast

           $NO_3(g) + NO(g) \rightarrow 2NO_2(g)$ fast

    (ii) $O_3(g) \rightleftharpoons O(g) + O_2(g)$ fast equilibrium, $K_{(ii)}$

           $NO(g) + O(g) \rightarrow NO_2(g)$ slow, $k_{(ii)}$

(a) Show that both mechanisms agree with the overall reaction stoichiometry.

(b) What is the rate law predicted for each mechanism? Do these results help you determine whether either of these mechanisms is plausible? Explain why or why not.

(c) Propose another mechanism that would agree with both the reaction stoichiometry and the experimental rate law.

Section 11.5. First Order Reactions

11.43. In Figure 11.6(b), the horizontal lines showing the fraction of radioactive nuclei present, 1.00, 0.50, and 0.25, are equally spaced on the ln(cpm) axis. What is the spacing? How do you explain this value and why is the spacing equal?

11.44. Phosphorus-32, ^{32}P, is used extensively in molecular biology and biochemical research. To save the expense of radioactive waste disposal laboratories often store their ^{32}P wastes until the activity has decreased to less than 1% of its starting value and then dispose of it as chemical waste (a safe and appropriate procedure). Suppose you are the laboratory safety officer and assume that you collect the ^{32}P wastes weekly. How many weeks would you have to keep each collection to assure that its activity is under 1% of the starting value? *Hint:* Use data from Figure 11.6.

11.45. A specimen of bone from an archaeological dig has a carbon-14 radioactivity of 2.93 counts·min^{-1}. If, under the same counting conditions, the activity found in samples of living plants and animals is 12.6 counts·min^{-1}, what is the age of the bone specimen. Explain. *Hint:* Use data from Check This 11.37.

11.46. An antibiotic, **A**, is metabolized in the body by a first order reaction:

$$-\frac{\Delta(\text{concentration of A})}{\Delta t} = k(\text{concentration of A})$$

The rate constant, k, depends upon temperature and body mass. At 37 °C, $k = 3.0 \times 10^{-5}$ s^{-1} for a 70-kg person.

(a) How frequently must a 70-kg person take 400-mg pills of the antibiotic, in order to be sure to keep the concentration of antibiotic from falling below 140 mg per 70 kg body weight? You can assume that, upon ingestion, the antibiotic is almost immediately distributed uniformly throughout the body mass. Explain the reasoning for your answer.

(b) At 39 °C, $k = 4.0 \times 10^{-5}$ s^{-1}. If the person in part (a) has a fever of 39 °C, how often must s/he take the antibiotic to maintain a concentration at or above the effective level of 140 mg per 70 kg body mass. Explain.

11.47. These data are for the high temperature (504 °C) decomposition of gaseous dimethyl ether by the net reaction: $(CH_3)_2O(g) \rightarrow CH_4(g) + H_2(g) + CO(g)$

time, sec	0	390	777	1195	3175
pressure $(CH_3)_2O(g)$, kPa	41.5	35.1	29.8	24.9	10.4

Is the reaction first order, as the net reaction suggests? If so, determine the rate constant for the reaction. If not, explain how you draw this conclusion.

11.48. When heated, cyclopropane, $C_3H_6(g)$, isomerizes to propene, $CH_3CH=CH_2(g)$, in a first order reaction with a rate constant of 6.0×10^{-4} s^{-1} at 773 K. If cyclopropane at a pressure of 56.7 kPa is placed in a reaction vessel at 773 K, what will be the pressure of propene in the vessel after 15.0 minutes? Explain your reasoning.

11.49. Bacteria in a nutrient medium cause it to be cloudy: the more bacteria, the cloudier the medium. The number (concentration) of bacteria in a nutrient medium is sometimes determined by measuring the absorbance of the medium, which is directly proportional to the number of bacteria present. These are data for a bacterial growth experiment.

time, min	absorbance
0	0.053
15	0.095
30	0.167
45	0.301
60	0.533

During the part of their growth called the "log phase," the rate law for bacterial growth is:

$$\frac{\Delta(\text{number of bacteria})}{\Delta t} = k(\text{number of bacteria})$$

This equation for the appearance (growth) of bacteria is identical in form to equation (11.35) for the first order rate of disappearance of a reactant, with the exception that the term on the left here does not have a negative sign.

(a) Why does this equation for growth not have a negative sign?

(b) Use the method outlined in the text (or calculus) to show that the equation above leads to the result: $\ln(\text{number of bacteria})_t = kt + \ln(\text{number of bacteria})_0$. *Hint:* For small values of f, $\ln(1 + f) \approx f$.

(c) Plot the data in the table first on an absorbance *vs.* time graph and then in a way that will permit you to determine the rate constant, k, for the growth. Why is this growth phase called the log phase? Explain your reasoning.

(d) First order decays or disappearances are characterized by their half lives. Growth curves are characterized by their doubling times. Show how to use the value of k from part (c) to obtain the doubling time for this bacterial growth.

(e) In another nutrient medium, the doubling time for these bacteria was about 13 minutes. What conclusion(s) can you draw about this medium relative to the first?

11.50. Regulations from the U.S. Public Health Service specify that milk may contain a maximum of 20000 bacteria per mL when it has been freshly pasteurized. At 5 °C, the usual temperature of a refrigerator compartment, these bacteria are reported to have a doubling time about 40 hours. After milk is stored in a refrigerator for 10 days, what would we expect the maximum bacteria count to be? Explain your reasoning. *Hint:* See Problem 11.49.

11.51. The decomposition of dinitrogen tetroxide has been studied in the gas phase and in solution, as a model system to see how solvents affect reactions. The stoichiometry of the decomposition reaction in carbon tetrachloride, $CCl_4(l)$ (tetrachloromethane), solution is:

$$2N_2O_5(CCl_4) \rightarrow 4NO_2(CCl_4) + O_2(g)$$

This is the same net reaction as in the gas phase. These are time and concentration, $[N_2O_5(CCl_4)]$, data for the solution phase reaction at 45 °C.

time, min	0	184	319	526	867	1198	1877
$[N_2O_5(CCl_4)]$, M	2.33	2.08	1.91	1.67	1.35	1.11	0.72

(a) How many days did this experiment take?

(b) Plot these data on a concentration *vs.* time graph and then in a way that will permit you to determine the numeric value of the rate constant, k, for the reaction, assuming it is first order with respect to $[N_2O_5(CCl_4)]$. Does your plot justify the assumption? Explain your reasoning.

(c) Is your conclusion in part (b) consistent with the gas phase results for this reaction from Problem 11.37? Explain why or why not.

(d) This reaction was followed by measuring the volume of oxygen produced from the reaction. If the reaction solution volume was 25.0 mL and the evolved gas was measured at a pressure of 95.5kPa at 45 °C, what volume of gas was measured after the first 184 minutes of reaction? What was the total volume of gas evolved? Explain.

11.52. Show that the time required for a first order reaction to go to 99.0% completion is twice as long as the time it takes for the reaction to go to 90.0% completion.

Section 11.6. Temperature and Reaction Rates

11.53. (a) If the rate of a reaction at 55 °C is 6.7 times the rate of the reaction at 35 °C, what is the activation energy for the reaction? Explain your reasoning.

(b) At what temperature would the rate of this reaction be twice that at 35 °C? Explain.

11.54. Rate constants for the reaction, $2NOCl(g) \rightarrow 2NO(g) + Cl_2(g)$, are 9.4×10^{-6} s^{-1} at 350 K and 6.9×10^{-4} s^{-1} at 400 K.

(a) What is the rate law for this reaction? How do you know?

(b) What is the activation energy for this reaction? Explain your reasoning.

(c) What is the Arrhenius frequency factor for this reaction? Explain.

11.55. Rate constants for the reaction, $2N_2O_5 \rightarrow 4NO_2 + O_2$, at two temperatures in the gas phase and in carbon tetrachloride solution are:

	25 °C	45 °C
gas phase	3.38×10^{-5} min^{-1}	43.0×10^{-5} min^{-1}
CCl$_4$ solution	4.69×10^{-5} min^{-1}	Problem 11.46(b)

(a) Determine the activation energy and Arrhenius frequency factor for the gas phase reaction. Explain your reasoning.

(b) Determine the activation energy and Arrhenius frequency factor for the reaction in solution.

(c) In Problem 11.51, we said that this reaction had been studied in the gas phase and in several solvents to see how the rate parameters compare. Does the solvent carbon tetrachloride have a large affect on the rate parameters calculated here? Explain the basis for your conclusion.

11.56. California ground squirrels provoke northern Pacific rattlesnakes to get them to rattle and reveal their temperature (related to rattle frequency) which is an indication of the speed with which they can strike. Cold snakes are more lethargic and their aim is poorer. Use the data from this figure to estimate the activation energy for the rattling process. Explain how you obtain your answer.

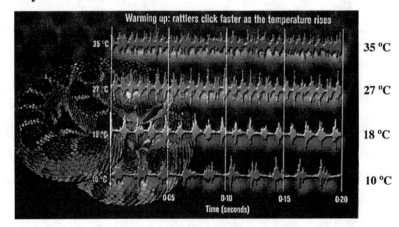

11.57. A study has been done to find out the optimum temperature to store milk, in order to get the best foam for espresso coffee drinks like cappuccinos and lattes. Among the conclusions of the study were that, "milk should be stored at no more than 40 °F. Every 5 °F increase in storage temperature cuts shelf life by half." What is the activation energy for the process that "cuts the shelf life," that is, spoils the milk? Explain clearly how you obtain your answer.

11.58. Raw (unpasteurized) milk sours in about 9 hours at 20 °C (room temperature), but it takes about 48 hours to sour in a refrigerator at °5 C.

(a) What is the activation energy for souring of raw milk? Explain your solution.

(b) The milk referred to in Problem 11.57 is presumably pasteurized milk. How does the activation energy for souring of raw milk compare with the activation energy for spoilage of pasteurized milk? What might be the cause of any difference?

11.59. The cyclopentadiene dimer dissociates in the gas phase to cyclopentadiene:

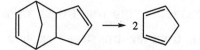

dimer   cyclopentadiene

The Arrhenius frequency factor and activation energy for this reaction are 1.3×10^{13} s^{-1} and 146 kJ·mol^{-1}, respectively.

(a) What is the rate constant for this reaction at 200 °C? Explain.

(b) What is the rate of the reaction at 200 °C, if the amounts are stated in pressure units, kPa? Explain.

(c) If a 15.0 kPa sample of the dimer is placed in a 200 °C reaction vessel, how long will it take for ten percent of the dimer to dissociate to cyclopentadiene? Explain your reasoning.

11.60. The hydrolysis of sucrose, reaction equation (11.66) is also catalyzed by [H$^+$(aq)]. Data for the temperature dependence of the rate constant for this reaction are:

temperature, °C	18	25	32	40	45
k, M^{-1}·s^{-1}	0.0022	0.0054	0.013	0.032	0.055

(a) Use a graphical method to find the activation energy and Arrhenius frequency factor for the acid-catalyzed reaction. Explain your method.

(b) What is the rate law for this reaction? How do you know?

(c) What is the half life for sucrose at 37 °C, in a solution with [H$^+$(aq)] = 1.0 M? Explain your reasoning.

11.61. Development of a photographic image is controlled by the kinetics of silver halide reduction by the developer. Data for development time for a commercial developer are:

temperature, °C	18	20	21	22	24
development time, min	10	9	8	7	6

Show how to use these data to estimate the activation energy for the reaction.

Section 11.7. Light: Another Way to Activate a Reaction

11.62. Hydrogen peroxide is always packaged in opaque containers. The container says to keep the hydrogen peroxide out of the light and away from high temperatures.

(a) What effect do you predict light to have on the rate of decomposition? Explain how you reach your conclusion.

(b) What effect do you predict high temperature to have on the rate of decomposition? Explain how you reach your conclusion.

(c) In Investigate This 11.6, you found that many substances catalyze the decomposition of hydrogen peroxide. Would it be a bad idea to dip your finger, or hair, or even a cotton swab into a bottle of hydrogen peroxide from the drugstore? Explain your reasoning.

11.63. Some molecules, **M**, after absorbing a photon that boosts them to an excited state, **M***(Chapter 4, Section 4.5), emit a photon to return to the ground state. Usually the emission is at a longer wavelength (lower energy) than the absorbed radiation. Excited molecules may also return to the ground state by transferring their energy to other molecules, **Q**:

(i) $M + hv_1 \rightarrow M^*$ rate = (constant)I[M]
(ii) $M^* \rightarrow M + hv_2$ rate = k_1[M*]
(iii) $M^* + Q \rightarrow M + Q^*$ rate = k_2[M*][Q]

In some cases, **Q*** also loses its energy by emission of light, but in many cases it dissipates the energy in other ways. If a system containing **M** and **Q** molecules is irradiated with a constant intensity, I, of light with a frequency v_1, a steady state is reached for which the intensity of the emitted radiation, I_e, at frequency v_2, is a constant.

(a) How does v_2 compare to v_1? Explain.

(b) If the intensity of emitted radiation, I_e, is directly proportional to [M*], what can you say about [M*] at the steady state? Explain your reasoning.

(c) Derive an expression for the [M*]/[M] ratio at the steady state in terms of the variables in the rate expressions.

(d) How would the intensity of emitted radiation, I_e, vary as [Q] varies? Be as specific as you can in your explanation. Explain why molecules like **Q** are often called "quenchers."

Section 11.8. Thermodynamics and Kinetics

11.64. In Chapter 8 we found that processes that resulted in a net entropy increase were spontaneous, that is, they could occur without any outside influence. Do spontaneous reactions always occur quickly? Explain your response.

11.65. Hydrogen-oxygen mixtures are highly explosive and dangerous. Yet mixtures of hydrogen and oxygen can be kept for long periods of time (perhaps millennia) without reacting. How can you explain this apparent contradiction?

11.66. The dissociation of the dimer of cyclopentadiene shown in Problem 11.59 also proceeds in the reverse direction to form the dimer:

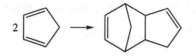

The Arrhenius frequency factor and activation energy for this reaction are 1.3×10^6 M^{-1}·s^{-1} and 70. kJ·mol^{-1}, respectively.

(a) Use the information here and in Problem 11.59 to sketch and clearly label an activation energy diagram (energy *vs.* progress of reaction) for this system.

(b) From your plot in part (a), what is your estimate of the standard enthalpy change. ΔH° for the above reaction? Explain.

(c) Show how to use bond enthalpies, Table 7.3, Chapter 3, Section 7.7, to estimate ΔH° for the above reaction. How does the result compare to your estimate in part (b) from the kinetic data? Is the difference, if any, between your estimates what you might expect on the basis of the molecular structures? Explain.

(d) Cyclopentadiene is a volatile liquid, b.p. 40 °C, with a density of about 0.8 kg·L^{-1}. What is the molarity of the pure liquid? What is the initial rate of the above second order dimerization reaction in the pure liquid at 25 °C? Explain your method.

(e) About how long will it take for 10% of the cyclopentadiene in the pure liquid to form dimer at 25 °C? Explain your reasoning and any assumptions you make.

(f) Cyclopentadiene is a useful reactant and starting material for many chemical syntheses, but it is not available from chemical suppliers as the pure liquid. How is your result in part (e) related to this lack of availability of the pure liquid?

(g) Chemists who need pure cyclopentadiene purchase the dimer, a viscous liquid, b.p. 170 °C, heat it in a distillation apparatus, collect the volatile cyclopentadiene product and use it immediately. Use your results from Problem 11.59 and this one to explain how this procedure works kinetically and thermodynamically.

11.67. Which of these statements about elementary reactions are correct and which incorrect? For the ones that are correct, show why they are correct. For the ones that are incorrect, show why they are incorrect and write a corrected statement.

(a) The equilibrium constant for a reaction is the ratio of the reverse to the forward rate at equilibrium.

(b) For a reaction with a very small equilibrium constant, the rate constant for the forward reaction is much smaller than the rate constant for the reverse reaction.

(c) For an endothermic reaction at equilibrium, an increase in temperature decreases both the forward and reverse reaction rates, but they are equal when equilibrium is reattained.

(d) For an exothermic reaction at equilibrium, an increase in temperature increases both the forward and reverse rate constants by the same proportion.

11.68. Consider the elementary reaction, $A \rightarrow B + C$, which is reversible and has been studied going in the reverse direction. The reaction, as written, is endothermic with $\Delta H^\circ = 57$ kJ·mol^{-1}. Species C is colored, so the rate of the reaction can be followed and the equilibrium constant determined colorimetrically. When equal concentrations of B and C, both 0.046 M, were reacted at 25 °C, the initial rate of reaction was 1.12×10^{-5} M·s^{-1}. When equilibrium was attained (no further change in the color of the solution), the concentration of C in the solution was 0.027 M. A series of temperature studies on the rate of the reverse reaction gave an activation energy of 43 kJ·mol^{-1}.

(a) Sketch and label an activation energy diagram for the reaction written above. In particular, show the activation energies and enthalpy of reaction. Explain how you use the data to draw your diagram correctly.

(b) Show how to calculate the equilibrium constant for the reaction written above.

(c) Show how to calculate the rate constant, k_r, for the reverse reaction.

(d) Show how to calculate the rate constant, k_f, for the forward reaction.

(e) Show how to calculate the Arrhenius frequency factors for the forward, A_f, and reverse, A_r, reactions.

(f) If the temperature is increased, will the equilibrium constant increase or decrease? Answer this question using both a thermodynamic argument and a kinetic argument.

(g) Show how to calculate the equilibrium constant at 40 °C for the reaction written above.

11.69. The rate of a first order reaction decreases with time as the reactant is used up, but the concentration of the reactant should eventually go essentially to zero (or below the limit of detectability). However, for some first order reactions the concentration of the reactant does not go to zero, but comes to some constant measurable value.

(a) What is probably happening in the case that the concentration comes to a constant value and does not go to zero?

(b) What experiments would you propose to test your explanation in part (a)? What would you expect the outcomes of the experiments to be, if your explanation is correct?

11.70. In your investigations with phenolphthalein and high concentrations of hydroxide ion, $[OH^-(aq)]$, Investigate This 11.1 and 11.4, you found that the color of the phenolphthalein dianion, $P^{2-}(aq)$, disappeared from the solution – it all reacted by reaction (11.2) going in the forward direction: $P^{2-}(aq) + OH^-(aq) \rightleftharpoons POH^{3-}(aq)$. At lower $[OH^-(aq)]$, say 0.10 M, observations on this system are different. Initially, the rate of disappearance of the color is first order, rate = $k_{expt}[P^{2-}(aq)]$, where the experimental first order rate constant is, $k_{expt} = k_f[OH^-(aq)]$, just as at the higher $[OH^-(aq)]$. However, as the reaction proceeds, k_{expt} gets smaller (even though $[OH^-(aq)]$ is not changing appreciably) and the solution never completely decolorizes but comes to an unchanging light pink color. Compare these observations with the behavior described in Problem 11.69.

(a) How would you characterize the reacting system thermodynamically and kinetically when the solution reaches its unchanging pink color? Be as specific as possible in your explanation.

(b) Assume that reaction (11.2) represents an elementary reaction that can go in both the forward and reverse directions. The rate law for the forward reaction is written above. Write the rate law for the reverse reaction with rate constant k_r.

(c) When $[OH^-(aq)] = 0.10$ M, $k_r \approx (1/10)k_f[OH^-(aq)]$. Under these conditions, what is the $[P^{2-}(aq)]/[POH^{3-}(aq)]$ ratio when the forward and reverse rates are the same? Show how you get your result.

(d) If $[OH^-(aq)] = 1.0$ M, what is the $[P^{2-}(aq)]/[POH^{3-}(aq)]$ ratio when the forward and reverse rates are the same? Does this result explain why the solution goes colorless at higher hydroxide concentrations? Explain why or why not.

(e) The reaction, as written, is exothermic. Why? Sketch and label an activation energy diagram for the reaction.

(f) If no other conditions change in the 0.10 M hydroxide solution, will the light pink color get darker or lighter if the temperature of the solution is increased? Explain.

Section 11.10.   EXTENSION – Enzymatic Catalysis

11.71. These data are initial rates (10^6 μmol = 1 mol) of an enzyme-catalyzed reaction measured with the same amount of enzyme and varying initial concentrations of substrate, S.

$[S]_0$, M	V_0, μmol·min^{-1}
2.0×10^{-2}	60
2.0×10^{-3}	60
2.0×10^{-4}	48
1.5×10^{-4}	45
1.3×10^{-5}	12

(a) What is V_{max} for this reaction? Explain your reasoning.

(b) Why doesn't V_0 continue to increase for $[S]_0$ greater than 2.0×10^{-3} M?

(c) What is the approximate concentration of free enzyme in the reaction with $[S]_0 = 2.0 \times 10^{-2}$ M? Explain.

(d) What is the numeric value of K (with units) in the Michaelis-Menten equation for this reaction? Explain your approach to solving this problem. (Note that K has units, because it is conventionally expressed, equation (11.70), in terms of concentrations, instead of dimensionless concentration ratios.) *Hint:* See equation (11.77).

11.72. Experiments on an enzyme-catalyzed reaction that followed Michaelis-Menten kinetics showed that $K = 1.2 \times 10^5$ M^{-1} and that, with the same amount of enzyme, the initial rate of the reaction, 45 μmol·min^{-1} (10^6 μmol = 1 mol), was the same for substrate concentrations of 0.10 M and 0.010 M.

(a) Use the Michaelis-Menten equation to show why the initial reaction rate is the same for both of these substrate concentrations.

(b) Predict the initial rate of reaction for this same amount of enzyme at a substrate concentration of 2.0×10^{-5} M. Explain your reasoning. *Hint:* See equation (11.77).

11.73. Suppose that a mutant enzyme in an organism binds the substrate 10 times more tightly than the native (normal) enzyme, that is, $K_{mutant} = 10K_{native}$.

(a) If nothing else about the catalytic mechanism is different, how will V_{max} for the mutant and native enzymes compare? Explain your reasoning.

(b) For a substrate concentration that is not saturating for either enzyme, how will the initial reaction rates compare? Explain.

(c) On the same graph, sketch and label V_0 *vs.* $[S]_0$ plots for both enzymes.

11.74. The enzyme urease catalyzes the hydrolysis of urea to ammonium and carbonate ions (or bicarbonate under acidic conditions), reaction equation (11.80):

$$NH_2CONH_2(aq) + 2H_2O \xrightarrow{\text{urease}} 2NH_4^+(aq) + CO_3^{2-}(aq)$$

Initial rate data as a function of substrate, $NH_2CONH_2(aq)$, concentration are:

$[NH_2CONH_2(aq)]_0$, M	V_0, M·sec^{-1}
0.00065	0.226
0.00129	0.362
0.00327	0.600
0.00830	0.846
0.0167	0.975
0.0333	1.03

(a) Plot these data on a V_0 vs. $[NH_2CONH_2(aq)]_0$ graph and, assuming that the reaction obeys Michaelis-Menten kinetics, estimate V_{max} and K for the reaction.

(b) Show that the Michaelis-Menten equation, in the form given in equation (11.77), can be transformed to this equation by inverting both sides.

$$\frac{1}{V_0} = \left[\frac{(1/K)}{V_{max}}\right]\left(\frac{1}{[S]_0}\right) + \frac{1}{V_{max}}$$

This is a linear equation for $1/V_0$ plotted as a function of $1/[S]_0$ – a "double reciprocal" plot. From the intercept we get V_{max} (= 1/intercept) and the slope divided by the intercept gives $1/K$. This transformed equation is often called the Lineweaver-Burk equation. Plot the data for the urease reaction on a double reciprocal plot and find the numerical values for V_{max} and K.

(c) How do your results from parts (a) and (b) compare?

(d) What initial concentration of urea is required to give an initial rate of reaction that is 99% of V_{max}? Explain.

11.75. The activity of an enzyme is usually defined in terms of the amount of substrate that reacts in a specified time under specified conditions. The unit of activity for a certain bacterial pyrophosphatase is hydrolysis of 10 μmol (10^6 μmol = 1 mol) of pyrophosphate in 15 minutes at 37 °C. A sample of the purified enzyme has an activity of 2750 units per milligram of enzyme. Under the test conditions, how many moles of substrate react per second per milligram of enzyme? Explain your reasoning.

11.76. Enzymatic reactions are affected by many factors, including molecules that bind to the enzyme at the site(s) where their substrate(s) must bind to undergo reaction. These molecules are often called "inhibitors," because their usual effect is to slow the enzymatic reaction. The initial rate data here are for an enzymatic reaction in the absence and presence of a constant concentration of an inhibitor.

$[S]_0$, M	V_0, μmol·min^{-1} inhibitor absent	V_0, μmol·min^{-1} inhibitor present
1.0×10^{-4}	28	18
1.5×10^{-4}	36	24
2.0×10^{-4}	43	30
5.0×10^{-4}	63	51
7.5×10^{-4}	74	63

(a) Assuming that the inhibitor binds reversibly to the active site of the enzyme, just as the substrate does, write a series of reactions that includes the inhibitor binding as well as the Michaelis-Menten reactions. Use equation (11.69) as a model and sketch a pictorial representation of your mechanism. Explain, with reference to your mechanism, why the enzymatic reaction is slower with the inhibitor present.

(b) Plot the data with the inhibitor absent and present on the same V_0 vs. $[S]_0$ graph. From your plots estimate numeric values for V_{max} and K in the absence and presence of a constant amount of inhibitor. Are your results consistent with what you would expect from your mechanism in part (a)? Explain why or why not.

(c) Plot the data on a double-reciprocal graph (see Problem 11.74(b)) and use the equations of the lines to calculate numeric values for V_{max} and K in the absence and presence of a constant amount of inhibitor. How do these results compare with those from part (b)? Do you need to revise any of your explanation from part (b)? If so, how?

11.77. The normal substrate for the enzyme aspartate transcarbamylase is aspartate. Succinate is an inhibitor (see Problem 11.76) of the normal reaction. Based on their structures, does this inhibition make sense? What kinds of interactions in the active site are probably responsible for the binding of succinate in place of aspartate? Explain.

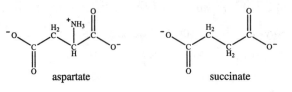

aspartate   succinate

11.78. The rate of almost all reactions, including enzyme-catalyzed reactions, increases with increasing temperature. For enzymatic reactions, however, the rate decreases rapidly beyond a certain temperature (which is dependent on the enzyme under study). How can you account for this behavior? *Hint:* See Chapter 1, Section 1.9.

11.79. In Problem 11.28, you found that the decomposition of ammonia on a hot tungsten wire goes at the same rate even as NH_3 decomposes and its pressure (concentration) decreases. The reaction rate is independent of the amount of ammonia left unreacted (at least for the data we have).

(a) Could this catalytic reaction system be displaying a saturation effect? Discuss this possibility and consider what the sites are that might be saturated, if this is such an effect. What experiment(s) could you do to test your ideas?

(b) Assuming that the results do represent a saturation effect, similar to an enzymatic reaction, sketch a curve showing the initial rate of ammonia decomposition as a function of initial pressure of ammonia from zero to 26 kPa.

General Problems

11.80. The rate law and reaction stoichiometry are important data we can use to propose a reaction pathway, but other information is also helpful. For example, consider the hydrolysis of phosphate esters, such as butyl phosphate (the phosphate and phosphoric acid product transfer protons to water, but these reactions are omitted here):

$$C_4H_9OP(O)(OH)_2 + H_2O \rightarrow C_4H_9OH + (HO)_3PO$$

When this reaction is carried out in water that has been enriched in oxygen-18 (that is, has a higher percentage of O-18 atoms than normal), the oxygen-18 is found in the phosphoric acid product, but not in the alcohol. What does the result suggest about the atoms that interact when the water and ester begin to react? How does this result help to establish a pathway for the reaction? *Hint:* Review Chapter 6, Section 6.7.

11.81. Reaction pathway (i) in Problem 11.37 proposes that dinitrogen pentoxide, O_2NONO_2, reacts by decomposing in a rapid reversible reaction to NO_3 and NO_2, which then can undergo this rate limiting reaction:

$$NO_3 + NO_2 \rightarrow NO_2 + O_2 + NO$$

(a) Write Lewis structures and calculate the formal charges on all the atoms in dinitrogen pentoxide, nitrogen trioxide, and nitrogen dioxide.

(b) Is there electron delocalization that could stabilize any of these molecules? Which one(s)? Could this factor account for any part of the proposed reaction pathway? If so, explain how. If not, explain why not.

(c) What would be a plausible activated complex for the rate limiting reaction written above? Explain why you think it is reasonable and why the reaction might be slow.

11.82. In Problem 11.35, a pathway was proposed for the net reaction:

$$(CH_3)_3CCl + OH^- \rightarrow (CH_3)_3COH + Cl^-.$$

This pathway is consistent with the rate law you derived, based on the information in Problems 11.15 and 11.25. Experiments with $(CH_3)_3CBr$ and $(CH_3)_3CI$ show that they undergo this same reaction with the same rate law, but at different rates.

(a) If the proposed reaction pathway is the same for all three halogen compounds, can it explain the different rates for the reaction? If so, show how. If not, show why not.

(b) Which of the three reactants, the chloro, bromo, of iodo compound, probably reacts most rapidly? Explain the reasoning for your choice. Hint: See Table 7.3, Chapter 7, Section 7.7.

(c) If studies at different temperatures are carried out, which reaction rate will be most affected by the temperature? Explain your reasoning.

(d) In all three reactant systems there is a side reaction that forms a small amount of 3-methyl propene, $(CH_3)_2C=CH_2$. The proposed pathway for this side reaction is a rapid elimination reaction: $(CH_3)_3C^+ \rightarrow (CH_3)_2C=CH_2 + H^+$. The ratio of the alkene product to the alcohol product is the same in all three reactant systems. Is this observation consistent with the proposed reaction pathway for these reactions? Explain why or why not.

Thermodynamic Data at 298 K (25 °C)

Substance	ΔH°_f, kJ·mol⁻¹	S°, J·K⁻¹·mol⁻¹	ΔG°_f, kJ·mol⁻¹
Aluminum — Al(s)	0	28.33	0
Al³⁺(aq)	−524.7	−321.7	−481.2
Al₂O₃(s)	−1675.7	50.92	−1582.3
Barium — Ba(s)	0	62.8	0
Ba²⁺(aq)	−537.64	9.6	−560.77
BaCO₃(s)	−1216.3	112.1	−1137.6
BaSO₄(s)	−1465.2	132.2	−1353.1
Bromine — Br₂(l)	0	152.23	0
Br(g)	111.88	175.02	82.40
Br₂(g)	30.91	245.46	3.11
Br⁻(aq)	−121.55	82.4	−103.96
HBr(g)	−36.40	198.70	−53.45
Calcium — Ca(s)	0	41.42	0
Ca²⁺(aq)	−542.83	−53.1	−553.58
Ca(OH)₂(s)	−986.09	83.39	−898.49
CaCO₃(s), calcite	−1206.9	92.9	−1128.8
CaCO₃(s), aragonite	−1207.1	88.7	−1127.8
CaC₂(s)	−59.8	69.96	−64.9
CaF₂(s)	−1219.6	68.87	−1167.3
CaCl₂(s)	−795.8	104.6	−748.1
CaBr₂(s)	−682.8	130.	−663.6
CaSO₄(s)	−1434.11	106.7	−1321.79
Carbon — (also see pages A-7—A-9)			
C(s), graphite	0	5.740	0
C(s), diamond	1.895	2.377	2.900
C(g)	716.68	158.10	671.26
CO(g)	−110.53	197.67	−137.17
CO₂(g)	−393.51	213.74	−394.36
CO₃²⁺(aq)	−677.14	−56.9	−527.81

Substance	ΔH°_f, kJ·mol^{-1}	S°, J·K^{-1}·mol^{-1}	ΔG°_f, kJ·mol^{-1}
Chlorine — $Cl_2(g)$	0	223.07	0
$Cl(g)$	121.68	165.20	105.68
$Cl^-(aq)$	−167.16	56.5	−131.23
$HCl(g)$	−92.31	186.91	−95.30
Copper — $Cu(s)$	0	33.15	0
$Cu^+(aq)$	71.67	40.6	49.98
$Cu^{2+}(aq)$	64.77	−99.6	65.49
$CuSO_4(s)$	−771.36	109	−661.8
$CuSO_4·5H_2O(s)$	−2279.7	300.4	−1879.7
Fluorine — $F_2(g)$	0	202.78	0
$F(g)$			
$F^-(aq)$	−332.63	−13.8	−278.79
$HF(g)$	−271.1	173.78	−273.2
Hydrogen — $H_2(g)$	0	130.68	0
$H(g)$	217.97	114.71	203.25
$H^+(aq)$	0 (assigned)	0 (assigned)	0 (assigned)
$H_2O(l)$	−285.83	69.91	−237.13
$H_2O(g)$	−241.82	188.83	−228.57
$H_2O_2(l)$	−187.78	109.6	−120.35
$H_2O_2(aq)$	−191.17	143.9	−134.03
Deuterium — $D_2(g)$	0	144.96	0
$D_2O(l)$	−294.60	75.94	−243.44
$D_2O(g)$	−249.20	198.34	−234.54

Substance	ΔH°_{f}, kJ·mol^{-1}	S°, J·K^{-1}·mol^{-1}	ΔG°_{f}, kJ·mol^{-1}
Iodine — $I_2(g)$	0	116.14	0
$I(g)$			
$I_2(g)$	62.44	260.69	19.33
$I^-(aq)$	−55.19	111.3	−51.57
$HI(g)$	26.48	206.59	1.70
Iron — $Fe(s)$	0	27.28	0
$Fe^{2+}(aq)$	−89.1	−137.7	−78.90
$Fe^{3+}(aq)$	−48.5	−315.9	−4.7
$Fe_3O_4(s)$, magnetite	−1118.4	146.4	−1015.4
$Fe_2O_3(s)$, hematite	−824.2	87.40	−742.2
$FeS(s)$	−100.0	60.29	−100.4
Lead — $Pb(s)$	0	64.81	0
$Pb^{2+}(aq)$	1.7	21.3	−24.43
$PbO(s)$	−217.86	69.45	−188.49
$PbO_2(s)$	−277.4	76.57	−217.33
$PbSO_4(s)$	−919.94	148.57	−813.14
$PbCl_2(s)$	−359.2	136.4	−313.97
$PbBr_2(s)$	−287.7	161.5	−261.92
$PbI_2(s)$	−175.1	177.0	−171.69
Lithium — $Li(s)$	0	28.0	0
$Li^+(aq)$	−278.46	14.2	−293.8
$Li_2O(s)$	−595.8		
$LiOH(s)$	−487.2	50.2	−443.9
$LiCl(s)$	−408.8		
$LiBr(s)$	−350.3		

Substance	ΔH°_f, kJ·mol^{-1}	S°, J·K^{-1}·mol^{-1}	ΔG°_f, kJ·mol^{-1}
Magnesium — Mg(s)	0	32.68	0
Mg^{2+}(aq)	−466.85	−138.1	−454.8
MgO(s)	−601.70	26.94	−569.43
Mg(OH)$_2$(s)	−924.66	63.1	−833.75
MgCO$_4$(s)	−1095.8	65.7	−1021.1
MgSO$_4$(s)	−1278.2	91.6	−1173.6
MgCl$_2$(s)	−641.8	89.5	−592.3
MgBr$_2$(s)	−524.3	117.2	−503.8
Mercury — Hg(l)	0	76.02	0
Hg(g)	61.32	174.96	31.82
Hg^{2+}(aq)			−164.38
HgO(s)	−90.83	70.29	−58.54
Hg$_2$Cl$_2$(s), calomel	−265.22	192.5	−210.75
HgCl$_2$(s)	−230.1		
Nitrogen — N$_2$(g)	0	191.61	0
N(g)			
N$_2$O(g)	82.05	219.85	104.20
NO(g)	90.25	210.76	86.55
NO$_2$(g)	33.18	240.06	51.31
N$_2$O$_4$(g)	9.16	304.29	97.89
NO$_3^-$(aq)	−206.57	146.4	−110.5
NH$_3$(g)	−46.11	192.45	−16.45
NH$_3$(aq)	−80.29	111.3	−26.50
NH$_4^+$(aq)	−132.51	113.4	−79.31
N$_2$H$_4$(l)	50.63	121.21	149.34
NH$_4$Cl(s)	−314.43	94.6	−202.87
NH$_4$NO$_3$(s)	−365.56	151.08	−183.87

Substance	ΔH°_f, kJ·mol^{-1}	S°, J·K^{-1}·mol^{-1}	ΔG°_f, kJ·mol^{-1}
Oxygen — $O_2(g)$	0	205.14	0
$O(g)$	249.4	160.95	230.1
$O_3(g)$	142.7	238.93	163.2
$OH^-(aq)$	−229.99	−10.75	−157.24
Phosphorus — $P(s)$, white	0	41.09	0
$P_4(g)$	58.91	279.98	24.44
$P_4O_{10}(s)$	−2984.0	228.86	−2697.0
$H_3PO_4(aq)$ or $(HO)_3PO(aq)$	−1288.34	158.2	−1142.54
$H_2PO_4^-(aq)$ or $(HO)_2PO_2^-(aq)$	−1302.48	89.1	−1135.1
$HPO_4^{2-}(aq)$ or $HOPO_2^{2-}(aq)$	−1298.7	−35.98	−1094.1
$PO_4^{3-}(aq)$	−1284.07	−217.57	−1025.59
Potassium — $K(s)$	0	64.18	0
$K^+(aq)$	−252.38	102.5	−283.27
$KOH(s)$	−424.76	78.9	−379.08
$KCl(s)$	−436.75	82.59	−409.14
$KBr(s)$	−393.80	95.90	−380.66
Silver — $Ag(s)$	0	42.55	0
$Ag^+(aq)$	105.58	72.68	77.11
$AgCl(s)$	−127.07	96.2	−109.79
$AgBr(s)$	−100.37	107.1	−96.90
$AgI(s)$	−61.84	115.5	−66.19
$AgNO_3(s)$	−123.1	140.9	−32.2

Substance	ΔH°_f, kJ·mol^{-1}	S°, J·K^{-1}·mol^{-1}	ΔG°_f, kJ·mol^{-1}
Sodium — Na(s)	0	51.21	0
Na$^+$(aq)	–240.12	59.0	–261.91
NaOH(s)	–425.61	64.46	–379.49
NaCl(s)	–411.15	72.13	–384.14
NaBr(s)	–361.06	86.82	–348.98
NaI(s)	–287.78	98.53	–286.06
Na$_2$CO$_3$(s)	–1130.9	135.98	–1047.67
NaHCO$_3$(s)	–947.68	102.09	–851.86
Na$_2$SO$_4$(s)	–1384.49	149.49	–1266.8
Sulfur —S(s), rhombic	0	31.80	0
S(s), monoclinic	0.33	32.6	0.1
SO$_2$(g)	–296.83	248.22	–300.19
SO$_3$(g)	–395.72	256.76	–371.06
HSO$_4^-$(aq) or HOSO$_3^-$(aq)	–887.34	131.8	–755.91
SO$_4^{2-}$(aq)	–909.27	20.1	–744.53
H$_2$S(g)	–20.63	205.79	–33.56
H$_2$S(aq)	–39.7	121	–27.83
Zinc — Zn(s)	0	41.63	0
Zn^{2+}(aq)	–153.89	–112.1	–147.06
ZnO(s)	–348.28	43.64	–318.30
ZnS(s)	–202.9	57.7	–198.3
ZnCl$_2$(s)	–415.89	108.37	–369.26
ZnSO$_4$(s)	–978.6	124.7	–871.6

Compounds of Carbon

Name	Formula(state)	ΔH°_f, kJ·mol^{-1}	S°, J·K^{-1}·mol^{-1}	ΔG°_f, kJ·mol^{-1}
Hydrocarbons				
methane	CH$_4$(g)	−74.81	186.26	−50.72
ethane	C$_2$H$_6$(g)	−84.68	229.60	−32.82
propane	C$_3$H$_8$(g)	−103.85	270.2	−23.49
butane	C$_4$H$_{10}$(g)	−126.15	310.1	−17.03
pentane	C$_5$H$_{12}$(g)	−146.44	349	−8.20
ethene	C$_2$H$_4$(g)	52.26	219.56	68.15
propene	C$_3$H$_6$(g)	20.42	266.6	62.78
1-butene	C$_4$H$_8$(g)	1.17	307.4	
1-pentene	C$_5$H$_{10}$(g)	−20.92	347.6	
ethyne	C$_2$H$_2$(g)	226.73	200.94	209.20
cyclopropane	C$_3$H$_6$(g)	53.30	237.4	104.45
cyclobutane	C$_4$H$_8$(g)	26.65	265.4	
cyclopentane	C$_5$H$_{10}$(g)	−77.24	292.9	
cyclohexane	C$_6$H$_{12}$(g)	−123.1	298.2	
	C$_6$H$_{12}$(l)	−156.4	204.4	26.7
benzene	C$_6$H$_6$(g)	82.9	269.31	129.72
	C$_6$H$_6$(l)	49.0	173.3	124.3
Alcohols				
methanol	CH$_3$OH(g)	−200.66	239.81	−161.96
	CH$_3$OH(l)	−238.86	126.8	−166.27
	CH$_3$OH(aq)			−175.23
ethanol	C$_2$H$_5$OH(g)	−235.10	282.70	−168.49
	C$_2$H$_5$OH(l)	−277.69	160.7	−174.78
	C$_2$H$_5$OH(aq)			−180.92
glycerol	C$_3$H$_5$(OH)$_3$(l)	−668.6	204.47	−477.06
	C$_3$H$_5$(OH)$_3$(aq)			−488.52

Name	Formula(state)	ΔH°_f, kJ·mol^{-1}	S°, J·K^{-1}·mol^{-1}	ΔG°_f, kJ·mol^{-1}
Aldehydes and Ketones				
methanal (formaldehyde)	HCHO(g)	–115.90	218.78	–109.91
	HCHO(aq)			–130.5
ethanal (acetaldehyde)	CH$_3$CHO(g)	–166.36	264.22	–133.30
	CH$_3$CHO(l)	–192.30	160.2	–128.12
	CH$_3$CHO(aq)			–139.24
propanone (acetone)	(CH$_3$)$_2$CO(l)	–248.1	200.4	–155.39
	(CH$_3$)$_2$CO(aq)			–161.00
Carboxylic Acids/Ions				
methanoic (formic)	HCOOH(l)	–424.72	128.95	–361.35
ethanoic (acetic)	CH$_3$COOH(l)	–484.5	159.8	–389.9
	CH$_3$COOH(aq)	–485.76	86.6	–396.46
ethanoate (acetate) ion	CH$_3$COO$^-$(aq)			–372.334
oxalic	(-COOH)$_2$(s)	–827.2	120	–697.9
succinic	(-CH$_2$COOH)$_2$(s)	–940.90	175.7	–747.43
	(-CH$_2$COOH)$_2$(aq)			–746.22
succinate ion	(-CH$_2$COO$^-$)$_2$(aq)			–690.23
pyruvic	CH$_3$COCOOH(l)	–584.5	179.5	–463.38
pyruvate ion	CH$_3$COCOO$^-$(aq)			–474.33
lactic	CH$_3$CHOHCOOH(s)	–694.08	142.26	–522.92
lactate ion	CH$_3$CHOHCOO$^-$(aq)			–517.812
Nitrogen Containing				
hydrogen cyanide	HCN(g)	135.1	201.78	124.7
	HCN(l)	108.87	112.84	124.97
	HCN(aq)	107.1	124.7	119.7
cyanide ion	CN$^-$(aq)	151.0	118.0	165.7
urea	(NH$_2$)$_2$CO(s)	–333.19	104.6	–197.15
	(NH$_2$)$_2$CO(aq)	–319.2	173.85	–203.84

Name	Formula*(state)*	ΔH°_{f}, kJ·mol^{-1}	S°, J·K^{-1}·mol^{-1}	ΔG°_{f}, kJ·mol^{-1}
Amino Acids	*side group*			
glycine	-H*(s)*	–537.2	103.51	–377.69
	-H*(aq)*			–379.9
alanine	-CH$_3$*(s)*	–562.7	129.20	–370.24
	-CH$_3$*(aq)*			–371.71
leucine	-CH$_2$CH(CH$_3$)$_2$*(s)*	–646.8	211.79	–357.06
	-CH$_2$CH(CH$_3$)$_2$*(aq)*			–353.09
cysteine	-CH$_2$SH*(s)*	–533.9	169.9	–343.97
	-CH$_2$SH*(aq)*			–340.33
aspartic acid	-CH$_2$COOH*(s)*	–973.37	170.12	–730.23
	-CH$_2$COOH*(aq)*			–719.98
aspartate ion	-CH$_2$COO$^-$*(aq)*			–698.69
Sugars				
glucose	C$_6$H$_{12}$O$_6$*(s)*	–1274.4	212.1	–910.52
	C$_6$H$_{12}$O$_6$*(aq)*			–917.47
fructose	C$_6$H$_{12}$O$_6$*(s)*	–1272		
	C$_6$H$_{12}$O$_6$*(aq)*			
galactose	C$_6$H$_{12}$O$_6$*(s)*	–1285.37	205.4	–919.43
	C$_6$H$_{12}$O$_6$*(aq)*			–924.58
sucrose	C$_{12}$H$_{22}$O$_{11}$*(s)*	–2222.1	360.2	–1544.65
	C$_{12}$H$_{22}$O$_{11}$*(aq)*			–1551.76
lactose	C$_{12}$H$_{22}$O$_{11}$*(s)*	–2236.72	386.2	–1566.99
	C$_{12}$H$_{22}$O$_{11}$*(aq)*			–1569.92

Appendix YY

Standard Reduction Potentials: $E°$ and $E°'$(pH 7)

Half cell contents	Half-cell reaction	$E°$, V	$E°'$(pH 7), V
	Strongest Oxidizing Agent		
$F_2|F^-$	$F_2 + 2e^- \rightleftharpoons 2F^-$	2.87	
H_4XeO_6, XeO_3	$H_4XeO_6 + 2H^+ + 2e^- \rightleftharpoons XeO_3 + 3H_2O$	2.38	
Co^{3+}, Co^{2+}	$Co^{3+} + e^- \rightleftharpoons Co^{2+}$	1.92	
H_2O_2, H_2O	$H_2O_2 + 2H^+ + 2e- \rightleftharpoons 2H_2O$	1.763	
$PbO_2, SO_4^{2-}|PbSO_4$	$PbO_2 + SO_4^{2-} + 4H^+ + 2e^- \rightleftharpoons PbSO_4 + 2H_2O$	1.658	
$HClO|Cl_2$	$2HClO + 2H^+ + 2e^- \rightleftharpoons Cl_2 + 2H_2O$	1.630	
MnO_4^-, Mn^{2+}	$MnO_4^- + 8H^+ + 4e^- \rightleftharpoons Mn^{2+} + 4H_2O$	1.507	
$HOI|I_2$	$2HOI + 2H^+ + 2e^- \rightleftharpoons I_2 + 2H_2O$	1.430	
$Cr_2O_7^{2-}, Cr^{3+}$	$Cr_2O_7^{2-} + 14H^+ + 6e^- \rightleftharpoons 2Cr^{3+} + 7H_2O$	1.36	
$Cl_2|Cl^-$	$Cl_2 + 2e^- \rightleftharpoons 2Cl^-$	1.359	
chlph(II)$^+$|chlph(II)	chlph(II)$^+$ + $e^- \rightleftharpoons$ chlph(II)		0.9
$MnO_2|Mn^{2+}$	$MnO_2 + 4H^+ + 2e^- \rightleftharpoons Mn^{2+} + 2H_2O$	1.230	
$O_2|H_2O$	$O_2 + 4H^+ + 4e^- \rightleftharpoons 2H_2O$	1.229	0.816
$IO_3^-|I_2$	$2IO_3^- + 12H^+ + 10e^- \rightleftharpoons I_2 + 6H_2O$	1.210	
$Br_2|Br^-$	$Br_2 + 2e^- \rightleftharpoons 2Br^-$	1.087	
$Fe(o\text{-}phen)_3^{3+}$, $Fe(o\text{-}phen)_3^{2+}$	$Fe(o\text{-}phen)_3^{3+} + e^- \rightleftharpoons Fe(o\text{-}phen)_3^{2+}$	1.06	
$NO_3^-|NO$	$NO_3^- + 4H^+ + 3e^- \rightleftharpoons NO + 2H_2O$	0.955	
$Hg^{2+}|Hg$	$Hg^{2+} + 2e^- \rightleftharpoons Hg$	0.908	
$Ag^+|Ag$	$Ag^+ + e^- \rightleftharpoons Ag$	0.799	
$Hg_2^{2+}|Hg$	$Hg_2^{2+} + 2e^- \rightleftharpoons 2Hg$	0.796	
Fe^{3+}, Fe^{2+}	$Fe^{3+} + e^- \rightleftharpoons Fe^{2+}$	0.771	
cytochrome a$_3$	Fe(III) + $e^- \rightleftharpoons$ Fe(II)		0.55
chlph(I)$^+$|chlph(I)	chlph(I)$^+$ + $e^- \rightleftharpoons$ chlph(I)		0.4
Qu, QuH$_2$ (quinhydrone)	$Qu + 2H^+ + 2e^- \rightleftharpoons QuH_2$	0.699	0.286
$O_2|H_2O_2$	$O_2 + 2H^+ + 2e^- \rightleftharpoons H_2O_2$	0.69	0.295
methylene blue	MB(ox) + $2H^+$ + $2e^- \rightleftharpoons$ MB(red)	0.54	0.01
$I_2|I^-$	$I_2 + 2e^- \rightleftharpoons 2I^-$	0.5355	
$NiO_2|Ni(OH)_2$	$NiO_2 + 2H_2O + 2e^- \rightleftharpoons Ni(OH)_2 + 2OH^-$	0.49	
$O_2|OH^-$	$O_2 + 2H_2O + 4e^- \rightleftharpoons 4HO^-$	0.403	
vitamin C (ascorbic acid)	dehydroascorbate + $2H^+$ + $2e^- \rightleftharpoons$ ascorbate + H_2O	0.390	
$Fe(CN)_6^{3-}, Fe(CN)_6^{4-}$	$Fe(CN)_6^{3-} + e^- \rightleftharpoons Fe(CN)_6^{4-}$	0.36	
$Cu^{2+}|Cu$	$Cu^{2+} + 2e^- \rightleftharpoons Cu$	0.337	
cytochrome a	Fe(III) + $e^- \rightleftharpoons$ Fe(II)		0.290

Half cell contents	Half-cell reaction	$E°$, V	$E°'$(pH 7), V
$Hg_2Cl_2 \vert Hg$	$Hg_2Cl_2 + 2e^- \rightleftharpoons 2Hg + 2Cl^-$	0.268	
$AgCl \vert Ag$	$AgCl + e^- \rightleftharpoons Ag + Cl^-$	0.222	
cytochrome c	$Fe(III) + e^- \rightleftharpoons Fe(II)$		0.254
hemoglobin	$Fe(III) + e^- \rightleftharpoons Fe(II)$		0.17
Cu^{2+}, Cu^+	$Cu^{2+} + e^- \rightleftharpoons Cu^+$	0.161	
S, H_2S	$S + 2H^+ + 2e^- \rightleftharpoons H_2S$	0.144	
Sn^{4+}, Sn^{2+}	$Sn^{4+} + 2e^- \rightleftharpoons Sn^{2+}$	0.139	
UQ, UQH_2 (ubiquinone)	$UQ + 2H^+ + 2e^- \rightleftharpoons UQH_2$		0.10
cytochrome b	$Fe(III) + e^- \rightleftharpoons Fe(II)$		0.077
myoglobin	$Fe(III) + e^- \rightleftharpoons Fe(II)$		0.046
fumarate, succinate	$^-O_2CCH{=}CHCO_2^- + 2H^+ + 2e^- \rightleftharpoons {}^-O_2CCH_2CH_2CO_2^-$		0.031
$HgO \vert Hg, OH^-$	$HgO + H_2O + 2e^- \rightleftharpoons Hg + 2OH^-$	0.098	
$S_4O_6^{2-}, S_2O_3^{2-}$	$S_4O_6^{2-} + 2e^- \rightleftharpoons 2S_2O_3^{2-}$	0.09	
(SHE) $H_3O^+ \vert H_2$	$2H_3O^+ + 2e^- \rightleftharpoons H_2 + 2H_2O$	0.000	
$Fe^{3+} \vert Fe$	$Fe^{3+} + 3e^- \rightleftharpoons Fe$	−0.04	
$Pb^{2+} \vert Pb$	$Pb^{2+} + 2e^- \rightleftharpoons Pb$	−0.13	
$Sn^{2+} \vert Sn$	$Sn^{2+} + 2e^- \rightleftharpoons Sn$	−0.14	
pyruvate, ethanol	$CH_3COCO_2^- + 3H^+ + 2e^- \rightleftharpoons CH_3CH_2OH + CO_2$		−0.14
$FAD, FADH_2$	$FAD + 2H^+ + 2e^- \rightleftharpoons FADH_2$		−0.18
pyruvate, lactate	$CH_3COCO_2^- + 2H^+ + 2e^- \rightleftharpoons CH_3CHOHCO_2^-$		−0.19
acetaldehyde, ethanol	$CH_3CHO + 2H^+ + 2e^- \rightleftharpoons CH_3CH_2OH$		−0.20
$Ni^{2+} \vert Ni$	$Ni^{2+} + 2e^- \rightleftharpoons Ni$	−0.23	
$S \vert H_2S$	$S + 2H^+ + 2e^- \rightleftharpoons H_2S$		−0.23
1,3-DPG, G3P, P_i	$^{2-}O_3POCH_2CHOHCOOPO_3^{2-} + 2H^+ + 2e^- \rightleftharpoons {}^{2-}O_3POCH_2CHOHCHO + HOPO_3^{2-}$		−0.29
$PbSO_4 \vert Pb$	$PbSO_4 + 2e^- \rightleftharpoons Pb + SO_4^{2-}$	−0.356	
$Cd^{2+} \vert Cd$	$Cd^{2+} + 2e^- \rightleftharpoons Cd$	−0.40	
$Fe^{2+} \vert Fe$	$Fe^{2+} + 2e^- \rightleftharpoons Fe$	−0.41	
$NAD^+, NADH$	$NAD^+ + H^+ + 2e^- \rightleftharpoons NADH$	−0.105	−0.32
$NADP^+, NADPH$	$NADP^+ + H^+ + 2e^- \rightleftharpoons NADPH$	−0.105	−0.32
$H_3O^+ \vert H_2$	$2H_3O^+ + 2e^- \rightleftharpoons H_2 + 2H_2O$		−0.414
ferredoxin	$Fe(III) + e^- \rightleftharpoons Fe(II)$		−0.43
gluconate, glucose	$C_5H_{11}O_5CO_2^- + 3H^+ + 2e^- \rightleftharpoons C_5H_{11}O_5CHO + H_2O$		−0.44
3PG, G3P	$^{2-}O_3POCH_2CHOHCO_2^- + 3H^+ + 2e^- \rightleftharpoons {}^{2-}O_3POCH_2CHOHCHO + H_2O$		−0.55
acetate, acetaldehyde	$CH_3CO_2^- + 3H^+ + 2e^- \rightleftharpoons CH_3CHO + H_2O$		−0.60
$Cr^{3+} \vert Cr$	$Cr^{3+} + 3e^- \rightleftharpoons Cr$	−0.74	
$Zn^{2+} \vert Zn$	$Zn^{2+} + 2e^- \rightleftharpoons Zn$	−0.763	
$Cd(OH)_2 \vert Cd$	$Cd(OH)_2 + 2e^- \rightleftharpoons Cd + 2OH^-$	−0.81	

Half cell contents	Half-cell reaction	$E°$, V	$E°'$(pH 7), V
$H_2O|H_2$	$2H_2O + 2e^- \rightleftharpoons H_2 + 2OH^-$	-0.8281	
$Cr^{2+}|Cr$	$Cr^{2+} + 2e^- \rightleftharpoons Cr$	-0.89	
$Zn(OH)_2|Zn, OH^-$	$Zn(OH)_2 + 2e^- \rightleftharpoons Zn + 2OH^-$	-1.25	
$Al^{3+}|Al$	$Al^{3+} + 3e^- \rightleftharpoons Al$	-1.66	
$Al(OH)_3|Al$	$Al(OH)_3 + 3e^- \rightleftharpoons Al + 3OH^-$	-2.33	
$Mg^{2+}|Mg$	$Mg^{2+} + 2e^- \rightleftharpoons Mg$	-2.38	
$Na^+|Na$	$Na^+ + e^- \rightleftharpoons Na$	-2.71	
$Ca^{2+}|Ca$	$Ca^{2+} + 2e^- \rightleftharpoons Ca$	-2.76	
$K^+|K$	$K^+ + e^- \rightleftharpoons K$	-2.92	
$Li^+|Li$	$Li^+ + e^- \rightleftharpoons Li$	-3.045	

Strongest
Reducing Agent

Index of Terms

α–helix	1-48	acidic	2-63
δ (a small amount)	1-28	activated complex	11-46
λ (wavelength)	4-14	activation energy	11-46
π (pi bonding orbital)	5-33	active site (enzyme)	11-71
π* (pi antibonding orbital)	5-78	active site	5-73
σ (sigma bonding orbital)	5-19	activity	9-17
σ (sigma orbital)	5-18	alcohol	1-64
σ bonding framework	5-22	alcohols	5-68
σ framework	5-22	aldehydes	5-69
σ_n (sigma nonbonding orbital)	5-20	alkali metal	4-6
ν (wave frequency)	4-15	alkali metals	2-21
ψ (electron wave function)	4-65	alkaline earth metal	4-6
Π (osmotic pressure)	8-64	alkaline earths	2-21
Δ [change (final – initial)]	1-53	alkanes	5-65
ΔE (energy change)	1-53	alkenes	5-66
ΔE (internal energy change)	7-54	alkyl	1-64
ΔG (free energy change)	8-45	allotrope	7-41
ΔG° (standard free energy change)	8-45	allotropes	7-41
		alpha particle	3-25
ΔH (enthalpy change)	7-22	ambiphilic	8-50
ΔH_f° (standard enthalpy of formation)	7-43	amino acid side group	9-40
		amino acid	1-47
3-d structure	5-48	amount	1-58
A (amperes)	10-12	amperes	10-12
A (frequency factor)	11-44	amplitude	4-14
A (mass number)	3-16	anion	2-18
absolute zero	8-22	anode	10-10
absorption spectrum	4-12	aqueous	1-5
acetic acid-water reaction	9-19	Arrhenius equation	11-42
acid anhydride	7-49	ascorbic acid	6-81
acid-base indicator	6-79	aspartic acid	9-44
acidic	2-61	atom	1-13